Legal Aspects of ENGINEERING

Sixth Edition

Richard C. Vaughn

KENDALL/HUNT PUBLISHING COMPANY
4050 Westmark Drive Dubuque, Iowa 52002

Cover images courtesy of Corel.

Library of Congress Catalog Card Number: 98-75481

ISBN 0-7872-5641-2

Printed in the United States of America
10 9 8 7 6 5 4 3 2 1

Contents

Table of Cases

♦ Preface ♦

Legal Aspects of Engineering is intended to introduce engineers, architects, and others interested in engineering projects to pertinent legal concepts. It is not intended to replace the services of an attorney. Rather than replace those services, it should make them more useful to the legal client. If the services of a member of the bar are sought by an engineer who is aware of basic legal principles pertaining to the engineering activities, the contacts with the attorney can be expected to be more efficient and satisfactory. In addition, an engineer who has some knowledge of the law as well as engineering specialty is better prepared to act as an expert witness if the need arises.

Earlier editions of this text evolved from classroom notes to teach engineering classes. Changes and additions to this 6th edition come from two main sources. The first is commentary from users of the text. The second is the result of various cases handled by one of the authors. Based on this experience, the chapter on intellectual property (patents, trade secrets, and so on) was greatly expanded. Although many of the old cases are retained, many were replaced to reflect changes in technology and legal philosophy; and many of the chapters were substantially rewritten.

Law cases are presented at the end of most chapters. The cases serve to illustrate the principles and concepts in the chapters, but more importantly, each case also provides an example of legal reasoning. Because legal reasoning is somewhat different from engineering problem-solving logic, the cases have been preserved virtually intact. The only deletions are the case references; while such references are of interest to an attorney, they can tend to break an engineer's trend of thought as the case is read.

R.C.Vaughn

Preface

Part One

OVERVIEW OF ENGINEERING AND THE LAW

The first jobs most engineers hold after they receive their bachelor's degrees are in the employment of others. They become members of management teams. Many of them rise to higher positions of management, where they continue to use their engineering backgrounds even though their titles may imply only management responsibilities. Other engineers find that solving engineering problems is so exhilarating that they go on to solve such problems for others as consultants. Still other engineers pursue careers in the military, academia, and in federal, state, or local governments.

In any of these endeavors, the engineer's relationships with others are prescribed by rules of law and ethics. Such rules provide the respective rights and responsibilities of the parties. A knowledge of these rules, then, is valuable to the engineer. In this first part we study these rules as well as relationships and controversies that spring from them.

Chapter 1

The Engineer in Management

Engineers are problem solvers. They possess unique tools for stating problems in such a manner that they can be solved and for providing solutions to such problems. Expertise in stating and solving problems results from screening, natural selection, and specialized training in problem-solving techniques. It is this ability to state and solve problems that employers hope to find when they hire engineers, and it is because of the many uses for this knowledge that there is presently a substantial demand for engineering graduates.

Most training offered to engineers equips them to solve problems of a mathematical nature—problems that may readily be reduced to symbolic form. However, not all problems lend themselves to such an attack. The stress resulting from a given force applied to a particular design of beam is easily stated in mathematical terms. It is a little more difficult, but usually still possible, to assume probabilities and solve for the number of parts to be run or for warehouse space required for next year's production. It is exceedingly difficult, though, to formulate or state laws governing the relationships between people in terms of x's and y's with proper coefficients and thereby solve legal and ethical problems. Most such problems involve the interpretation of human laws and the use of discretion and judgment in determining rights and responsibilities.

Often, the solutions of legal problems and problems involving human relations are no less important to a successful engineer than the solutions of mathematical problems. In most engineering jobs, the engineer is part of a so-called management team. Before turning to the aspects of law with which engineers should be familiar, consider certain management skills they should strive to acquire.

ENGINEERING MANAGEMENT

For the vast majority of engineering graduates the first job secured is merely a stepping stone to higher positions. Most people, including engineers, are ambitious. It is only natural that engineering neophytes should raise their sights toward positions that offer greater rewards.

Normally, entry-level engineering jobs require a large amount of technical skill. As they move up their career ladders, the percentage of time in which seasoned engineers use their technical skills usually decreases. Regardless of the ladder the engineer has chosen to climb—research, manufacturing engineering, consulting, sales, or any other job—progression to higher levels depends on at least four factors in addition to engineering ability.

♦ Communication Skills

An idea initiated by an engineer may have very great latent value, but until it is used or communicated in some way, the idea is worthless to the engineer and to others. In addition, the mental work necessary to develop the idea in a detailed description or a diagram is in itself of value. Nearly everyone has had the experience of gaining new insight or of discovering added features of an idea when faced with the task of trying to explain it to someone else.

♦ People Management

A promotion from a strictly technical position to one of greater responsibilities almost always leads to handling people. Being "boss" isn't easy. People can be forced to work under threat of being deprived of their paychecks, but such threats usually stifle initiative. The manager who takes time to explain to subordinates "why" and to keep them informed is likely to be more successful than one who does not.

♦ Sense of Cost

Most operations are undertaken with a profit motive. Even in situations when operations that are not expected to make a profit (such as a service department), cost is usually important. If the selling price of a company's product is unchanged, money saved in manufacturing or raw material cost represents added profit; conversely, added cost decreases profit. Many engineers have won promotions and many consultants earn their livelihoods on their ability to analyze operations and reduce costs.

♦ Knowledge of Law

Engineers are not expected to become attorneys from an exposure to one survey course in law, any more than attorneys could become engineers by taking one survey course in engineering. However, engineers should be cognizant of the probable effects of carelessness in dealing with others. Engineers should know when they need the advice of an attorney. Legal background is a "preventive" asset; that is, with a basic knowledge of law, engineers should be equipped to prevent costly lawsuits against their company. Meticulous reading of contracts before signing is an important preventive measure. It is often surprising how little attention is paid to contracts and supporting documents, particularly in view of the fact that these documents outline the rights and responsibilities of the parties.

EXECUTIVE QUALITIES

In recent years there has been an increasing trend toward filling top management or executive positions with engineers. Companies have recognized the value of the engineer's analytical approach to executive problems.

Although there is a good deal of truth to the often quoted comment that "there is always room at the top," those who get there usually possess special abilities. Engineering training is beneficial to the executive aspirant, but so is a knowledge of many other fields.

What makes an executive? Why does one person achieve this goal while many others strive and fail? At first glance, the behavior of one successful executive appears to have little in common with that of another who is equally successful. One may be the brusque bull-of-the-woods; another may be as smooth as silk. However, closer examination reveals certain similar behavior patterns. Each usually possesses the four qualities just mentioned for successful managers, often to a high degree. Other qualities, too, are seemingly common to most top executives and deserve consideration.

♦ Leadership

The quality known as leadership is difficult to define. It is clearly evident in one person and strangely lacking in another. Psychologically, leadership indicates an identification of the group with the one who leads—it is necessary that the leader be considered by the group as one of them. It is also required that the leader be somewhat superior to the others in the group in one or more qualities esteemed by them.

It is doubtful that anyone is truly a "born leader." It is more probable that leadership qualities result from training acquired both consciously and subconsciously—study and observation so ingrained that the leader's responses to various situations are almost as natural as breathing. Thus, the term "born leader" has come to be used in reference to the leader who seems to do everything right at the right time in a very natural way. Leadership qualities seem to be enhanced by practice.

Opportunities to practice the poise and purposefulness of leadership occur in virtually limitless ways. One of the main reasons job applications nearly always contain space for listing organizational activities is to determine the amount of practice in leadership the candidate has had.

Two outstanding leadership characteristics are the ability to keep the ultimate goal uppermost in mind and the ability to pursue it enthusiastically. Enthusiasm is infectious—it rubs off on others. The relative success of dictators and would-be dictators attests to this. A speech delivered in a monotone makes for dull listening; however, the same speech using virtually the same words but delivered enthusiastically can move people to action.

A leader does not need a leaning post—either literally or figuratively. Leadership stems partly from the ability to stand firm on principles. There is a popular misconception that a leader should not admit mistakes. Few people make perfect decisions all the time, however. Not only must leaders admit their own mistakes; they must also take responsibility for the mistakes of their subordinates, because those actions result from the leaders' direction or lack thereof. Scapegoating is a popular art, but few effective leaders in top management will stoop to it to avoid criticism.

♦ Delegation

A characteristic most top executives share is the facility for delegating authority and responsibility to others. It is virtually impossible for anyone to rise to the top of a modern industrial organization without the ability to delegate. There is just not time for one person to effectively and thoroughly perform the requirements of a top management job. Executives who delegate few tasks rob themselves of time needed for adequate thought before making decisions. Also, the failure to delegate routine tasks to others can be a barrier to executives' promotions; if no one can be found who has performed a portion of the executives' tasks with the authority necessary for that performance, it is natural to leave the executives where they are.

Specialization is an inherent advantage of effective delegating. No one is a specialist in everything. By assigning some of their tasks to others, executives can obtain the advantage of specialized treatment.

Delegation as the term is used here (and in most businesses), means more than merely assigning tasks to be performed. Delegation includes clothing the delegee with the necessary authority to carry out the assigned function. It is this parting with a portion of authority that causes shortsighted executives—consciously or unconsciously—to oppose delegation to others. It is this very aspect of delegation, however, that contributes to the growth of assistants. The able executive realizes this and takes full advantage of it in helping others develop.

♦ Decision-Making

All of us make decisions involving choices from among alternatives. Our choices are not always correct. One attribute that seems characteristic of those people who reach top management is their ability to be right a higher percentage of the time than the average person. Of course, top management decisions are decisions about particularly difficult problems. Decisions run all the way from a single-variable problem (like checking a part with go/ no-go gauge) to multivariable problems where little, if anything, is fixed or known. Generally, routine decisions are delegated to others; the top manager is the one who makes the decision when major uncertainties exist. People that make decisions such as these are venturers-people who will assume risks in their decisions. Generally, the greater the risk undertaken, the greater the possible reward. The decision to expand a plant or install new production facilities based on an apparently expanding market is such a decision. No one can predict the future—the further into the future the planning, the more inaccurate it is likely to be.

Many of the assumptions can be reduced to probabilities. If enough of this can be done, the problem can be programmed for a computer, which will then give the executive some answers. However, the answers are based on assumptions and probabilities, and the executive must decide whether to go ahead. The risk is still the executive's, not the computer's.

A few top executives possess such vast knowledge and the ability to analyze and synthesize that they can make rapid-fire policy decisions that are nearly always right. However, such people are rare. Generally, people in top management do not make hurried policy decisions. There is often grumbling from below because of apparently undue procrastination. Despite the grumbling, such delay is usually the course of wisdom, because the risks are frequently sizable. A decision based on inadequate facts or erroneous assumptions is hazardous, and delay in waiting for more facts is often in-

Top management decisions generally consist of five elements, dealt with in sequence:

1. a gathering of facts
2. a recognition of limiting conditions
3. assumption of facts and conditions as they are expected to be and recognition that these are assumptions
4. analysis of the facts, limits, and assumptions
5. decision

escapable. Even the rare management genius who makes correct decisions rapidly usually has had many years of experience in more methodical decision making that has equipped him or her for this present role.

♦ Discipline

Discipline is a necessary component of any well-run organization. People must be taught; old habit patterns must be changed. Most top executives are masters of the use of reward and punishment in changing the behavior of subordinates. To be effective, executive orders must imply some form of reproof for disobedience; rewards of some sort must follow outstanding performance if the effort required for the performance is to be continued.

The extent to which reward and punishment are necessary depends to a great degree on the personal stature of the executives. If they are held in high regard by their subordinates, a word or so of reproof is often the equivalent of the proverbial 10 lashes.

In addition to people's drive for food, water, and the means of satisfying other basic needs, they have a whole host of derived needs, not the least of which is the need of recognition. Every person needs recognition or respect from others-lack of it causes loss of self-respect and, eventually, diminished effort. Recognition can be either tangible or intangible, and both forms are required. Verbal praise sounds hollow after awhile if it is not accompanied by some material reward. Similarly, material rewards without praise for accomplishments are incomplete.

It has often been stated that rewards should be public, with criticism or punishment private. The truth in the statement is inescapable. Most top managers observe this principle in the interest of preserving their organizations.

These are only a few of the principles that guide top executives in their management of discipline. Most of these guidelines are understood and observed without conscious thought when disciplinary occasions rise.

Many qualities can make a person successful in top management. Only a few have been mentioned here. Nevertheless, these few are basic and must be mastered by executive aspirants. The purpose of a business and its management is to produce something. Converting time and raw material into goods and services requires production facilities. Assembly of the machines and equipment required to produce a product is normally undertaken as an engineering project. Not only must the original facilities be planned and built but every design change or functional change of the product also requires changing machines and equipment. The job of setting up production facilities becomes, then, not a "one-shot" enterprise, but an almost continuous replanning and rearrangement.

The burden of deciding when and how much to change—and what to change—falls on top management; the job of planning and carrying out the details of rearrangement is usually assigned to the engineering department.

ENGINEERING PROJECTS

A large proportion of the capital wealth of our country has resulted from engineering projects. Civil engineering projects-roads, bridges, buildings, and the like-are most familiar to the public. As a result, whenever the term *engineering project* is used, visions of a dam or expressway cloverleaf are likely to come to mind. The value of civil projects cannot be denied, but contributions by other engineering fields are also significant, even though the public is not as aware of these activities or results.

Since the development and adoption of mass-production methods in the United States, a new combination of engineering talents has taken place. People are needed who can apply knowledge of civil, mechanical, electrical, chemical, industrial, and other engineering fields to manufacturing problems. This combining of engineering talents to solve manufacturing problems has come to be known by many names, but the term *manufacturing engineering* seems more appropriate than most. Typically, manufacturing engineering is concerned with the process required to mass-produce some product. It starts with an analysis of someone's brain child and continues as long as there are engineering problems to be solved. The following discussion covers the various stages of an engineering project in a manufacturing engineering context. However, the same basic concepts apply to other types of engineering projects, too.

♦ Project Phases

Any engineering project involves three phases or stages of development:

1. conception of the idea
2. reduction of the idea to practice
3. refinement of the idea and ensuring that the project works.

The stages are fairly distinct, and a particular engineering group may have responsibilities in one or more of the stages.

Conception of the Idea

Just about every product and convenience we enjoy started as someone's idea or "screwball notion." Neither products nor the processes by which they are manufactured can be developed without someone's original idea. Not all ideas are practical, however. A large number of those that are adopted require substantial alterations before they are acceptable. Many ideas appear attractive in the beginning only to be demonstrated as impractical by objective examination. This objective examination of a possible engineering project is known as a *feasibility study*.

A feasibility study is a preliminary examination of a proposed idea. It is meant to answer such questions as: What will it cost to produce various quantities per year? Can we market enough to make a reasonable return on the required investment? How many can be sold at a given price? What processes will be better in the long run? The answers given determine whether it is desirable to proceed to the next stage—actually setting up to produce.

Reduction to Practice

Turning someone's idea into a reality can be quite complex in a manufacturing situation. Planning is necessary and requires imagination—a vision of the future—and it continues until all the pieces are firmly in place. Even then, maintenance should be planned. Changes are made easily in the planning stage—it costs little to erase a machine location on a layout and place the machine in another location. Even rearrangements of the entire process are inexpensive at this point. It is here that questions pertaining to equipment sizes, locations, and added features must be answered and the answers justified if the process is to be successful. Layout changes after the process equipment has been placed are very expensive. For this reason, questions that should have been raised in the planning stage but were never brought up reflect on the process engineer's ability. A member of a manufacturing engineering department cannot be omniscient, but it would help.

The process engineer designs a layout of the process, complete with machines and equipment, and writes specifications for the machines to function as desired. The specifications are then sent out, proposals received, and contracts awarded to the successful bidders. Engineers are the owner's agents; as such they often must supervise the building of machines or other structures to fit the layout and then supervise the installation. Engineers also must control the times of completion of the elements of layout. Rarely is a process completed and functioning properly within the original timeframe. There is nearly always at least one contractor who is late. A wise engineer will allow some time for this in the schedule.

Refinement and Oversight

It is probably safe to state that in every manufacturing process ever installed there were special problems to be solved before full-scale production could begin. The presence of "bugs" in a newly installed process is about as normal as any expectancy can be. The bugs must be removed before the process can be considered complete. The engineer who set up the process is the logical person to remove these bugs before the operation is turned over to the production people.

LAW AND ENGINEERS

In any engineering project, engineers act as professionals; they are the representatives, or agents, of the owner. Their function is to act in the best interests of the owner—to get the best possible results with a minimum of delay and problems. Engineers must deal with the rights of others.

> Engineers' actions affect others' property rights and personal rights—rights they have due to ownership of property, contractual obligations, torts, or crimes.

An engineer is a guardian of the owner's rights and, in a manner of speaking, of the rights of others with whom the owner deals. Because court proceedings are costly in both time and money, engineers generally should avoid entanglements that would lead to litigation. And, because violation of the rights of others is likely to lead to court proceedings, engineers must know the characteristics of these rights if their preventive job is to be well done.

The relationships between the owner and contractor are set forth in a series of documents drawn up by the engineer. Documents such as instructions for bidders, requests for proposals, proposals, general terms and conditions, specifications, drawings (and sometimes purchase orders, order acknowledgments, invoices, and the like) compose parts of the contract. Careless errors in the preparation of these documents can cause legal controversies or place the owner and engineers in indefensible positions when controversies arise. Engineers must formulate the documents in such a way that the owner's position is protected. Imposing an undue hardship on the contractor may lead to unnecessary litigation. Similarly, ambiguities in the terminology or leaving too much to future agreement can lead to unnecessary litigation. Hence, specificity and realistic goals are what the contract usually requires.

In some respects the engineers' position is between the owner and the contractor. When disputes arise, engineers are likely to be called on to mediate or at least enter into the controversies. To do a reasonable job in this position, engineers must be acquainted with the legal rights and responsibilities of both parties. Engineers don't have to be attorneys, but some knowledge of the law is essential. Engineers should be able to recognize situations in which it is necessary to consult an attorney. Some knowledge of the law is required even for this; engineers can't very easily recognize legal troubles unless they have some knowledge of the rights involved.

As with other human endeavors, obtaining expert advice as early as possible can avoid problems or serve to provide damage control. There is a second reason for engineers to acquire a knowledge of the law. Besides their professional activities, engineers are citizens as well. The law controls many of our private day-to-day dealings with others. When we buy insurance or sign a chattel mortgage for the purchase of a refrigerator, our rights and responsibilities should be clear to us. At the very least, the idea of reading the document before we sign it should occur to us.

Engineers are members of society as well as professionals who possess technical skills. As educated members of society and professional people, their knowledge and abilities should extend well beyond their technical skills. One popularly accepted criterion of the cultured person is the ability to analyze and discuss news events with some perception. Much of the news presented to us by radio, television, newspapers, and magazines has legal significance. If engineers are to be accepted and respected as learned people in their community, their interests and knowledge must be broad enough to justify this acceptance. An acquaintance with legal issues is a step in this direction.

LEGAL ANALYSIS

Engineers have backgrounds in analysis of scientific things. In the design of a bridge or automotive component, for example, an engineer analyzes forces and reactions to them, statically and dynamically. Using the results of analyses, engineers adjust the designs to serve both their employers and the public appropriately.

In the legal setting, engineers are often still concerned with engineering analysis in tort (particularly product liability) cases. They may conduct investigations known as *failure analyses* to find why some malfunction occurred. Scientific analysis may be required in a variety of criminal case settings. Engineers are accustomed to this kind of analysis where most things are reasonably precise and even the provision for error is usually predictable. Legal analysis is somewhat different.

When individuals reason legally they reason on what is generally a much broader playing field. They have not only the incident or event of immediate concern, but also the setting in which it occurred. Many other events or properties usually have a bearing on it and may determine the outcome if the case goes to court. A homicide may be felonious, justifiable, or excusable. Which it is depends on other events near or at the time of the homicide, as well as the developed law regarding such things. In addition to these considerations, individuals have the effect of testimony, physical evidence, and even the social and political setting in which the event and trial occurred. The judge and/or jury has this complex morass to wade through to make a decision and administer justice. Very little is dependably fixed except, perhaps, the statutes and constitution, but even here there is a problem. The writings in a statute use words of a language (English, in our case) and, despite dictionary definitions, these words and combinations of them can take on a variety of meanings. Much of legal education is devoted to review of cases, what the jurists wrote to justify the courts' decisions and contrary opinions by dissenting judges.

Frequently, engineers find the transition from scientific analysis to legal analysis frustrating; such a reaction may be beneficial in keeping one out of trouble or in smoothing the road to relief of legal problems.

REVIEW QUESTIONS

1. Why should an engineer have some knowledge of the law?
2. Name at least three more qualifications an engineer should possess for success in management. Name at least three additional qualities of successful executives.
3. What are the stages of an engineering project? What would each stage be composed of in the proposed manufacture of, say, table lamps?
4. What is manufacturing engineering?

Chapter 2

Ethics

Engineers are professionals; the occupation they have chosen is one of the newer professions. Until quite recently, the only callings of sufficient dignity and dedication to public service to be termed *professions* were law, medicine, and theology. These three "learned professions" are still looked on by the public as the pinnacle of professions. Because of the intimate contact between members of these professions and the public, this view may never change.

Engineering has existed as a separate calling for nearly two centuries. Considering the span of its existence, the progress of engineering toward top professional standing is quite striking. Certainly there was engineering prior to 200 years ago, as the ancient pyramids, the aqueducts of Rome, and other engineering works mutely testify. The designing and building of these structures was not known as engineering, however.

Civil engineering was the first type of engineering to be recognized as a separate calling. Around 1750, John Smeaton, an English engineer, made the first recorded use of the term civil engineering. In 1818 the Institution of Civil Engineers was founded in Great Britain. That organization defined engineering as "the art of utilizing the forces of nature for the use and convenience of man." With the development of new areas of technological knowledge, many other fields of engineering have been established.

FEDERAL LAWS

A separate, recognized field of learning and the presence of societies of its members do not make a "profession." People do not have "professional" status merely because they have graduated from a school and joined a society. Perhaps it is best to consider for a moment the meaning of the term *profession*. It is not difficult to find definitions of the term. Webster's dictionary says it is "a calling requiring specialized knowledge and often long and intensive academic preparation." *Black's Law Dictionary* calls it "a vocation or occupation requiring special, usually advanced, education, and the labor and skill involved in a profession is predominantly mental or intellectual rather than physical or manual."

Although each of these definitions serves a purpose, each is brief at the expense of completeness. A more complete definition was given by the U.S. Congress in the Labor Management Relations Act. In this act, Congress defined the term *professional employee* as follows:

The term "professional employee" means:

a. any employee engaged in work (1) predominantly intellectual and varied in character as opposed to routine mental, manual, mechanical, or physical work; (2) involving consistent exercise of discretion and judgment in its performance; (3) of such a character that the output produced or the result accomplished cannot be standardized in relation to a given period of time; (4) requiring knowledge of an advanced type in a field of science or learning customarily acquired by a prolonged course of specialized intellectual instruction and study in an institution of higher learning or a hospital, as distinguished from a general academic education or from an apprenticeship or from training in the performance of routine mental, manual, or physical processes; or

b. any employee who (1) has completed the courses of specialized intellectual instruction and study described in clause (4) of paragraph (a); and (2) is performing related work under the supervision of a professional person to qualify himself to become a professional employee as defined in paragraph (a).

Besides the four requirements just stated, various other criteria are frequently added to the list:

a. registration requirements for practicing the professions;
b. representation of members and control of activities by a professional society;
c. the public service nature of the occupation; and
d. adherence to a code of ethics.

STATE REGISTRATION LAWS

The state laws of each of the United States require engineers to be registered before they are allowed to practice professional engineering in the state. Registration in any one state does not give engineers the right to act as professional engineers in another state. However, many states have reciprocal agreements whereby registration is much simplified if engineers are already registered in another state.

The primary purpose of the state engineering registration laws is to protect the public from shoddy engineering practices.To this end, it is necessary that prospective licensees convince a board of examiners that they are qualified to practice professional engineering. The usual method for this is a scrutiny of the candidates' past engineering work and training and a qualifying examination. This examination commonly lasts two days and is either oral and written or entirely written. It usually covers the basic sciences and specialization in a particular field of engineering.

Full registration as professional engineers allows licensees to act as professional engineers within the state and to resort to the courts to collect fees for their services. Penalties in the form of fines and /or confinement are usually specified for practicing without a license.

According to at least one court decision,[1] engineering became a profession on the enactment of the state registration law.

However, professions do not magically spring into existence on the day a law is passed. The acceptance of a field as a profession requires continuing efforts to maintain high standards of service to the public.

PROFESSIONAL SOCIETIES

Each of the recognized branches of engineering has formed at least one society of its members. In addition to these bodies, there are three organizations that represent and serve all engineers.

♦ Accreditation Board of Engineering and Technology

In 1932, the Engineers Council for Professional Development (ECPD) was formed. For 47 years it was concerned with accrediting engineering curricula and with other nonaccrediting activities such as guidance, ethics, and the development of young engineers. In 1979, ECPD restructured itself into an accreditation board and joined with the American Association of Engineering Societies (AAES) for its other functions.

Accredited engineering and technology curricula are examined at least once every six years by an examiner from one of two commissions of the Accreditation Board of Engineering and Technology (ABET): the Engineering Accreditation Commission examines engineering curricula while the Technology Accreditation Commission examines technology programs. In 1992, there were 1,416 accredited engineering programs and 766 accredited technology programs.

1. *H. C. Downer and Associates, Inc. v. The Westgate Realty Company,* No. 4892, Court of Appeals, Ninth Judicial District, Ohio, November 25, 1959.

♦ American Association of Engineering Societies

The American Association of Engineering Societies (AAES) was founded in 1979, superseding the Engineers Joint Council (founded in 1945). Its objectives are to advance the science and practice of engineering in the public interest; and to act as an advisory, communication, and information exchange agency for member activities. As noted earlier, AAES also acquired some of the functions of the Engineers Council for Professional Development when it restructured.

♦ National Society of Professional Engineers

The National Society of Professional Engineers (NSPE), formed in 1934, is concerned with the social, economic, political, and professional interests of all engineers. NSPE activities were largely responsible for the passage of engineer registration laws in the various states, public recognition of engineering as a profession, and recognition of the value of engineering activities.

CODE OF ETHICS

♦ Definition

According to Webster's dictionary, ethics is "the discipline dealing with what is good and bad and with moral duty and obligation."

Ethics are the ground rules of our moral conduct. They consist of our attitudes toward honesty, integrity, trust, and loyalty; they are exhibited in our day-to-day contacts with others. No laws compel engineers to take an interest in community affairs or to give a completely unbiased report of the results of an investigation.The manner in which engineers act depends on their own moral code or ethics.

♦ Establishment of Moral Patterns

A person does not acquire a code of ethics or sense of moral duty by reading a passage in a textbook and then deciding to abide by what was stated. Rather, a personal code grows out of the experiences and observations of one's life.

Perhaps we each inherit a predisposition toward a certain type of moral behavior. It's doubtful whether this sense is inherited, though. Certainly we are influenced by what happens after birth—by the environment in which we mature. Examples set by parents, friends, classmates, teachers, and professors all con-

Professional Lifestyle of an Engineer

A degree in engineering is a foundation; a life's career in the engineering profession can be built on it. Engineers can reasonably expect to be treated as professionals by their clients and superiors. However, members of the public are often unaware of contributors made by members of the engineering profession. The service rendered by a doctor, an attorney, or a member of the clergy is obvious to the public. A person experiencing pain seeks out a doctor for diagnosis and treatment. One who has been accused of a crime requires personal contact with an attorney. A family experiencing domestic difficulties may turn to a member of the clergy for aid.

The public has little knowledge, though, of the engineer whose work makes crossing a bridge or riding an airplane safe and convenient. Engineers' work is just as vital and their contributions are as great as those of other professions—a person can be a victim of an accident caused by mechanical malfunction. Engineers' contributions, however, are often unseen; it is for this reason, among others, that engineers must act professionally if they are to establish and maintain the respect accorded members of other professions!

An engineering education is never finished. Engineers have a duty to their clients and profession to learn new developments in the engineering field. Unquestionably, professional people's day-to-day bread-winning efforts are important to them; but so is the extra time in the evenings and on weekends that is required for reading technical periodicals and attending engineering society meetings. For dedicated professionals, the day does not end when the office doors close behind them. There is always more to learn.

tribute. Punishment for censured acts and praise for achievements are the building blocks for individual codes of ethics. All of these factors contribute to engineers' moral structure; to these will be added his experiences and observations on the job.

In July of 1996, the NSPE revised its code of ethics presented here. This code is not meant as a body of inflexible laws, to be observed "or else." It is meant as a guidepost, to be worked into and be the basis for changes in the engineer's moral standards. It has been in existence in various stages of development for several decades and has been found to be of value in dealing with others. It provides professional standards.

NSPE CODE OF ETHICS FOR ENGINEERS

♦ Preamble

Engineering is an important and learned profession. As members of this profession, engineers are expected to exhibit the highest standards of honesty and integrity. Engineering has a direct and vital impact on the quality of life for all people. Accordingly, the services provided by engineers require honesty, impartiality, fairness and equity, and must be dedicated to the protection of the public health, safety, and welfare. Engineers must perform under a standard of professional behavior that requires adherence to the highest principles of ethical conduct.

♦ I. Fundamental Canons

Engineers, in the fulfillment of their professional duties, shall:

1. Hold paramount the safety, health, and welfare of the public.
2. Perform services only in areas of their competence.
3. Issue public statements only in an objective and truthful manner.
4. Act for each employer or client as faithful agents or trustees.
5. Avoid deceptive acts.
6. Conduct themselves honorably, responsibly, ethically, and lawfully so as to enhance the honor, reputation, and usefulness of the profession.

♦ II. Rules of Practice

1. Engineers shall hold paramount the safety, health, and welfare of the public.
   a. If engineers' judgment is overruled under circumstances that endanger life or property, they shall notify their employer or client and such other authority as may be appropriate.
   b. Engineers shall approve only those engineering documents that are in conformity with applicable standards.
   c. Engineers shall not reveal facts, data or information without the prior consent of the client or employer except as authorized or required by law or this Code.
   d. Engineers shall not permit the use of their name or associate in business ventures with any person or firm that they believe are engaged in fraudulent or dishonest enterprise.
   e. Engineers having knowledge of any alleged violation of this Code shall report thereon to appropriate professional bodies and, when relevant, also to public authorities, and cooperate with the proper authorities in furnishing such information or assistance as may be required.
2. Engineers shall perform services only in the areas of their competence.
   a. Engineers shall undertake assignments only when qualified by education or experience in the specific technical fields involved.
   b. Engineers shall not affix their signatures to any plans or documents dealing with subject matter in which they lack competence, nor to any plan or document not prepared under their direction and control.
   c. Engineers may accept assignments and assume responsibility for coordination of an entire project and sign and seal the engineering documents for the entire project, provided that each technical segment is signed and

sealed only by the qualified engineers who prepared the segment.

3. Engineers shall issue public statements only in an objective and truthful manner.
   a. Engineers shall be objective and truthful in professional reports, statements, or testimony. The shall include all relevant and pertinent information in such reports, statements, or testimony, which should bear the date indicating when it was current.
   b. Engineers may express publicly technical opinions that are founded upon knowledge of the facts and competence in the subject matter.
   c. Engineers shall issue no statements, criticisms, or arguments on technical matters that are inspired or paid for by interested parties, unless they have prefaced their comments by explicitly identifying the interested parties on whose behalf they are speaking, and by revealing the existence of any interest the engineers may have in the matters.
4. Engineers shall act for each employer or client as faithful agents or trustees.
   a. Engineers shall disclose all known or potential conflicts of interest that could influence or appear to influence their judgment or the quality of their services.
   b. Engineers shall not accept compensation, financial or otherwise, from more than one party for services on the same project or for services pertaining to the same project, unless the circumstances are fully disclosed and agreed to by all interested parties.
   c. Engineers shall not solicit or accept financial or other valuable consideration, directly or indirectly, from outside agents in connection with the work for which they are responsible.
   d. Engineers in public service as members, advisors, or employees of a governmental or quasi-governmental body or department shall not participate in decisions with respect to services solicited or provided by them or their organizations in private or public engineering practice.
   e. Engineers shall not solicit or accept a contract from a governmental body on which a principal or officer of their organization serves as a member.
5. Engineers shall avoid deceptive acts.
   a. Engineers shall not falsify their qualifications or permit misrepresentation of their or their associates' qualifications. They shall not misrepresent or exaggerate their responsibility in or for the subject matter of prior assignments. Brochures or other presentations incident to the solicitation of employment shall not misrepresent pertinent facts concerning employers, employees, associates, joint venturers, or past accomplishments.
   b. Engineers shall not offer, give, solicit or receive, either directly or indirectly, any contribution to influence the award of a contract by public authority, or which may be reasonably construed by the public as having the effect of intent to influencing the awarding of a contract. They shall not offer any gift or other valuable consideration in order to secure work. They shall not pay a commissions, percentage, or brokerage fee in order to secure work, except to a bona fide employee or bona fide established commercial or marketing agencies retained by them.

♦ III. Professional Obligations

1. Engineers shall be guided in all their relations by the highest standards of honesty and integrity.
   a. Engineers shall acknowledge their errors and shall not distort or alter the facts.
   b. Engineers shall advise their clients or employers when they believe a project will not be successful.
   c. Engineers shall not accept outside employment to the detriment of their regular work or interest. Before accepting any outside engineering employment they will notify their employers.
   d. Engineers shall not attempt to attract an engineer from another employer by false or misleading pretenses.
   e. Engineers shall not actively participate in strikes, picket lines, or other collective coercive action.
   f. Engineers shall not promote their own interest at the expense of the dignity and integrity of the profession.

2. Engineers shall at all times strive to serve the public interest.
   a. Engineers shall seek opportunities to participate in civic affairs; career guidance for youths; and work for the advancement of the safety, health, and well-being of their community.
   b. Engineers shall not complete, sign, or seal plans and/or specifications that are not in conformity with applicable engineering standards. If the client or employer insists on such unprofessional conduct, they shall notify the proper authorities and withdraw from further service on the project.
   c. Engineers shall endeavor to extend public knowledge and appreciation of engineering and its achievements.
3. Engineers shall avoid all conduct or practice that deceives the public.
   a. Engineers shall avoid the use of statements containing a material misrepresentation of fact or omitting a material fact.
   b. Consistent with the foregoing, Engineers may advertise for recruitment of personnel.
   c. Consistent with the foregoing, Engineers may prepare articles for the lay or technical press, but such articles shall not imply credit to the author for work performed by others.
4. Engineers shall not disclose, without consent, confidential information concerning the business affairs or technical processes of any present or former client or employer or public body on which they serve.
   a. Engineers shall not, without the consent of au interested parties, promote or arrange for new employment or practice in connection with a specific project for which the Engineer has gained particular and specialized knowledge.
   b. Engineers shall not, without the consent of all interested parties, participate in or represent an adversary interest in connection with a specific project or proceeding in which the Engineer has gained particular specialized knowledge on behalf of a former client or employer.
5. Engineers shall not be influenced in their professional duties by conflicting interests.
   a. Engineers shall not accept financial or other considerations, including free engineering designs, from material or equipment suppliers for specifying their product.
   b. Engineers shall not accept commissions or allowances, directly or indirectly, from contractors or other parties dealing with clients or employers of the Engineer in connection with work for which the Engineer is responsible.
6. Engineers shall not attempt to obtain employment or advancement or professional engagements by untruthfully criticizing other engineers, or by other improper or questionable methods.
   a. Engineers shall not request, propose, or accept a commission on a contingent basis under circumstances in which their judgment may be compromised.
   b. Engineers in salaried positions shall accept part-time engineering work only to the extent consistent with policies of the employer and in accordance with ethical considerations.
   c. Engineers shall not, without consent, use equipment, supplies, laboratory, or office facilities of an employer to carry on outside private practice.
7. Engineers shall not attempt to injure, maliciously or falsely, directly or indirectly, the professional reputation, prospects, practice or employment of other engineers. Engineers who believe others are guilty of unethical or illegal practice shall present such information to the proper authority for action.
   a. Engineers in private practice shall not review the work of another engineer for the same client, except with the knowledge of such engineer, or unless the connection of such engineer with the work has been terminated.
   b. Engineers in governmental, industrial, or educational employ are entitled to review and evaluate the work of other engineers when so required by their employment duties.
   c. Engineers in sales or industrial employ are entitled to make engineering comparisons of represented products with products of other suppliers.
8. Engineers shall accept personal responsibility for their professional activities, provided, however,

that Engineers may seek indemnification for services arising out of their practice for other than gross negligence, where the Engineer's interests cannot otherwise be protected.

a. Engineers shall conform with state registration laws in the practice of engineering.

b. Engineers shall not use association with a nonengineer, a corporation, or partnership as a "cloak" for unethical acts.

9. Engineers shall give credit for engineering work to those to whom credit is due, and will recognize the proprietary interests of others.

a. Engineers shall, whenever possible, name the person or persons who may be individually responsible for designs, inventions, writings, or other accomplishments.

b. Engineers using designs supplied by a client recognize that the designs remain the property of the client and may not be duplicated by the Engineer for others without express permission.

c. Engineers, before undertaking work for others in connection with which the Engineer may make improvements, plans, designs, inventions, or other records that may justify copyrights or patents, should enter into a positive agreement regarding ownership.

d. Engineers' designs, data, records, and notes referring exclusively to an employer's work are the employer's property. Employer should indemnify the Engineer for use of the information for any purpose other than the original purpose.

"By order of the United States District Court for the District of Columbia, former Section ll© of the NSPE Code of Ethics prohibiting competitive bidding, and all policy statements, opinions, rulings or other guidelines interpreting its scope, have been rescinded as unlawfully interfering with the legal right of engineers, protected under the antitrust laws, to provide price information to prospective clients; accordingly, nothing contained in the NSPE Code of Ethics, policy statements, opinions, rulings or quotations or competitive bids for engineering services at any time or in any amount."

STATEMENT BY NSPE EXECUTIVE COMMITTEE

In order to correct misunderstandings which have been indicated in some instances since the issuance of the Supreme Court decision and the entry of the Final judgment, it is noted that in its decision of April 25, 1978, The Supreme Court of the United States declared: "The Sherman Act does not require competitive bidding."

It is further noted that as made clear in the Supreme Court decision:

1. Engineers and firms may individually refuse to bid for engineering services.
2. Clients are not required to seek bids for engineering services.
3. Federal, state, and local laws governing procedures to procure engineering services are not affected, and remain in full force and effect.
4. State societies and local chapters are free to actively and aggressively seek legislation for professional selection and negotiation procedures by public agencies.
5. State registration board rules of professional conduct, including rules prohibiting competitive bidding for engineering services, are not affected and remain in full force and effect. State registration boards with authority to adopt rules of professional conduct by adopt rules govering procedures to obtain engineering services.
6. As noted by the Supreme Court, "nothing in the judgment prevents NSPE and its members from attempting influence governmental action..."

NOTE: In regard to the question of application of the Code to corporations vis-a-vis real persons, business form or type should not negate nor influence conformance of individuals to the Code. The Code deals with professional services, which services must be performed by real persons. Real persons in turn establish and implement policies within business structures. The Code is clearly written to apply to the Engineer and items incumbent on members of NSPE to endeavor to live up to the provisions. This applies to all pertinent sections of the Code.

♦ Law and Ethics

Not all unethical actions are illegal; indeed, many unethical acts do not involve any fine or imprisonment, as do criminal acts. Neither the examining boards nor the courts have any rights or responsibilities in the ethical practice of engineers. (That is not to suggest, however, that the courts are not concerned with the negligence of engineers. As will be discussed, the courts indeed hear such cases.) Whatever formal reproof there is for moral misbehavior must come from the engineering societies.

In the short history of the engineering profession there has not yet been developed a society equivalent to the American Medical Association or the local and state bar associations for handling ethical infractions. Nearly all the engineering societies are concerned with unethical practices, but few seem deeply concerned.

At present, enforcement efforts vary from apparent tacit condonation of the unethical in some sections of the United States to very active policing in others. However, the need has been generally recognized and it is likely that systems for detecting, investigating, and holding hearings on unethical engineering practices will soon be developed and vigorously maintained throughout the United States.

Despite the present lack of uniformity in formally policing ethical practices, informal penalties for infractions exist everywhere. The ultimate formal penalty is removal of membership in an engineering society. Although loss of membership in a society can be quite harmful to an individual member's professional status, the informal penalties are often much more severe. Loss of a job on a weak excuse or social censure within a community demonstrates the presence and effectiveness of informal sanctions.

Codes of professional ethics are not necessarily meant to be of practical value to an individual, yet in most cases they are. People who are honest and loyal in their adherence to a code of ethics in dealing with others often find that, as a result, others are honest and loyal in dealing with them. Adherence to ethics tends to inspire confidence and admiration of colleagues, clients, and employers.

♦ Gifts and Favors

A strict interpretation of the NSPE Code of Ethics indicates that anything offered to an engineer by a present or prospective contractor should be shunned.This would seem to include all manner of gifts, favors, and evidences of hospitality. If a ball point pen or a cigar is acceptable, why wouldn't also a set of golf clubs or a silver tea service? If there is no stigma attached to a free lunch, then why not also an evening of nightclubbing at the vendor's expense? If an inexpensive favor is to be condoned, where should the line be drawn?

It must be recognized that the "ivory tower" approach is unpopular in many circles. In numerous companies buyers and engineers seldom buy their own lunches, and pen and pencil sets and other gifts are accepted with no qualms. In such places engineers' refusal of such "advertising" would mark them as a bit odd. Perhaps minor gifts and favors should be accepted; maybe even those larger than "minor" are acceptable. However, engineers should consider well what is at stake when they do accept these favors.

Whenever engineers accept a gift, favor, or hospitality from a contractor or potential contractor, their freedom of action is inhibited. The obligation to deal with the particular contractor may not be very evident—many times it acts only as a subtle reminder. Nevertheless, engineers are human beings capable of being persuaded even against their best engineering judgment. Influence of this nature is against the best interests of the engineers' company. In recognition of this, many large companies have adopted policies restricting or eliminating receipt of gifts from vendors.

It has been said that everyone has a "price." Perhaps this is true. However, those in the engineering profession feel that an engineer's integrity cannot be bought by the people with whom the engineer deals as the employer's agent—it has no price tag. If engineers cannot afford to buy their own lunches or pocket knives, they should look to a new employer for economic improvement, not to the present employer's vendors.

♦ Recruiting Practices

In recent years engineers have been in short supply. It is occasionally contended that anyone with a reasonable pulse and a diploma certifying him to be a graduate of an engineering curriculum is eligible to be hired as an engineer. This demand for engineers is not likely to decline substantially in the foreseeable future.

As might be expected, this unprecedented demand for engineers has led to some peculiar and even reprehensible recruiting practices. Proper use of the talent and problem-solving ability of qualified engineers can make or save money for a company. Therefore, pressure is brought to bear on those whose duty it is to acquire such staff. The results are not always in keeping with the highest ethical practices. For instance, many consider it quite unethical for a company to contact engineers working for another company in an effort to lure them away. If the engineers make the first move, though, the resulting job change usually is considered to be above reproach. Sometimes the efforts used to cause engineers to make this first move can be ludicrous.

The blame for a company's loss of engineers may sometimes be laid at its own doorstep. A distressingly large number of companies have a tendency to overlook contributions made by their engineers. If engineers' method or design changes result in a sizable saving to the company, the engineers may feel that there should be some recognition of them as a result—after all, they could have accepted paychecks and performed only as their managers required. The company, on the other hand, may feel that engineers are paid both for periods in which their contributions are outstanding and for many other periods in which seemingly little is accomplished.

As a result of these conflicting views, engineers often become dissatisfied and are easy marks for competing offers. Most companies find that some form of tangible recognition (salary increase or bonus) along with words of praise for outstanding jobs will inhibit such dissatisfaction.

Engineers, however, owe a duty of full service to their employer. During the first few months after new engineers are hired their contributions are not likely to be great, yet their employer has invested time and money in them. If engineers quit before they have repaid this investment, the company loses money. Engineers who have frequently "job-hopped" may find that even though a demand for engineers exists, they have a rough time getting another job.

The NSPE's code of ethics addresses the professional activities of engineers, not the activities of students or others in seeking employment. To cut down on the more flagrant abuses in the recruiting context and to aid in the development and maintenance of high ethical standards in college recruiting of engineers, the American Society for Engineering Education formulated a code of ethics. This code of ethics, "Recruiting Practices and Procedures—1959," was endorsed by the ECPD and has been distributed to placement offices on engineering college campuses. The code outlines the responsibilities of the employer, the college, and the student applicant in an attempt to secure fair treatment of each by the other two.

Engineers' ethics will be tested many times during their careers. For example, consider an engineer who knows of a design problem with the steering system for a new car. What does the engineer do if management decides the $15 per unit required for the alternative design is too much? Have a discussion with the supervisor about the design? Send a memorandum questioning the decision and emphasizing that lives are at stake? Provide the press with information about the chosen design and the costs of alternatives? At what stage has the engineer done enough to discharge his or her ethical obligations?

Other situations may test an engineer's ethics. Suppose the engineer knows of a discharge of a carcinogenic chemical into a nearby river that is used for drinking water. Does the engineer have a duty to report the discharge? If so, to whom—the engineer's supervisor, the Environmental Protection Agency, the local authorities using the river for drinking water, or the press? No clear answers exist. Hence, the engineer must base such decisions on personal ethics and values.

♦ The Ethics Tool

Professional engineers are taught to be ethical. Their code of ethics is as much a tool as is their knowledge of the grain structure of steel or the deflection of a beam. Proper use of the tools engineers possess will give them a rewarding career; improper use may lead to frustration and disaster.

CRIMINAL LAW

In the United States we have many rights and freedoms. We can do as we please up to the point where the things we do infringe on the rights or freedoms of others. A crime is an antisocial act, but, of course, not all antisocial acts are crimes. Telling true but unflattering stories about another to secure a promotion would probably be considered antisocial by most, but it would not be a crime. Society is injured when a crime is committed. The main purpose of trying a person accused of a crime and punishing him or her when guilt is determined is to prevent recurrences of the criminal act.

One main distinction between a criminal action and a civil action is that for the criminal action the state undertakes prosecution, whereas a citizen undertakes civil action. A criminal action is undertaken to punish the wrongdoer; a civil action is pursued to get compensation for a loss suffered or to prevent a loss from being suffered.

The same act may constitute both a tort and a crime—as such it gives rise to both a civil and a criminal action against the person who committed the act. In fact, it takes some thought to conceive of a *tort* (an injury to another's person or property) that may not be a crime and a crime that would in no way be a tort.

A *crime* is an act prohibited by statute. As time passes and society becomes more complex, we tend to increase the number of laws defining certain acts as crimes. When a criminal statute is made, it is generally necessary to answer at least three questions in the statute:

1. What is the act (or omission) that is to be prohibited?
2. Who can commit the crime (or, conversely, who cannot)?
3. What is the punishment for commission of the act?

Often there is both a physical and a mental component of a crime. The mental element is known as *criminal intent*. Many state statutes eliminate a requirement for this proof on the basis that the act speaks for itself and that a person will intend the natural results of such acts.

Punishment for a criminal offense ordinarily takes one of the following forms: death; imprisonment; fine; removal from office; or disqualification to hold and enjoy any office of honor, trust, or profit under the constitution or laws of the state.

It should be recognized that basic morals and ethics form much of our society's criminal laws. Current criminal laws prohibiting homicide are not too different from the Old Testament commandment "Thou shall not kill." At the same time, it is important to recognize that criminal laws usually reflect the lowest common denominator of society's standards. Such laws basically set the low standard; if you violate these laws, you may be punished by society. For example, many consider a breach of contract unethical, yet it is rarely, if ever, a criminal act.

Degrees of Crime

Treason

The highest crime is treason. The United States Constitution states: "Treason against the United States shall consist only in levying War against them, or in adhering to their Enemies, giving them Aid and Comfort. No Person shall be convicted of Treason unless on the Testimony of two Witnesses to the same overt Act, or on Confession in open Court."

Felony

Felonies compose the second level of crime. Early common law punished felonies with a sentence to death. Today a **felony** is generally defined as an act that is punishable by death or imprisonment in a penitentiary for a term of longer than one year.

Misdemeanor

The lowest level of crime is known as a *misdemeanor*. It consists of all prohibited acts less than felonies. Traffic violations are misdemeanors, as are zoning law violations and breaches of the peace. Punishment for a misdemeanor usually consists of a fine or jail sentence or, generally, anything less than death or imprisonment in a penitentiary for a term of longer than one year.

REVIEW QUESTIONS

1. A process engineer was about to recommend the purchase of equipment for a new manufacturing process for his employer. A vendor calling on him told him of a new and apparently cheaper means of accomplishing the same result. The engineer was familiar with the type of work in question but had never heard of the new process. The vendor invited him to go (at the vendor's expense) to several sites where the new process was being used. Should the engineer go to see the new process? At the vendor's expense? What canons or rules have a bearing on the situation?
2. If the engineer took the trip mentioned above and the vendor suggested an evening at a local night club to avoid the boredom of a hotel room (with the vendor picking up the night club tab), should the engineer accept or decline? Why or why not?
3. In going through the files on a process in which her employer finds himself in trouble, an engineer finds several instances of poor judgment and miscalculation by her predecessor who set up the process. Most of the present problems in the process are caused by the previous engineer's errors. The previous engineer left the company and is now working for another firm. The present engineer's assignment is to improve the process. How should the improvements be justified to her employer? What, if anything, should the engineer say or do about her predecessor's mistakes? What canons or rules apply?
4. About a month ago an engineer made an outstanding improvement in a process. His company produces approximately 300,000 parts per year through the process, and the direct labor savings alone amounts to approximately $0.40 per part. The supervisor of the department in which the process is located has complimented the engineer on his achievement, but no one else in the plant has done anything more than mention it to him. What, if anything, should the engineer do?
5. What are some of the characteristics of a "professional person?"
6. Just before Christmas an engineer receives a package from a vendor with whom she has dealt in the past. The engineer is now concerned with work entirely outside the field of the vendor's interest. The package contains eight place settings of sterling silver. Should the engineer keep the gift or return it? Why?
7. An engineer is approached by a friend who argues stoutly for joining an engineer's union. The friend points out that promotions will be based largely on seniority, that wages will be paid according to the class of work undertaken (which is likely to improve the engineer's economic situation), and that overtime will be paid for all work in excess of 40 hours per week or 8 hours per day. Should the engineer join the union? What are the arguments for doing so and for not doing so?
8. Describe and give an example of each of the three degrees of crime under United States law.

Chapter 3

Development of Law

Our activities are regulated by laws. As we live and work we become familiar with many laws, particularly those concerning the physical world—the laws of nature. We know that if we are near the earth when we drop an object, it will fall to the earth in accordance with the law of gravity. We can even predict accurately how fast the object will fall and where it will strike the earth if we consider the laws of motion and the retarding forces. Such natural laws form a particular kind of universe of laws. They are not human-made laws, only human-discovered. These laws would exist even if we passed a legislative act against them. It is interesting to note that such an act was once attempted; a state legislature tried to set the value of π as 3.000 to make calculations involving the diameter and circumference of circles more convenient.

In this text we are concerned with human-made laws, the laws governing relationships between people. As we will talk about it here, *law* refers to a set of rules and principles set up by society to restrict the conduct and protect the rights of its members.

A person living alone who had no contacts with others would have no need for human-made laws. Add another person and the need for law would become apparent. Each of the two would have rights that might be infringed on by the other. In fairness, each must control his or her behavior in such a manner that the other's rights are protected. In such a simple society the relationships would not be complex; simple rules would be sufficient.

People are gregarious; our social instincts are highly developed. Judging from the steadily increasing percentage of urban residents in our population, our social tendencies appear to be increasing, due either to the strength of the "pull" or to the ease of succumbing to it. As we become a more urban society we require more laws and restrictions of greater complexity to govern our behavior.

BEGINNINGS OF LAW

Laws began as social customs. It was considered proper to behave in certain manners in particular circumstances. At first the tribal chief and later the priest was charged with the preservation of these customs, including punishment for infractions. In the hands of the priests, the idea that laws were of divine origin was fostered. Thus many of our early laws, as well as the present Islamic system, were said to have resulted from divine manifestations.

♦ Divine Laws

Two codes of divine laws have made major contributions to the laws of Western civilization. The first of these is the Code of Hammurabi (about 1750 B.C.), based on the idea "an eye for an eye, and a tooth for a tooth." Early justice in the United States was sometimes not far removed from this concept, and the sentence of death for first degree murder is similar.

The second set of divine laws that influenced our law is the Ten Commandments. Present influence of these laws in our legislation and court decisions is easily found. As rules, these Mosaic laws have an outstanding feature—they are short and simple. Each person is presumed to know the laws under which he or she lives. It is necessary that this presumption be made; if it were not, anyone could plead ignorance of the law and thereby avoid it. Yet how valid is this presumption? How many U.S. citizens know the local ordinances under which they live? Or the state statutes? Or the federal laws? Or even the rights guaranteed or the restrictions imposed by the Constitution and Bill of Rights? Children can be taught the Ten Commandments, on the other hand, and they will be able to recall many of them when they reach adulthood, even though they may have little contact with the commandments in the intervening years.

♦ Civil Law

Two great systems of law are used in the Western nations. *Common law* is used in most of the English-speaking countries; *civil law* is used in the remainder of Western societies. Civil law (sometimes called the *Continental system*) originated about 450 B.C. as the Law of the Twelve Tables. This was the law of the Roman Empire as it expanded and contracted during the next 10 centuries. During that time statutes were passed and meanings clarified in court decisions. Under Emperor Justinian in the sixth century these laws were all boiled down to their essentials and published as the *Corpus Juris Civilis*, the body of the civil law. This civil law spread to other countries and became the foundation of the legal systems of continental Europe.

In operation, Roman civil law and English common law are quite different, though in practice the results are usually the same. Civil law is based on the idea of comprehensive and complete written codes or statutes, the court's task being to apply the correct statute to the particular set of facts in the case. Common law is built on the use of cases—prior decisions in similar factual situations.

Most of the law brought to the United States by its early settlers was common law. However, in the states settled by French and Spanish settlers, remnants of civil law may be found. Thus, in Louisiana, Texas, and California, principles of civil law have had and continue to have some influence.

The term *civil law* has come to have a dual meaning. It means a code of law based on the Roman codes, but also it has come to be used today to describe our system of private law as opposed to criminal law. When the term civil law is used in the remainder of the text it will refer to the system of *private law*.

♦ Enforcement

In governing the behavior of individuals in a society, a human-made rule has little practical value unless it can be enforced. Enforcement takes one of three forms:

- *Punishment.* Fines or imprisonment are the usual means of punishing someone who has committed a crime.
- *Relief.* In actions involving private rights, the relief sought is usually money damages for the person who has been harmed at the hand of another, or a court order preventing future harm.
- *Social censure.* Frequently the strongest enforcement factor is the fear of social ostracism—the fear of public opinion. Although, strictly speaking, public opinion is not a recognized legal means of enforcement, its existence and strength cannot be denied.

Few of us possess a formal knowledge of the laws that govern our actions. Yet we obey them. Even the worst criminals obey nearly all the laws almost all the time. We comply unconsciously. The laws have become a part of each of us—they have been a part of our lives since earliest childhood. Thus, even though we may not know the specific laws to which we conform, we do conform, and we are aware when others do not. Often we are not aware of the definite rule violated or the rights infringed. An attempt is made here to bring such an awareness to the reader.

COMMON LAW AND EQUITY

♦ Types of Laws

Our laws are of four basic types:

- constitutional
- statute
- common law
- equity

♦ Constitutional Law

Constitutional law sets up the operation of a government, including its powers and limitations. It states fundamental principles in the relationships between citizen and state, including rights which may not be infringed. Rights may arise under both state and federal constitutions.

♦ Statute Law

A *statute* is a law stating the express declaration of the will of a legislature in the subject of the law enacted. A state law prohibiting gambling and setting forth a maximum penalty of $500 or six months in jail for violations would be an example of statute law. Similarly, federal laws, such as the Interstate Commerce Act, and municipal laws (or ordinances) regulating traffic are statutes. Administrative regulations, such as those of the Federal Trade Commission, have the same force and effect as statutes. Treaties also have generally the same force as statutes passed by Congress.

One of the many reasons for passage of a statute is dissatisfaction with the law in a particular field. Frequently, when decisions have for some reason become muddled in dealing with a problem, an appeal is made to the legislature for a law to clarify the issues involved. Passage of a statute voids the common law covering the same point within the legislature's jurisdiction. The statute must not, of course, conflict with the U.S. Constitution or with the state constitution involved. If, when a case arises, the courts find that a statute conflicts with the constitution, the constitution is protected, and the statute is held invalid.

Another reason for passing a statute is the legislature's ability to develop a comprehensive set of rules to take care of a specific problem. Suppose that the problem of low-level radioactive waste disposal is being considered in a given state. That state's legislature can enact a statute dealing with all the relevant questions of who makes the decisions, who handles the waste, where waste storage is to be located, when should it be done, how the waste should be handled, who is liable to whom and what amounts of liability are to be imposed, and what remedies are available. Needless to say, it would be quite likely that many years and several complex decisions at the least would be needed before the judiciary could develop similar rules.

♦ Common Law

U.S. law originated as the common law brought over from England by the colonists. It is still our law; it is used by a court when no statute or constitutional law covers the particular legal problem involved.

♦ Origin

The law that existed before the Norman conquest of England was local law. Each town or shire had its own law, and each town's body of law differed somewhat from those of other towns. The king took an interest in the law only in an exceptional case.

The Norman conquerors were organizers. Under them a national *king's council* was established to make laws and decide cases. Eventually, the council evolved into *parliament* and a system of king's courts.

In the first century or so following the Norman conquest, what we know as common law began. It arose from the practice of judges to write their opinions, giving the general principles and the reasoning they followed in deciding cases. When the facts were similar, judges tended to follow earlier opinions of other judges. These decisions were followed by the courts throughout England. Hence, the term *common law* was used to describe these decisions.

Stare Decisis

Abiding by previous decisions is known as *stare decisis*. The main feature of common law is that the law itself is built on case decisions. When a case is decided, that decision becomes the law for that court and other courts within its jurisdiction[1] in deciding similar future cases. (In the cases in this text you will see many references to previously decided cases. Many other references have been omitted to allow easier reading of each case.) A judge uses earlier cases as the foundation for decision. About 200 years after the Norman conquest a jurist named Henry Bracton compiled the decisions that had been rendered under the king's court systems. This was the beginning of case reporting. These and later decisions formed the common law of England and, eventually, of nearly all of the United States.

1. A court's jurisdiction, in the manner used above, indicates the area in which the court may operate. For example, the jurisdiction of the Supreme Court of Ohio is the State of Ohio, and a common law decision by the Ohio Supreme Court would be binding on all other courts in Ohio.

Shortly before the American Revolution, Sir William Blackstone, an English jurist, completed his *Commentaries on the Laws of England*. In this law-book—the first major contribution since the time of Bracton—Blackstone clarified and made intelligible the English common law. An American jurist, James Kent, made a similar contribution in this country about 60 years later. Common law is defined in Kent's *Commentaries* as "those principles, usages, and rules of action applicable to the government and security of persons and property, which do not rest for their authority upon any express or positive declaration of the will of the legislature." In both books the principles of common law were extracted from recorded decisions. These two books provided the basis for further development of common law in both countries.

It is frequently stated that courts do not make the laws, that they merely enforce them. The statement is largely true of constitutional law and statutory law, but not for the common law. As decisions are made for new types of cases, new interpretations of law are made by the courts; the law is amplified in this way continuously.

Business Custom

In deciding cases the courts often will make use of a relevant business custom. For instance, as will be discussed later, silence on the part of one to whom an offer is made usually cannot constitute acceptance. However, if there has been a history of dealings between persons or if there is a practice in a particular business such that silence constitutes acceptance, the court will consider this and decide accordingly. Terminology peculiar to a trade is given its trade usage interpretation in court.

Business customs played an extremely important role in legal evolution, because these customs developed into what is often called the "law merchant." During the Middle Ages and then the Renaissance, commerce and trade flourished in certain parts of Europe. Merchants often needed some certainty about the rules that would be applied to their disputes by courts in various locations. Eventually, a body of law developed out of the business customs, and this set of laws was used by various courts in deciding commercial cases between merchants.

Changes in the Law

Frequently one hears the complaint that the law is behind the times, that it is slow to change in a rapidly changing world. The complaint is fairly well founded; law *is* slow to change. However, it is usually far better to have a law or fixed principle on which one can depend than to have a law or principle that may this time be decided one way and the next time, another way.

The law does change, slowly, to reflect changes in society and changes in technology. By way of illustration we can look back a few years to the changes that became necessary when the automobile took over personal transportation from the horse and buggy. Similarly, we can look ahead to changes that are likely to be needed when space travel becomes a commercial reality.

Laws usually are changed in two ways: by overruled decisions and by the enactment of statutes.

Overruled Decisions

When a case goes to court, the attorneys for both parties in the case have usually done some research on the law involved. It is likely that both will be armed with decisions in previous cases on which the judge in the present case is expected to base the case's decision. Let us assume that one of the cases used is based on facts quite similar to those in the case at hand. Let us further assume that the decision in the earlier case was handed down by the state supreme court. If the present case is in a lower court and the facts of the case are in all ways the same, the judge should follow the prior decision. However, because there are nearly always differences in the facts of two cases, let us assume that the judge, on the basis of slightly different facts, did not follow the earlier case and that an appeal resulted. The case at hand is finally taken to the state supreme court. If the supreme court justices see the facts in the two cases as being essentially the same but render a decision different from the decision in the precedent case, the earlier case has been either distinguished on the factual differences or overruled. The law has been changed. There is no effect as to the parties in the earlier case; that case was decided by the law of that time. The law was not changed until the state supreme court changed it in the process of overruling its earlier precedent.

In the interest of retaining stability in the law, courts are quite reluctant to overrule prior decisions. At the same time, they recognize that nothing is as permanent in this world as change itself.

♦ Equity

In the beginning the common law was administered by royal writs. These writs were orders, in written form, to a sheriff or other officer to administer justice in a particular way or to summon a defendant before the royal justices. We still use writs—written and sealed court commands or mandates ordering some specified action to be done. Following a court judgment, a writ of execution, for instance, may be issued to an officer telling the officer to take possession of and sell some of the loser's property to satisfy the judgment.

During the first two or three centuries following the inception of the king's court in England, the law underwent a hardening process. The writs accepted by a common-law court and the remedies offered became standardized and limited.

The limited remedies of common law brought about equity. Consider an example. Black, a contractor, is hired to build a structure for White. During the excavation, Gray, next door, notes an impending separation of her house from its foundation. The walls start to crack, and the ceiling begins to bulge. At this point, at common law, Gray has no remedy. With common law she would have to await whatever damage might be forthcoming and sue to get compensation for it. However, in equity she can do something about it now. She can obtain a temporary injunction or restraining order to prevent the excavation next door until something more appropriate can be done to prevent her house from sliding into the hole.

Origin

With the limiting of common-law writs, certain injustices could take place without any relief being offered by the courts.

Common law could not and would not do any of the following:

1. Prevent a wrong from taking place
2. Order persons to perform their obligations
3. Correct mistakes

It soon became apparent that these gaps in the law had to be filled.

Courts of Chancery

With the King's court system the king had become established as the "fountain of justice." As such, he could offer redress beyond that available in the common-law courts. In unusual cases the king was petitioned and he settled the cases with "unusual" remedies. When the load became too heavy for the king, the task of hearing the unusual cases was delegated to his chancellor and, eventually, to vice-chancellors. From this the new courts took on the name of *courts of chancery*.

In the United States, there used to be separate court rooms and judges for equity cases. In a few states, the court rooms are the same and the judges are the same ones who deal with other types of law, but the equity procedure is different. In most states and in the federal court system, law and equity have been completely combined.

Unusual Remedies

The remedies available in a court of equity are quite different from those offered in common law. They are *in personam* remedies; that is, they are directed to a person, whereas common law acts *in rem*, or on an object. Probably the most common and well-known equity remedies are the injunction and specific performance. The list of equity remedies, though, is quite long and includes divorces, mortgage foreclosures, accountings, reformation of contracts, and many others. In fact, an equity court can act in any way necessary to secure a right or remedy a wrong. The very word *equity* implies that justice will be done, and if a remedy must be invented to serve the purpose, that will be done.

The injunction exists in two general forms: temporary and permanent. A *temporary injunction* (a restraining order or injunction *pendente lite*) is readily obtainable for cause. An attorney can request a temporary injunction from the court. As soon as the judge signs the temporary injunction, it is an act in contempt of court for anyone who has knowledge of the order and who is subject to the order to fail to obey it. The object of a temporary injunction is to hold the status quo until a hearing can be held on the merits of the case. The complaint may be dismissed at the hearing, some other type of remedy may be given, or a *permanent injunction* may result when the facts are heard. If a permanent injunction is issued, the order is effective as long as the cause of the injunction exists.

Specific performance usually arises in connection with contracts involving land. Land is considered to be unique; no one piece of land is exactly like another. When a contract is made for the sale of a piece of land and the seller refuses to deed the land to the buyer, the seller may be forced to do so by an order for specific performance from a court of equity. Property other than real estate is treated the same way only when it is recognized as being unique, for example, an antique or a rare painting.

Equity is reluctant to give a remedy that will require continuous supervision over an extended period. In some cases such remedies have been given, but where another remedy will suffice, the other remedy is preferred.

The remedies afforded in equity courts are not limited to equity remedies. Once a cause of action is legitimately in an equity court, the court will settle all the issues involved, including the remedies afforded at common law and the equity remedy. For instance, an equity decree might include money damages for injuries already suffered as well as an injunction against further injury. Equity, however, will not give a remedy that is directly contrary to common law; neither will equity act when an adequate remedy exists under a statute or under common law.

Features

Certain features of equity law are different from those of common law.

Requirements

To obtain equitable relief, the plaintiff is required to meet the applicable standard. Different courts have different standards for various types of equitable relief. To obtain a preliminary injunction, one court requires that the person seeking relief show the following:

1. a substantial likelihood of success on the merits
2. irreparable injury
3. that the harm if the relief requested is denied outweighs the harm if the relief is granted
4. that the public interest is not disserved by granting the relief.

Irreparable injury usually refers to the court's inability to completely remedy the situation through an award of money damages, such as when such damages cannot be calculated with accuracy or could not provide a complete remedy. A *nuisance*—such as smoke or noise that makes a home untenantable—would be a basis for getting into equity jurisdiction with this reason.

Speed

A court of equity generally acts more rapidly than does a court of law. As previously noted, a temporary injunction may be obtained with little delay. In addition, there is usually no jury involved inequity cases, which eliminates the time consuming selection of jurors and deliberation over the evidence.

Privacy

Because there is no jury involved, a case in equity may be decided more privately. This feature is particularly significant when a case involving a trade secret is tried. If a jury were to hear the facts, there would be 12 more people to hear the secret. The secret would become virtually public information.

Injustices

On occasion, equity decisions appear to work injustices on defendants. Consider this example. Black owns an industrial plant bordering a stream. The plant discharges waste materials into the stream. White owns property downstream from the plant, which also borders the stream. White's property is far enough downstream so that the water purifies itself before it gets to his property. White has never been injured as a result of the waste discharge into the stream and does not use the stream except as a location of the boundary of his property. White brings suit to enjoin waste disposal into the stream by Black. A court of equity would be likely to enjoin the waste disposal, even though no one has been injured by it. If the court were to allow the waste disposal to continue, it is possible that, with the growth of the plant, harmful contamination could result. Also, the possibility exists that someone else might later obtain and use property between Black's plant and White's property and be injured by the stream's pollution. If the court dismissed the suit and the contamination became harmful, the dismissal could stand in the way of future relief, giving Black an apparent right to continue. Therefore, the court would have to find some solution other than outright dismissal of the case. Any solution other than dismissal is likely to add to Black's cost of operation.

Principles

Many equity principles and maxims form the background for equity decisions. Five prominent equity maxims are the following:

- For every right a remedy.
- He who seeks equity must do equity.

- He who comes to equity must come with clean hands.
- Equity regards substance rather than form.
- Equity aids the vigilant, not those who sleep on their rights.

HOLZWORTHV. ROTH
101 N.W. 2d 393 (S.D. 1960)

This suit in equity commenced on July 18, 1958, claimed a breach of the covenants of title in a series of standard form warranty deeds by which the title to Lot 37, Harmony Heights Addition in Spearfish, South Dakota, went from the Lampert Lumber Company to the Holzworths, as ultimate grantees. Those named as defendants are the grantors in this series of deeds. They interposed numerous objections to the suit, one of them being that in the circumstances of this case a suit in equity did not lie.

During the time that these named defendants successively owned the property, between June 26, 1956 and April 26, 1957, the construction of a residence thereon was commenced and carried on. Eight of the laborers and materialmen contributing to this improvement were not paid so they filed mechanic's liens[2]. Six of them were filed for improvements commenced during the period when these defendants were the owners, and two for improvements started after the plaintiffs acquired their title. Subsequently on October 22, 1957, two of these lien claimants instituted an action to foreclose their liens naming the other six lien claimants and Martin Holzworth as defendants. This is urged by the plaintiffs as a breach of the covenants of title. None of the grantors in these various deeds were made parties to this foreclosure action nor does it appear that they had notice of it. Whether it was further prosecuted and with what results does not appear in this record. However, plaintiffs still occupy the premises.

In the suit here involved the Holzworths as the ultimate covenantees asked that the defendants be required to remove the mechanic's liens or pay the amount thereof. The defendants by motion for judgment, by their answer, and by a requested conclusion of law directed the trial court's attention to their objection that a suit in equity was not the proper remedy. The trial court overruled this contention and entered judgment requiring the defendants to remove the various liens or pay to the plaintiffs an amount equal to any judgment entered in the lien foreclosure action. From this judgment the defendants appeal.

In this state the distinction between actions at law and suits in equity is abolished by statute.... All relief is administered through one proceeding termed a civil action. However, this statutory abolition of distinctions applies only to the form of action, and not to the inherent substantive principles which underlie the two systems of procedure.... In other words, the essential and inherent differences between legal and equitable relief are still recognized and enforced in our system of jurisprudence.... One of these principles is that if the primary right which is the foundation of the litigation is legal in nature and there is a remedy at law, the action is one at law.... Equity has jurisdiction in such cases only if the legal remedy is not full, adequate, and complete.

When property is conveyed by our standard form warranty deed, SDC 51.1403 writes into it these covenants on the part of the grantor, his heirs, and personal representatives:

> that he is lawfully seized of the premises in fee simple, and has good right to convey the same; that the premises are free from all incumbrances; that he warrants to the

---

2. Note: A *mechanic's lien* is a claim established against property to secure priority of payment for work done to improve that property; it may be established by almost any umpaid person who had a hand in the improvement.

> grantee, his heirs, and assigns, the quiet and peaceable possession thereof; and that he will defend the title thereto against all persons who may lawfully claim the same.

They spell out the obligation on the part of the grantor, his heirs, and personal representatives arising out of the agreement between the parties. A breach of any of them is in effect a breach of their contract for which an action for damages will lie.

Significantly the chapter of our statutes concerned with damages for breach of contract,...prescribes the measure of damages to be allowed on breach of the covenants contained in our standard form warranty deed.... It seems to us that an action for breach of these covenants is clearly an action at law.... We think it follows that equity has no jurisdiction in this case if the remedy at law is full, adequate, and complete....

Concerning the jurisdiction of equity the question of adequacy of the remedy at law appears in two aspects. Where the suit is properly cognizable in equity the existence of an adequate remedy at law may justify a court of equity in refusing to entertain the matter.... On the other hand, where the matter is legal in nature the absence of an adequate remedy at law is necessary to confer equitable jurisdiction. This is the aspect in which it is here involved. Accordingly, the burden of establishing this is on the plaintiffs.

We are unable to find that such inadequacy was urged by them or that it exists. On this feature the record is silent except as it is referred to in the contentions of the defendants. The complaint does not plead the inadequacy of an action at law nor are there any factual allegations from which such conclusion is inferable. In the findings and conclusions proposed by plaintiffs and adopted by the court this matter is not mentioned. Nor did the plaintiffs in either their brief or argument in this court make a claim of such inadequacy. In the absence of fraud or some other unusual circumstances rendering the remedy at law inadequate equity will not interfere in this type of case.... Since it does not appear that plaintiffs' remedy at law was inadequate we must hold that equity is without jurisdiction.

Their remedy at law would have been even more efficient if Holzworth, who was a defendant in the lien foreclosure proceeding, had given these defendants notice of it and requested them to come in and defend the title they had warranted. This is called "voucher to warranty."... After such notice and request the judgment in that proceeding would be binding on them if they did not defend it.... That he neglected to utilize this privilege is of no help in getting this matter into equity.... The trial court should have dismissed this suit.

Reversed.

REVIEW QUESTIONS

1. Why do we need human-made laws? What purpose do they serve?
2. What are the four basic types of laws in our legal system? What does each consist of?
3. What is the meaning and significance of the term *stare decisis*?
4. How do changes in common law take place?
5. What factors led to the establishment of equity as a separate system of law?
6. What types of remedies are offered by common law? By equity?
7. Party A owns a factory that emits large quantities of foul-smelling smoke. People in a nearby housing development are annoyed by the odors whenever the wind shifts to a certain direction. Party B, a resident in the housing development, has lodged a complaint against A for public nuisance. In what kind of court would the case be likely to be tried in your state? Why? What would be the probable result of the legal action?
8. What are the general requirements for equity jurisdiction? How might these requirements be met?
9. In *Holzworth v. Roth* the decision of the trial court was reversed with the statement that the case should have been dismissed. Does this mean that Holzworth must give up his attempt to recover for the apparent injustice he has suffered? What else can he do?

Chapter 4

Courts, Trial Procedure, and Evidence

The fact that one person accuses another of a wrong does not mean that the accusation is proper. Since the beginning of civilization the problem of determining the truth of an accusation has existed. Court trials with a judge, a jury, and attorneys for both plaintiff and defendant have not always been used. Probably the earliest form of trial was trial by battle. Accused and accuser (or their representatives) faced each other in battle, with the outcome determining the justice of the accusation.

In many of the early civilizations trial by battle was replaced either by trial by ordeal or trial by jury. The codes of Hammurabi required trial by ordeal. For certain acts the accused was to be thrown in the divine river. If the river held him (if he couldn't swim), the guilty verdict and punishment were delivered at the same time. Trial by jury, in one form or another, existed in many ancient civilizations. Our present jury trial system developed with the common law. Its continuance in criminal and civil cases is guaranteed in the sixth and seventh amendments of the U.S. Constitution.

COURTS

In the United States we have a system of courts for each state as well as a system of federal courts. The systems are somewhat similar. Both the federal and state systems provide an ultimate tribunal, a supreme court. At the next lower level are the appellate courts (in the federal system and some state systems) that handle appeals from the lower courts. The lower courts are the district courts in the federal system, and usually county courts in the state systems.

♦ Federal Courts

Supreme Court

The United States Supreme Court is our highest tribunal. It is our final court of appeal. Article 3 of the U.S. Constitution provides for our Supreme Court and whatever inferior federal courts Congress may from time to time require. The jurisdiction (for types of cases) is limited in the Constitution to nine categories. The only cases that may originate in the Supreme Court are ones involving ambassadors, public ministers, and consuls, or in which a state is a party. In these the U.S. Supreme Court has original jurisdiction; all other cases go to the Supreme Court by appeal.

Cases are appealed to the Supreme Court from either U.S. courts of appeals or from state supreme courts in the normal course of events. However, cases may go directly from any court to the U.S. Supreme Court if the question to be settled involves the U.S. Constitution or is of very great public interest. Appeal is usually made by a petition to the court for a *writ of certiorari*. If the petition by the appellant is successful, a writ of certiorari will be issued to the lower court demanding that the case be sent up for review. Only a very small portion of such petitions are successful—something like 1 out of 20.

The nine justices who sit on the U.S. Supreme Court have the final say about what our law shall be. They do not, of course, make formal statutes—this is the function of the legislative branch of the government. Their highly significant function is to interpret what is meant by the wording of the Constitution or the wording of a statute. The interpretation given by the Supreme Court determines the lawful interpretation to be used in future cases in lower courts.

Courts of Appeals

Thirteen United States courts of appeals exist. Appeals on federal questions are normally settled by three justices who then form a panel of each court. These courts were first established by Congress in 1891 because of the burden of appeals on the U.S. Supreme Court. The courts of appeals function as appellate courts only and do not conduct trials. Issues between parties are appealed on the basis of a conflict in the law, the facts of the issue having been decided previously in a lower court.

District Courts

The trial courts of the federal court system are the United States district courts. The districts presided over by the 100 or so U.S. district courts are formed in such a way that no state is without a federal district court. The number of justices in a U.S. district court is determined by statute and is based on the number of federal cases arising in the district.

For a case to be tried in a federal court, it must meet at least one of these criteria:

- Arise from the U.S. Constitution, federal laws, or treaties of the United States
- Affect ambassadors, other public ministers and consuls
- Arise in admiralty or maritime jurisdictions
- Involve two or more states as parties
- Involve the United States as a party
- Be between citizens of different states with the amount in controversy greater than $50,000
- Be between citizens of a state who claim land grants in another state
- Be between a state and a foreign country
- Be between a United States citizen and a foreign country or its citizens

Most of the controversies handled by the district courts involve federal statutes or the Constitution (that is, "federal question" cases) or "diversity" cases (that is, those between citizens of different states).

Common examples of cases tried in United States district courts are admiralty, patent, copyright and trademark, restraint of trade, tax cases, and infringement of personal rights. Many other cases arise from decisions of administrative boards (such as the National Labor Relations Board) that operate as quasi-judicial entities and look to the federal courts for enforcement of their orders. Claims against the federal government may be filed either in a U.S. district court or in the U.S. Court of Claims. The U.S. Court of Claims is a special court in the federal system. Other special courts include the U.S. Tax Court.

♦ State Courts

State court systems are far from uniform throughout the United States. Not only do the systems differ, but the names of the courts differ as well. A few generalities can be stated.

Each state has a final court of appeal, usually called the supreme court of the state. Nearly always the highest court confines its work to appeals of cases tried in lower courts. Courts of intermediate appellate jurisdiction are interposed between the supreme court and lower courts in many states.

The next lower tier of courts consists of the trial courts of general jurisdiction, known variously as *circuit courts, courts of common pleas, county courts, superior courts*, or, in New York, the *supreme court.* Probate courts or surrogate courts handle cases of wills, trusts, and the like. Such courts are limited in geographical jurisdiction to a particular county, district, or other major political subdivision of the state.

At the lower end of the judicial hierarchy are various municipal courts. Police courts, justices of the peace, small claims courts, juvenile courts, and recorders' courts are common examples of these courts of very limited jurisdiction.

♦ Jurisdiction

The jurisdiction of a court means its right or authority, given either by a legislature or constitution, to hear and determine causes of action presented to it. A court's jurisdiction relates to geographical regions (for the location of persons and subject matter), to types of cases, and possibly to the amount of money or types of relief concerned.

Courts have jurisdiction over property, both real and personal, located within their assigned geographical limits. Even though the owner may not be available, his or her property may be taken in satisfaction of a judgment. A court has no authority over property lying outside its territorial limits. Let us assume that a Tennessee court has awarded Black $5,000 as a result of a damage action against White. If White does not pay and cannot be made to pay in satisfaction of the judgment, justice for Black is rather hollow. The Tennessee court could take and sell that part of White's property that could be found in Tennessee until the judgment was satisfied. However, it could not touch any of White's property in, say, Georgia.

A court has jurisdiction over all persons who have had sufficient contacts with the state such that the exercise of jurisdiction does not offend traditional notions of fair play and substantial justice, whether the persons are residents or not. Jurisdiction over a person is exercised by serving that person with a summons or other legal process—a defendant would be served with a summons; a witness with a subpoena. How service may be made is a matter clarified by each state's statutes. Generally, a sheriff or other officer is directed to serve the process on the person; however, if the person cannot be found there is usually an alternate means of service. In certain types of cases, such as quieting title to real property, divorce, probate of a will, and others in which the thing involved is within the court's jurisdiction, a process may be served even though the person on whom it is to be served is outside the court's jurisdiction.

The means by which this is done is known as *constructive service*. This service on an absentee is made by publication of the process in a local paper. In most damage actions, though, absence of both defendant and property acts as a serious obstacle—action against the defendant would be pointless unless some recovery could be anticipated.

Jurisdiction of courts is also limited as to types of cases that they have authority to handle. A probate court, for instance, ordinarily has no authority to handle criminal cases; criminal courts usually do not handle civil suits.

A monetary limitation is placed upon many of the courts. Justice of the peace courts and small-claims courts are limited to cases involving no more than the statutory limit—usually $150 to $1,000, but up to $5,000 in some states.

PRETRIAL PROCEDURE

As with many areas of law, the rules relating to procedural issues vary. In the federal district courts, the Federal Rules of Civil Procedure control the case.

♦ Pleadings

A lawsuit begins when a complaint is filed with the appropriate court. A complaint should set forth the facts giving rise to the cause of action asserted and should set forth the relief requested by the plaintiff.

Following its receipt of a complaint, the court issues a summons to the defendant. The summons is simply a notice to the defendant that he or she has been sued. A copy of the complaint may or may not accompany the summons, depending on the jurisdiction. If the complaint does not accompany the summons, it is made available to the defendant by the clerk of the court. If the defendant is served a summons but does not answer it, the result may be a judgment against the defendant by default.

Once a complaint has been filed and the defendant has been served according to the statutes or rules relating to service of process, the defendant has a set period of time in which to answer the plaintiff's allegations. Usually, the defendant files an answer that responds to the allegations in the complaint. Generally, most answers deny at least some of the complaint's allegations (or there would be no dispute) and include affirmative defenses, which amount to reasons why the defendant should not be held liable. Often, a defendant also files one or more counterclaims, by which the defendant seeks a recovery from the plaintiff. (In federal court, the plaintiff then needs to file the plaintiff's "answer" to the defendant's counterclaim.)

A defendant also often includes various motions with the answer. For example, a defendant may contest the court's jurisdiction over him or her with a motion to dismiss for lack of personal jurisdiction. A defendant may test the legal sufficiency of the plaintiff's complaint by filing what is called a *motion to dismiss* for failure to state a claim on which relief may be granted. Such a motion, as one might guess, argues that the law does not recognize the claim asserted in the plaintiff's complaint.

Generally, the rules covering pleadings allow the allegations to be fairly general and do not require a great deal of specificity. Sometimes, however, a pleading will be so vague that it is difficult to formulate a response. In such situations, a motion for a more definite statement is appropriate.

♦ Discovery

After the lawsuit begins and before trial, both sides usually engage in *discovery*, in which each side learns about the other's evidence, versions of the facts, and legal theories relating to the case.

Depositions

The Federal Rules of Civil Procedure provide a number of mechanisms for each side to learn about the other's case.

One of the most commonly used discovery procedures is that of depositions on oral questions. A *deposition* is simply a procedure in which the witness, whether a party to the case or not, orally answers questions that are put to the witness by an attorney for one of the parties in the case. The questions, answers, and any objections by other attorneys, are all taken down and transcribed by a court reporter. Before the deposition begins, the witness takes an oath to tell the truth. The transcription thus provides a complete written record of the witness' testimony. More recently, litigants have begun to use videotape depositions. Videotape depositions have the advantage of providing a record of not only the questions and answers but also the facial expressions of the witness and any expressions of nervousness, such as sweating, excessive blinking, and fidgeting.

Request for the Production of Documents

Another method of discovery is the use of a request for the production of documents. To use this method, one party prepares a request that lists the categories of documents that it wants the opponent to produce for inspection and copying. The opponent then reviews the list of categories and responds whether such documents will be produced. The use of such requests for the production of documents allows each side to see the other's documents, which are often quite revealing. Moreover, documents do not forget, as real people sometimes do. Thus, documents are often critical in discovery.

Interrogatories

Yet another type of discovery is the use of interrogatories. *Interrogatories* are essentially written questions that are submitted to the other side. After receiving a set of interrogatories, the other side must respond with any appropriate objections and with answers that are sworn to by the party or an officer of the party. Generally, interrogatories are much cheaper to use than taking the deposition of the other side, but the amount and quality of information received through interrogatories is generally much less and not as good as that received through depositions. Besides the types of discovery procedures just outlined, other procedures are available.

♦ Pretrial Motions

Many cases are resolved short of a full trial. Perhaps the most common method of resolving a case before trial is through a *motion for summary judgment*. Under the Federal Rules of Civil Procedure, a summary judgment is to be rendered by the court if the pleadings, depositions, answers to interrogatories, and any affidavits on file with the court, show that "there is no genuine issue as to any material fact" and that the moving party is entitled to a judgment as a matter of law. To determine whether there are any genuine issues of material fact, the court must look to the substantive laws that relate to the claims asserted in the pleadings. For example, in a case involving a complaint that asserts breach of contract and a counterclaim that asserts fraud, the court will look to contract law and tort law regarding fraud to determine what the issues are. Once the court determines the issues, it looks to the evidence on file (much of which has been developed through discovery) and then determines whether there is evidence sufficient to raise issues of fact for each and every issue of the claim or counterclaim that is the subject of the motion. If there is no genuine issue of material fact for each of the elements of the breach of contract and fraud claims the court grants summary judgment for both.

Another motion that is often used to dispose of some or all of a case is the *motion to dismiss for failure to state a claim* on which relief may be granted. Essentially, this motion is used to take the position that, as a matter of law, there can be no recovery on the claim asserted, even if the facts alleged are true. This motion is often used when new legal theories are being asserted, such as when a court in one jurisdiction allows a certain type of claim that has not yet been recognized in the jurisdiction in which the case is pending.

♦ Joining Additional Parties and Claims

The rules of procedure that apply to cases in the federal courts allow for joining additional claims and additional parties besides those involved in the plaintiff's complaint and the defendant's answer and counterclaims. The rules enable either the plaintiff or the defendant to assert as many claims as each party may have against the other, whether or not these claims are related to the main case filed by the plaintiff. The reason for allowing the parties to add multiple claims is the idea that no inconvenience can result from joining such matters in the pleadings but only from the trial of two or more matters that have little or nothing in common. Accordingly, the rules also provide for separate trials of the different claims or issues to avoid prejudice, to expedite the case, or merely for convenience.

The rules also allow either party to bring in additional parties. For example, suppose you burn your hand on a toaster. If you decide to sue the store that sold you the toaster, the store may then decide to bring the manufacturer of the toaster into the case as an additional party. The store will want to do this to hold the manufacturer responsible and to gain reimbursement for any liability to you. Likewise, the manufacturer may then proceed to bring in still more parties, such as other manufacturers who manufactured the heating control element or the wiring of the toaster. The manufacturer originally named may try to point the finger at other manufacturers of such component parts, arguing that those manufacturers were instead at fault. To pursue the toaster example even further, suppose that a manufacturer of the heating control component has a breach of contract claim against the manufacturer of the toaster. The component manufacturer can then pursue a breach of contract claim against the manufacturer of the completed toaster even though this claim has practically nothing to do with the original claim brought by you for personal injuries due to the toaster.

TRIAL PROCEDURE

♦ The Jury

Parties to a case can choose whether they will have a jury decide the facts of the case. Both parties may agree to submit all the issues, including issues of fact, to the judge. If there are questions of fact (for example, "Did the plaintiff act reasonably?"), both parties must be in agreement if they are to dispense with the jury,

A trial jury is known as a *petit jury*. It usually consists of 12 persons but may consist of fewer people for civil cases or for crimes less serious than capital offenses.

Selection of the individual members of the jury involves both attorneys and the judge. Grounds for challenging prospective jurors are established by statute. The two opposing attorneys may use these grounds to disqualify prospective jurors being examined. For example, grounds for disqualification include a financial or blood connection between the prospective juror and one of the litigants. In addition to disqualifying a juror on the basis of some bias, each attorney may usually disqualify a limited number of prospective jurors arbitrarily by exercising the *right of peremptory challenge*. The judge supervises the qualification proceedings. When the jury has been impaneled the case is ready for trial.

♦ Courtroom Procedure

When the pleadings are complete, the case has come up on the court's docket, and the jury has been chosen and sworn in, the trial begins. Following opening statements by the attorneys, witnesses are sworn in and the evidence is examined. Each attorney sums up the case to the jury, the judge charges the jury, and the jury retires to reach a verdict. In the judge's charge to the jury, he or she sums up the case and instructs the jury about the issues to be decided. After reaching a verdict, the jury returns to the courtroom and the foreman announces the decision. The judge then enters the judgment that is the official decision of the court in the case.

♦ New Trial or Appeal

Within a certain time after the judgment has been entered, a new trial or an appeal may be requested. Generally, a new trial is concerned with error in the facts of the case, whereas an appeal is concerned with a misapplication of the law.

A successful motion for a new trial may be made on the basis of an almost unlimited number of circumstances. It is argued that if a new trial is not granted, there will be a miscarriage of justice. The following are examples of reasons given for requests for new trials:

- Unfairness or bias in selection of members of the jury
- Prejudice stemming from financial or blood relationship between a jury member and a party to the trial
- Misconduct by a jury member, for example, talking with a witness or an attorney in the case
- Error by the judge in failing to allow evidence that should have been admitted or in admitting evidence that should have been excluded
- False testimony (perjury) of a witness
- Unforeseen accident preventing the appearance of a witness

If the motion to the trial court for a new trial is unsuccessful, appeal may be made to a higher court to order a new trial.

Either party may appeal a decision in a civil case; only the defendant may appeal from an adverse criminal judgment. Appeals are based on questions of law—the trial court has decided issues of fact.

Reasons for appeal are presented to the appellate court based on objections or exceptions to the ruling of the trial court. These objections are usually concerned with objections to and rulings on the admissibility of evidence, errors in the conduct of the trial, and instructions by the judge. Evidence presented and testimony taken are usually included along with the judge's instruction to the jury.

The *appellant* (the appealing party) is usually required to post an *appeal bond* when the case goes to a higher court. The purpose of the appeal bond is to insure that the appellant will pay court costs and the judgment to the appellee if the trial court decision is upheld.

The appellant prepares and files with the appellate court a brief of the case. The *brief* contains a statement of the case from the appellant's position and a list of errors forming the basis of appeal. The *appellee* (the party not appealing) then prepares and files a reply brief.

♦ Equity Suits

An equity procedure may be different from that of common law. In an equity suit the judge decides both questions of fact and questions of law. There is no jury unless the judge specifically requires a jury recommendation on a question. Even then, the verdict of the jury is only a recommendation and the final conclusion as to fact rests with the judge. If factual questions are long and involved, the judge may appoint a "master" to take testimony and make recommendations.

Equity acts *in personam*. The decree directs a person to do or not to do a certain thing. The decrees are either interlocutory or final. An *interlocutory decree* reserves the right of the court to act again in the case at some later time. A temporary injunction is an example of such a decree.

♦ Cases

In both common law and equity the doctrine of *stare decisis* is followed. Similar prior cases are followed in deciding present issues. Because cases must be known to be followed, it may be desirable to consider for a moment how cases are recorded and reported.

The West Publishing Company, St. Paul, Minnesota, publishes reports of all state cases that reach the appeal courts and all federal cases. Consider the two cases at the end of this chapter. *Globe Indemnity Co. v. Highland Tank and Mfg. Co.* is reported in 345 F. Supp. 1290. This indicates that the Globe Indemnity Co. and Highland Tank and Mfg. Co. were plaintiff and defendant, and the case was reported in the 345th volume of the Federal Supplement series of case reporters, starting on page 1290 of that book. The case of *Archer Daniels Midland Co. V. Koppers Co.*, 485 NE 2d 1301 (Ill. App. 1 Dist. 1985), is a case in which Archer Daniels Midland Co. and Koppers Co. were plaintiff and defendant. The case was reported in the 485th volume of the second series of the North Eastern Reporters, starting on page 1301.

The case reporters used by the West Publishing Company for reporting state cases divide the United States into seven districts. Each district covers several states. In addition to the *North Eastern Reporter*, there is the *Pacific Reporter, the South Western Reporter, the Southern Reporter, the South Eastern Reporter, the North Western Reporter,* and the *Atlantic Reporter*. The cases in these reporters are state cases that have been appealed from lower court decisions. Such cases are important because appellate decisions become controlling law for that type of case in that state's courts. A particular decision in an appealed case may even be the basis for decisions on similar cases in other states or in the federal courts.

Another set of reporters covers federal case decisions. The *Federal Reporter* reports U.S. circuit court of appeals cases; U.S. district court opinions are found in the *Federal Supplement*; Supreme Court decisions are reported in the *Supreme Court Reporter*. Federal special court cases and decisions of administrative boards are found in other series of West volumes.

EVIDENCE

Evidence is used to prove questions of fact. The facts, as presented by the contestants in a law case, often stray somewhat from the truth. The truth may be shaded a bit in the presentation. Each contestant must be willing to prove that what he or she said is true. The judge or jury then has the task of determining the true situation. Evidence is the means of establishing proof.

In a criminal case the evidence must prove guilt "beyond a reasonable doubt" for the defendant to be found guilty. A civil case, by contrast, is won or lost on the comparative weight of the proof.

The burden of proof in a criminal case is always assumed by the state. In a civil case the burden usually rests on the plaintiff to prove his or her charge but, under certain circumstances, the burden may shift to the defendant. One such circumstance occurs when a counterclaim is made.

♦ Real Evidence and Testimony

Evidence may be classified in many ways. It is classed as *real evidence* if it is evidence the judge or jury can see for themselves. For example, a fire extinguisher shown in court to be defective would be real evidence; so would a defective cable or a ladder that broke because of a defective rung. *Testimony* consists of statements by witnesses of things that have come to their knowledge through their senses. Testimony might be used to prove that a driver was operating a car unsafely and thus contributed to the cause of an accident.

♦ Judicial Notice

Certain facts are so well known and accepted that the court will accept them without requiring proof. The court takes such *judicial notice* of logarithm tables, provisions of the federal or a state constitution, Newton's laws of motion, and the like. Evidence would not be required to prove that gasoline is combustible, but the presence of gasoline in a particular situation might require proof.

♦ Presumptions

Conclusive evidence is evidence that is incontestable. It is evidence that can in no way be successfully challenged. The existence of a written contract is conclusive evidence that someone wrote it.

Prima facie evidence is something less than conclusive. It is rebuttable. It is capable of being countered by evidence from the opposing side but, if allowed to stand, is sufficient to establish some fact. A signature on a written document would be evidence of this nature.

♦ Direct and Circumstantial Evidence

Direct evidence goes to the heart of the matter in question. It is evidence that, if uncontested, would tend to establish the fact of the issue. A witness to a signature could establish the fact of signing by a particular party.

By contrast with direct evidence, circumstantial evidence attempts to prove a fact by inference. *Circumstantial evidence* might be used to show that the person now denying the contract would not have acted as he or she did at some past time unless he or she had entered into the contract. A net of circumstances is often woven to show a high probability that a particular version of an issue is true.

♦ Best or Primary Evidence

The court will require the best evidence possible in a particular case; that is, the highest and most original evidence available. *Secondary evidence* will not be used unless, for some reason, the primary evidence is not obtainable. In a case involving a written document, for instance, the document is the *best evidence* of its existence and provisions. If the original document were destroyed or lost, a copy of the document or, if no copy exists, testimony as to its existence and contents probably would be admissible. The document, itself, would be best or primary evidence; testimony would be secondary.

♦ Hearsay Evidence

Hearsay evidence is second hand. It is testimony about something that the witness has heard another person say. Most hearsay evidence is objectionable to the court for three main reasons:

- The person whose observation is quoted is not present in the court to be seen by the jury.
- The original testimony was not under oath.
- There is no opportunity for the original testimony to stand the test of cross-examination.

There are a few exceptions to the exclusion of hearsay evidence by the court. The rules vary somewhat in this respect from state to state. Hearsay is sometimes permitted where other evidence is completely lacking. Hearsay testimony about dying declarations is usually admissible. Business records are often admitted. A prominent exception to the hearsay rule is the *res gestae* statement.

♦ Res Gestae

In common English, the term *res gestae* means "things done." In courtroom procedure *res gestae* refers to allowable hearsay testimony about spontaneous utterances closely connected with an event. Consider an industrial accident situation in which the victim was killed. If, just before the accident occurred, the victim shouted that the safety device didn't work, testimony to that effect by a fellow worker might be allowed as *res gestae*. The statement explains the cause.

♦ Parol Evidence Rule

A contract or statement that has been reduced to writing is the best evidence of the meanings involved. Testimony (*parol evidence*) that would tend to alter these meanings is objectionable. Only when the terminology is ambiguous or when unfamiliar trade terms are used will testimony be allowed to clarify the meaning, and then an expert witness may be called on for an interpretation.

The parol evidence rule is confined to interpretations of wording of a document; it does not apply to a question of the validity of the instrument. In a question about the reason why a person entered into a contract—for instance, a claim of duress—oral evidence would be allowed. Similarly, an attack on the legality of consideration offered would permit testimony.

♦ Opinion Evidence

During the taking of testimony in a trial, the objection is occasionally heard that the "counsel is asking for a conclusion of the witness." An ordinary person appearing on the witness stand generally is not allowed to express opinions or conclusions in evidence. Such is the rule, but, as with many other rules of law, there are exceptions. In some instances the nature of the testimony requires opinions to be given—otherwise, the evidence will not be clear. In fact, just about any perception of anything is a conclusion based on sensory responses and experience.

The main objection to opinion evidence is that logical deduction or reasoning is being required of a witness. If such reasoning and conclusions are required in the progress of a trial, an expert from the field of knowledge involved should be called on to express an opinion about the facts or the meaning of a series of facts. If an opinion is necessary it is desirable to have the best possible opinion.

Engineers are qualified by training and experience to act as expert witnesses in certain types of cases. Occasionally, it is by giving expert testimony that members of engineering faculties obtain fees that enable them to afford to remain teachers.

♦ Witnesses

The determination that a witness is competent to testify is part of the court's (judge's) function. Competence is usually questioned on the basis of the witness's mental capacity or mental ability. The witness is required to testify only as to information having some bearing on the case at hand. If the testimony given under examination or cross examination gets too far afield, either the opposing attorney or the judge may object. Wandering by the witness under examination is likely to result in an objection from the other counsel on the basis that the testimony is "irrelevant."

Irrelevancy refers to the lack of relationship between the issues of the case and testimony requested or given. In a case involving machinery specifications, a question pertaining to an engineer's home life is hardly relevant.

The right of a witness not to testify on certain matters is known as *privilege*. According to the Fifth Amendment to the U.S. Constitution, no person may be compelled to testify against himself or herself. Communications between certain people need not be revealed in court. Examples of such privileged communications are those between husband and wife, doctor and patient, and attorney and client. Neither party may be made to testify unless the privilege is waived by the party affected by the trial.

THE ENGINEER AS AN EXPERT WITNESS

Most opinion evidence is excluded from a trial; the opinion of an average person acting as a witness is inadmissible. Opinion implies conjecture, and the law looks with disfavor on indeterminate factual situations. Nevertheless, such factual situations do arise and the truth in them must be determined as closely as possible. Such questions as the adequacy of design of a structure or the capabilities of a specially designed machine often must be answered.

♦ Expert

Attorneys and judges are usually quite learned people. Knowledge of the law requires a broad general knowledge of many specialized fields to understand the factual situations presented in cases. Most lawyers, though, would admit that their knowledge of a particular technical field is quite general and that, therefore, they would be incapable of drawing intelligent conclusions on complicated technical questions. For such purpose an expert is needed.

An expert was once facetiously defined as "any person of average knowledge from more than 50 miles away." Such a criterion would hardly stand up in a court of law. An *expert* is a person who, because of technical training and experience, possesses special knowledge or skill in a particular field that would not be possessed by an average person.

Qualification before the court as an expert witness is part of the expert testimony. The prospective expert witness can expect to be asked questions about his or her background, projects in which he or she has been involved, training, whether he or she is registered, and other similar questions. The court then determines whether that prospective witness is qualified to give expert testimony.

♦ Expert Assistance

There are three important ways in which an expert may assist an attorney in a case involving a technical matter:

- advice and consultation regarding technical matters in the preparation of the case;
- assistance in examining and cross-examining technical witnesses
- expert testimony on the issues involved

Advice and Consultation

In engineering, as in other technical fields, it is necessary for a person who is thoroughly familiar with technical concepts to explain them to laypeople. A jury is made up of a cross section of a community (or some approximation to that). Part of engineers' functions as expert witnesses are to present to their attorney and to the judge and jury the facts of the case in such a manner that laypeople can understand them. The minimum requirements for this communication are technical proficiency and the ability to effectively express the knowledge.

The basis of any effective presentation is investigation. Engineers, as expert witnesses, should be so thoroughly familiar with the facts of the case that nothing the opposition can propose will come as a surprise. Drawings and specifications may have to be carefully read, materials tested, and building codes or other laws examined. All should be analyzed for the presence of flaws if engineers are to do an effective job. Assistance in the preparation of the attorney's brief requires the best of engineers' investigative powers.

An engineer's obligations to client and attorney requires an objective approach to the case. Not only should the facts to substantiate a client's claim be present but opposing facts should also be shown. There are two sides to any controversy. In the interest of winning the case, opposing arguments must be considered and rebuttals prepared.

Trial Assistance

In a case involving technical fields, both parties normally obtain experts to aid them and to testify in their behalf. In addition to assistance in preparation of a case and testimony in court, the expert may also be valuable in suggesting questions the attorney should ask. Direct examination questions are prepared in advance, but most cross-examination questions and redirect and recross questions must be planned during the course of the trial.

A flaw in the technical argument posed by the opposition might escape an attorney's notice, but it should not escape the notice of technical experts in that field. The questions suggested may be aimed at the opposition by way of cross-examination; or they may be in the form of direct questions to be asked of the experts when they are put on the stand to counter the opposition's evidence.

Testimony

The judge, attorneys, and jurors are laypeople as far as the technical expert's field is concerned. Experts, therefore, must present their information in a simplified manner so that it will be readily understood by laypeople. Such presentation is quite akin to teaching. A simple foundation must first be laid and then the complexities built on it. The preparation of such a presentation is not always easy.

Such testimony necessitates comparison of technical principles with everyday occurrences and the use of pictures, slides, models, and drawings to make a meaning clear. Many hours of preparation are required for an hour of effective presentation. Each part of the presentation must be as nearly perfect as possible—incapable of being successfully questioned by opposing counsel.

It is almost essential for expert witnesses to be present at the entire trial if they are to be effective on the witness stand. Prior evidence established by opposing counsel may require alterations in the expert's presentation and changes in the attack on the opposing witnesses.

Honesty in answers is one of the prime requisites of any witness under examination. The opposing attorney will look (with expert assistance) for any point on which he or she can attack the expert's testimony. Once found, and properly worked on, a small loophole in the presentation can destroy the effect of laboriously developed testimony. If engineers do not know the answer to a particular question, the least damaging answer is a simple "I don't know." If the question involves prior testimony, witnesses may request the reading of that prior testimony before answering the question. If the question requires calculations or a consultation between witnesses and the attorney or other experts, time for such calculation or consultation may be requested from the court.

The demeanor of expert witnesses on the stand is important. Their appearance and answers to questions posed should inspire confidence in their ability. A professional appearance, professional conduct, and a professional attitude toward the entire proceedings come through clearly to the others in the courtroom. During cross-examination the opposing counsel usually tries to belittle or pick apart testimony damaging to the opponent's case; usually the more damaging the testimony, the greater the effort to reduce it to a shambles. Failing this, the attorney may attempt merely to enrage witnesses in hopes that an opening in the testimony may occur.

Calm and considered answers by witnesses are the best defense against the opposing counsel's attack. Courtesy and self-control must be exercised.

Reference to writings by acknowledged authorities in the technical field is advisable. Often it seems that a quotation excerpted from a textbook has more weight than an oral statement on the witness stand by the author.

Depositions

If a witness is unable to attend a trial to testify, then testimony may be taken in some place other than the courtroom prior to the trial. The testimony is taken under oath and recorded for use in the trial. Both opposing attorneys must be present, and the rights to direct examination, cross-examination, redirect, and recross are the same as they would be in the courtroom. At the time of the trial the deposition is read into the court record and becomes part of the proceedings. A deposition is generally considered to be less effective than testimony given in open court where jurors can see and hear the presentation even though it is videotaped. However, there are times when it is the only means available and is far preferable to the alternative of omission.

Deciding Whether to Take the Case

Engineers who are asked to be expert witnesses should decide whether they really believe in the prospective client's position. The quality of support rendered by engineers in the case often depends on how firmly they believe in the case of their client. On the witness stand engineers will be required to tell the truth of the case as they see it. Conviction that their client is right is often apparent in the manner in which the testimony is given.

Fees

The fee engineers charge for acting as expert witnesses in a case should correspond to the fees they would normally charge for other consulting work. Charges should be based on the amount of time required and the relative importance of the engineers' role in the proceedings. In no case should the fee be contingent on the outcome of the case. "Double or nothing" is closer to gambling than it is to payment for services. Contingency fees are seriously frowned on by the court as tending to cause inaccurate testimony; opposing counsel will not hesitate to take full advantage of this arrangement.

The fees for professional engineering services vary depending on where the engineer's business is located, the engineer's expertise, and the services involved. A typical minimum charge in 1998 ranges from $150 to $250 per hour or so in addition to such expenses as required travel and hotel accommodations.

GLOBE INDEMNITY CO. v. HIGHLAND TANK & MFG. CO.

345 F. Supp. 1290 (1972)
MEMORANDUM AND ORDER

Newcomer, District Judge

The plaintiff in a previously tried indemnity and/or contribution action seeks to gain a new trial based on many grounds. The Court seeks to direct its consideration on the disqualification of plaintiff's expert witnesses as the only ground which may have any arguable merit in granting a new trial, all other grounds being without merit.

It is well settled law that the trial court must be left to determine the qualifications of an expert witness.... The decision upon the fitness of the individual witness is based upon the circumstances of the expert's experience and the factual setting in which the expert's testimony is necessary. Further, the expert's qualifications must be expressly shown....

This Court is well aware that the witnesses which the plaintiff attempted to qualify have distinguished themselves in the academic world and has no difficulty in accepting and recognizing their credentials. However, their credentials alone do not qualify them as experts to testify on the factual setting which was in issue—the design of a molasses storage tank.

An analogy involving expert testimony would be as follows: A properly licensed driver with ten years of driving experience being called upon to testify as an expert witness involving the circumstances of a collision in an auto racing mishap. The witness would be knowledgeable in the general rules of driving, i.e., staying to the right of the road, passing on the left, etc., but would not be qualified to testify as an expert on the more specific area of driving in an auto race just because he was a licensed driver, without the showing of other specific qualifications. Therefore, the trial court would be acting within its discretion in disqualifying the witness as an expert on auto racing, and still permitting the witness to testify on the general rules of driving. This Court within its discretion felt that the analogy put forth applied to Mr. Maurer, an electrical engineer, and Mr. Bradley, an industrial hygienist.

Mr. Maurer was generally qualified as an engineer. However, Mr. Maurer was asked to testify on the design criteria necessary to insure the safe use of molasses holding tanks in an industrial setting. Such a question is not within the expertise or experience of every qualified engineer. One needs specific training or experience to be able to comment on this question. Some engineers may have this training or experience. Indeed, in the opinion of this Court, it would not even be necessary to be an engineer in order to make expert statements concerning the criteria to be set up for engineers to follow to insure a safe design in a particular industrial situation. There are people in the world who would qualify to give expert testimony on this question. However, there was no evidence that Mr. Maurer had any experience or expertise regarding the proper formulation of safety criteria to be followed in the design of molasses tanks in this particular industrial setting. The defective safety criteria in the design of the tank was a kingpin in the plaintiff's case, and there are experts in the field of holding tank design with an expertise in safety criteria available. Therefore, Mr. Maurer was not allowed to testify.

The Court also ruled that Mr. Bradley was qualified in the areas of industrial toxicology, environmental health, air pollution, but not safety unless qualified within the areas of his expertise, nor safety design.... The above analogy also applies to Mr. Bradley and that was the reason for his disqualification as a safety design expert.

In making these decisions as to the qualifications of the plaintiff's experts, the Court was well aware that an expert witness may not have personally, and cannot be expected to have personally, experienced the factual situation in question about which he is to testify in order to qualify. The plaintiff does not have to get the best possible expert available on the subject, but should have produced an expert more specifically and expressly qualified in the safety aspects of the design of storage tanks. Neither expert here had any prior experience or observational knowledge to testify as to the proper design of a molasses storage tank under the factual setting presented. Therefore, the disqualification of plaintiff's experts in this area was proper and resulted in plaintiff failing to make out a prima facie case. All other grounds for a new trial are without merit, and plaintiff's Motion for a New Trial is denied.

Archer Daniels Midland Co. v. Koppers Co.
485 N.E. 2d 1301 (Ill. App. 1 Dist. 1985)

McNamara, Justice.

Plaintiff, Archer Daniels Midland Company, brought this action to recover damages allegedly sustained on July 7, 1979, when its structure, located in Galesburg, Illinois and used for the storage of soya meal, collapsed. One of the defendants, A.O. Smith Harvestore Products, Inc., designed, manufactured, and sold the product. During the course of pre-trial discovery, A.O. Smith refused to produce an employee's report as ordered by the court after several hearings and an *in camera* inspection of the report, (This court has also made an *in camera* inspection of the report.) Trial counsel for A.O. Smith requested the trial court to find him in civil contempt to facilitate appeal. The trial court did so and A.O. Smith appeals, contending that the report is protected from disclosure by the attorney-client privilege.

Plaintiff purchased the structure in November, 1976. The unit was shipped to its Galesburg plant in December 1976, and construction was completed in April 1977. Repairs were required in June 1977 and October, 1978. In July 1979, the unit collapsed and caused damage to 350 tons of soya meal stored in the structure. Plaintiff filed suit, and subsequently initiated discovery.

In August 1982, plaintiff asked A.O. Smith for any documents relating to the storage tank and any documents regarding examination or reports made by experts which touched on the issue raised in the suit. After numerous denials that such documents reflecting expert analysis existed, plaintiff took the deposition of A.O. Smith's senior product engineer, James E. Gordon.

Gordon stated that he performed a study in 1980 at the request of A.O. Smith's in-house counsel. The study, according to Gordon, "was purely an analysis of the structural strengths of each model structure that we had in use for soybean storage. It was a relative comparison of the strength of one model versus the strength of another model, and a projection of what might constitute potential danger versus what might not." The law department had requested the report, according to A.O. Smith, to "provide them with some technical information on structural integrity as it depended upon structure application." A.O. Smith's in-house counsel further described the report as a "review of the structural parameters of all [A.O. Smith] structures and to compare those structural parameters to the structural parameters of particular units for non-agricultural purposes that had resulting problems. Further advice was requested on technical modifications that might be required in the non-agricultural [A.O. Smith] structures."

The report by Gordon was dated November 20, 1980, and was directed to the then acting in-house counsel for A.O. Smith. It was marked "confidential" and copies were given to four individuals in A.O. Smith who had some involvement with product safety. The report apparently was used in making a decision to notify the owners of certain types of A.O. Smith structures of potential dangers in using the structure.

Plaintiff's production request was renewed repeatedly. Numerous hearings were held. The trial court made an *in camera* inspection of the report and found that there was no attorney-client privilege applicable to the report.

The Illinois courts maintain a broad discovery policy, looking to the ultimate ascertainment of truth.... The courts, therefore, narrowly construe the attorney-client privilege in order to avoid trammeling upon the broad discovery policy. The need for this narrow construction demands particular attention in a corporate context, where the privilege has the potential of posing an absolute bar to the discovery of relevant and material evidentiary facts because of the large number of employees who frequently contact the corporation's lawyers, and the masses of documents used in business today. (Consolidation Coal Co. V. Bucyrus-Erie Co.).... That holding mandated that we must strive to deter such extensive insulation of these vast amounts of materials from the truth-seeking process by limiting the privilege for the corporate client to the extent reasonably necessary to achieve the basic purpose of the privilege. Yet we must remember that the purpose underlying the attorney-client privilege is to encourage and promote full and frank consultation between the attorney and client, and this is done by removing the client's fear that the information will be disclosed by the attorney.

...The threshold requirements include a showing that the communication originated in a confidence that it would not be disclosed; was made to an attorney acting in his legal capacity for the purpose of securing legal advice or services, and remained confidential. The burden of showing these facts is on the party claiming the exemption....

In the present case, the report was marked "confidential" and only distributed to the in-house counsel and four other people at A.O. Smith. The attorney used the report to make a recommendation regarding A.O. Smith's potential legal liability. The exact contents of the report have evidently remained confidential, notwithstanding the fact that the end result was to notify customers of potential dangers. The customers were not sent copies of the report.

When a corporate client claims the attorney-client privilege, the corporation must go beyond these threshold requirements and show that the employee involved falls within the control group of the corporation, as defined in *Consolidation Coal.* An employee's communications receive the protection of the attorney-client privilege under the umbrella of the "control group" when: 1) the employee is in an advisory role to top management, such that the top management would normally not make a decision in the employee's particular area of expertise without the employee's advice or opinion; and 2) that opinion does in fact form the basis of the final decision by those with actual authority.... Employees not within the control group include those whom top management merely relies upon for supplying information....

A.O. Smith argues here that Gordon advised the legal department as to what should be done regarding the product safety problems because legal counsel lacked the necessary technical and scientific expertise. No corporate decision maker, however, makes decisions in a vacuum. They necessarily depend upon information from other persons. When a manufacturer faces potential liability due to product design defects, the corporation typically goes through a decision-making process. It first must gather information from various sources. Hypothetically, the decision makers might ask computer operators to find relevant data in the computer files; clerks to summarize in a report all of the relevant correspondence, testing, examinations, and reports gathered by the corporation over the years; engineers to review previously approved designs and comment and recommend in light of new information showing the designs or manufacturing is causing safety problems; or other scientific experts to analyze technical aspects underlying the product's defects. None of these employees are necessarily within the *Consolidation Coal* definition of a control group, regardless of how expert their skills are, or how inexperienced the decision makers are in those same skills. A.O. Smith's argument that its legal counsel lacked technical expertise to develop a recommendation fails to convince us that Gordon was rendering the type of opinion required by *Consolidation Coal*.

In defining "control group," *Consolidation Coal* refers to decision makers or those who "substantially influence" corporate decisions. The court, however, focused on individual people who substantially influenced decisions, not on facts that substantially influenced decisions. Here, Gordon supplied technical data and opinions to people within A.O. Smith who then decided what to do with that information. The decision to notify owners of A.O. Smith structures of potential dangers was not a decision in which Gordon participated except to provide technical data and his analysis of that data. His opinions were technical opinions in regard to the designs and manufacture of those structures, not opinions about the corporate policy. As an engineering expert, Gordon supplied information, technical recommendations, and opinions to employees, such as the corporate in-house counsel, whose legal opinions were sought and relied upon by others, such as A.O. Smith's top management who ultimately made the decision to send notices to customers. Gordon, therefore, was not part of the control group as defined by *Consolidation Coal*. Thus, we hold that Gordon's report is not privileged and must be made available for inspection.

For the foregoing reasons, the judgment of the circuit court of Cook County is affirmed.

Judgment affirmed.

REVIEW QUESTIONS

1. Describe the federal court system. Describe the court system in your state.
2. Assume that you have been the unfortunate victim of someone's negligence. What steps will you or your attorney take in attempting to get compensation for the loss you have suffered?
3. Distinguish between the proof required in a criminal case and the proof required in a civil case.
4. How would you prove the existence and terms of an oral agreement if there were no third parties present to overhear the agreement?
5. Why does a court hold in disfavor:
   a. Oral testimony as to the meaning of a written document?
   b. Hearsay testimony?
   c. Expression of an opinion by a witness?
6. Why is circumstantial evidence so often used in criminal trials?
7. In the case of *Globe Indemnity Co. v. Highland Tank and Mfg. Co.*, an engineer and an industrial hygienist were disqualified as expert witnesses. What were the grounds for their disqualifications? Based on this case, who would be the best possible expert witness in a case of an automotive brake failure allegedly causing a crash which resulted in the plaintiff becoming a paraplegic?
8. Briefly outline the requirements for communications between an engineer and a corporate representative to be considered "privileged" according to Judge McNamara's discussion in Archer Daniels Midland Co. V. Koppers Co.

Part Two

CONTRACTS

Chapter 5 *Introduction to Contracts*

6 *Parties*

7 *Agreement*

8 *Reality of Agreement*

9 *Consideration*

10 *Lawful Subject Matter*

11 *Statute of Frauds*

12 *Third-Party Rights*

13 *Performance, Excuse of Performance, and Breach*

14 *Remedies*

15 *Sales and Warranties*

The world of the engineer is an environment of serious communications, nearly all of which have contractual implications. A contract to build a bridge, road, or building may necessitate numerous subcontracts—contracts with material suppliers, labor contracts, leasing contracts, easements, utility contracts, and many others. A contract to manufacture and ship parts to an appliance manufacturer on a continuing basis may similarly precipitate additional contracts. The day-to-day engineering management of such contracts requires a very large number of activities and communications, almost all of which involve contracts. For example, the quality of a final product is often spelled out or implied in the contract for that product. The product's quality is determined by the production processes used, the manner in which they are used by the production work force, and the materials on which the processes operate. All of the design and operating decisions involved with such production have quality overtones and for that reason, if for no other, are important contract considerations. The next few chapters briefly examine the law of contracts, Then, the following section looks at engineering contracts.

Chapter 5

Introduction to Contracts

Modern civilization is a world of contracts. Each one of us depends on them. Every purchase is a contract whether it is a pair of socks, a restaurant meal, or a battleship. When you turn on your television set to watch your favorite program, it is done in execution of a contract. The utility company has agreed to furnish electric power and you have agreed to pay for it. You go to work as your part of a contract with your employer. On the job you do what your employer wants you to do, and at the end of a week or a month your employer pays you for it. It is all part of the same contract. If your employer pays you by check it is because your employer has a contract with a bank to safeguard its money and give it out on order. If your employer pays you in cash, the currency represents a contract between the bearer and a Federal Reserve Bank. When you die and are buried, the security of your last resting place may depend on the terms of the contract under which the land was obtained. You can't avoid contracts, even by dying.

Even a simple action such as leaving your watch at the jeweler's for repair involves you and the jeweler in a complicated legal situation. Everything is resolved painlessly when you pick up the watch and pay the jeweler a week later. In the meantime, the relationship between you and the jeweler involved the following:

- Personal property-the watch
- Agency (quite likely a clerk represented the owner)
- The law of bailment (personal property was left with another for repair)
- Insurance law (had the watch been lost)
- A contract (the jeweler's agreement to repair and your agreement to pay for the service)

All contracts are agreements and, morally at least, all agreements are contracts, but moral duties are not always enforceable under the law. Suppose that Black accepted Dr. White's invitation to dinner on Tuesday, then forgot the engagement and did not appear. Black broke a social contract and in doing so committed a serious breach of social ethics. But Black's contract was not an enforceable one in the sense that a court of law would award damages or order Black to appear and, no matter how much pain and suffering Black caused Dr. White, the doctor could not collect damages.

Now, suppose White is a practicing engineer and Black is an attorney looking for help, that is, a prospective client. Black calls and makes an appointment with White for 3:00 p.m.on Thursday. In breaching this appointment, Black may be breaching a contract. The commodity in which White trades is time and his ability as an engineer. Many jurisdictions would hold that unless White could otherwise gainfully use the time set aside for Black's appointment, White could collect for that time. In deciding the case, the court would examine the understanding of the parties and also business practices in the area, and these factors would largely determine the case's outcome.

If these two parties entered into an agreement whereby Black agreed to purchase a new computer network from White for $6,000, such an agreement would be a contract and enforceable in a court of law. Each party to the contract would have an action at law available if the other failed to perform as agreed.

In modern America, we take individual freedom for granted. Many of our opportunities and freedoms can be traced to the law of contracts and a strong public policy of freedom of contract. In effect, the courts allow the parties a great deal of freedom to agree to the terms and conditions by which goods are made and sold or by which services are rendered. Such freedom, however, is not without restraint; an agreement to perform a criminal act, for example, will not be enforced. Such freedom of contract helped to eliminate the strict rules governing the status of individuals that developed and existed in England during medieval times. For example, the ability of an earl or baron to enter into a binding agreement was vastly different from the ability of a medieval serf.

In the following chapters, consider whether the policy of allowing the parties to freely contract is or should be subject to other public policies. Also consider whether you agree with the cases in which a court found the parties' agreement to be void as against a public policy (other than freedom of contract).

DEFINITION OF CONTRACT

The only kind of an agreement the law recognizes as a contract is defined in *Black's Law Dictionary* as "an agreement between two or more persons which creates an obligation to do or not to do a particular thing." The Uniform Commercial Code (UCC), which has been adopted by every state, defines a *contract* as "the total legal obligation that results from the parties' agreement as affected by [applicable] law." The term *agreement*, in turn, is defined as "the bargain of the parties in fact as found in their language or by implication from other circumstances including course of dealing or usage of trade or course of performance.... *Whether an agreement has legal consequences is determined by the applicable law of contracts,"* (Emphasis added.)

Stated even more simply, a contract could be defined as "an agreement enforceable at law."

SOURCES OF CONTRACT LAW

The area of law dealing with contracts developed under common law through the decisions of English and, later on, American judges. Many "modern" rules of contract law have their roots in cases that are centuries old. The common law rules were distilled into a multivolume treatise called the *Restatement of the Law of Contracts*. Later, a second Restatement was published. The Restatement (Second) remains an important authority for the common law on contracts. Certain types of contracts, however, are governed by distinct statutes and treaties. For example, the UCC controls contracts for the sale of goods. The UCC was written and enacted fairly recently in terms of legal history—within the last 40 years. Another more recent example of a source of contract law is the *United Nations Convention on Contracts for the International Sale of Goods*. This Convention was signed by the United States in 1980 and took effect in 1989. Certain contracts involving the international sale of goods with buyers and sellers located in different countries will now be governed by the Convention's rules (instead of the common law or the UCC). Most of the following discussion focuses on the general common law rules and the UCC.

ELEMENTS OF A VALID CONTRACT

Analysis of valid contracts reveals that they are made up of the following five elements, each of which must exist according to law:

1. Agreement
   a. Offer
   b. Acceptance
2. Competent parties
3. Consideration
4. Lawful purpose
5. Form

Each element is only briefly introduced here but is treated in more detail in the next six chapters.

♦ Agreement

The agreement consists of an offer and acceptance. The *offeror*[1] states the terms of the proposed contract. The *offeree*[2] must accept the terms as they are proposed to complete a binding contract.

1. An *offeror* is the person who makes an offer to someone else.
2. An *offeree* is the person to whom an offer is made.

♦ Parties

For a contract to be thoroughly binding it must be made by at least two parties, neither of whom will be able to avoid the contractual duties by pleading as a defense an incompetence to contract.

♦ Consideration

Consideration in a contract consists of the money, promises, and /or rights given by each party in exchange for the money, promises, and/ or rights that each party receives from the other.

♦ Lawful Purpose

The purpose and consideration in the contract must be lawful if the contract is to be enforceable in a court of law. A contract to perform an unlawful act will not be upheld in a state that imposes a penalty for committing the same act.

♦ Form

Difficulties in proving the existence, validity, and terms of contracts have prompted the requirement for certain types of contracts to be in written form. Failure to comply with this requirement renders such contracts unenforceable by the court.

As soon as the offeree expresses acceptance of the terms offered, a contract is created. Usually, the contract at this stage is *executory*, and as long as something remains to be done by either or both of the parties under the terms of the agreement, it remains so. Even when one of the parties has performed that party's obligations completely and the other party has not yet completed performance, the contract is *partially executed*, but in a legal sense it is still executory. After all parties to the contract have performed all the actions required according to the agreement, the contract is fully *executed*.

A contract to purchase a refrigerator is executory when the sales agreement is made between the buyer and the seller. It is partially executed (but legally executory) when the refrigerator is delivered to the buyer and fully executed when the buyer has finished paying for it.

♦ Parties

The "garden variety" contract involves two parties as promisor and promisee. However, it is more difficult to determine the liabilities of the individuals when they concern several parties and the individual interests of some have been merged. When there are merged interests, the relationships of the parties are treated as *joint, several,* or in some cases as *joint and several.* Any legal entity can be a party to a contract. Thus, an individual, corporation, partnership, or joint venture can enter into a contract. For example, an employment agreement (written or not) exists between an engineer (an individual) and the engineer's employer, such as a corporation. The federal, state, and local governments also enter into contracts for a vast array of different goods and services.

Individuals who merge their interests into a joint relationship to form a party to a contract may be thought of as partners in the promises made. If the individuals together agree to "bind themselves" or "covenant" to do a certain thing in the terms of an agreement, it is treated as a joint contract. The liability is much the same as it is in a partnership. That is, should the parties breach the contract, each is liable for complete performance of the contract until the full value has been satisfied.

Say that Black and White agree jointly to hire Gray, a contractor, to make alterations in an existing building and to pay him $4,000 for the alterations. If, after completion of the alterations, payment is refused by Black and White, Gray must sue them jointly. If Black cannot pay part of the debt, White may have to pay the entire amount. If either White or Black dies, the survivor will be liable for the full amount of the debt.

Restriction of the liability of individuals merging their interests is a feature of the several contract. If, in the contract just described, Black and White had agreed to "bind themselves severally," or "covenant severally" to pay $4,000 for the alterations that Gray was to make (restricting themselves to, say, $2,000 each), the liability of Black and White each would be limited to the amount stated. In case of default of payment, Gray could take only separate actions against Black and White.

If merging individuals agree to "bind themselves and each of them" or to "covenant for themselves and each of them," the contract is treated as a joint and several contract. If the Black and White contract just described were joint and several, and Gray found court action necessary, such action could be taken either jointly or severally, but not both ways.

In a *third-party beneficiary* contract, the purpose of the contract is usually to benefit a third party. If the intent to benefit the third party is clear from the agreement, the third party can enforce those benefits in a court of law. If, however, the consideration under the contract benefits one of the parties to the contract and only incidentally benefits a third party, no right of court action is available to the third party. Probably the most common example of a thirdparty beneficiary contract is life insurance.

♦ Means of Acceptance

A *unilateral* contract is a promise of a consideration made without receiving a promise of consideration from another party. A common example of the unilateral contract is the reward notice. White publishes a promise of a reward of $500 for the return of a diamond-studded wrist watch. Black, on finding the watch, accepts the offer of a reward by returning the watch and then claims the reward. To obtain the reward in a legal action, Black must have been aware of the reward promised when Black returned the watch. A *bilateral* (or *reciprocal*) contract is a mutual exchange of promises, present consideration, or both, by two or more parties.

DIFFERENT CLASSIFICATIONS OF CONTRACTS

Express contracts involve overt statements and communications to reach an agreement by the parties, the terms being expressed between them orally, in writing, or both.

Implied contracts are based on implications of fact. If the parties, from their acts or conduct under the circumstances of the transaction, make it a reasonable or necessary assumption that a contract exists between them, it will be held that such a contract does in fact exist.

Quasi-contracts are contracts implied in law. They are generally based on the theory of *unjust enrichment,* which in effect says that it would, in all justice and fairness, be wrong to allow one person to be enriched by another without having to pay for it. The theory amounts to a legal fiction created by the courts to permit recovery in cases where, in fact, there would be no recovery otherwise. Black contracts orally to pay $50,000 for White's services for a period of 18 months. The Statute of Frauds (described in Chapter 11) requires such a contract to be in writing if it is to be enforceable. In ignorance of this requirement, White performs the services for Black and now sues for payment. White cannot recover on the original contract but may recover the reasonable value of the services under a quasi-contract on the theory that it would be unjust to allow Black to be enriched by White's services without having to pay for them.

In construction projects, it is not unusual to see a contractor or subcontractor recover for goods and services previously provided despite a breach of contract. Such a recovery is for the value of the goods and services, not for the contractually agreed on amount.

♦ Legal Status

Contracts also can be classified as to legal standing into valid, unenforceable, voidable, and void.

A *valid* contract is an agreement voluntarily made between competent parties, involving lawful consideration, and in whatever form may be prescribed by law for that particular kind of subject matter. It contains all the essential elements previously described. Courts will enforce such a contract.

An *unenforceable* contract is one that creates a duty of performance that may be recognized in a court of law but that, because of some defect in the contract, may not be enforced by the court.

A detailed contract with a number of different provisions may have one or two of them found to be unenforceable. For example, suppose the contract requires the payment of money on the delivery of goods and, if the payment is late, the principal amount plus interest at a rate of 20 percent per year is due. Many states have laws restricting the rate of interest charges; charge of 20 percent in some states would constitute "usury" and would not be enforced. However, depending on the applicable state statute on interest, the contract's provisions regarding the required payment of the principal amount would still be enforceable.

Sometimes, however, the whole contract may be unenforceable. A common example of such a contract is one that is not made in accordance with the Statute of Frauds. An oral contract that, because of the subject matter, should have been in writing is an unenforceable contract.

A *voidable* contract is one in which one or both of the parties may avoid the contract if either so desires. Black, a minor, purchases a bicycle from White, an adult. Black may return the bicycle and demand the money back or keep the bicycle and pay the price. White has no similar right of avoidance.

A *void* contract is, strictly speaking, not a contract. A contract to commit a crime is an example of a void contract. The courts ordinarily treat contracts for an unlawful purpose as nullities.

♦ Formality

Contracts also may be classified as *formal* and *informal*. Most contracts are informal in nature. Contracts to perform services or to sell goods to another are usually informal contracts. One particular type of formal contract (so called because the contract derives its validity or effect from its form) is the negotiable instrument.

Negotiable Instrument.

A *negotiable instrument* is a contract for the payment of money. It derives its negotiable characteristics from its form. The requirements for a contract to be a negotiable instrument are set forth in the Uniform Commercial Code. Briefly, the agreement must include the following features:

- Be in writing and signed
- Offer an unconditional promise or order
- Agree to pay, in money, an ascertainable amount
- Be surrendered on demand or at a fixed or determinable future time
- Be payable to order or bearer
- Identify a drawee, if there is one

Consider these features in detail. The instrument must be in writing and signed. It is obviously impossible to have an oral negotiable instrument. The writing may be in long-hand or printed; a negotiable instrument could be written on wrapping paper. The signature may be written or stamped (as is frequently the case with payroll checks) and may appear anywhere on the instrument, although it is commonly placed in the lower right-hand corner.

The negotiable instrument must be an unconditional promise or order. An acknowledgment of an existing debt is not a promise to pay it unless such a promise is made or implied in the instrument. A mere request or authorization is not considered to be a promise or order to pay. The promise or order must not be conditioned on any future contingency other than the passage of time.

The instrument must call for the payment of a definite amount of money in legal tender. The amount to be paid must be fixed or be capable of being determined from the instrument itself. Payment may actually take the form of goods or services, but the instrument must give the holder the right to call for payment in money.

The negotiable instrument must provide for payment on demand or at a fixed or determinable future time. If no time of payment is stated, the instrument is assumed to be payable on demand. The instrument is also considered to be payable on demand if it is past due. A "determinable" future time may be tied to some event that is certain to take place, but about the only future event certain enough for the courts is death.

The instrument must be payable to order or bearer. Certain words of negotiability are required in a negotiable instrument. These words are *order* or *bearer*. The negotiable instrument must contain one or both of these words used in a way that indicates that the maker intended the instrument to be capable of being negotiated. Such phrases as "pay to bearer" and "pay to the order of *X*" serve this purpose.

Finally, if there is a *drawee,* the instrument must identify that party. If the negotiable instrument calls for payment by a third party (as a check or other bill of exchange), the party who is to make the payment (the drawee) must be indicated with reasonable clarity.

Promissory Note

Negotiable instruments are of two general types—promissory notes and bills of exchange. A *promissory note* is a two-party negotiable instrument in the form of a promise by one person to pay money to another. The two parties to the instrument are the *maker,* who undertakes the obligation stated in the instrument, and the *payee,* the person to whom the promise is made. The classification of promissory notes includes conditional sales notes, chattel mortgage notes, real estate mortgage notes, and the coupons on coupon bonds.

Bill of Exchange

The bill of exchange is a three-party negotiable instrument. The *drawer* (corresponding to the maker of a promissory note) orders a third party (the drawee, usually a bank) to pay money to another party, the payee. Common forms of the bill of exchange include checks, bank drafts, trade acceptances, sight drafts, and time drafts.

Negotiation

A negotiable instrument *(bearer paper)* made payable to "bearer" is negotiated by delivery alone. Possession by itself entitles the bearer to be paid. If the instrument is made payable to "order" (called an *order paper*), it is

No. 1001

Sylvania, Ohio August 11, 19 –

Pay to the order of Jed Black $ 100.00

One hundred and no/100 -- Dollars

For P.O. No. 1001

Bank of Sylvania
Sylvania, Ohio

Sam White

Bill of Exchange (Check)

$100.00 Gainesville, Florida August 11, 19 –

Ninety days after date I promise to pay to
the order of Jed Black
One hundred and no/100 -- Dollars
at Graytown Bank, Graytown, Florida

Value received.

No. 1001 Date Nov. 9, 19 – Sam White

Promissory Note

negotiated by *endorsing* (the holder of the instrument writes his or her signature) on the back of it.

There are three general types of endorsement of negotiable instruments: endorsement in blank, special endorsement, and restrictive endorsement. An *endorsement in blank* consists of the signature of the payee on the back of the instrument. Such an act makes the instrument, in effect, bearer paper in that it may be further negotiated by nothing more than delivery (anybody can cash it).

A *special endorsement* names the endorsee. "Pay to the order of Jed Black, Sam White" is a special endorsement and must be endorsed by Jed Black if he wants to negotiate it. If the endorsement had read just "Sam White," it would have been an endorsement in blank.

A *restrictive endorsement* restricts future negotiation on the instrument. "Pay to the Graytown Bank for deposit only" is an example of a restrictive endorsement.

A *holder in due course*[3] of a negotiable instrument has certain rights superior to those of the original parties. The holder is entitled to payment on the negotiable instrument in spite of certain personal defenses available to the original parties. If, for instance, the maker was induced by fraud to draw up the note, this is no defense against a subsequent holder in due course, although the maker still has an action available against the person who defrauded him or her.

Certain so-called real defenses may, however, be used successfully against a holder in due course. The most prominent of these defenses is forgery of your signature. If your signature has been forged, you should not be held liable.

3. A *holder in due course* is a party other than the original parties to the negotiable instrument, to whom the instrument has been negotiated in good faith and for value and without notice of defects in the instrument, such as its earlier "dishonor" or any defenses or claims relating to it.

Why should our laws allow a holder in due course special rights? By doing so, banks and others will be more likely to honor such instruments and, as a result, persons providing goods and services (as well as those who make loans) will be more likely to provide credit to the person who is to pay.

KUENZI v. RADLOFF
34 N.W. 2d 798 (Wis. 1948)

The plaintiff is licensed under provisions of sec. 101.31 as a professional engineer. He has 40 years experience in designing buildings and was licensed in 1932.

In December 1945 the defendants were considering the erection of a building in the city of Waupun to be occupied by bowling alley and a tavern. The defendant Radloff consulted the plaintiff and as a result of this consultation the plaintiff wrote the following letter:

Dec. 29, 1945

Mr. Harold Radloff
Waupun, Wisconsin

Dear Sir:

I wish to confirm our conversation of some time ago wherein I named you a fee of 3% of the estimated value of the project for services in making up plans for the construction of a proposed bowling alley to be built at Waupun, Wis. This also includes the services of securing a full approval of the Industrial Commission.

Respectfully submitted
Yours truly,
Arthur Kuenzi

Accepted
O. A. Krebsbach
H. Radloff

Upon receipt of the signed proposal from the defendants the plaintiff proceeded with the design and preparation of the plans for the proposed building. The defendants from time to time during the preparation of the plans consulted with the plaintiff and his associates and changes were made in accordance with the suggestions made by the defendants. The plans were completed early in March 1946 and were presented to the Industrial Commission and duly approved by it and then promptly delivered to the defendant, Radloff.

On April 11, 1946, an application for the allocation of construction materials, to which application was attached a copy of the plans, was made on behalf of Radloff and Krebsbach to the Civilian Production Administration. A second application to the Civilian Production Administration, signed by both defendants, was submitted to the Civilian Production Administration on April 25, 1946. In each of these applications the cost of the structure, including fixtures and building service, is stated to be $80,000. The following statement was made in the application.

"A site has been obtained and an architect engaged for the construction of such building and the plan submitted to a contractor who has in turn ordered various materials for the construction thereof. All of said obligations and commitments made previous to March 26, 1946." Both applications were denied on May 1, 1946.

On April 28, 1946, the plaintiff sent to the defendant Radloff an invoice for $1,350 based on the estimated value of the building of $45,000. The plaintiff also demanded payment from the defendant Krebsbach before the commencement of this action. The plaintiff received the following letter from the defendant Radloff:

Monday morning

Dear Sir:

Sorry to keep you waiting but we are still working through Washington to get started building.

We will make payment to you just as quickly as possible. Milan Nickerson, one of our partners, dropped out, didn't want his money laying idle so he went into the cement block business.

We are picking out another good partner and will get in touch with you or write when we have the partners lined up.

This Nickerson was undecided for some time and that's why we didn't send you any money, until we have the other party lined up.

How is the steel coming? We will have to pay you and have it on hand when it comes and wait for the permit to start to build. You said in your last letter of quite a while ago that the steel would be here within a couple of weeks.

Yours very truly,
H. Radloff

P.S. Keep this under your hat about a third party and if you should happen to know of someone who has 15 or 20 thousand and wants to put it in a good business of a bowling alley and tavern let us know.

The letter was undated; neither the plaintiff nor the defendant, Radloff can fix the date on which it was sent. Evidently it was sent after the receipt of the invoice from the plaintiff because payment is promised.

Upon notice of the denial of their application by the Civilian Production Administration on May 1, 1946, the defendants abandoned the project, and the building for which the plans were prepared has never been erected.

Upon the facts it appears as a matter of law that the plaintiff had a contract with the defendants for the making of plans for the construction of the proposed building; that he proceeded to carry out his part of the contract by preparing the plans, procuring their approval by the Industrial Commission, and delivering them to the defendants; that they were accepted by the defendants and used by them in their efforts to procure a priority order from the Civilian Production Administration.

The defendants seek to defeat the plaintiffs' claim on a number of grounds....

♦♦♦

It will be observed that the contract provided that the fee should be 3% of the estimated value of the project for services in making up plans for the construction of the proposed bowling alley. The defendants place a great deal of emphasis on the term "estimated value." The defendants made the claim, and the court sustained it, that this meant the value of the building after its erection. In so holding it is considered that the trial court was in error. The basis on which plaintiffs' fee was to be computed was the estimated value of the project, not of a building that might never be erected. In its opinion the trial court said: "The court refused to allow this testimony as to cost due to the fact that the letter that comprised the basis for the case said value of the project as the basis rather than cost of construction," and excluded all testimony as to cost except that of plaintiff, and restricted the evidence to what the building would be worth on the site after it was constructed. Just how the value of a building that has never been constructed can be determined does not appear. "Value," in the sense in which the trial court used the term, means market value, what the property could be sold for after the building was erected. Even if that were the test, evidence of the cost would be relevant and material, but it is not the test in this case.

It is considered that this case is ruled by *Burroughs v. joint School District,* 155 Wis. 426, 144 N.W. 977. It was there held that if when the term "value" is applied to a particular contract, or conditions growing out of it, it leads to results clearly not contemplated by the contract read as a whole, and it is susceptible of another meaning that harmonizes with all the provisions of the contract, such other meaning should be given to it. In that case a building contract provided for payment in each month of a sum equal to 90% of the value of the work done and material furnished during the preceding month, as assessed by the architects. In that case the word "value" was construed to mean not market but contract value. It is considered that in this case the term "estimated value" of the project referred to the estimated cost of the material and services necessary to complete the building according to the plans. There is not a scintilla of evidence in this case that plaintiff was to wait for his compensation until the completion of the building and an appraisal thereof. The idea that the base on which the fee was to be computed can be established by the sale of a mythical building owned by a mythical owner and sold to a mythical buyer, is too elusive and indefinite a standard to apply to practical affairs. So in this case we hold that the term "estimated value of the project" means the estimated cost of completing it.

Upon this point the plaintiff testified as follows: "I made an estimate of $45,000 being the value of this building, which is the same amount as the estimated cost. My estimate was the sum of $45,000 as I recall it. I made a charge against the defendants based on $45,000 and charged 3% of that amount or $1,350."

Two witnesses were called on behalf of the defendant, who testified that the value of the completed building would be fifteen to twenty thousand dollars, around twenty thousand dollars. This was on the theory that the building when completed would have to be rebuilt if used for any other purpose than a bowling alley. One witness testified: "You would have quite a job getting $20,000 for it, because it would have to be torn apart and fixed over for something else." This sort of evidence comes far short of establishing the estimated value of a project.

Considerable evidence was received in regard to the income tax returns of the plaintiff. So far as we are able to ascertain this evidence was immaterial. Whether the proper income tax returns were made or not is a concern of the Department of Taxation, and has no relevancy on the question raised in this case. It was introduced in an effort to establish that the real plaintiff was a co-partnership, a matter that has already been considered.

It is considered that, having fully performed the contract between the parties, including the procurement of the endorsement of the Industrial Commission, and defendants having accepted and acted on the plans as delivered to them, the plaintiff is entitled to compensation on the basis of the lowest estimated cost of the project appearing in evidence, to-wit the sum of $45,000 with interest and costs.

The judgment appealed from is reversed and the cause is remanded with directions to the trial court to enter judgment for the plaintiff as indicated in the opinion.

♦

REVIEW QUESTIONS

1. Name the essential elements of a valid contract. Identify these elements in *Kuenzi v. Radloff.*
2. Distinguish between a joint contract and a several contract. Give an example of each.
3. What is a quasi-contract? Who creates it? For what purpose?
4. Draw up a valid negotiable instrument with John Doe as payee and Richard Roe as maker or drawer.
5. Distinguish between unenforceable contracts and voidable contracts. Give an example of each.
6. Consider the word *value.* How many ways are there to estimate the value of a project that exists only in the planning stage? How many ways are there to estimate the value of something that has been in existence for a period of time—for example, a five-year-old punch press?

Chapter 6

Parties

There must be at least two parties to a contract. There may be more than two; in fact the only ceiling for parties is the number it is practicable to identify.

No person may make a contract with himself or herself. Consider Black, who is executor of White's estate and an ordinary citizen. If Black, as executor of White's estate, agrees with Black, as an individual, to do something, no valid contract will result.

The law offers its protective shield to those who, due to their immaturity or for some other reason, are held to be incompetent to contract. The law does not usually offer its protective shield unless it is asked to do so, however. A person must plead incapacity to contract to receive the protection.

Among those whose ability to contract is in some way limited are infants, married women, insane persons, intoxicated persons, corporations, governments, and professional people. To illustrate the rights and defenses of the parties to a contract, infants' contracts will be treated at a greater length than the contracts of other incompetent parties.

INFANTS

♦ Age of Infancy

According to common law, infancy ends when a person reaches his or her twenty-first birthday. There is no legal adolescence. On one day a person is an infant according to law; on the next day that person is an adult. The only status that might come close to a legal adolescence is that of *emancipated minor*. Emancipation of a minor (the legal meaning of *infant* is the same as for *minor*) takes place when a minor's parents or guardians surrender their rights to the minor's care, custody, and earnings. Such an emancipated minor is usually treated somewhat more sternly by the courts than a nonemancipated minor. The minor's ability and knowledge are considered to be less than those of an adult. A day later (at common law) the minor is vested with the full powers, rights, and responsibilities of an adult.

Statutes in many states have altered the age that ends infancy or have provided for removal of the incapacity under other circumstances. Several states provide that legal majority is reached on a person's eighteenth birthday. Other states make the age 19. Many statutes provide that marriage removes the infant's incapacity. Statutes of various states provide means by which an infant may remove the incapacity by request to a court of law. This is often done when an infant undertakes a business venture, such as is the case with young film or recording stars.

♦ Partnership

When an adult becomes a partner, the adult stands to lose some or all of his or her personal fortune as well as any investment in the business if the partnership becomes bankrupt. This is because all of the partners remain personally liable for the debts of the entire partnership. However, when a minor becomes a partner in a business venture, the minor can lose in bankruptcy only whatever value the minor has contributed.

♦ Agency

Generally, an agency contract in which a minor is the principal is void (not voidable) under common law. A recent trend, however, is to consider such contracts as voidable, which seems more logical. Infancy is no bar to acting as an agent for another, however, because it is the principal who is bound rather than the agent. An adult, therefore, may enter into a binding contract with a third person through an infant agent.

♦ Disaffirmance

In general it is true that if one party to a contract is not bound, neither is the other. However, the law recognizes certain exceptions to this generality. Infants' contracts form such an exception.

An infant may avoid the obligations under almost any contract. The minor needs merely to notify the other party of a disaffirmance of the contract. The infant's right to *disaffirm* is a personal right; no one else may disaffirm for the minor.

If a minor elects to disaffirm a contract that is still completely executory, no major problem is involved. Because no consideration has changed hands, none needs to be returned. The problems arise when consideration has been given. If a minor disaffirms a contract with an adult after having given the adult consideration, return of the consideration to the minor is mandatory. The minor must also return whatever consideration the minor has received if it is possible to do so. It has been stated, though, that a minor's right to avoid a contract is a higher right than the adult's right to get back the consideration.

If the consideration received by the minor has been demolished or depleted in value in the minor's hands, the minor may still return it and demand the return of his or her consideration. Destruction of the subject matter in the minor's hands is only further evidence of incapacity. Even if the minor cannot return any of the consideration received, the minor may still be successful in getting back what was given. White, a minor, buys a used car from Black Auto Sales. White pays $1,500 down and agrees to make a series of monthly payments for the car. Two weeks later White loses control of the car while driving it and, when it hits a tree, the car becomes a total loss. White could return the wreck and get the $1,500 back. If the car were not insured for theft and White lost it to a thief, White could still disaffirm and get the downpayment back.

Although the preceding example is representative of the rulings in a number of courts, other courts require the minor to return the consideration received or its equivalent in value to disaffirm. The minor retains the right to avoid the contract, but the courts will not allow the minor to harm the other party by disaffirming. In one such case, a minor had taken a car to a garage for repairs. When the repairs were finished, the minor refused to pay for them and demanded the car.[1] In requiring the minor to pay for the repairs, the court compared the legal protection offered to minors to a shield. The judge stated, in effect, that although a defensive shield is afforded by law, it is not intended to be used as a sword against another. There was nothing defensive in the acts of the minor in this case. In fact, it represents a bald swindle more than it does a contract. If the repairs had amounted to replacement of rod bearings, main bearings, and piston rings, for instance, the inherent difficulty of returning the consideration to the adult would be quite apparent.

If the infant poses as an adult to get the adult to deal with him or her, the court may demand that the adult be left unharmed or, at least, that the harm be minimized. It would seem unreasonable to allow an infant to harm another by deceit and then to protect the infant.

If the subject matter of the infant's contract is something other than real property (real estate) the infant may disaffirm the contract at any time before the infant becomes an adult. In addition, the infant has a reasonable period of time after reaching adulthood to disaffirm the contract. The infant may disaffirm either by an express statement or by implication. If the contract is executory and the infant does nothing about it within a reasonable time, it will be implied that the infant has disaffirmed it.

If the subject matter is real property, the infant must wait until the infant becomes an adult to disaffirm the contract. There seems to be good reason to distinguish between real and personal property in infants' contracts. Land sold by the infant will always be there. This is not necessarily true of personal property with which the infant has parted.

♦ Ratification

An infant's contract can be ratified as well as disaffirmed. However, to agree to be bound by the terms of the agreement, the infant must wait to become an adult. This seems reasonable: If an infant, in the eyes of the law, is incapable of making a binding contract, the in-

---

1. Egnaczyk v. Rowland, 267 N.Y.S. 14 (1933).

fant could hardly be expected to ratify one previously made. Contracts involving both real and personal property must await ratification until the infant reaches adulthood. Ratification may be either *express* or *implied*. The law merely requires the infant to indicate the intent to be bound if the infant so elects.

♦ Binding Contracts

An infant cannot avoid certain contracts he or she has made. Most prominent among such contracts are agreements by which the infant obtains the necessities of life. If a minor is not provided with such things as food, clothing, or lodging by a parent or guardian, the infant may make contracts for such things. If the minor is already supplied with the necessities, the minor cannot be made to pay for an additional supply of them. When payment is enforced by law, it is payment for the reasonable value of the goods or services, thus putting the liability on a quasi-contractual basis. The goods and services to be considered necessities vary according to the infant's station in life. Food, clothing, lodging, medical and dental care, vocational education, and tools of a trade are necessities for anyone. But conceivably such things as a car or a university education could be considered as necessities for an infant from a wealthy family.

If the infant has reached the minimum age to enter the armed forces, an enlistment agreement is not avoidable. Marriage cannot be avoided by a minor who is old enough to marry according to the state law.

♦ Parents' Liability

Ordinarily a parent (or guardian) is not liable for an infant's contract unless the parent has been made a party to it. For this reason, many who deal with infants insist that the adult responsible for the infant's welfare also agree to be bound. Only in cases where the parent (or guardian) has failed to provide the infant with necessities will the parent be required to pay reasonable value to third parties for needs supplied.

♦ Minors' Torts

There is a popular misconception about parents' and guardians' responsibility for the torts[2] committed by infants in their charge. The usual position taken by the law is that an infant is responsible for his or her own torts. Unless the infant is either acting under the parents' direction or should have been restrained by the parents, the parents are free of liability. The law seems reasonable in this stance. An adult can choose not to contract with a minor and thereby avoid difficulty.

However, it is often impossible to avoid tortious harm instigated by a minor (for instance, from an object propelled toward one's back). Knowledge that the *tort-feasor*[3] was a minor rather than an adult doesn't help much after the injury occurs. When the occurrence of tortious injury is inevitable, knowing the age of the tortfeasor is of little value to the victim. Therefore, the minor must answer for torts or crimes committed by the minor, even though the courts will allow the minor to avoid contracts.

MARRIED WOMEN

Under common law, a woman lost her right to contract independently when she married. A man and woman were one after marriage and the husband had all the contract and property rights. Today, however, a married woman's right to contract and own property is pretty much the equivalent of her husband's. A diminishing number of states, though, have remnants of limitations that have not yet been removed. For example, a wife may not be able to enter into a *binding surety contract*[4] or to make a contract with her husband once he has become her legal spouse.

INSANE PERSONS

Early court decisions involving contracts in which one party was insane declared such contracts to be void. The courts agreed that there could not be a meeting of the minds if one of the minds did not exist legally *(non compos mentis)*. Court decisions today usually hold that such contracts are voidable at the option of the insane person when that person becomes sane.

The law today recognizes that there are various forms and degrees of insanity. An insane person may have sane intervals in which the person is capable of contracting. A person may be sane in one or more areas

2. A *tort* is a wrongful injury to another's personalrights or property rights.

3. A *tort-feasor* is a person who commits a tort.

4. A *binding surety contract* is a contract in which one person agrees to pay another's debt if the other does not pay it.

of activity and insane in others. If a contract was made during the person's rational moments, providing the person had not been adjudged insane, it is binding despite previous or subsequent irrational behavior. Quite obviously, there can be some difficulty in establishing proof of sane behavior of an insane person or the limits of an area of activity in which the person is rational. These are questions of fact, though, and usually are left for a jury to decide on the basis of expert testimony of psychiatrists and psychologists.

Although it is generally true that courts will not examine the value of the consideration exchanged in a contract, they appear to make an exception where an insane person's contract is involved. If it appears that the other party to the contract took advantage of the insane person's mental condition, the courts will allow the insane person to avoid the contract. If it can be shown that the other party had no knowledge of the condition of the other bargainer's sanity and did not take advantage of the insane party, the courts will usually let the contract stand. This is especially true where the parties cannot be returned to their previous status (such as with contracts for services performed where the work has been completed).

INTOXICATED PERSONS

If a person tries to avoid a contract by pleading intoxication, that person must prove he was so inebriated that he could not be expected to understand the consequences of entering into a contract. A minority of the courts take a very dim view of a suit pleading intoxication. They hold that intoxication is a voluntary state and that the person should have had foresight enough to avoid getting drunk.

In most jurisdictions, however, intoxication of a party at the time the party entered into a contract is a sufficient ground to avoid the contract. If, however, an innocent third person would be harmed by avoidance of the contract, the contract usually will be allowed to stand. Even in jurisdictions where intoxication is frowned on by the courts as a means of avoidance of a contract, it can be used to show susceptibility to fraud. An intoxicated person, of course, remains liable for his or her torts.

CORPORATIONS

State laws provide means whereby an organization may become incorporated. The purpose and scope of its activities are set forth in the organization's articles of incorporation or charter. The corporation is an artificial person and possesses full power to act within the limits of its charter. If a contract of the corporation goes beyond its charter limits, it is an *ultra vires contract.* Before the corporation is given the right to do business in a state, it has no enforceable capacity to contract. Contracts entered into by the not-yet-existent corporation cannot form the basis of a suit by the corporation. Moreover, the individuals supposedly acting on the corporation's behalf (the "incorporators") remain liable, not the corporation. Until a corporation is formed (or registered as a foreign corporation if it was formed in another state), someone must make contracts for it. The corporation may take over these contracts after it is permitted to operate lawfully. The capacity to contract is also lost if the corporation's charter or its right to operate as a foreign corporation is suspended for some reason.

♦ Ultra Vires Contracts

There appears to be considerable variation in court decisions in cases involving *ultra vires* contracts. They can, however, be classified in a general way and some general statements can be made about them.

It is probably best to consider *ultra vires* contracts from the standpoint of their stage of completion. If an *ultra vires* contract is entirely executory, the courts probably will not enforce it. If it has been completely executed, the courts will tend to leave it alone. When the contract performance has been completed by one of the parties but not by the other (that is, the contract is partially executed), the courts generally give a remedy. Some courts allow recovery on the contract, and others place the remedy on a quasi-contractual basis, holding that the contract itself is void but allowing recovery on the basis of unjust enrichment.

♦ Torts and Crimes of Corporations

Because a corporation is an artificial person, it cannot personally commit a tort, but its agents can. Through the laws of agency, then, a corporation may be held liable for tortious injury to another by one of its employees. The employee must, of course, be engaged in the corporation's business at the time the tort is committed. As a general rule, the corporation is still liable even though the injury arose from an *ultra vires* act.

A corporation, as an artificial person, can commit crimes for which it may be held responsible. For example, the corporation is liable for criminal acts of its agents carrying out the business of the corporation, for antitrust law violations and other crimes. The penalty for a crime may be a fine or loss of the corporation's right to do business, either in a particular state or in the United States.

♦ The Corporate Veil

The corporation's status as an artificial, yet legally recognized, entity (often called the "corporate fiction" or the "corporate veil") affords limited liability to stockholders and the corporate officers. A stockholder generally risks losing only the amount invested in the stock. The same is true of corporate officers (president, vice-presidents, secretary, treasurer, and—perhaps—some managers) if the corporation has been fair and honest in its dealings with others.

The "corporate veil" protects these officers. But there is a limit to this protection and a reason to pierce the corporate veil when the people who run the corporation have purposely deceived, defrauded, and injured others. When such a situation occurs, these corporate officers (who are charged with the knowledge of the corporation's affairs) may be held personally liable for the liabilities of the corporation. The title of engineering vice-president may sound sweet, but with that title comes the joint and several responsibilities for the acts of the corporation.[5]

OTHER BUSINESS ENTITIES

Besides the corporate form, a number of other legal forms are used to conduct business. The sole proprietorship is probably the most common. In a *sole proprietorship*, the owner and the business are treated as one and the same. Other forms include *partnerships*, which can be general or limited, and *joint ventures*. In a *general partnership*, all partners are individually liable for the debts of the partnership.

A *limited partnership*, however, includes both "general" partners and "limited" partners. The general partners run the business of the partnership and remain individually liable for the partnership's debts. A limited partner, however, is not personally liable for the partnership's debts; at the same time, however, the limited partner cannot get very involved in the business operations, because to do so can result in personal liability. Hence, limited partnerships are somewhat like corporations in terms of allowing individuals to invest in the business and share in the profits while not subjecting themselves to individual liability (so tong as there is no active involvement in the day-to-day operations of the business).

GOVERNMENTS

People who do business with federal, state, or municipal governments have a practical need to know the contracting restrictions imposed. For example, when a city charter indicates the manner in which contracts are to be made, the city will not be bound to contracts made in some other fashion. If the Greenville city charter requires that certain contracts must be accepted by a public works board, acceptance by Mayor Brown will not bind the city. Someone performing a contract on the basis of the mayor's acceptance might find himself or herself an unwilling donor to the municipality. Federal and state statutes also may require that a given contract be submitted for competitive bids and that specified performance and payment bonds be given. If the bidding procedures are not followed, a court may declare the contract void or voidable.

PROFESSIONAL PERSONS

Nearly all recognized professions have restrictions placed on them by state statutes. The statutes customarily require professional persons to be registered or licensed as such before they are allowed to contract for their professional services. One who performs professional services without being registered or licensed may not resort to the courts to collect fees for such services, and such a nonprofessional may even be penalized for making such a contract. White, an engineer working for Black Manufacturing Company, designs a product that is built by Black and later involved in an injury to Brown. Brown brings a negligence action, alleging that faulty design of the product caused his injury. If the design is proved to be faulty, against whom does Brown have a right to recover? The general answer is that Black Manufacturing Co. would likely bear the loss. The reason for the liability is the concept of *respondeat superior*—the employer is responsible for the

5. See the case of *Mobridge Community Industries* v. *Toure* at the end of this chapter.

acts of employees during their employment. However, if White made the design while working as a consultant for Black, White would be liable for the design faults. In a similar situation, if White were an employee of Green Consultants in making the design for Black, Green is then liable, again under respondeat superior.

Sprecher v. Sprecher
110 A.2d 509 (Md. 1955)

This appeal is from a decree of the Circuit Court for Washington County setting aside a deed from the appellee to Martin L. Ingram, Trustee, dated July 28, 1950, and a deed of the same date from Ingram, Trustee, to Myron A. Sprecher[6] and Teresa 1. Sprecher, as joint tenants,[7] with right of survivorship. The ground of the Chancellor's action was that the deeds were executed while the appellee was still an infant and were disaffirmed by her within four months after she became of age. The appellant contends that the Chancellor erred in refusing to find that the appellee ratified and confirmed the deeds, or at least failed to seasonably disaffirm them.

The property conveyed, a six-room bungalow and a lot known as 345 S. Cleveland Avenue, in Hagerstown, was purchased as a home by the appellant and her then husband, Frank B. Sprecher, on May 16, 1939. Title was taken in their names as tenants by the entireties.[8] The price was $3,750, and they executed a mortgage to the First Federal Savings and Loan Association of Hagerstown in the amount of $2,700. The appellee, their daughter, had been born on July 14, 1932, and lived with her parents in the home until they separated on July 12, 1948. At that time the mortgage had been fully paid out of the husband's wages or perhaps out of their joint earnings since they were both employed, but on July 29, 1948, a new mortgage was obtained in the amount of $1,725. The proceeds were paid to Mrs. Sprecher, who testified that she used the money in part to repay a loan of $1,500 made to her by her mother, who resided in the home with them. On September 10, 1948, the parents and the daughter, who was then sixteen years of age, met with their attorneys, and it was agreed that the property be deeded unconditionally to the daughter, subject to the mortgage. On the same day the parents executed a deed of the property to trustees, who were in fact their respective attorneys, which recited that the conveyance was "upon trust to convey the same by deed to Teresa 1. Sprecher." On September 13, 1948, Frank B. Sprecher filed a bill for divorce against Mrs. Sprecher, and obtained a decree of absolute divorce on October 8, 1948. On November 13, 1948, the trustees executed a deed to the daughter.

The Chancellor found as a fact that at the time of the meeting to discuss a property settlement the father was unwilling that the mother should retain an interest in the property, but agreed that both should convey to the daughter. While the subsequent decree of divorce did not provide for custody, the daughter continued to live with her mother in the property, and the father contributed to the daughter's support until she was eighteen years of age and obtained employment. The Chancellor rejected the appellant's contention that she did not understand that she was parting with her interest in the property, in view of the plain language of the deed and the circumstances of its execution. The validity of these conveyances is not questioned on this appeal, and it is conceded that the infant was competent to take title to the property....

---

6. The wife and mother in this case.
7. Joint tenancy—an estate in land or other property, arising by purchasing or grant to two or more persons in which the parties own an equal and undivided interest.
8. Tenancy by the entirety is essentially a "joint tenancy," modified by the common law theory that husband and wife are one person, and survivorship is the predominant and distinguishing feature of each.

It appears, however, that on June 4, 1949, the appellant took the appellee to the office of the appellant's attorney and had her sign an agreement whereby Myron A. Sprecher agreed to support Teresa I. Sprecher during the period of her infancy and to maintain a home for her and cause her to attend school and high school. Teresa agreed, "as soon as practical after she shall attain the age of eighteen years," to reconvey the home property to Myron and herself as joint tenants. The appellee was sixteen years old at the time. On July 28, 1950, shortly after her eighteenth birthday, there was another visit to the attorney's office, and the execution of a deed to the attorney, as trustee, and a reconveyance to her mother and herself as joint tenants.

The appellee had graduated from high school in June 1950, secured permanent employment in August 1950, and thereafter paid her mother $10 a week board. She testified that she also paid three or four hundred dollars on the mortgage out of her earnings, but this was denied by the mother. On January 25, 1951, the appellee married Francis A. Pheil and shortly thereafter moved to her husband's apartment. Mrs. Sprecher continued to live in the property in question, and is still there. In 1952 she obtained an improvement loan on the property in the amount of $747 and spent it on improvements. The daughter was not asked to join in the application for this loan, or to execute any papers in connection therewith. On September 23,1953, Mrs. Sprecher married Leroy S. Hite, who moved into the property with his child. The appellee became twenty-one years of age on July 14, 1953. In August 1953, the appellant asked the appellee to join her in placing another mortgage on the property. The appellee declined to do so and a quarrel ensued. In October she consulted an attorney and on November 10. 1953, she wrote a letter to her mother formally disaffirming the deed she had executed on July 28, 1950. Shortly thereafter she filed the present bill.

At common law the period of infancy extended to the age of twenty-one years in the case of both sexes.... By statute in Maryland certain disabilities were removed in the case of females at the age of eighteen. One of the earliest of these statutes was Chapter 101, Acts of 1798, now codified as Code 1951, Art. 93, sec. 206, requiring a guardian to distribute the personal property of a female ward when she became sixteen (later changed to eighteen).... But in *Davis v. Jacquin & Pomerait,* 5 Har. & J. 100, it was held that although she could receive the property she could not dispose of any of it until she attained the age of twenty-one years.... Likewise, in *Greenwood v. Greenwood,* 28 Md. 369, 385, it was held that the right of a father to services of a female minor continued until she was twenty-one, despite a statutory limitation to eighteen in the case of apprenticing a female child. Statutes have been passed dealing with the right of females between the ages of eighteen and twenty-one to release dower...to make a will...to release an executor, administrator, or guardian...to release a trustee...to execute a release for any money paid, property delivered, or obligation satisfied...to make a deed of trust of her property, real, personal, or mixed, provided the same is approved and sanctioned by a court of equity.... We do not find any statute altering the rule stated in *Davis v. Jacquin,* supra. A conveyance made by an infant under twenty-one years of age is not void but is voidable, if disaffirmed, within a reasonable time after he or she attains the age of twenty-one years.... We think four months is a reasonable time under the circumstances.

The appellant contends that the appellee ratified the deed prior to her disaffirmance. The appellant testified that in August 1953, the appellee and her husband, Mr. Pheil, both said they didn't want to live in the property, that the appellee said: "I don't see why you don't let me sign the property over to you, my share of it, then you won't have to have anyone else to sign a deed, or a mortgage, either one." The appellant also testified that the Pheils wanted $5,000 for their interest; she offered them $3,000, but they would not take it. The appellant's husband, Mr. Hite, her sister, and her brother-in-law testified that Teresa, in the presence of her husband, made statements in August 1953 to the effect that the property belonged to her mother and she did not want it. The appellee and her husband denied that such statements were made. We cannot hold that the Chancellor was clearly wrong in finding these statements, if made, did not amount to ratification under the circumstances. The offer to convey her share for a consideration did not ripen into an agreement. At most, it was no more than a tacit recognition of the fact that she had previously conveyed a half interest to her mother. As pointed out by the Chancellor, the appellee had not then consulted counsel and was unaware of her rights. Indeed, it seems to have been assumed by everyone that the appellee was absolutely bound by her conveyance when she was eighteen years of age. Some of the authorities hold that ratification may be effective although made in ignorance of one's legal rights.... But the rule in Maryland seems to be to the contrary.... In any event, mere acquiescence or inaction, if not continued, is not enough. All of the authorities seem to recognize that there must be some positive act or declaration of an unequivocal nature in order to establish ratification.... We think the general statements testified to in the instant case fall short of this.

The appellant also contends that the appellee is barred or estopped from disaffirming her conveyance by the acceptance of benefits during her infancy which she is unable to restore. If we assume, without deciding, that the right of disaffirmance and recovery of the consideration paid may be barred under some circumstances...we think the principle cannot properly be invoked in the instant case. The appellee was under no legal or moral duty to reconvey a one-half interest to her mother. The prior conveyance to her had been made as a result of a property settlement between her father and mother, by which the mother was bound. The conveyance was upon a good consideration and in the nature of an advancement that would negate the implication of a resulting trust.... Nor did the mother undertake to perform any services for the appellee that she was not legally obligated to render. As the natural guardians of the infant child, she and the father were jointly and severally charged with its support, care, nurture, welfare, and education. . . . The appellee graduated from high school before she was eighteen, and thereafter until the time of her marriage in 1951, when she left the home, was employed and paid board to her mother. Until she was eighteen her father contributed to her support. These facts distinguish the case of Wilhelm v. Hardman, 13 Md. 140, where necessaries, in the form of support and education, were supplied by a third party.

The decree appealed from set aside the deeds and declared the property to be the sole and absolute property of Teresa I. Sprecher Pheil, subject to a balance of $408.92 due on the mortgage of $1,725 placed on the property July 29, 1948. The Chancellor held that the appellant was not entitled to reimbursement for mortgage payments made by her during the period from that date to March 1953, when she discontinued making payments. The daughter did not become twenty-one until July 14, 1953. We see no occasion to disagree with the Chancellor's finding. If we assume, without deciding, that reimbursement might be allowed under some circumstances, the appellant here failed to make out a case for equitable relief.... The entire proceeds of the mortgage were paid to her and not to the infant, and most of the payments made by the mother were made during the period when the property stood in the name of the daughter alone. The appellee claimed to have made payments of more than $300 in 1951. The daughter is now required to assume payment of the balance due. Moreover, the mother has occupied the premises rent-free during the whole period.

The Chancellor impressed an equitable lien on the property in the amount of $897.28, representing sums expended by the appellant in improvements to the property, and to pay the balance due on her personal loan for that purpose, and required the appellee to pay the costs below. The appellee has not appealed from that part of the decree, so its propriety is not before us. Under all the circumstances, we think the decree appealed from should be affirmed.

Decree affirmed, with costs.

MOBRIDGE COMMUNITY INDUSTRIES v. TOURE
273 N.W.2d 128 (S.D. 1978)

This case involves a breach of contract action brought by Mobridge Community Industries, Inc. (MCI), against Toure, Ltd. (Toure), an Illinois corporation, and its board of directors. The trial court ruled that MCI was entitled to recover the sum of $250,000 plus interest, advanced insurance premium, and cost from Toure and directors Cook, Bisbee, Spruck, and Zwald, jointly and severally. Directors Cook and Bisbee appeal. We affirm the judgment.

The breach of contract centers around an agreement dated November 25, 1975, between MCI and Toure. The agreement provided for the sale to Toure of personal property and equipment located in a plastics plant in Mobridge, South Dakota, together with a lease and option to buy the plant building. The agreement required the payment of $250,000 in five equal annual installments, insurance on the property, and rental payments on the building. Toure further agreed that if any personal property or equipment were sold or removed from the plant building it would be replaced with personal property or equipment of equal value to retain the overall value of the plant.

At the time the agreement was executed, MCI was fully incorporated in the State of South Dakota and Toure was duly incorporated in the State of Illinois. Toure, however, had failed to register and qualify to do business as a foreign corporation in South Dakota prior to or subsequent to the execution of the agreement.

Toure failed to make the first annual installment of $50,000 for personal property and equipment in the plant. The entire property was not insured as required by the agreement. Toure also failed to make several rental payments to MCI. Over a period of time, various items of personal property and equipment were removed from the plant building and sold without replacement of items of equal value as agreed.

The fact that the agreement was breached is apparent from the record and is not in serious dispute. The real question is the attachment of liability for the breach. The trial court found Toure and its directors Cook, Bisbee, Spruck, and Zwald to be jointly and severally liable for the breach[9] on three basic theories as follows: (1) statutory liability under SDCL 47-2-59; (2) liability for negligent and wrongful acts resulting in piercing the corporate veil; and (3) the trust fund doctrine.

On our review of the appeal, the successful party is entitled to the benefit of his version of the evidence and of all inferences fairly deducible therefrom that are favorable to the judgment of the trial court.... The findings of the trial court upon conflicting evidence are presumed to be correct, and we will not set such findings aside unless they are clearly erroneous.... In applying the clearly erroneous standard of review, the question is not whether we would have made the same findings that the trial court did but whether, on the entire evidence, we are left with a definite and firm conviction that a mistake has been committed....

With the proper standard of review in mind, we address the trial court finding of liability under SDCL 47-2-59, which reads as follows:

9. The trial court entered judgment against Toure, Spruck, and Zwald by default and dismissed the case as to defendant Krohn pursuant to stipulation. The trial court found that the record did not support recovery against defendant Raby by virtue of his association with Toure. Raby, however, did leave his position as manager of the Mobridge plant and removed some equipment because Toure had failed to pay his wages. The trial court found that MCI would be entitled to judgment against Raby upon a proper showing of the value of the equipment taken.

> All persons who assume to act as a corporation without authority so to do shall be jointly and severally liable for all debts and liabilities incurred or arising as a result thereof.

This statutory provision has been thoroughly analyzed in a well-reasoned opinion written by Federal District Judge Bogue in which he held that SDCL 47-2-59 does not require that the board of directors of a foreign corporation be held personally liable for the obligations of the corporation undertaken while doing business in South Dakota without a certificate of authority to do so.... The terms "corporation" and "foreign corporation" are defined separately in SDCL-47-2-1 and context of Title 47 does not require the first term to be read as including the second term.[10] SDCL 47-2-59 refers only to persons who purport to act as a corporation without incorporating and has the effect of negating the possibility of a de facto corporation....

The trial court attempts to distinguish the Cargill case on the fact that American Pork Producers, Inc., was a bona fide Iowa corporation and remained a recognized foreign corporation, even though its certificate of authority to do business in South Dakota had been revoked for a period of time, whereas Toure never did exist or function as a bona fide foreign corporation and never sought a certificate of authority to do business in South Dakota. The record contains a letter from the Illinois secretary of state to MCI's counsel stating that Toure was duly incorporated in that state on May 29, 1974, and had designated its registered agent along with his address pursuant to statutory dictates. Since SDCL 47-2-59 applied only to persons acting as a corporation without incorporating and Toure was duly incorporated in Illinois during all times pertinent to the misrepresentations made in this case, the trial court finding that appellants were personally liable under SDCL 47-2-59 is clearly erroneous.

The second theory of liability relied upon by the trial court is that of piercing the veil of the corporate entity. The general rule is that the corporation is looked upon as a separate legal entity until there is sufficient reason to the contrary. Such reason exists when retention of the corporate fiction would "produce injustices and inequitable consequences." ...In order to promote the ends of justice in appropriate cases, the corporate veil will be pierced and the corporation and its stockholders will be treated identically.... In deciding whether the corporate veil will be pierced, we recognize that "each case is sui generis and must be decided in accordance with its own underlying facts."

Disregarding the corporate entity or piercing the corporate veil may result from the occurrence of numerous factors. The primary factor considered by the trial court was the misrepresentation of Toure's financial condition by the directors during negotiations prior to execution of the agreement. Five basic requirements for the recovery from individual directors on the theory of fraudulent representation of financial condition are listed as follows: (1) a false representation of a material fact made by the directors; (2) the directors making the representation knew the fact was not true or made the statement recklessly with no reasonable grounds for believing it to be true; (3) the misrepresentation was made with the intent to induce MCI to act upon it; (4) MCI took action or refrained from acting in reliance upon the misrepresentation; and (5) MCI incurred damage from such reliance....

The trial court found that various representations were made by the directors as to the financial ability of Toure to improve and operate the Mobridge plant and the investment money that was forthcoming. The Toure financial statement exhibited by the directors to MCI during the negotiations showed a net worth of $90,000, and a second statement showed a net worth of

10. This is true in part because SDCL 47-2-59 is in the chapter dealing with the formation of domestic corporations, and foreign corporations are treated separately in Chapter 47-8.

$65,000. This discrepancy was explained as the good will factor of approximately $25,000. At the time the statements were exhibited to MCI, Toure had approximately $62.08 in its corporate account. Thus, the first requirement is satisfied. Regarding the second requirement of knowledge on the part of the directors, the trial court found that the directors were experienced in business affairs, were accompanied by a banker, and, considering outstanding obligations, certainly knew that Toure was "broke." We find that this requirement is further satisfied due to the fact that directors of a corporation are "presumed to have knowledge of the financial affairs of their corporation when making statements concerning its financial condition."...The third requirement is satisfied by the very nature of the facts and circumstances surrounding the negotiations between Toure and MCI regarding the Mobridge plant. The financial statements were padded or inflated to induce MCI to close the transaction. The final two requirements are met in that MCI acted in reliance on the inaccurate financial information given by the directors in entering into the agreement for sale and lease and incurred damages as a result of its reliance, as Toure was not in any financial condition to fulfill its side of the agreement. Therefore, the trial court finding that the theory of fraudulent representation of financial condition rendered the directors liable jointly and severally is supported by the record and is not clearly erroneous.

A further factor justifying a disregard of the corporate entity in this case is inadequate capitalization. An obvious inadequacy of capital, measured by the nature and magnitude of the corporation's undertaking, is an important factor in denying directors and controlling shareholders the corporate defense of limited liability.... The rationale is more completely explained in *Briggs Transp. Co. v. Starr Sales Co.,* supra, as follows:

> If a corporation is organized and carries on business without substantial capital in such a way that the corporation is likely to have no sufficient assets available to meet its debts, it is inequitable that shareholders should set up such a flimsy organization to escape personal liability. The attempt to do corporate business without providing any sufficient basis of financial responsibility to creditors is an abuse of the separate entity and will be ineffectual to exempt the shareholders from corporate debts. It is coming to be recognized as the policy of the law that shareholders should in good faith put at the risk of the business unencumbered capital reasonably adequate for its prospective liabilities. If capital is illusory or trifling compared with the business to be done and the risks of loss, this is a ground for denying the separate entity privilege....

The record supports the conclusion that Toure was inadequately capitalized in view of the magnitude of its undertaking embodied in the agreement, and therefore the corporate entity must be disregarded.

The directors of Toure made several false representations to MCI in addition to that of financial ability to carry out the agreement. Cook's banker accompanied him during the negotiation process, and Cook's role as treasurer of Toure was emphasized to the point that he would "ride herd" over and be "watchdog" of the funds. There were also misrepresentations regarding marketing networks that were waiting for Toure's products. The fact that appellants were involved in the venture provided considerable credibility to the representations and negotiations. Further misrepresentations came in the form of a Toure management resume that included several reputable individuals in the business community who were to join Toure management within a month's time. There were other misrepresentations as to startup time, worldwide patents owned or pending, availability of molds, plant improvements, and an influx of investment money.

Appellants argue that representations as to future events are not actionable and that the false representations must be of past or existing facts. Although this is the general rule, we have stated that an exception comes into existence when the misrepresentation of a future event is in regard to a matter that is peculiarly within the speaker's knowledge.... For example, the directors of Toure collectively exhibited a particular expertise in the products involved and represented that some of the products were so new and superior to others on the market that the Mobridge plant was ideal so that production would remain a secret in the industry and the product ideas would not be stolen prior to patenting and marketing. These representations as to the innovative nature of the products were also tied into the anticipation of the market, addition of management personnel, and influx of investment money. We conclude that the exception applies under these circumstances and the representations as to future events are actionable along with the representations of past or existing facts.

Appellants further argue that they acted in good faith and were not aware of what was going on at all times pertinent to this case. This is without merit, because corporate officers and directors are held to a high degree of diligence and due care in the exercise of their fiduciary duty to shareholders. By accepting such duty, each director is charged with monitoring the heartbeat of the business and knowing where the corporation stands in regard to finances, obligations, goals, policies, etc. Where such a duty exists, ignorance due to neglect of that duty "creates the same liability as actual knowledge and a failure to act thereon."...Since appellants failed to exercise the duty imposed on corporate directors, they can hardly claim ignorance and lack of awareness "by emulating the three fabled monkeys, hearing, seeing, and speaking no evil."...

We conclude that the conduct of the directors constitutes the kind of a dealing which would make it inequitable for the trial court to recognize the corporate entity of Toure and thus permit the individual directors to escape personal liability. While we have held that a corporation existed during the material time when the misrepresentations were made in regard to the assets of this corporation, we are aware of the fact that the Toure board on February17, 1976, voted to dissolve the corporation by March 15, 1976, and that the corporation was officially dissolved by the State of Illinois on December 1, 1976. It was during this dissolution process that most of the property was removed from the Mobridge plant. This may not meet the requirements of SDCL 47-2-59 for purposes of holding the members of the board jointly and severally liable, but it certainly adds weight to our decision that this is a proper case to pierce the corporate veil because the legal corporation was of short duration (two and one-half years), was never used for any purpose except the personal immunity of the members of the board in their questionable activities, and was in the process of being dissolved when the assets were actually looted from the Mobridge plant. We are not compelled to say that the trial court finding of sufficient facts to warrant piercing the corporate veil of limited liability is clearly erroneous, and we will not disturb such finding on appeal.

The third theory of liability relied upon by the trial court is the trust fund doctrine that once Toure became insolvent the directors were charged with the fiduciary responsibility of holding the corporate assets in trust for the benefit of creditors. On the basis of our conclusion that the theory of piercing the corporate veil as discussed above warrants a finding of personal liability of the directors of Toure, it is unnecessary to discuss the third theory that the trial court used to charge the directors with personal liability on the breach of contract.

♦

REVIEW QUESTIONS

1. Black, a minor, sells a piece of real estate that Black inherited. Black spends the money rather foolishly and, by the time of adulthood, Black is deeply in debt. Black's creditors force Black into bankruptcy. The creditors find that the land Black sold as a minor is quite valuable; in fact, the value is sufficient to pay Black's debts and leave a substantial remainder. Black refuses to disaffirm the sale. May the creditors disaffirm for Black? Why or why not?
2. What is an *ultra vires* contract? Give an example of such a contract.
3. How could a married woman's capacity to contract be determined?
4. Gray, a salesperson, contracted with Green to sell Green a piece of equipment at a price about 10 percent below the usual market price for such equipment. The contract occurred in a local restaurant during an extended lunch hour. Gray had two or three martinis before eating lunch. Gray now claims he was drunk at the time the contract was made and, on this basis, seeks to avoid the contract. Green has established that all of Gray's other acts (including paying both restaurant tabs) gave no indication that Gray was under the influence of alcohol. Is it likely that Gray will be held to the contract? Why or why not?
5. Are contracts for professional services voidable by the recipient of the services if the professional is not licensed?
6. In the case of *Sprecher v. Sprecher*, what acts by the daughter would have amounted to ratification under the circumstances?
7. In the case of *Mobridge Community Industries v. Toure*, suppose there is another "White" who could show that his or her membership on the board of directors of Toure was nominal only and that he or she was unaware of Toure's fraudulent activities. Would White still be liable? Suppose further that White is the only person on the board with sufficient assets to pay the court's judgment (all the other directors are "broke"). How much would White have to pay?

Chapter 7

Agreement

A contract is an agreement, although not all agreements are contracts. According to *Black's Law Dictionary*, an *agreement* is "the coming together in accord of two minds on a given proposition." An agreement consists of an offer and an acceptance of that offer. Refer back to the definitions of a "contract" and an "agreement." The Uniform Commercial Code (UCC) speaks in terms of the parties' "bargain." The UCC does not define what constitutes an offer or an acceptance. Instead, the common law remains effective as to such issues. Hence, this chapter's discussion addresses the common law rules on the formation of contracts.

CONTRACT FORMATION

To form a contract, there needs to be an offer and an acceptance of that offer. What amounts to an offer or an acceptance is considered here, as are issues regarding just when and how an offer may be accepted.

♦ Offer

An offer contemplates a future action or restraint of action. It is a proposal to make a contract. In general, an offer states two things: what is desired by the offeror and what the offeror is willing to do in return.

The Restatement (Second) of Contracts defines an *offer* as follows:

> [An offer is] the manifestation of willingness to enter into a bargain, so made as to justify another person in understanding that his assent to that bargain is invited and will conclude [the bargain].

The offer does not have to be a formal statement to be binding on the offeror. When the intent to offer is conveyed from the offeror (the person making offer) to the offeree (the recipient of the offer), an offer occurs. Under certain circumstances the law will consider a series of acts by the parties to be an offer and acceptance of the offer resulting in a binding contract. A man entering a hardware store where he is well known, grasping a shovel marked $14.95, motioning to the owner who is busy with another customer, and then leaving the store with the shovel, has made an offer to purchase. Although no word was spoken, an offer to buy the shovel at the suggested price of $14.95 was made. By allowing the man to leave with the shovel, the hardware store owner accepted the offer, and a contract resulted.

Intent

The usual test of both offer and acceptance is the *reasonable man* standard. Would a reasonable man consider the statements and acts of the party to constitute a binding element of the agreement necessary in a contract? The circumstances surrounding the statements and acts of the parties are examined to determine the existence of an intent to contract. In considering the circumstances, the courts gauge the parties' intentions objectively, rather than each party's "subjective" view of what they meant and understood the other to mean. In one case,[1] a blacksmith shop owner, enraged at the loss of a harness, stated in his irate ravings about the alleged thief that he would pay anyone $100 for information leading to the capture of the thief and an additional $100 for a lawyer to prosecute him. The court held that the language used, under the circumstances, would not show an intention to contract to pay a reward. In another case,[2] Justice Holmes stated, "If, without the plaintiff's knowledge, Hodgdon did understand the transaction to be different from that which his words plainly expressed, it is immaterial, as his obligations must be measured by his overt acts."

It is often quite difficult to determine by the language used by the parties whether a valid offer and acceptance have occurred. In the haggling, "horse-trading," or bargaining that so often occurs in attempting to reach an agreement, words are used that, removed from context, sound very much as though the parties had formed a contract. One person may say to another, "I would like to sell my car for $1,500," or "Would you give me $1,500 for my car?" These are not offers. The courts, in recognition of business practices, view such statements as solicitations to offer and not as binding offers.

Advertisements

When making an offer, the offeror must be prepared for acceptance of the offer. Acceptance would complete the contract, and, if the offeror could not perform the contract, the offeree would have a legal action for damages available against the offeror. For this reason, advertising—prices marked on merchandise in stores and similar publicity—is considered by the courts to consist of mere invitations or inducements to offer to purchase at a suggested price. The courts generally do not view such conduct as amounting to an offer. An expression by a customer of willingness to buy the merchandise at the price stated does not form a contract. The customer, not the store, has made the offer. The store has the right to accept or reject the offer. It is possible for the store to have sold the last of its merchandise in that line and thus to find performance to be impossible. If an offer of the price by a customer were to be construed as acceptance and the store could not perform, it would be liable for breach of contract. To hold that the store has made a contract in all cases in which a customer tenders the price of certain merchandise would therefore be unreasonable.

Of course, some advertisements are worded (whether purposely or accidentally) so as to constitute an offer. Acceptance by anyone will then form a binding contract. A common example of an advertisement that constitutes an offer is an ad offering a reward. To accept, anyone with knowledge of the advertisement may perform the act required in the ad, thus completing the requirements for a valid contract.

An advertisement (or solicitation or request) for bids on construction work generally does not constitute an offer. Unless the advertisement specifically states that the lowest bid will be accepted, the advertiser has the right to reject any or all bids submitted. Similarly, a quotation of a specific price usually does not amount to an offer. Such requests for bids and price quotations are usually viewed as preliminary negotiations only.

Continuing Offer

An offer generally provides the offeree with a continuing power to accept the offer until the offer is revoked. When the offeror agrees not to "revoke" the offer (that is, to keep the offer open for some set period of time) and such an offer is based on consideration, it is an option, a contract in itself. The offeror basically agrees to perform or not to perform certain acts on acceptance by the offeree, providing only that the offer be accepted within the specified time limit.

Termination of Offer

An offer may end in various ways: by acceptance, by revocation or withdrawal, by rejection, by death of a party, or through the passage of time. The expiration of an option supported by consideration, however, ends only on expiration of the time period for which it was

1. *Higgins v. Lessig*, 49 111. App. 459 (1893).
2. *Mansfield v. Hodgdon*, 147 Mass. 304,17 N.E. 544 (1888).

to be held open. An option cannot be revoked, and rejection of the offer or death of one of the parties usually does not end its life.

The offeror has the right to revoke the offer at any time before acceptance. The withdrawal or revocation must be communicated to the offeree before the offeree's acceptance. If the offeree has knowledge of the sale to another of the goods offered on a continuing offer, the offeree has effective notice of withdrawal of the offer.

Rejection by the offeree ends the offer. The rejection of the offer often takes the form of a qualified acceptance[3] of the offer or a request for modification of the offer. Either act usually amounts to rejection. Mere inquiry as to whether the offeror will change the terms, though, does not constitute a rejection of the offer.

Generally, the death of either party before acceptance of an offer terminates the offer. This is the rule even when the offeree has not received notice of the death. The Restatement (Second) of Contracts describes this rule as a "relic" of the view that it is necessary for a meeting of the minds to occur for a contract to be formed. If one of the minds is no longer among the living, so the reasoning goes, no contract can occur.

If a time limit is specified in a continuing offer, the offer terminates on the expiration of the period of time in which the offer was to remain open. If no time was specified in the offer, the offer will be open for acceptance for a reasonable time, the reasonableness of the time depending on the circumstances and subject matter of the offer. One would expect that a reasonable time to decide whether to purchase a truckload of ripe bananas (about to lose their value) would be much shorter than a reasonable time to decide whether to buy furniture.

♦ Acceptance

Generally, for a contract between two parties to exist there must have been an acceptance of an offer. However, an offer may be accepted and a contract thus formed only during the life of the offer. With the exception of an option contract, as mentioned earlier, once an offer has been rejected, revoked, or its time has run out, it cannot be accepted.

Similarly, once an offer has been accepted, it is impossible for the offeror to withdraw the offer. The rules sound simple, but in many situations it is often difficult to determine whether an acceptance has taken place. Such uncertainty often arises because of various ambiguities in the parties' statements, letters, or conduct.

An offer may be accepted only by the person to whom it was made. A person other than the offeree who obtains the offer cannot make a valid contract by accepting. The best that such a person can do is to make a similar offer to the original offeror for acceptance. An offer must be communicated to the offeree by the offeror or the offeror's agent to constitute an acceptable offer. Black tells Gray, who is a friend of both Black and White, of Black's intent to offer White a television set for $500. Gray reveals this to White. White cannot at this point create a valid contract with Black by stating White's acceptance. White must either wait for Black to communicate the offer to her or offer Black $500 for the TV set. At this point there is no offer to be accepted. However, Black could have used Gray as his agent, requesting Gray to give the offer to White. Black's comments then would have constituted a valid offer.

Acceptance of an offer is held to consist of a state of mind evidenced by certain acts and/or statements. The mere determination to accept is not sufficient; it must be accompanied by some overt act that reveals that the offeree accepts the offer. Without such an overt act, it is impossible for the offeror or the court to determine whether the offer has been accepted, despite the offeree's secret intent to accept, unless previous dealings have established the offeree's silence or inactivity as acceptance. Generally, the offeror may specify the manner in which the offer may be accepted, such as by letter, telecopy, signature of the contract document, or the like. If the offeror does not specify the mode of acceptance, the offeree may accept in any manner and by any medium reasonable under the circumstances.

The communication of offer and acceptance creates a contract regardless of how it is done. When the offer and acceptance are oral, the contract is formed as soon as the offeree speaks the words of acceptance.

One sticky problem often arises in regard to the offeree's communication of acceptance. Suppose the parties are not in voice contact with each other. For example, suppose they are communicating by mail. Then let us suppose the acceptance is made in timely good order but gets lost in transit. Is there a contract or not?

3. The phrasing of such an acceptance resembles this: "I accept your offer providing you will do [an additional condition] in addition." Such a statement may not constitute acceptance, but rather a counteroffer. The Uniform Commercial Code is more likely than the common law of contracts to lead to a finding of acceptance in situations such as this.

This is rather important information for the offeror, because the same offer probably will be made to someone else if the original offer is rejected.

After some years of confusion on this point the law is now pretty well settled—there is a contract. It sprang into being when the letter of acceptance was mailed. The offeror can protect himself or herself from a possible breach of contract action by including a requirement that acceptance be made in a writing received in the offeror's office by some fixed time to be effective.

♦ Time of Revocation

An offer may be withdrawn by the offeror up until the time it is accepted. The offeror's revocation, however, is never effective until it has been communicated to the offeree. When the negotiations are carried on by mail or telegram, a question may arise about whether a contract exists. To point up this difficulty, consider the following example. Black and White have been negotiating by mail for the sale of White's land. White finally makes an offer that Black finds acceptable, and Black sends a letter at 12:00 noon to White accepting the offer. Previous to this, White has sent a letter to Black revoking the offer, but the withdrawal is not received by Black until some time later than noon on the day Black mailed his acceptance. The courts would hold that there is a contract between the parties, and even the loss of the letter of acceptance would not change the holding, providing only that Black could prove sending such a letter and that the letter constituted acceptance of the contract. As you might conclude, these rules encourage offerors to convey a revocation of the offer quickly, lest the offeree accept in the meantime.

♦ Silence as Acceptance

Unless there is a customary practice or agreement between the parties to the contrary, silence generally cannot be construed as acceptance. Wording an offer to the effect that "If I do not hear from you, I will consider my offer accepted by you" has no legal effect upon the offeree. Mere silence generally does not form a binding contract.

♦ The "Battle of the Forms"

In many situations, the parties informally negotiate only a very few terms of a contract, such as price, quantity, and time of delivery. The parties then exchange preprinted forms—the buyer sending a purchase order and the seller sending an order acknowledgment or invoice. The preprinted forms usually contain the buyer's terms of purchase and the seller's terms of sale. Almost always, these forms are written with a very one-sided view in favor of the party sending the form. Thus, the buyer's terms almost never match the seller's terms. Each party wants its form to apply rather than the other's. The question arises as to whether a contract was ever formed and, if so, what were its terms? Because the parties often at least partially perform the contract by delivering goods or by accepting or paying for all or some of the goods, it is usually clear that there was some contract, whatever its terms.

Because the common law applied a "mirror image" rule (requiring the acceptance to "mirror" all of the terms and conditions of the offer), the party sending the last form usually retained the advantage. Why? The buyer's purchase order usually amounts to an offer. The seller's invoice or order acknowledgment, however, usually acts simultaneously as a counteroffer and a rejection of the original offer. (Remember, the seller's terms almost always vary substantially from the buyer's terms.) If the buyer sends no further forms and makes no further offers, and the seller then delivers goods that are accepted by the buyer, the buyer will be deemed to have agreed to the seller's terms by way of performance.

The UCC approaches the problem involving the battle of the forms differently than does the common law. The UCC rules are as follows:

§ 2-207. Additional Terms in Acceptance or Confirmation

1. A definite and seasonable expression of acceptance or a written confirmation which is sent within a reasonable time operates as an acceptance even though it states terms additional to or different from those offered or agreed upon, unless acceptance is expressly made conditional on assent to the additional or different terms.
2. The additional terms are to be construed as proposals for addition to the contract. Between merchants such terms become part of the contract unless:
   a. the offer expressly limits acceptance to the terms of the offer;
   b. they materially alter it; or
   c. notification of objection to them has already been given or is given within a reasonable time after notice of them is received.
3. Conduct by both parties which recognizes the existence of a contract is sufficient to establish a contract for sale although the writings of the parties do not otherwise establish a contract. In such case the terms of the particular contract consist of those terms on which the writings of the parties agree, together with any supplemental terms incorporated under any other provisions of [the UCC].

Notice that section 2-207(2) creates a division between the rules applicable to contracts "between merchants" and contracts not between merchants. (Generally speaking, a merchant is one who regularly deals in goods of the kind that are the subject of the contract.) If the parties are not merchants, either party's added terms on the paperwork (e.g., order or order acknowledgement) are treated as "proposals" for addition to the contract. If, however, the purchase order states that the seller must accept all the buyer's terms to complete the contract, then an order acknowledgement cannot alter those terms without defeating the contract. On the other hand, the parties' subsequent conduct in performing the contract may establish the existence of some contract, probably governed largely not by either of the printed forms but by the UCC. If the parties are merchants, the new terms presented by the seller's order acknowledgment apply unless any of the following occurs:

- The offer limits acceptance.
- The new terms materially alter the buyer's terms.
- The buyer objects.

A careful buyer using a printed form usually limits the terms of the acceptance and also objects to the addition of any new terms. Generally, such new terms will not be added and the buyer's terms often will control. Thus, the UCC adopts a rule favoring buyers in many situations (at least in those situations involving merchants).

The complexity of modern business often leads to situations in which it is unclear exactly what terms govern the parties' relationship. For example, assume that the buyer and seller agree over the phone on price, quantity, and date of delivery, then exchange their conflicting forms. Whose terms govern? Consider the cases at the end of the chapter. Were the results fair? Is there a better way to handle this problem? At least one scholar has noted that it is lucky that most business persons do not read the other's forms and that most transactions do not develop into disputes.

MELVIN v. WEST

107 So. 2d 156
(Fla. Dist. Ct. App. 1958)

West, a real estate broker, sued Melvin, a property owner, to recover a real estate commission allegedly earned by West for effecting the sale of certain of Melvin's lands to one Buchanan and his associates. By stipulation, the case was tried before the Circuit Judge without a jury and judgment was for the plaintiff. The defendant has appealed from the judgment, assigning as grounds for reversal that the evidence does not support the judgment and that the evidence shows that a nonlicensed employee of the plaintiff actively participated in procuring the listing of the land and in procuring the purchaser, and hence the plaintiff is barred by law from recovering a commission.

According to the evidence most favorable to the plaintiff, the defendant owned a piece of land which plaintiff understood contained approximately 3,000 acres and was for sale at $1,100 an acre. The defendant orally listed this property for sale with the plaintiff, and several other brokers, with the specific understanding that "the first registered real estate broker who brought in a check and a signed sales contract would be the successful negotiator of the deal."

About a month after West had been given the listing, he informed Melvin that he had a prospect who was interested in buying the 3,000-acre tract at $1,100 an acre. In reply, Melvin told West that the price was not $1,100 an acre for 3,000 acres, as understood by West, but was $3,500,000 whatever the acreage. West conveyed this information to his prospect, Buchanan, and thereafter and with full knowledge of the fact that the price of the property had been fixed at $3,500,000, West and Buchanan inspected the property.

Shortly after viewing the property, West and Buchanan conferred with Melvin about the purchase of the property by Buchanan and two associates, but came to no definite agreement in regard to a sale. A few days after this conference, Buchanan informed West that he had ascertained that the tract did not contain 3,000 acres, as had been originally represented by West to him, but a lesser acreage, and hence that he and his associates were not interested in buying the property at the quoted price of $3,500,000 but might be willing to pay $2,750,000, and that if the counter offer was not acceptable they "would have no further interest in the matter."

After receiving this counter offer from Buchanan, West phoned Melvin and told him that if he was interested "in helping save the deal" he and West should "get (their) heads together and see if (they) could work out some way to work the situation out...(that) Mr. Buchanan was still interested in the property as such...and if the price and terms could be worked out to (Buchanan's) satisfaction...(he, West) felt the deal could still be made."

The evidence shows that a few hours after this conversation, West, Melvin, and L. V. Hart, one of West's employees, met in West's office at Ft. Lauderdale to discuss what might be done to "save the deal." The meeting ended when Melvin decided, with West's approval, that since he and Buchanan "were the two principals in the situation" he, Melvin, would negotiate directly with Buchanan, if someone would drive him to Miami Beach for the purpose. Thereupon Hart, West's employee, agreed to drive Melvin to Miami Beach where Buchanan was found and a conference was held, which began at approximately 5 o'clock in the afternoon and ended about three hours later, when Melvin finally agreed, after much discussion, to reduce the property from $3,500,000 to $2,900,00, provided $600,000 was paid in cash and the remainder was paid in specifically agreed annual installments, with interest, over a 10-year period; and also agreed that since Buchanan had no authority to commit his associates to the purchase of the property at the new price and terms agreed on at the conference he, Buchanan, might have "until two o'clock the next day (Friday, March 16, 1958) to obtain the permission of (his) associates to continue with the business on the basis of this present agreement and to authorize the contract to be drawn up on this basis"; that in the event permission was obtained "to continue with the business," Buchanan would inform West and would "deposit" with him a check in the sum of $50,000 as a "binder payment"; and that Mr. and Mrs. Melvin would meet with West and Buchanan, at Buchanan's office, on Saturday morning, March 17, 1956, to execute a contract of purchase, to be prepared by West's attorney, which would bind the parties to the sale and purchase of the property.

The evidence is to the effect that on Friday morning, March 16, Buchanan got in touch with his two associates and obtained their permission to purchase the property at the price and terms agreed to by Melvin at the Thursday afternoon conference; that Buchanan called West, at about 11:30 Friday morning, to advise him that he had had "the matter confirmed and...had a check available"; and that West immediately phoned the Melvin residence to inform Melvin of what Buchanan had said, but was told by the person answering the phone that Melvin was away from the premises. West continued calling the Melvin residence at frequent intervals throughout the remainder of the day but was never able to contact Melvin. That night, at a few minutes past midnight, Melvin called West to inform him that on that day he had sold the property to two prospects produced by a real estate salesman, one Harris, who also had an open listing on the property; that a downpayment check had been offered and accepted, and that a sales contract had been executed by seller and buyers and hence that he could not sell the property to Buchanan.

The testimony of plaintiff's witness, Harris, the real estate salesman who produced the purchasers to whom the property was sold, was to the effect that on Thursday, March 15, 1956, he had interested two prospects in buying the property at the price of $3,500,000 that was being asked by Melvin; that around 6 o'clock Thursday evening his prospects had offered him a "binder" check that he had refused because "they had not stepped on the property"; that he thereupon attempted to call Melvin between 8:45 and 9 o'clock on Friday morning to inform him that he had the property sold and to request Melvin to meet with the purchasers so that the transaction might be closed without delay; that he met with Melvin one hour and ten minutes later and introduced him to the purchasers; that Melvin then went with Harris and the purchasers to the property, where a purchase contract was executed and a check was delivered between 2:45 and 3 o'clock in the afternoon.

The evidence shows, that after receiving this information late Friday night, West went to Miami Beach early Saturday morning to inform Buchanan that Melvin had sold the property to another prospect. While there West presented a contract of purchase for Buchanan's signature, had Buchanan execute it, and received from Buchanan his personal check in the sum of $50,000 that was dated March 16, 1956, and was made payable to the order of "Wm. H. West, Realtor." According to West, he accepted the check because "Buchanan felt that there might still be a possibility that Mr. Melvin would change his mind and that the other transaction would not go through...(and) he wanted to be on record that he had carried out the part of the deal that he had made with Mr. Melvin."

About thirty days later, after it appeared certain that Melvin had no intention of honoring his oral promise to Buchanan, West returned the check to Buchanan and instituted the present suit to recover a commission for selling the property.

The first question on the appeal is whether the foregoing evidence, when viewed in the light most favorable to the plaintiff, is sufficient to support the judgment rendered in his favor.

Two types of brokerage contracts are generally used in the business of selling real estate. Under the first type, the seller employs a real estate broker to procure a purchaser for the property, and the broker becomes entitled to his commission when he produces a purchaser who is ready, able, and willing to purchase the property upon the terms and conditions fixed by the seller, leaving to the seller the actual closing of the sale. Under the second type, the seller employs a broker to effect a sale of the property, and the broker, to become entitled to his commission, must not only produce a purchaser who is ready, able, and willing to purchase the property upon the terms and conditions fixed by the seller but must actually effect the sale, or procure from the purchaser a binding contract of purchase upon the terms and conditions fixed by the seller....

Where a broker employed to effect a sale procures a purchaser who is ready, able, and willing to purchase the involved property upon the terms and conditions fixed by the seller and before he can effect the sale or procure a binding contract of purchase the seller defeats the transaction, not because of any fault of the broker or purchaser but solely because he will not or cannot convey title, the broker will be entitled to his commission even though the sale has not been fully completed, if the buyer remains ready, able, and willing to purchase the property upon the terms and conditions fixed; the strict terms of the contract between the seller and the broker requiring him to actually complete the sale or procure a binding contract of purchase being deemed, in such case, to have been waived by the seller....

As pointed out in *Hanover Realty Corp. v. Codomo,* Fla., 95 So. 2d 420, the rule excusing the broker from complete performance where performance has been made impossible by arbitrary acts of the seller is simply an application to brokerage contracts of the rules relating to contracts generally to the effect that "where a party contracts for another to do a certain thing, he thereby impliedly promises that he will himself do nothing to hinder or obstruct that other in doing the agreed thing," and "one who prevents or makes impossible the performance as happening of a condition precedent upon which his liability by the terms of a contract is made to depend cannot avail himself of its nonperformance."

It is vigorously contended by West that when Melvin went to Miami Beach, with West's acquiescence, and made the oral promise to Buchanan to reduce the price of the property from $3,500,000 to $2,900,000 and to allow Buchanan until 2 o'clock the following day to ascertain from his associates whether or not they would be interested in buying the property at the new price, West thereupon became entitled to his commission, because, in contemplation of law, he became "the first broker to sell, because he fully performed his contract to sell when (Melvin) took over and made the oral agreement with Buchanan." To support his contention, he relies upon what is said in *Knowles v. Henderson,* supra, to the effect that where a broker employed to sell property finds a purchaser who is ready, able, and willing to buy at terms fixed by the seller, but before he can effect the sale or procure a binding contract of purchase the seller arbitrarily refuses to go through with the transaction, the broker will be entitled to his commission, if the purchaser remains ready, able, and willing to buy at the terms fixed.

We fail to see how the facts of the case at bar bring the case within the exception stated in *Knowles v. Henderson,* supra.

It is perfectly plain that when Melvin went to Miami Beach to confer directly with Buchanan and thereby attempt "to save the deal," West had not become entitled to a commission, because he had not found a purchaser who was ready, able, and willing to purchase the property at the price demanded by Melvin. Since Buchanan and his associates had flatly refused to buy the property at $3,500,000 and Melvin had rejected Buchanan's counter offer of $2,750,000, the most that can be said for West's legal position at the time was that he had three possible prospects, one of whom was willing to confer further about the property but had no power to bind his associates beyond the counter offer that had already been made and rejected.

It is equally plain that since West had utterly failed to perform his nonexclusive oral contract within the terms of his listing, Melvin was not legally obligated to West to negotiate with Buchanan, at the conference, in any particular manner, except to refrain from entering into an arrangement with West's prospects for the purpose of fraudulently preventing West from earning a commission.

So far as can be ascertained from the record, nothing was said or done by Melvin throughout his conference with Buchanan that could be deemed a waiver of the terms of West's original brokerage contract, or that would have, or could have, prevented West from becoming entitled to a commission if a sale had been made to his prospects at the new purchase price, prior to the time some other broker had effected a sale of the property within the terms of his listing. For, as has already been stated, when the conference ended between Melvin and Buchanan no sale had been made of the property. Melvin had orally promised to reduce the purchase price of the property from $3,500,000 to $2,900,000—which Buchanan had been unable to accept because he had no authority from his associates to do so—and had given Buchanan until 2 o'clock the following afternoon to ascertain from his associates whether or not they were interested in purchasing the property at the reduced purchase price.

Even though the oral promise made by Melvin to extend the time for Buchanan to talk with his associates was not enforceable as an option, since no consideration was given for the promise...it is impossible to understand how, if West had been negotiating with Buchanan in place of Melvin, he could have accomplished as much as, or more than, was accomplished by Melvin in an effort to effectuate a sale. Manifestly, West could not have lawfully made the concession as to price that was made by Melvin, since he had no authority, in respect to price, except to offer the property for sale at $3,500,000—a figure that had already been presented to Buchanan and rejected by him. And, assuming for the sake of argument that West had been given authority, prior to the conference, to make concessions in regard to price and terms and thereby bind Melvin, it is plain that Buchanan could not have accepted a single concession on behalf of his associates, since at the time of the conference he had no authority to bind his associates beyond the amount of the counter offer of $2,750,000 that had already been made to, and turned down by, Melvin. Consequently, when the conference between Melvin and Buchanan ended, West occupied no weaker position, because of anything said or done by Melvin during the conference, than he had occupied prior to that time; indeed, it appears that he was in a stronger position. Before the conference, all that West had were three prospects whom he had been unable to interest in the property at the price and terms fixed by the seller; hence he had failed to fulfill the terms of his nonexclusive contract. After the conference, West not only still had his three prospects but also had, in effect, an oral nonexclusive contract to sell that authorized him to sell the property to Buchanan and his associates for $2,900,000 (while other brokers had to sell, under the terms of their contracts, for $3,500,000) provided, of course, he actually effected a sale of the property or procured a binding contract of purchase, prior to the time some other broker did, and provided the sale was effected prior to the time the gratuitous offer to sell the property at the reduced price terminated or was lawfully withdrawn by the seller.

As we have already stated, Harris, who also had a nonexclusive contract to sell the property on a "first come, first served" basis, called Melvin not later than 9 o'clock on Friday morning March 16, 1956, to inform him that he had sold the property at the fixed price of $3,500,000; and at 10:10 on the same morning Harris introduced his purchasers to Melvin. All of this occurred more than an hour before Buchanan advised West that he and his associates were willing to buy the property and hence occurred prior to the time that West could have notified Melvin of the fact, even if the latter had chosen to remain at home throughout the whole of Friday morning. The contract of sale was executed by the seller and the purchasers produced by Harris at 2:45 or 3:00 Friday afternoon. As a matter of law, the execution of this contract terminated all nonexclusive brokerage contracts that were then outstanding. . . . We find nothing in the oral concessions made by Melvin at the conference that could be construed as an agreement on his part to pay West a commission regardless of whether a sale was made in the meantime by some other broker; nor was there anything said or done by Melvin to indicate that Melvin intended to give West an exclusive agency to sell the property at the new price stated at the conference.

We conclude, therefore, that the trial court erred in giving judgment to the plaintiff; and consequently, that the judgment should be reversed and the cause remanded for further proceedings according to law.

The conclusions we have reached make it unnecessary for us to consider the second point urged for reversal by the appellant.

Reversed.

LORANGER CONST. CORP. v. E F. HAUSERMAN CO.
384 N.E.2d 176
(Mass. 1978)

The plaintiff, a contractor, was preparing its bid for construction at the Cape Cod Community College. It received an "estimate" of $15,900 for movable steel partitions from the defendant and used the estimate in preparing the bid it submitted. The construction contract was awarded to the plaintiff, the defendant refused to perform in accordance with its estimate, and the plaintiff engaged another company to supply and install the partitions for $23,000. The Appeals Court upheld an award of damages to the plaintiff, we allowed the defendant's petition for further appellate review, and we affirm the judgment for the plaintiff.

♦♦♦

The Appeals Court held that the plaintiff was "foreclosed from recovery on any traditional contract theory" but could "recover on the theory of promissory estoppel, a basis for recovery not previously explicitly accepted in the courts of this Commonwealth."...The defendant argues that "the adoption of this new theory of law is procedurally unfair, unwarranted by the facts in the case, and contrary to the statutory policy of the Commonwealth."

We summarize the evidence most favorable to the plaintiff. On May 20, 1968, the plaintiff was preparing its bid to become general contractor on the construction project. The specifications called for movable metal partitions from the defendant or one of two other suppliers, "or equal." About fifteen days earlier, a sales engineer employed by the defendant had prepared a "quotation" or "estimate" of $15,900 for supplying and installing the partitions. The figure was based on information received from the architect's office, and the engineer knew that the general contractor would submit a bid based on such estimates from subcontractors. The estimate was given to the plaintiff by telephone on May 20, 1968; it was also given to other general contractors. The engineer waited until shortly before bids were due on the general contract to prevent the general contractor from shopping for a lower price from other subcontractors. The plaintiff received no other quotations on the partitions and used the defendant's quotation in preparing the bid on the general contract, submitted the same day.

The general contract was awarded to the plaintiff on June 21 or 26, 1968. Sometime in August or September, the plaintiff informed the defendant that it was getting ready to award the partition contract and asked whether it had the defendant's lowest price. Thereafter, on September 12, 1968, the plaintiff sent the defendant an unsigned subcontract form based on the $15,900 figure. The defendant rejected the subcontract, and the plaintiff engaged another company to supply and install the partitions for $23,000. The partition work was not scheduled to begin until the summer of 1969; in fact, work began in the summer of 1970, and the last payment for it was made in 1972.

♦♦♦

The questions argued to us relate to the question whether the evidence made a case for the jury.

1. *The offer or promise.* The defendant argues that the "quotation" or "estimate" made by its sales engineer was not an offer or promise, but merely an invitation to further negotiations, citing *Cannavino & Shea, Inc. v. Water Works Supply Corp.* 361 Mass. 363, 366, 280 N.E.2d 147 (1972). But the Cannavino case involved the circulation of a price list without specification of quantity. Here there was more; the defendant was to do a portion of the work called for by the plans and specifications. Of course, it was possible for the engineer to invite negotiations or offers.... But it was also possible for him to make a commitment. His em-

ployer stated in answer to interrogatories that it was "unable to determine whether or not an employee of the defendant spoke with any of the plaintiff's employees on or about May 20, 1968," and the only direct evidence of the estimate was the testimony of the engineer. We think the jury were warranted in resolving ambiguities in his testimony against the defendant, and in finding that the estimate, in the circumstances, was an offer or promise....

2. *Reliance on the promise.* It seems clear enough, as the Appeals Court held, that the evidence made a case for the jury on the basis of the plaintiff's reliance on the defendant's promise. "An offer which the offeror should reasonably expect to induce action or forbearance of a substantial character on the part of the offeree before acceptance and which does induce such action or forbearance is binding as an option contract to the extent necessary to avoid injustice."... This doctrine is not so novel as the defendant contends.... When a promise is enforceable in whole or in part by virtue of reliance, it is a "contract," and it is enforceable pursuant to a "traditional contract theory" antedating the modern doctrine of consideration.... We do not use the expression "promissory estoppel," since it tends to confusion rather than clarity.

3. *Procedural unfairness.* The defendant contends that the decision of the Appeals Court, resting on "the new theory of promissory estoppel," departed from the pleadings and from the theory on which the case was tried. So far as the pleadings are concerned, count 1 of the declaration alleged an exchange of promise for promise and also the submission of a bid by the plaintiff in reliance on the agreement between the parties. If either allegation was sustained by proof, the other could be treated as surplus. The pleadings could have been amended to conform to the evidence, even after judgment; failure so to amend does not affect the result of the trial....

   In view of the defendant's claim of procedural unfairness, we requested and received a transcript of the judge's charge to the jury. The defendant does not assert any error with respect to the charge, and did not include the charge in its record appendix. We do not treat the charge as the "law of the case".... But we find that the case was presented to the jury on the basis of offer, acceptance, and consideration; there was no reference in the charge to reliance on a promise. We therefore cannot attribute to the jury a finding that the offer or promise of the defendant induced action "of a substantial character" on the part of the plaintiff. We consider the case on the basis on which it was submitted to the jury.... Pursuant to the charge and on the evidence before them, the jury might have found that the defendant's offer was accepted in any one of three ways. First, there might have been an exchange of promises in the plaintiff's telephone conversation with the defendant's engineer, before the plaintiff's bid was submitted. Second, the offer might have been accepted by the doing of an act, using the defendant's estimate in submitting the plaintiff's bid. Acceptance in this way might be complete without notification to the offeror.... Finally, the offer might have remained outstanding, unrevoked, until September 1968, or it might have been renewed or extended when the plaintiff asked whether it had the defendant's lowest price; in either case it might have been accepted when the plaintiff sent the defendant a subcontract form on September 12.

   The evidence warranted the jury in finding that the defendant invited acceptance in any one of the three modes, and in finding that the plaintiff's promise or act furnished consideration to make the defendant's promise binding.

   "In the typical bargain, the consideration and the promise bear a reciprocal relation of motive or inducement: The consideration induces the making of the promise and the promise induces the furnishing of the consideration."... In the present case, the jury could infer that the defendant's engineer intended to induce the plaintiff's promise or action in the hope that defendant would benefit, and thus that his offer or promise was induced by the hoped-for acceptance. Even more clearly, the jury could find that the plaintiff's promise or action was induced by the defendant's offer or promise. Such findings would warrant the conclusion that there was a "typical bargain," supported by consideration..... Indeed, review of the cases suggests that many decisions based on reliance might have been based on bargain.... Once consideration and bargain are found, there is no need to apply sec. 90 of the Restatement, dealing with the legal effect of reliance in the absence of consideration.

4. *Statutory policy.* The defendant did not argue any question of statutory policy to the Appeals Court. It argues to us that the decision of the Appeals Court is contrary to the policy of G.L. c. 149, secs. 44A-44L, regulating bidding on contracts for the construction of public works. The argument seems to relate primarily to subcontract bids described in sec. 44C. Such bids must be listed in the general contractor's bid under sec. 44F and must be filed with the awarding authority under sec. 44H. The defendant was not in any of the trades to which those provisions apply. In any event, the argument relates only to the reliance doctrine on which the Appeals Court based its decision. We decide on a different basis.
5. *Other issues.* Several other matters argued by the defendant to the Appeals Court are discussed in the opinion of that court: unreasonable delay by the plaintiff in notifying the defendant that it was to be the subcontractor, "bid shopping" by the plaintiff, and application of the statute of frauds, G.L. c. 106, sec. 2-201, and c. 259, sec. 1, Fifth. The defendant has not emphasized these matters in its argument to us. The Appeals Court held that they did not bar recovery based on reliance, and they have no more force to bar recovery based on bargain plus reliance. We therefore do not consider them.

Judgment of the Superior Court Department affirmed.

REVIEW QUESTIONS

1. Black is in need of a turret lathe similar to a used one White is trying to sell. White offers Black the lathe for $2,500. Black: "I accept, providing you will deliver the lathe and have it in my shop tomorrow noon." White: "Sorry, but my truck will not be back in town before tomorrow night." Black: "Then I will pick it up myself, because I need it immediately." White: "I believe I will wait a while longer before selling it." Black: "You can't; we made a contract." Was there a contract? Why or why not?

2. On January 3, Gray sends Brown the following offer by first class mail: "I offer you my vertical milling machine in the same condition it was when you last saw it for $4,200. This offer will remain open for your air mail acceptance until January 15." On January 9, at 10:00 a.m., Brown sends the following telegram to Gray: "I accept the offer of your vertical milling machine for $4,200. Am sending truck for it immediately." Is there a contract? Why or why not?

3. Expanding on question 2, assume that Brown received a withdrawal from Gray at 9:00 a.m., January 9, but, because the original letter stated that the offer was open until January 15, Brown sent an air mail acceptance an hour later. Was there a contract? Why or why not?

4. Referring again to question 2, assume that Brown's air mail acceptance on January 9 was lost in the mail and not delivered until January 16. Was there a contract? Why or why not?

5. Green required parts for an appliance he intended to manufacture. He sent out blueprints and specifications to several manufacturers, including the White Manufacturing Company, requesting bids on 100,000 of such parts. White Manufacturing Company submitted what turned out to be the lowest bid. The bid was refused, however. Can White Manufacturing Company demand and get the job on the basis of its bid? Why or why not?
6. Brown lost an expensive watch that had his name engraved on it. He placed an advertisement in a local paper offering a $100 reward for its return. Black found the watch, noticed Brown's name on it, found Brown's address in a telephone book, and returned the watch to him. Later, he read the reward notice and demanded the reward. Is he entitled to the $100? Why or why not?
7. The price of White's product is $50. Newspaper advertising inadvertently lists the price as $30, precipitating a deluge of orders. White honors the advertised price even though his cost is $35. He has begun an action against the newspaper for $20 for each unit sold. Who will win? Why?
8. Referring to the case of *Melvin v. West,* summarize the two types of brokerage contracts commonly used in selling real estate. Which type of contract existed between Melvin and West?
9. In the case of *Loranger Const. Corp. v. E. F. Hauserman Co.,* in what ways could you change the facts so defendant must win?
10. In the third paragraph of its opinion in *Loranger Const. Corp. v. E. F. Hauserman Co.,* the court says that it summarizes the "evidence most favorable to the plaintiff." Why should the appellate court adopt a standard of review that tends to favor the party that won below?

Chapter 8

Reality of Agreement

Chapter 7 stated that there must be an agreement or meeting of the minds between contracting parties. The meeting of minds must be voluntary and intentional. It is implied that neither of the parties has been prevented from learning the facts of the proposed contract. In short, the assumption is made that all contracting parties have entered into the agreement freely and "with their eyes wide open."

This assumption especially applies when the contract is reflected by a written and signed instrument. Generally, an agreement in writing binds the parties to the agreement according to the wording on the paper. A person is held to have read and understood the terms set forth. If the person did not understand the document, he or she should not have signed it. If one wishes to enter into a written agreement with another but disagrees with some of the terms, that person has the right to eliminate these terms, get the other party to the contract to agree to the change, and obtain the other's signature to the change before signing. Most written contracts are either typed or hand printed. If, in reaching an agreement, one of the parties changes the contract in the party's handwriting with the other's knowledge of the handwritten change, this writing will stand in lieu of the preexisting printed or typed part with which it is in conflict.

If the assumption that the contracting parties freely and voluntarily entered into the contract is not true, the contract may be held void or unenforceable. Although the law will uphold a person's right to contract as part of that person's freedom and will hold that person, as well as the party with whom that person bargains, to the agreement they have made, it would be unjust to do so in the absence of free and voluntary action. Where one person knowingly attempts to take advantage of another, it would not seem right to enforce the attempt.

The rules and principles of contracts are very practical and logical as, of course, they should be when they govern practical events. Probably there is no better demonstration of the straightforwardness of contracts than in cases where assent by one of the parties is not voluntary and intentional.

Contracts in which the *reality of agreement* is questionable fall into five general categories: (1) mistake, (2) fraud, (3) innocent misrepresentation, (4) duress, and (5) undue influence. These are treated separately in this chapter. After considering these situations, the available remedies are briefly discussed.

SITUATIONS IN WHICH ASSENT IS NOT VOLUNTARY

♦ Mistake

A mistake may be either unilateral or mutual. A *unilateral mistake* is a mistake that one party to a contract makes about the material circumstances surrounding the transaction. Usually the law does not allow relief for unilateral mistakes. A *mutual* or *bilateral mistake* occurs when both of the parties were mistaken about some material fact. When both are mistaken, the law will not bind either of them.

Unilateral Mistake

When a mistake results from poor judgment of values or negligence by the injured party, the law usually will hold the injured party to the bad bargain. Black sells White a used automatic screw machine for $15,000 and White finds later that a similar machine could have been purchased for $10,000. White cannot avoid the contract because of the difference in value.

In certain instances of unilateral mistake, a party may rescind the contract. If one party to the contract, with full knowledge of the other's mistake, took advantage of the situation, the courts usually do not hold the injured party to the bargain. Such a situation might easily fit into the picture of fraud given later in this chapter. Certain types of clerical or calculation errors will also allow rescission. If a contractor, in preparing a bid on a job, makes a clerical error or a mathematical miscalculation, a court may hold that there is no binding contract. This is particularly likely if the error is an obvious one and the other party must have known or reasonably suspected an error in the quotation.

Mutual Mistake

If both parties to a contract are mistaken about a material fact regarding a matter, the courts generally hold the contract to be voidable. Two types of situations commonly occur: mistakes about identity and mistakes about existence. Suppose Black agreed to purchase White's computer network for $5,000, and the two parties had two different types of computer networks in mind. The courts would say that the minds of the two parties had not met, and that therefore, no enforceable contract resulted. In this example, the two parties may have contracted for the sale of the computer network at some location away from the place where the computer network was kept. If the computer network had in fact been destroyed by a fire, the contract would not be enforceable. The subject matter would have ceased to exist (at least in the manner assumed by the parties). The assumed existence of the subject matter would have been a mutual mistake.

When the value is something different from that assumed by both of the parties, though, a different view is taken by the law. For instance, Black purchases farmland from White at a reasonable value for farmland. Later, a valuable mineral deposit is discovered in the land. If Black had no prior knowledge of such deposits, the sale stands, and it is Black's gain. There are, of course, many instances of purchases of valuable paintings and antiques in which the parties involved knew nothing of the true value of the subject matter. Such sales are generally quite valid.

When the mistake concerns the legal rights and duties involved, the mistake generally is not a ground for rescission of the contract. Every person is supposed to be familiar with the law and is charged with that knowledge.

♦ Fraud

Five basic elements must be established to prove fraud. There must have been (1) a false representation (2) of a material fact (3) made with the intent that it be relied on and (4) reliance on the misrepresentation by the injured party (5) to his or her detriment.

A false representation may, but need not, be an outright misstatement. It may, instead, take the form of an omission of information which should have been passed along to the injured party. A false representation also might be an artful concealment, such as using S.A.E. 90 oil in the crankcase of a car to conceal the need for an overhaul. In certain instances, particularly where a previous relationship of trust has existed between the parties, there may be a duty to speak and reveal all information concerning the subject matter. The courts have been quite general in describing how fraud may arise, preferring not to set up a fixed blockade for the sharp operator to attempt to avoid. The false representation must be of a "material" fact that is not obviously untrue. For instance, if you are told that a car you are thinking of buying is in perfect condition when, in truth, it has a broken rear window that anyone could see, the statement would not support an action for fraud.

The false representation must be material. If the seller represented the car as having brand-X spark plugs when the spark plugs were really another brand of equivalent quality, the statement would hardly be considered material. Also, the representation must falsely represent a fact. When you stop to think of it, it may be a little difficult to define what is fact. Certainly, your actions yesterday and today may be stated as facts. But, aside from a few accepted certainties, it is often impossible to state something about the future as fact. In attempting to get Black to buy land, White tells Black, "I intend to start a housing development in a section very near this land." If Black buys in reliance on the statement and no housing development is begun, there might be some question about whether White ever intended to build. Was the statement a statement of fact? It could very well be, because it is a statement of a present intention.

If a statement is made in such a way as to express an opinion, it is not a statement of fact and cannot support a claim for fraud. Sales talk, puffing, statements that a certain thing is the "best," or is "unsurpassed," or a "good buy" are not fraudulent statements at common law. An exception to this occurs when a person maintains that he or she is qualified to give an expert opinion on the subject. If a party is injured in reasonable reliance on such an opinion, the injured person may be able to recover for fraud.

Generally, then, the representation of material fact must be a false representation, other than opinion, having to do with something past or present. It is not necessary for the statement to be the sole inducement to action by the injured party. It is only necessary that, if the injured party had known the truth, that party would not have taken the action actually taken.

Knowledge of the truth or falsity of a statement made does not necessarily determine the existence of fraudulent representation. If the person making the representation knows it to be false, it is fraudulent. If the person does not know whether it is true but makes the representation in reckless disregard of the truth and to persuade the other party, the party making the statement may be charged with having acted fraudulently.

A misrepresentation for the purpose of fraud is normally made with the intent that the party to whom it is made will act on it. If the other party does not act on it there is no fraud. If a third party (for whom the representation was not intended) overhears it and acts on it to his or her detriment, it is nevertheless not fraud. Action must be taken by the person for whom the false representation was intended.

The injured party must have acted in reliance on the representations made. People are assumed by the courts to be reasonable, and the courts assume that they will take ordinary precautions when dealing with others. One is charged with the knowledge or experience that anyone in the same trade or profession would normally possess. For instance, an auto mechanic should be far less gullible than a lingerie salesperson when purchasing a used car. If White has conducted her own investigation before entering into an agreement rather than relying on another's representations, it is far more difficult for her to sustain a charge of fraud.

The false representations of material fact for the purpose of fraud may be proved; they may have been made with full knowledge of their falsity and with intent to deceive; action may have been taken in reliance on the misrepresentations, but if no injury resulted there can be no recovery for fraud.

Black is induced by White, as a result of various fraudulent misstatements, to purchase 1,000 shares of "Ye Olde Wilde Catte" oil stock at $1 per share. On discovery that Ye Olde Wilde Catte stock isn't worth the paper the shares are printed on, Black sues, but before the case comes up on the court docket the oil - company strikes oil and the stock price increases to $2 per share. Black quite obviously would not continue the action, but even if Black wanted to, Black couldn't recover damages because there was no injury.

♦ Misrepresentation

To point up the distinction between fraud and misrepresentation, consider the following example. Black contracts to buy a hunting dog from White for $200. During the negotiations, White tells Black that the dog is 3 years old. A few months later the dog becomes ill and Black takes him to a veterinarian. After treatment the vet tells Black that Black has a fine 10-year-old dog. Black has been a victim. Black has either been defrauded or has been victimized by an *innocent misrepresentation.*

The distinction depends on the presence or absence of White's knowledge and intent. If White was present 10 years ago when the dog was born, White can certainly be charged with knowledge of the dog's age. If White knew the dog's age, the statement to Black indicating that the dog's age was 3 years could have been made only with an intent to deceive Black. If, however, the dog was acquired by White 6 months ago from Gray, who stated the dog's age to be 2 years, the statement made by White was probably an innocent misrepresentation (or mistake, in some courts).

Although no fraud results from an innocent misrepresentation, the injured party has the right to rescind the contract, giving up the consideration that was received. However, in situations involving an innocent misrepresentation there is usually no independent right to sue for damages, as there is in cases of fraud.

♦ Duress

Duress is forcing a person to consent to an agreement through active or threatened violence or injury. The threat must be accompanied by the apparent means of carrying it out. To be duress, the threat must place the victim in a state of mind such that the person no longer has the ability to exercise free will.

In the early court decisions, a holding of duress was limited to cases in which the victim's life or freedom was endangered or in which the victim was threatened with bodily harm. Modern courts take the view that duress is the deprivation of a person's free will; the means of accomplishing this is immaterial as long as unlawful injury is involved. Threat of injury to other persons, even to property, could conceivably be the means employed to gain the desired end. White is an art lover and has a collection of paintings. Black, to obtain White's signature on a promissory note, threatens to cut up one of White's paintings with a pocket knife. This threat could constitute duress.

Some courage is inferred by the courts. It must be shown not only that threats of unlawful violence occurred, but that they were sufficient to overcome the person's free will. When duress has been used, the contract is voidable and the victim can disaffirm the contract or affirm the contract.

♦ Undue Influence

Duress and undue influence are alike in at least one respect. When either has been successfully exercised, the free will of the victim has been overcome. The means of accomplishing the objectives are somewhat different. Duress requires violence or a threat of violence or harm. *Undue influence* usually results from the use of moral or social coercion, most often arising where a relationship of trust or confidence has been established between two people. Common relationships of this nature are those between husband and wife, parent and child, guardian and ward, doctor and patient, and attorney and client. It also may arise where a person is in dire need, or in some physical or mental distress that would put the potential victim at the mercy of the other party.

To establish undue influence, all of the following must be shown:

- There was a person who, because of a special relationship, was in position to be influenced.
- Improper pressure was exerted by the person in the dominant position.
- The pressure influenced the victim—that the victim acted on it and was injured as a result.

REMEDIES

In situations involving a mutual mistake or an innocent misrepresentation, the most common remedies used by the courts are rescission and reformation. *Rescission* involves placing the parties in the respective positions they would have enjoyed if there had been no contract. This usually means a return of the amounts paid and/ or the goods delivered. Where the contract involves services already rendered, rescission is sometimes not practical. *Reformation* involves the modification of the contract to reflect the intentions of the parties. For example, a contract for a road might be understood to involve an asphalt surface. If the written agreement specified a concrete surface, however, that portion of the contract would be subject to reformation to clearly reflect the parties' agreement.

Several remedies are available to a person who has been defrauded:

- That person may wish to continue the contract with full knowledge of the fraud.
- If the contract has been completed, that person may wish to retain what was received and institute a case for what damages have been suffered.
- The defrauded person may rescind the contract, return the consideration received, and get back what was given.
- If the contract is still executory, the defrauded party may merely ignore the obligations under it, relying on the proof of fraud if the other party brings an action to enforce the contract.

In any event, if the defrauded party wishes to bring a court action after discovering the fraud, the injured party must do so in a reasonable time. If the injured party does not take action within a reasonable length of time, the right to take action will be lost. If a party waits too long, the state's statute of limitations may prevent bringing the claim. As in cases involving fraud, a party subjected to duress may disaffirm the contract. Moreover, fraud and duress often amount to torts (discussed in a later chapter); thus, a victim of fraud or duress also may sue for damages.

MCMULLEN v. JOLDERSMA
435 NW 2d 428 (Mich. 1988)
Per Curiam

Plaintiffs appeal as of right a December 20, 1986, circuit court judgment of no cause of action against defendants Paul and Mary Joldersma and separate circuit court orders granting defendants Buehler Realty, Inc., and James M. Parrette's motion for summary dispositions. Plaintiffs present a myriad of issues for our review; However, we find that none require reversal.

This case arose out of dispute over alleged fraudulent representations and omissions surrounding the sale of a party store located in Newaygo County, Michigan. On July 23, 1981, defendants Paul and Mary Joldersma sold their party store to plaintiffs on a land contract. In September, 1984, plaintiffs filed suit against the Joldersmas, Buehler Realty, Inc., and realtor James Parrette, alleging that defendants fraudulently concealed the material fact that the State of Michigan had plans to construct a highway bypass as part of M-37 running north from Newaygo which would substantially divert all traffic from both M-82 and M-37 away from the party store. Plaintiffs further alleged that the construction of the bypass, which was completed in 1984, destroyed the value of their business. Plaintiffs sought rescission of the land contract, restitution, and exemplary damages for mental distress.

On April 18,1985, plaintiffs filed an amended complaint alleging the following: fraud against the Joldersmas, Parrette, and Buehler (Counts I and IV); innocent misrepresentation against the Joldersmas, Buehler, and Parrette (Counts II and V); and breach of contractual duty of care against Buehler and Parrette (Count III).

On May 8, 1985, defendants Buehler Realty and Parrette moved for summary disposition pursuant to MCR 2.116(C) (7), (8), and (10). On June 14, 1985, defendants Paul and Mary Joldersma followed suit and filed their motion for summary disposition on the basis of MCR 2.116(C) (7) and (10).

On June 18, 1985, an initial hearing on all defendants' motions was held before Judge Terrence Thomas. Judge Thomas took the various motions under advisement and, while a decision was pending, removed himself from the case because of a potential conflict of interest. On February 24, 1986, a hearing was held before Judge Charles Wickens on the various summary disposition motions. On March 13, 1986, Judge Wickens issued his opinion granting summary disposition to defendants Buehler Realty, Inc., and its agent, James Parrette. In the same opinion, defendants Joldersmas' motion was denied because the court concluded that a factual issue was presented.

Plaintiffs then proceeded to a three-day bench trial against the only remaining defendants, Paul and Mary Joldersma, in which plaintiffs primarily sought rescission of the land contract and damages for fraud. About six months after trial, Judge Wickens issued his opinion of no cause of action and awarded costs to the Joldersmas.

Plaintiffs argue first that the court erred in granting summary disposition in favor of Buehler Realty, Inc, and James M. Parrette. As there is no mention whatsoever in the court's opinion regarding the statute of limitations as a basis for summary disposition, we decline to address that aspect of plaintiffs' argument.

In granting summary disposition in favor of these defendants, the court reasoned: "There is no privity of contract between the Plaintiffs and the real estate agency that would justify this action. There is no activity by either of these defendants growing out of the scope of their agency relationship with the sellers which would warrant an action against the real estate agency separate from the sellers."

Although not expressly stated, the tenor of the Court's ruling indicates that it was based on MCR 2.116 (C) (8).

In Count IV of the first amended complaint, plaintiffs alleged "silent fraud" or fraudulent concealment of a material fact against Parrette and Buehler, In Count V, the theory of innocent misrepresentation was alleged.

We first address the former theory, i.e., "silent fraud" or fraudulent concealment of a material fact. As correctly noted by defendants, Michigan courts have held a seller liable to a buyer for failing to disclose material defects in the property or title.... However, to date there are no cases holding that a real estate agent is similarly liable for such a fraud. We do not believe that, by virtue of their agency relationship as real estate agents for the sellers, defendants were duty bound to disclose the pending bypass plans to the buyers. Neither Parrette nor Buehler was a party to the underlying business transaction nor was this a situation where the challenged information was subsequently acquired rendering previous misstatements untrue or misleading.... Under the circumstances, the imposition of such a duty would necessarily conflict with the duty defendants owed to the seller. Moreover, we find It noteworthy that plaintiffs were not without representation. Indeed, they were represented by an attorney and, also, a certified public accountant, Roy Heppe. By her own testimony, Virginia McMullen testified that prior to concluding the transaction she asked Heppe "to check out the area and about the flow of traffic and so forth." Thus, we find defendants owed no duty, either legal or equitable, to disclose the fact that there existed plans for the reconstruction project. Consequently, plaintiffs would not be able to prevail on this claim.

Although plaintiffs focus on Count IV, we also find that Count V, innocent misrepresentation, was properly dismissed. In order to properly state a claim under this theory, privity of contract must be established.... We find that such a requirement was lacking as between plaintiffs and Parrette and Buehler and, thus, summary disposition was properly granted as to this claim as well, Similarly, plaintiffs also claim that the trial court erred in determining that defendants Paul and Mary Joldersmas' actions in failing to disclose the reconstruction bypass plan constituted actionable fraud. We disagree.

In *Taffa v. Shacket,*....the elements for establishing fraud or silent fraud were set forth: "(1) a material representation which is false; (2) known by defendant to be false, or made recklessly without knowledge of its truth or falsity; (3) that defendant intended plaintiff to rely upon the representation; (4) that, in fact, plaintiff acted in reliance upon it; and (5) thereby suffered injury.... The false material representation needed to establish fraud may be satisfied by the failure to divulge a fact or facts the defendant has a duty to disclose. Such an action is one of fraudulent concealment."

After hearing all the evidence, the trial court concluded there was "no fraud or innocent or intentional concealment or misrepresentation on the part of Defendant sellers" The trial judge found no existing facts that created for the sellers a duty to disclose the possibility that the relocation of M-37 might take place in the future, pending federal funding. Additionally, the pertinent facts with respect to the proposed project were of public record and plaintiffs employed Roy Heppe, a CPA, to investigate these facts for them. Finally, the court concluded that even if the relocation of M-37 would have been material to plaintiffs' store operation, they did not rely on the sellers to disclose this information but, rather, on Heppe, in whom they had placed their reliance.

We are not persuaded, after thoroughly examining the entire record...and giving due consideration to the trial court's opportunity to judge the credibility of the witnesses...that the court erred in this instance. Testimony clearly established that approval for the project was still ongoing several months after plaintiffs purchased the store. Indeed, final approval from the federal government did not obtain until September 10, 1982, well over a year after the purchase (July 23, 1981) Moreover, such information was a matter of public record and had been so for years.

Not insignificantly, plaintiffs wrote to the state Department of Licensing and Regulation on June 20, 1983, requesting a full inquiry regarding Heppe's alleged malpractice with respect to the purchase. Excerpts from this letter further evidence plaintiffs' reliance on Heppe regarding the decision to purchase the property: "In summary: *Our purchase was contingent upon the findings and recommendations of Mr. Heppe, whom we paid, for his professional expertise.* Because of this, we now find ourselves with a party store that was over priced—which we cannot sell because of the land contract balances and the fact it is to be by-passed by the Michigan State Highway Department.

"When construction of the by-pass is started, I can think of no conceivable way a payment of $1167 per month plus expenses can be made. Mr. Joldersma knew this—Mr. Parrette knew this—we paid Mr. Heppe—*now* we know it. *We trusted Mr. Heppe completely*—placed our life-savings, assets and future in his hands. In addition, I terminated my employment at Ford Motor Company after 24 years (30 year retirement and benefits) to purchase this store.

"In conclusion, may I ask—is not this C.P.A. responsible? Do we have recourse? In our personal opinion, we feel he may be guilty of breach of ethics." (Emphasis added.)

We are unpersuaded that the court erred in finding that plaintiffs did not in fact rely on the statements or omissions of the Joldersmas, James Parrette, or Bonnie Robbins. Rather, plaintiffs relied upon the recommendations of their agent, Roy Heppe. As reliance upon a false representation or omission of fact is a requisite element,...plaintiffs simply cannot assert that a fraud was perpetrated upon them by parties upon whom they did not rely. For this reason we similarly find no merit to plaintiffs' claim that the Joldersmas are responsible for any fraudulent concealment by their agent James Parrette and listing agent, Bonnie Robbins.

Plaintiffs next claim that they should not be charged with knowledge of the public record regarding the bypass plans.... Significantly, plaintiffs admit that prior to entering into the business transaction, they took note of the "rickety old bridge." A reasonable inquiry would have revealed the existence of the bypass project.

Finally, with respect to this issue, plaintiffs rely upon the following rule quoted in their brief: "The law of constructive notice can never be so applied as to relieve a party from responsibility for actual misstatements and frauds, and to prevent a representee from having a right to rely upon representations under such conditions. *Thus, one to whom a misrepresentation is made is not held to constructive notice of a public record which would reveal the true facts.*" (Emphasis added by plaintiffs.)

However, in this case defendants did not make "actual misstatements" regarding the bypass plans. Indeed, the very essence of this case is their failure to say anything at all.

We have carefully reviewed plaintiffs' next argument—that a seller can be liable for a misrepresentation of a future event and find it has no merit. At the time of the purchase, the bypass project was a future possibility—contingent upon federal approval and funding. Because of this fact, the Joldersmas' failure to inform plaintiffs did not constitute a fraudulent omission.

We have similarly reviewed plaintiffs next allegation—that defendants failed in their burden of proving that plaintiffs did not rely upon defendants' misrepresentations or omissions of fact—and find no reason to set aside the verdict on this basis.

Plaintiffs also claim that a defrauded party to a contract can rely on more than one representation and, thus, reliance on Heppe's recommendation does not relieve the Joldersmas of liability because plaintiffs also relied on a continuation of traffic as represented by the Joldersmas. However, the court expressly found plaintiffs did not rely upon statements or omissions of the Joldersmas. Having previously concluded that the court did not err in its findings, we find no basis for this claim.

Plaintiffs also contend that the Michigan Department of Transportation's adoption of Proposal B (Bypass Plans) was a material fact. Be that as it may, materiality is but one element of fraud.

The court concluded, and we do not disagree, that plaintiffs failed to establish all of the requisite elements necessary to prevail.

As to plaintiffs' next issue, we agree that in the context of a rescission action, as here, it is irrelevant to what extent plaintiffs attempted to mitigate their damages.

Plaintiffs also assert that they have demonstrated the requisite legal damage suffered as a result of their reliance upon the alleged misrepresentation of defendants. However, defendants' expert, Eric Adamy, testified that in his opinion the store was not nearly as profitable after plaintiffs assumed the business for reasons other than the relocation of M-37. According to Adamy, plaintiffs did not adequately promote the store, they did not maintain an adequate inventory, there was decreased emphasis on electronic sales, there was an economic recession, and finally, the number of party stores in Newaygo County increased. The court apparently gave considerable weight to this testimony and concluded that plaintiffs had not been damaged by the highway bypass. We do not find error.

We cannot lose sight of the fact that during the last full year the Joldersmas operated the store, the gross sales were $365,727.02. After the first full year plaintiffs operated the business, gross sales dropped to approximately $200,234 (year ending June 30, 1982), followed by another decline to $163,730 for the next year (ending June 30, 1983). By the close of the next fiscal year (ending June 30, 1984), sales had increased slightly to $174,104. By the end of the fiscal year, sales had declined to $152,000. Significantly, however, the bypass project was not constructed until the first six months of 1984 which lends credence to the court's conclusion that the relocation of M-37 had no significant impact on the decline in sales. Rather, it was owing to factors other than the bypass.

Plaintiffs further contend that the trial court erred in excluding a witness, Calvin Deitz, from testifying in plaintiffs' case in chief. However, by failing to state any authority for their contention that the court's ruling was error, this issue is not properly before us. As we have consistently stated, a party may not leave it to this Court to search for authority to sustain or reject a position....

We have carefully reviewed plaintiffs' final allegation of error and are unpersuaded that the court abused its discretion by admitting testimony concerning financial operations for the party store after the date of closing. As the parties agree, plaintiffs sought rescission, an equitable remedy. Not only is this information relevant as to plaintiffs' fraud claim, the remedy of rescission returns the parties to the status quo, i.e., it places the parties in the position they occupied before the transaction in question....

Rescission of a land contract should not be granted where the result would be inequitable. The decision rests within the sound discretion of the court and each case must be decided on its own particular facts.... Here, it was necessary for the court to examine all the facts to determine the appropriateness of rescission. After doing so, the court concluded, inter alia, that the plaintiffs could not restore defendants to the status quo and were not entitled to rescission. We cannot say, given all the circumstances of this case, that the court erred so as to require reversal by allowing the challenged testimony. Moreover, and contrary to plaintiffs' claim, we find no evidence that the court was biased in defendants' favor.

We are not unmindful of the hardship that has befallen plaintiffs with respect to acquisition of the party store and we express no opinion, beyond the scope of our appellate review, as to the origin. However, this case was heard on the merits and we find no basis in law or equity to upset the final verdict.

AFFIRMED.

COMMUNITY BK., L. OSWEGO, OR. v. BANK OF HALLANDALE & T. CO.
482 F. 2d 1124 (5th Cir. 1973)

The critical question on this appeal is whether the plaintiff bank in Oregon, in making a personal loan, relied on misrepresentations of the trust officer of the defendant bank in Florida regarding the borrower's financial condition. Although the District Court found the representations to be false, it held that the plaintiff's tort claim for damages must fail for want of reliance. We think that the evidence compels a finding of justified reliance and reverse.

On May 31, 1971, Jerome A. Lurie executed a $50,000 promissory note payable to plaintiff, The Community Bank, Lake Oswego, Oregon. The loan had not yet been approved and was to be secured on a two-for-one basis by I.B.M. grade securities in Lurie's trust account with the Bank of Hallandale.

On June 2, 1971, Maldyn C. Evans, Community Bank's president, telephoned Craig G. Kallen, the Bank of Hallandale's sole trust officer, introduced himself, and inquired about Lurie. Evans indicated that the Oregon bank was considering lending Lurie $50,000 and sought information concerning Lurie and his trust account with the Florida bank. Kallen replied that he had discussed this matter with Lurie, who was an extremely good customer of the bank, had a high seven-figure trust with the bank, and had good balances on his accounts. Kallen offered, if Evans would send him a copy of the promissory note, to set aside securities held by Lurie at the defendant bank as collateral for the loan. All of this information was in the hands of the Florida bank, and Kallen neither qualified these statements nor raised any suspicion about his representations.

The facts reported by Kallen were false. As of June 2nd, Lurie had a zero balance at the Bank of Hallandale. At no time, either before or after that date, were the securities referred to by Kallen in the Florida Bank's possession.

On June 3rd, during an audit, it was discovered that certain securities that the Florida bank held as collateral for its own loans to Lurie were in street name, had been stolen and were subject to stop orders. Consequently, Kallen was relieved of his duties but remained in his office and received telephone calls as a bank officer until his employment was terminated on June 11th. The Court found, despite denials by Kallen, that in a second telephone conversation with Kallen, on June 10, 1971, Evans asked if collateral had been set aside for the proposed loan and Kallen responded, "I told you already they had." When asked if he would forward a list of securities, Kallen replied, "I told you I would."

After the June 10th call, Evans approved the loan. On his instructions, the cashier and vice-president of Community Bank telephoned Lurie at Kallen's office in the Florida bank. Kallen answered the telephone, indicated that Lurie was in his office, and then turned the telephone over to Lurie, who authorized disbursement of the $50,000. The funds were then disbursed.

On June 16, 1971, Evans wrote Kallen a letter which the District Court held evidenced an intent to obtain future documentation before disbursement. Since the Court apparently relied exclusively on this documentary evidence to find no reliance, it is set out in full:

> Dear Mr. Kallen:
>
> In connection with the loan transaction we have with your client, Dr. Jerome A. Lurie, we enclose herewith a certified photocopy of the note executed by Dr. Lurie and payable to us in accordance with the terms therein.
>
> It is our understanding that this copy, together with Dr. Lurie's instructions to you, will permit you to collateralize his loan at this bank on a two-for-one value ratio. We enclose our form of collateral agreement, which we presume will be executed by

your Department for us, and that a collateral receipt describing the securities pledged (or the equivalent thereof) will be forthcoming for our files.

When replying, we would appreciate it if you would include either a photocopy of Dr. Lurie's financial statement or sufficient narrative information to indicate his general financial strength and reputation.

Thank you for your assistance.

Very truly yours,
M. C. Evans, President

We think that this letter cannot serve the Bank of Hallandale as a defense. This action is based not on an agreement or an understanding between the parties, but on the tort claim of misrepresentation. Plaintiff predicated this action upon Kallen's fraudulent telephone misrepresentations and the doctrine of apparent authority in regard to the Bank of Hallandale. . . .

We conclude the evidence compels a finding that, in authorizing the Lurie loan, Community Bank's president did rely on Kallen's factual misrepresentations that Lurie had securities in proper form in his trust account with the Bank of Hallandale and that these could be used as collateral for the loan and had been set aside for this purpose. The June 16th letter could not preclude recognition of Community Bank's reliance on earlier telephone conversations regarding Lurie's financial condition. Although paperwork needed to be completed to conclude the collateralization, it was not the Bank of Hallandale's failure to comply with the collateral agreement, but the fact that the Florida bank never possessed the securities in proper form that was fundamental to the damage to Community Bank.

It is the general law, as well as the controlling law in Florida, that recovery may be had for misrepresentation as to a third party's financial condition where a person, for the purpose of inducing another to lend money to said third person, misrepresents the financial responsibility or solvency of such third person. . . . The District Court, in its final judgment, listed elements essential to proving misrepresentation: (1) a false statement of fact, (2) known by the defendant to be false at the time it was made, and (3) made for the purpose of inducing the plaintiff in reliance thereon; (4) action by the plaintiff in reliance on the correctness of the representations; and (5) resulting damage to the plaintiff. . . . The court recognized in this fact situation each of these elements except reliance.

For reliance to be proved,

> (i)t is enough that the representation has had a material influence upon the plaintiff's conduct and been a substantial factor in bringing about his action. It is not necessary that the representation be the paramount, or the decisive, inducement which tipped the scales, so long as it plays a substantial part in affecting the plaintiff's decision. . . .

It is clear from the record that Evans sought information from Kallen, as the Bank of Hallandale's trust officer, as to facts concerning Lurie's financial position, the existence of the subject securities, and their segregation in his trust account. Thus, Kallen's misrepresentations had a "material influence" on and were a "substantial factor" in Community Bank's loan decision. No stricter standard of reliance is required. A false representation is actionable if it substantially contributed to the formation of plaintiff's determination to act. . . .

To constitute remediable fraud, however, it must appear not only that there was reliance on the misrepresentations, but that the reliance was justified under the circumstances. . . . For the reliance to be justified, the misrepresentation must concern some material thing unknown to the plaintiff either because he did not examine it, because he had no opportunity to become informed, or because his entire confidence was reposed in the defendant. . . .

In the light of Kallen's position as trust officer of the Florida bank, the fact that essential information regarding Lurie's financial condition was in the hands of the Bank of Hallandale and the fact that Kallen obviously intended for Evans to believe the information given, Evans was justified in relying on Kallen's telephone communications.

On remand, the District Court need consider only the question of damages.

Reversed and remanded.

On Petition for Rehearing Per Curiam:

It is ordered that the petition for rehearing filed on behalf of Bank of Hallandale and Trust Company, Appellee in the above entitled and numbered cause be and the same is hereby denied. Although the Court's opinion indicated that the District Court need consider on remand only the question of damages, that reference applied only to Community Bank's action against Bank of Hallandale and would not preclude consideration of Bank of Hallandale's third party action against its former trust officer, Craig Kallen, or any other related actions.

OTTAWA STRONG & STRONG v. McLEOD BISHOP SYSTEMS

676 F. Supp. 159 (N. D. Ill. 1987)
Memorandum Opinion and Order

Aspen, District Judge:

Plaintiffs Ottawa Strong and Strong, Inc. brought this action in the Illinois Circuit Court of LaSalle County against defendants McLeod Bishop Systems, Inc. ("McLeod") and Gordon Schlagel. Defendants petitioned to remove to the United States District Court for the Northern District of Illinois on the basis of diversity jurisdiction. Plaintiffs petition for remand to the Illinois court. For the following reasons, we remand this action to the Illinois Circuit Court.

I

This four-count action against McLeod and Schlagel revolves around McLeod's sale of a computer system and lease of a software program to plaintiffs. Count I alleges breach of the sales and product license agreements. Counts II and III allege breaches of warranty. Count IV alleges fraudulent misrepresentation in the negotiation of the agreements.

♦♦♦

II

For Schlagel to have been properly joined, Count IV must state a fraudulent misrepresentation cause of action against Schlagel.... To state such a claim under Illinois law, [1] a plaintiff must plead with particularity that defendants knowingly made false representations of material fact intended to induce plaintiffs' reliance, and plaintiffs actually and justifiably relied on the misrepresentations to their detriment.... Defendants contend that rather than alleging misrepresentations of fact, plaintiffs alleged false promises to perform a future act, which do not constitute actionable fraud in Illinois[2]....

Looking at Count IV in its entirety, we believe plaintiffs have pled misrepresentations of fact, and there is a "reasonable basis for predicting that state law might impose liability of the non-diverse defendant. ...The gravamen of Count IV is that Schlagel misrepresented that the software program could be quickly and cheaply modified to accommodate plaintiffs' business requirements. Specifically, Schlagel stated that "it would not be difficult" to customize the computer software program at issue to plaintiffs' business needs, and that any modifications "could be made without a great deal of time expended by the Defendants and without the Plaintiffs having to pay a great deal in fees," knowing "that such modifications...would be extremely difficult and entail a great deal of modification time and fee expense." Complaint, Count IV, secs. 13–16. Plaintiffs also allege in Counts I through III that defendants' written proposal estimated a modification time of ten days at $400 per day....

Plaintiffs' position in the negotiation of these agreements is not unusual in the world of computer software purchasing and leasing. Vendors of large software application packages often market products that can be customized to the needs of a variety of businesses. Potential purchasers of these software programs must rely on software vendors' representations that the products can be adequately modified to fit their unique business requirements. They purchase or lease a product expecting that it can be so modified and often obtain the contractual promise of the vendor to make the modification. If the product cannot be modified as represented, the product may be useless to the purchaser or require significant delay and expense in making the modification.

There is no principled basis for denying those who purchase software on these representations of the protections of a fraudulent misrepresentation cause of action. Defendants contend that their promises to quickly and cheaply modify the software converted the representations of fact to a promise. We disagree. Representations as allegedly made by the defendants are representations of fact, regardless of a vendor's accompanying or subsequent promise to perform the modifications. A vendor's promise should not relieve the vendor of liability for inducing the purchase through misrepresentations. Defendants have not provided, and we have not found, any support in Illinois law for relieving a party of liability under the tort of fraud solely because the party guaranteed the representations in a contract.

III

Conclusion

Having alleged that Schlagel made misrepresentations of fact, plaintiffs have stated a cause of action against Schlagel in Count IV. Accordingly, Schlagel was properly joined and diversity is incomplete. The petition to remand to the Illinois Circuit Court of LaSalle County is granted. It is so ordered.

---

1. We agree with the parties that Illinois provides the governing law.
2. Defendants appear to contend that Count IV fails to state a claim against both McLeod and Schlagel.

REVIEW QUESTIONS

1. Black represents a company that makes computers. Black tells the White Manufacturing Company that installation of a system for cost control and the rental of the machines to process them will save the company $6,000 per month. As a result, the White Manufacturing Company rents the machines and installs the system. After six months of operation, it is determined that the system is losing money for the company at the rate of $3,000 per month. The White Manufacturing Co. sues Black's employer for the $18,000 lost by use of the system to date and the $36,000 that would have been saved if Black's claim had been correct. White Manufacturing Co. charges fraud. How much can White Manufacturing Co. recover? On what basis?
2. List the elements required for fraud. Which of these elements are required for misrepresentation?
3. Gray, an engineer for a manufacturer of die cast parts, inadvertently used the weight of the parts rather than the labor cost in calculating the bid price. As a result the price bid per part was about 25 percent above the cost of the raw die cast material rather than about 80 percent above the raw material cost (which would have resulted from proper bid calculation). The manufacturer to whom the bid was submitted immediately accepted. Gray's employer claims that he should be allowed to avoid the contract since an error was made in the calculation. Will the court be likely to uphold the contract or dissolve it?
4. Distinguish between duress and undue influence.
5. Brown, a machine operator in Green's plant, lost her left hand in the course of her employment. Because of the character of the work, the state workers' compensation laws do not apply. Brown threatens to sue Green for $80,000 for damage to her ability to earn a living for the remainder of her working life. Brown claims, with apparent good reason, that the machine should have been guarded at the point where the injury occurred. Green offers to pay $20,000 to Brown if Brown will drop the court action. Brown agrees, and Green pays her $5,000 on account. It soon becomes apparent that this is all Green intends to pay. Brown starts a court action for the remaining $15,000. Green sets up as a defense that the contract resulted from duress. Who would win the court action? Why?
6. The defendants won in the case of *McMullen v. Joldersma.* Suppose the plans were complete for the M-37 bypass and the financing was in place when the McMullens bought the party store. How would this change the outcome of the case?
7. Considering the case of *Community Bk., L. Oswego, Or. v. Bank of Hallandale & T. Co.:*
   a. Why was the Hallandale bank named as defendant rather than Kallen?
   b. What do you suppose Kallen intended to gain by the deception?
   c. What policy should Community Bank adopt in the future to prevent a recurrence of such frauds?
8. Ordinarily, allegation of fraud requires evidence that there was a false representation of a material *fact.* In *Ottawa Strong and Strong v. McLeod Bishop Systems,* Schlagel offered an *opinion* that customizing the software would not be difficult or expensive. How could such an *opinion* be the basis for a fraud action?

Chapter 9

Consideration

Consideration is the price (in terms of money, goods, or services) given by one party to a contract in exchange for the other party's promises. Consideration may take the form of a benefit to one of the parties. Consideration also can consist of a detriment to a party or a forbearance by a party. Whatever form it takes, if the price is freely bargained by the parties, it is construed by the courts as their consideration.

Generally, the courts will consider the question of whether the consideration was "sufficient" to serve as a basis for a binding contract. In the vast majority of situations, the promises exchanged constitute sufficient consideration; the sum of $10 may suffice. An example of insufficient consideration is a promise to perform an undisputed legal duty already owed to the other party.

Although the courts will consider the question of the sufficiency of the consideration, the courts very rarely consider the "adequacy" of the consideration. Adequacy refers to the relative values of the promises, goods, or services exchanged. Usually, the courts are not concerned with whether someone made a wise bargain; the courts only consider whether the bargain amounted to an enforceable contract.

ESSENTIALS OF CONSIDERATION

The consideration given by each party to a contract must meet certain requirements. The following requirements are examples:

1. The consideration *must have value*. Except for money given in satisfaction of a debt—the amount of which has been acknowledged by both parties and is not in dispute—the values exchanged by the parties need not be equal. Usually it is assumed that in the free play of bargaining, the parties have arrived at what each believes to be a reasonable consideration. If the consideration exchanged appears grossly unfair, this disparity in value may be used to establish a case of fraud or other lack of reality of agreement, but inequality of value usually is not sufficient proof by itself.

   Black promises an uncle, White, that Black will not smoke until attaining the age of 22. White, in turn, promises to pay Black $5,000 for not smoking until Black is 22. Each consideration has value. Black's consideration is a forbearance of a legal right to smoke before age 22. White's consideration is the $5,000. It may seem to an outsider that $5,000 is a high price to pay for the value received. Or it might seem to others that no price would be sufficient to pay for giving up the use of tobacco. However, in this contract, Black and White set the price—the value of Black's forbearance of a legal right to them was $5,000. A court would not question their determination of values.
2. The consideration *must be legal*. The courts usually do not enforce a contract in which the consideration given by one of the parties to the contract is contrary to an established rule of law. If, in the previous example, a statute pronounced

the act of smoking before age 22 a crime, Black's forbearance of smoking would not have been valid consideration; legally, Black couldn't smoke anyway. To be valid consideration, Black's forbearance would have to be forbearance of something that Black was lawfully entitled to do. Similarly, if it is a positive act that is promised as consideration, the act must be lawful. If Black's consideration had been a promise to use illegal narcotics or, for that matter, not to use them, the contract would not be enforced.

3. The consideration *must be possible* at the time the contract is made. If the impossibility of performance was known to either or both of the parties at the time of the offer and acceptance, the courts ordinarily do not grant damages for nonperformance. The situation is a little more difficult, though, if the contract calls for performance that later becomes an impossibility. Occasionally, the subsequent destruction of an item essential to the contract, through no fault of either party, renders performance of the contract impossible. Impossibility of performance is discussed further in Chapter 13.
4. The consideration *must be either present or future.* A present contract cannot be supported by a past consideration. Suppose that a number of students decide to paint several classrooms. After the painting has been completed, the dean calls the students together, compliments them, and then promises to give them each a $50 credit toward tuition next year. The only possible consideration for the dean's promise is the painting, a past act. Hence, there is no consideration.

   An exception to this rule may exist in certain situations where it appears that a gift was made. For the past consideration to support a contract, it must not have been intended as a gift when it was given to the other party. If the past act or forbearance was not intended as a gift but was done at the other's request, then an *unliquidated obligation* (that is, an obligation of some undetermined amount) was created by it. A present promise to pay for such a past act or forbearance is binding on the promisor. The unliquidated obligation has become liquidated. Past relationships between the parties concerning similar acts or forbearance often explain their intent either to make a gift or to create an unliquidated obligation. But valid past considerations form a rather minor exception to the general rule.

The vast majority of contracts generally look to the future. The exchange of a future promise for a similar promise has virtually unquestioned standing at law. If Green orders 10,000 die castings of a particular design to be delivered by August 31 and agrees to pay for them on delivery, both parties' exchanges of consideration are to take place in the future. Either this or present payment for future delivery sets up an obligation that the other party must carry out or risk action for breach of contract.

♦ Seal as Consideration

In ancient times an impression was made on a piece of clay or (later) a blob of wax attached to a document. The seal took the place of the signature. Certain documents of exceptional importance or high security were required to be sealed. The seal came to have such significance that the common law courts would not inquire whether any consideration existed for a contract if the contract bore a seal. The formality of the act of sealing the document made it legal and binding.

In private transactions the seal is no longer an impression made on wax. Now it may be composed of the word seal or any other notation intended by the parties as a seal, following the signature. Many government institutions (such as the courts) have formal seals. When a notary notarizes a document, the notary usually adds the notary's seal. In most states, the use of private seals to formalize contracts has no real legal effect. Most states have abolished the old common-law rules on seals. However, many states still require that certain documents (such as deeds for real property) be notarized; as noted, this usually involves the notary's seal.

Suppose Black promises in a letter to White to make White a gift of $1,000. At this point White has received nothing, because the present promise of a future gift either orally or in writing is unenforceable if it is unsupported by consideration from the other party. However, if the promise of the $1,000 had been sealed, it would be supported as a contract in those states that still recognize the use of seals.

♦ Theories of Consideration

Certain theories of consideration form the basis of court opinions in contract cases. A primary one is the *bargain theory,* which, in effect, holds that the parties have bargained in good faith for the consideration to be received. Each of the parties has determined the value of the consideration. If the agreement to exchange consideration was voluntary and intentional, the values involved—regardless of how dissimilar they may appear—are not subject to question.

The bargain theory is the basis of court decisions in which it appears that the parties negotiated freely. But what happens when one party deals from a position of great strength with a much weaker party? Is it fair when a dominant party may dictate terms on a "take-it-or-leave-it" basis to one who is practically forced to take it? Should landlord Brown, for example, be allowed to hold tenant Green to a very one-sided contract that Green was forced to take because Green had no alternative?

Recent cases have indicated that gross inequities of this nature may not be enforced. For example, if the contract wording would require the weaker party to assume responsibility for not only the weaker party's own negligence but also the dominant party's negligence, such terms may not be given their literal effect. These *adhesion* contracts have gradually come to be less enforceable. The modern trend seems to favor a balancing of the equities in such situations rather than blindly following the contract language.

Another theory is the *injurious reliance theory,* which is based on the principle of estoppel. Courts occasionally enforce promises that are not supported by consideration under a doctrine referred to as *promissory estoppel.* Essentially, this doctrine prevents a promisor from asserting the absence of consideration as a defense in situations in which a person relied on the promisor's promise to his or her detriment. Where one of two innocent parties must suffer, so the reasoning goes, the party whose act occasioned the loss should bear that loss.

Suppose Gray promises to contribute $10,000 to a charitable organization's building fund. In reliance on Gray's promise, the charitable organization contracts with others for a structure to be built. Gray's promise to contribute will be enforced in the courts if this is necessary to pay for the building.

A third theory, the *moral consideration theory,* is generally confined to those cases in which a preexisting debt has been discharged by law and the debtor, subsequent to the termination of the obligation to pay the debt, agrees to pay it anyway. The basis of the argument here is that though the *obligation to pay* the original debt has been relieved by law, the debt still exists. The debtor, in reaffirming this obligation to pay the debt, has transformed the moral obligation to pay into a legal duty to pay. This moral obligation is sufficient consideration to sustain a court action to recover on the debt. Cases of this nature often arise from bankruptcy proceedings and creditor's compositions.[1]

♦ Mutuality

Generally, both parties must be bound to a valid bilateral contract or neither is bound. The promises constituting consideration in a contract must be irrevocable. If either party to a contract can escape the duties under it as desired, there is no contract between the parties. The contract must fail for lack of consideration and mutuality.

In one case[2] a contract was made to sell and deliver 10,000 barrels of oil. The price per barrel was stipulated in the contract. Delivery, though, was to take place in such quantities per week as the buyer might desire. Payment was to be made upon delivery. No minimum quantity per week was agreed on; the buyer apparently could refuse delivery of any oil each week during the life of the contract. The buyer, in other words, really had not agreed to do anything. A contract may be sustained, however, if the quantities are ascertainable with reasonable certainty. If, in the oil example, the purchases had been tied to the buyer's needs or the seller's output, the contract probably would have been upheld.

In a similar case[3] the contract was to furnish the coal required for the steamers of a certain line. Although, in one sense, the quantity was indefinite, it was limited to the requirements of the steamers. Because the quantity was ascertainable at the end of any period of time, the contract was sustained.

♦ Legal Detriment

Consideration for a contract may consist of a legal detriment. *Legal detriment* means that the promisor agrees, in return for the consideration given by the other party, either to give up some right that is lawfully the promisor's or to do something that the promisor might otherwise lawfully avoid doing. Agreeing, for consideration in the form of a discount in price, to re-

♦———

1. See Chapter 13, "Performance, Excuse of Performance, and Breach," for a discussion of bankruptcy and creditors' compositions.
2. *American Co. v. Kirk,* 68 F.2d 791, 15 C.C.A. 540.
3. *Wells v. Alexander,* 130 N.Y. 642.

strict this year's purchases of raw material requirements to a particular source is an example of legal detriment. A company agreeing to so restrict its purchases would be giving up its legal right to choose to purchase its raw materials from any other source for the year's time.

♦ Forbearance

Forbearance, or a promise to forbear or give up a legal right, is sufficient consideration to support a contract. One instance of forbearance occurs when one party gives up the right to sue another. A person has a legal right to resort to the courts to have a claim adjudicated. For such forbearance to sue to constitute consideration, the claim on which the suit would be based must be a valid one. The party that forbears must have at least reasonable grounds on which to bring court action. If the claim is reasonably doubtful, or would be a virtual certainty in favor of the plaintiff, the promise to forbear is sufficient consideration.

In a small minority of cases, some courts have extended this reasoning even further, allowing as sufficient consideration forbearance to sue when only the party who would be plaintiff thinks, in good faith, that he or she has a valid claim. However, if a person brings a court action maliciously (for example, only as a nuisance to the defendant), forbearance of court action fails to constitute consideration on which a contract may be based.

One type of case in which forbearance of court action often occurs is that in which the amount of a debt is in dispute. If, in a compromise, the disputing parties agree on the amount to be paid, this agreement is binding on them. Each has given up the right to have the court decide the correct amount of the debt. In fact, most court cases are settled by compromise before they ever go to trial.

It is considered to be a different situation when the amount of debt is certain and due or past due. When such is the case, the acceptance by the creditor of a lesser sum may not discharge the complete debt. The creditor may bring an action at law later for the remainder of the debt. However, time has value. Prepayment of a lesser sum may, by agreement, discharge a fixed debt. Also, payment by means other than money may discharge a liquidated debt.

♦ Preexisting Duty

The law allows you to collect only once for what you do. If a person already has a duty to perform in a certain way, he or she cannot use that same performance as consideration in another contract. For example, assume that the house next door is on fire. The city's fire department is attempting to extinguish the blaze. An offer by you to pay a firefighter $500 extra to spray water on your roof to keep the fire from spreading to your house would not create a valid contract on acceptance. The firefighter is already paid for doing everything to keep the fire from spreading. The firefighter can give nothing that has not already been paid for in exchange for the $500.

An exception exists to the rule that consideration must not be something that one already is bound to do. Assumptions are usually made before parties enter into a contract. The actual conditions may not exist in the manner that the parties assumed. When this is found to be true and the performance of the resulting contract is thereby made more difficult, the added difficulty may be treated as consideration to support a modification of the contract price.

If, under these conditions, the parties agree to added compensation, the courts will often enforce the added payment. To be enforceable, though, the difficulty must be something that could not be reasonably anticipated by normal foresight.

Black, a contractor, agrees to build a structure for White for a fixed price. Test borings have shown the soil to be mainly clay. However, during excavation for the foundation and basement Black discovers that the test borings somehow missed a substantial rock layer. If White agrees to pay extra because of the necessity of excavating through the rock, the modified agreement should be enforceable. This is true even though the original contract did not require White to make such payments if unusual conditions were met. White might successfully refuse to pay extra, but once White has agreed to pay, White is bound to do so. If the price increase had been based on a factor normally considered foreseeable, such as a increases in labor or materials costs, there probably would not have been lawful consideration. Such contingencies could be anticipated and covered in the original price; such changes would not support added consideration.

MEADOWS v. RADIO INDUSTRIES
222 F.2d 347 (7th Cir. 1955)

Plaintiff sued in the District Court to recover damages occasioned, as he averred, by defendant's wrongful termination of his contract of employment. At the conclusion of his evidence, the court directed the jury to return a verdict for defendant and entered judgment against plaintiff. On appeal plaintiff contends that the court erred in directing a verdict.

On the 30th day of April, 1950, plaintiff was, and for some years prior thereto had been, a mechanical engineer residing in Wisconsin. In response to defendant's advertisement, he contacted defendant's officials and, as averred by plaintiff, the parties entered into a parole contract whereby he was to undertake production of resistors for defendant in its plant in Chicago. The agreement rests entirely upon the parole discussion between the parties, as related by the plaintiff. He testified that the parties made "an agreement and I (the plaintiff) was supposed to go with them," first, to make cold molding resistors, and then, to develop hot molding devices, for which defendant agreed to pay him $10,000 a year, and one-half of 1% of the gross sales; that he was supposed to be plant manager in exclusive charge of operating that part of the enterprise relating to resistors; that he went to work with this understanding without any further contract, was paid $190 per week, and stayed with the company for almost a year, when he was discharged. In the meantime he had twice threatened to quit but on each occasion had been prevailed upon to stay. He admitted that nothing was said about whether his employment was to be for a year, two years, or for life. He "supposed" it was permanent. There is nothing in his testimony to indicate that he ever promised to remain in employment for any certain period of time, or that there was any definite term of employment. He expressly denied that he was hired for a year and said that he did not at any time before he began working for defendant, offer to work for one year, two years, three years, or "whatever the time was."

His work during this period was largely the designing and building of machinery for the purpose of making hot mold resistors. At the time of his discharge, defendant told him the company was no longer going to continue the process that he claimed to have designed, and, upon his inquiry as to what that meant with regard to him, he was told "you're through." The evidence discloses that the defendant engaged in the resistor business for a few months after terminating plaintiff's employment and then dropped out of it entirely. On appeal plaintiff insists that his employment was to run until the resistor campaign had been fully launched and as long as it continued in operation.

It is apparent, therefore, that, if there was a contract of any character, it was of nebulous substance only, with absolute uncertainty as to duration and with lack of mutuality of promises upon the parts of the respective parties. In that situation, it seems clear (first) that the contract was void for lack of mutuality, and (second) that if any agreement existed it was one wholly at will, which either party had the right to terminate at any time.

It is well settled in Illinois that whenever a contract is incapable of being enforced against one party, that party is equally incapable of enforcing it against the other. . . . As the Appellate Court has said, in *Farmers' Educational and Cooperative Union* v. *Langlois . . . :* "Mutuality of obligation means that both parties are bound or neither (is) bound. In other words there must be a valid consideration. Without a valid consideration, a contract cannot be enforced in law or in equity." In various Illinois cases contracts have been held void for lack of mutuality. . . .

In the present case, a careful examination of the record discloses that there was not the slightest bit of evidence that plaintiff ever agreed that he would continue in the employment of defendant for any specified time. In other words, he had a right to terminate his employment at any time and did not promise to perform for any definite length of time. Therefore, the contract could not have been enforced against him and was lacking in mutuality. Consequently, he cannot enforce it against defendant. Plaintiff is in no better position if we consider his alleged agreement from the point of view of its term of existence, for the period of its continuance is indefinite. It is a contract at will, which may be terminated at the option of either party at any time.

♦♦♦

Nor can it avail plaintiff that his contract was, in his own words, permanent. . . . The Supreme Court of Illinois has expressly held that a contract for "permanent employment" is one at will. This is in accord with the decisions of other jurisdictions that contracts not expressly made for fixed periods may be terminated at the will of either party. . . .

The judgment is affirmed.

PARIS CONST. CO. v. RESEARCH-COTTRELL, INC.

451 F. Supp. 938 (W.D. Pa. 1978)

Paris Construction Company (Paris), plaintiff in the above action was a second-tier subcontractor in a contract with Nick Istock, Inc. (Istock). Istock in turn was a subcontractor to Research-Cottrell which was the prime contractor for construction of a facility for Jones and Laughlin Steel Company in Aliquippa, Pennsylvania.

Paris has not been paid for the work performed on the job and has sued Research-Cottrell for these items.

Research-Cottrell has moved for summary judgment on the claims of Paris and has filed evidentiary matters in support thereof. Paris' response to the motion relies to a very great extent on the same evidentiary materials, largely excerpts from depositions. Plaintiff's Amended Complaint alleges that its claim against Research is based on the following facts: (1) plaintiff's bid as a subcontractor became part of Istock's bid, which was accepted by the general contractor; (2) pursuant to the direction of Research-Cottrell and Istock, plaintiff entered into performance of its work; (3) at the specific request of agents and/or employees of Research-Cottrell, plaintiff agreed to do additional work; and (4) the additional work or "overages" performed by plaintiff were authorized by plaintiff's invoices, which were initialed by the onsite supervisor of Research-Cottrell, who acted in the capacity of agent of Research-Cottrell.

Defendant's contention, supported by evidentiary materials presented and not refuted by any evidence produced by plaintiff, is that there was no contractual relationship between Research and Paris; that the subcontract between Research-Cottrell and Istock prohibited any subcontracting by Istock; and, that any relationship between Istock and Paris was an arrangement of their own. Research-Cottrell's position is further fortified by the testimony that all extra work authorizations for the work claimed by Paris in this suit were on Istock forms (prepared by Paris) rather than on Paris forms, as alleged in the Amended Complaint. The Paris forms were used with respect to matters outside the Istock contract.

Fundamental to this lawsuit is the fact that Paris has filed an identical suit in the Common Pleas Court of Beaver County, Pennsylvania, against Nick Istock, Inc., for the same work and the same damages as those sought from Research-Cottrell. The sworn Complaint in that case alleges that Paris' contract on this job was with Nick Istock, Inc.

We conclude that there is not a genuine issue of material fact and that the plaintiff here, Paris Construction Company, had no express contract with the defendant Research-Cottrell.

Plaintiff also claims that it is entitled to recover from Research-Cottrell on a quantum meruit and/or implied contract theory in that it performed extra work at the request of the supervisor for Research- Cottrell. The primary inquiry in determining this matter is whether or not any circumstances exist which would reasonably have led Paris to believe that Research-Cottrell rather than Istock would pay for Paris' performance.

The evidence produced shows that all extra work performed by Paris on written orders signed by Research-Cottrell's on-site supervisor, Mulholland, were processed through Istock and that Paris billed Istock for such extra work, as well as for regular progress payment and payroll accounts.

There is no countervailing evidence to create an issue of fact here. Research-Cottrell had a contract with Istock which provided for the mechanism of payment for extra work ordered. The fact that Research-Cottrell ordered such work performed and signed authorizations for it is not inconsistent with the conclusion that Research-Cottrell was dealing at all times with the Nick Istock, Inc., contract and Istock personnel. The mere fact that the person to whom the request was made was an employee of Paris does not affect the result. Mulholland was dealing with Istock on the job. Research-Cottrell had no contractual relation at any time with Paris.

Count IV of the Amended Complaint asserts a claim based on detrimental reliance on the representations made to Paris by Research-Cottrell.

The elements of detrimental reliance are: (1) a promise to a promisee (2) which the promisor should reasonably expect to induce action by the promisee (3) which does induce such action by the promisee, and (4) which should be enforced to prevent injustice to the promisee. . . .

Unlike a claim in quasi-contract or implied contract, detrimental reliance does not establish a relationship giving rise to legal liability in absence of express contract. The doctrine of detrimental reliance is "not so much one of contract, with a substitute for consideration, as an application of the general principles of estoppel to certain situations." . . . Before detrimental reliance can be successfully alleged, a promise inducing action or forbearance by the promisee in reasonable reliance thereon must form the basis of the claim.

Again we find no evidence here creating an issue of fact. No promise or arrangement between Research-Cottrell and Paris existed upon which Paris could reasonably rely to induce it to perform. Any promises made with respect to work performed at the Aliquippa project were between Paris and Istock. Therefore, with respect to plaintiff's claim based on detrimental reliance we must grant summary judgment in favor of defendant.

REVIEW QUESTIONS

1. Does a conditional promise, such as a promise to resell a piece of property if you are able to buy it, constitute a legal consideration? Why or why not?
2. Black agreed to pay White, a member of the state house of representatives, $1,000 to do everything "reasonable and lawful" to defeat a bill that was to be presented to the legislature. Can White collect the $1,000 whether the bill passes or not? Why or why not?
3. What is the effect of sealing a contract in your state?
4. Black hires White Automation to build a piece of automatic machinery for $150,000. During the time that the machinery is under construction, union pressures result in a wage increase for the electrical suppliers. As a result, electrical equipment costs more than anticipated. White demands $8,000 more for the machinery and Black agrees to pay. After installation, Black refuses to pay more than $150,000 for the machine. White sues. What is the result? Why?
5. Gray owed Brown $600 on an old debt. Gray refused to pay the debt several times, claiming lack of funds each time. Brown finally attempted to collect the debt at a time when Brown knew Gray had sufficient money. Gray offered to pay $400 and give Brown a wrist watch worth about $30 if Brown would consider the debt paid in full. Brown accepted. Later Brown brought an action for the remainder of the debt. Can Brown get it? Why or why not?
6. In *Meadows v. Radio Industries,* would the outcome of the case have been different if the employment contract had been made for 10 months and Meadows was fired after 8 months?
7. In the usual employment contract between an engineer and his or her employer, how much notice is required for termination of the employment by either party?
8. In the case of *Paris Constr. Co. v. Research-Cottrell, Inc.,* suppose Paris lost its case against Istock as well as its case against Research-Cottrell, Inc. In this circumstance, would Paris appear to have a case against Jones and Laughlin Steel Company? Could Paris perhaps establish a mechanic's lien against the property improved?

Chapter 10

Lawful Subject Matter

Freedom to deal or to refuse to deal with others is one of our guaranteed rights. Just as some limitations must be placed on freedom of speech to protect the public, limitations also are imposed on the right to contract. The necessity that any contract have a lawful purpose is fundamental in the law. Generally, courts treat contracts based on illegal subject matter as void. Subject matter considered by the courts to be illegal may be classified in three categories: contrary to statutes (federal, state, local), contrary to common law, or contrary to public policy. More generally, illegal consideration may consist of any act or forbearance, or promise to act or forbear, that is contrary to law or morality or public policy. Ignorance by the parties as to what constitutes illegality cannot make valid a contract that is void because of its illegality.

INTENT

The intent of the parties at the time the contract was made often determines its legality. If both parties intended the contract to be performed illegally it usually will not be enforced by the courts. This is true even though actual performance was lawful. If only one of the parties intends to perform illegally, the courts will usually uphold the contract.

A very large number of contracts are capable of being performed either legally or illegally. A contract to drive a truck, for instance, may be performed either according to the law or in violation of it. If there is no evidence of an intent by both parties to perform the contract illegally, the truck-driving contract would be valid. Generally, the contract must be incapable of being performed in a legal manner for the courts to declare it void on the basis of illegality.

Knowledge that the subject matter of a contract is intended for later unlawful use by one of the parties usually will not void the contract. However, where the subsequent unlawful use involves a heinous crime (treason or murder, for instance), an exception is made to this rule. An action to collect for the sale of a gun where the seller knows the buyer intends to commit murder with the gun would not be successful. If the seller had no knowledge of the intended use of the gun and complied with the law in all other respects in the sale, however, the contract of sale would be valid and enforceable.

CONTRACTS CONTRARY TO STATUTE OR COMMON LAW

If the subject matter of a contract is contrary to an existing statute, the contract is usually void. The statute concerned may be a federal or state statute or a local ordinance. It is not necessary for the contract to violate the wording of the statute; the contract may be legally void if it conflicts with the implied meaning of the statute or the intent of the legislature when the law was passed.

♦ Law Passage Following Contract *Formation*

If a continuing contract[1] is formed and later a law is passed making its performance unlawful, the entire contract is not necessarily rendered void. Performance of the contract subsequent to the passage of the statute cannot be recovered for in the courts, but the contract can be enforced for the parties' performance prior to passage of the statute.

Shortly after the United States entered World War II, Congress passed several regulations outlawing the sale of various items and restricting the sale of others. In a case resulting from these restrictions where a new car sales agency had leased premises for the sole purpose of display and sale of new cars (so stipulated in the lease), it was held that because of the government restrictions on the sale of new cars, the lease contract was terminated. If the lease had been written so that the use of the premises had not been restricted to the sale of new cars, the governmental restrictions would have had no effect.

♦ Type of Statute

In considering the legality of a contract's subject matter, the courts sometimes distinguish between laws passed for the protection of the public and laws that have as their primary purpose the raising of revenue. If a contract violates a statute passed for the protection of the public, it is treated as void and unenforceable. If the contract violates a revenue statute, though, it is usually enforceable, subject to penalties for avoidance of the statute. An excellent example of this is found in the state licensing laws, such as those for business establishments and professional people. If the purpose of the licensing law is to protect the public and safeguard the lives, health, property, and welfare of citizens (for instance, licensing of professional engineers), a contract for professional services made by one who is not licensed in the state will not be enforced by the courts. If the primary purpose of the law is to collect revenue only (for instance, state gasoline taxes), the courts usually enforce the contract and impose a penalty on the violator.

1. A contract the performance of which will take place over some length of time.

♦ Harm to Third Person

If a third person would be harmed by a crime or tort in connection with the performance of a contract, the subject matter of the contract probably will be held unlawful. A court will not punish for the commission of a crime or tort on one hand and support a contract to commit such a crime or tort on the other. Where the contract is an inducement to commit a crime or tort, but the commission of such crime or tort is not necessarily required to perform the contract, the courts will examine the strength of the inducement and the general nature of the contract. Life insurance may be an inducement to murder; fire insurance an inducement to commit arson. For this reason, an insurable interest often must be shown before insurance can be obtained. A contract, in the performance of which a party to the contract must breach an existing contract, is usually unenforceable in the courts.[2]

♦ Contracts That Restrain Trade

Our economy is based on the principles of free enterprise and freedom of competition. Under the Sherman Antitrust Act, the Clayton Act, and state statutes similar in scope, contracts that tend to create a monopoly or maintain price levels or in other ways restrain trade may be unlawful. The penalty for a violation of these statutes is up to triple the amount of damages shown, plus payment of the court costs and reasonable attorney's fees for the plaintiff's attorney.

An example of an agreement that would probably be unenforceable as a violation of the antitrust laws would be an agreement by two competitors to fix prices at a set minimum or to divide the market.

Suppose two concrete manufacturers together have 90 percent of the business in a given state. They agree not to underbid each other in several counties, and each agrees not to bid at all for jobs in several other counties. Such an agreement would violate the antitrust laws and should be unenforceable. Other business practices, however, are not so clear cut. Should it be unlawful for a computer software manufacturer to require its customers to also buy maintenance services for the software? If so, are both the maintenance services

2. See *Reiner v. North American Newspaper Alliance*, 181 N.E. 561 (1930), in which a contract to breach the contract of passage on the Graf Zeppelin was held illegal.

agreement and the underlying software license agreement unenforceable? The law is not always clear in these areas.

In connection with their employment, many people sign written agreements in which they agree that, for some period of time after the termination of their employment, they will not compete against their former employer. Such "noncompetition" clauses (sometimes called "covenants not to compete") are considered restraints of trade and were strictly construed against enforcement by the common law. In some states, such covenants are prohibited in certain employment relationships. Generally, such noncompetition provisions are enforceable to the extent necessary to protect the legitimate business interests of the employer, such as the employer's trade secrets or good will. If the provisions are reasonable as to the restrictions on territory, scope of the former employee's activity, and the time period of the restrictions, then the provisions will be upheld and enforced.

Such provisions are also common in agreements involving the sale of a business. Despite the seeming restraint of trade present in such agreements, the courts will protect the purchaser of a business and the "good will" that goes with it. By such an agreement the seller states that the seller will not compete with the buyer of the business. Typically, the agreement states that the seller agrees not to enter into the same type of business in the same market area for a given period of time. If the restrictions about type of business, market area, and time are all reasonable, the restrictions probably will be upheld in court. It is only when these restrictions are unreasonable that a court will hold them to be in violation of the law.

Engineers are often asked to sign preprinted employment agreements that detail their obligations as employees. Besides noncompetition clauses, such employment agreements often include provisions dealing with the employee's duties regarding confidentiality and inventions. Occasionally, such forms provide that all inventions of an employee—whether or not they relate to the employer's business and whether or not they are made during business hours and at work—belong to the employer. Some forms provide that any inventions made by an employee within a year or so after the employment relationship ends are presumed to belong to the employer. A few states have statutes that limit the enforceability of such provisions; in some other states, such provisions are considered unenforceable as contrary to public policy. In still other states, however, such contracts can be enforced.

♦ Usury

Most states have passed laws limiting the rate of interest that may be charged for the use of money. Considerable variation exists in the attitudes of the various states when these laws are violated. In some states, when usurious interest is charged, the courts will not aid in collection of principal or interest; in other states, all interest may be forfeited as a result of usury; and in still other states, collection of principal and maximum allowable interest results.

♦ Wagering Contracts

In most states wagering contracts are illegal. It is often difficult for the courts to determine whether a particular contract is a legitimate business transaction or a wager. The test used by the courts in making their determination is centered about the creation of the risk involved. Is a risk created for the purpose of bearing that risk? If so, the contract is deemed a wager. In a dice game, on the first throw, a total of seven or eleven on the two dice pays off for the holder of the dice. The two bettors have created a risk of loss by placing their bets before the dice are thrown, with the score on the dice determining who shall lose the money. In a business contract calling for future delivery of a commodity for the payment of a present price, risk of loss resulting from a change in price of the commodity between the present and the delivery date is assumed by at least one of the parties. Has a risk been created for the purpose of assuming it? The courts answer no, providing that future delivery is actually intended. Manipulations in the stock and commodity markets are sometimes criticized on the grounds that the final result of the contract is, and was originally intended by the parties to be, a transfer of money rather than commodity.

♦ Other Types of Violations

A contract to withhold evidence in a court case is probably unlawful; so are contracts that tend to promote litigation. This is one reason why contingency fees for attorneys or others connected with a trial are sometimes frowned on. If a would-be plaintiff stands to lose nothing in a court case, this is an incentive to undertake the litigation. Contracts having the effect of compounding a crime are against public policy; for example, tax evasion agreements are unlawful.

A large category of unlawful contracts is included in those that either restrain marriage or promote divorce. The courts generally do everything within their power to promote matrimony and to discourage its dissolution.

Contracts that are immoral in subject matter or tend to promote immorality are sometimes declared by the courts to be unlawful and against public policy.

CONTRACTS CONTRARY TO PUBLIC POLICY

Contracts must conform to the common law and the applicable statutes and regulations to be lawful and enforceable. However, these are not the only limitations imposed by the courts on a contract's subject matter. In addition to the requirement that contracts conform to the written and unwritten laws of the community, the courts require conformity to public policy. Public policy is rather difficult to define because it is continuously changing. It must change to conform to changing ideas and changing technology, just as city planning, for example, has had to change to incorporate cloverleafs and jet transportation.

Generally, those contracts that would be held contrary to public policy are those that, if enforced and thereby encouraged, would be injurious to society in some way. It is not necessary that someone, or the public in general, be injured by the performance of the contract. Courts have held that a contract may be unenforceable because of an *evil tendency* found to be present in it.

Unenforceability because of the "evil tendency" of a contract is particularly characteristic of contracts in which the judgment or decision of public officials might be (or has been) altered by the contracts. In fact, any contract that tends to cause corruption of a public official may be subject to censure in the courts. This brings up a rather sensitive problem with which the courts often are faced: What acts of lobbying are to be condoned?

- The end sought to be accomplished by the lobbying must be a lawful objective—one which, if accomplished, would improve the public welfare, or at least would do it no harm.
- The means used must be above reproach. Generally, a lobbyist must be registered as such; his or her lobbying practices must not involve threats or bribery or secret deals.
- Courts generally treat contingent fees based on the passage of legislation as being against public policy.

Assume Green is hired as a consultant for an engineering design job, with payment to depend on his ability to obtain the acceptance of the design by political officials. Green, being human, might be tempted to go beyond what is considered right and ethical to obtain the fee. In the court's eyes, contingent fees are an inducement to the use of sinister and corrupt means of gaining the desired objective. Although the court may in certain cases overlook the presence of contingent fees, it nearly always detracts from the case presented by the proposed recipient of the fee. Contingency fees are allowed for attorneys in certain cases. The courts allow such fees on the idea that, by providing an incentive to the lawyer, more lawyers will make their services available to persons who otherwise could not afford the lawyer's fees.

♦ Fraud or Deception

A contract that has the effect of practicing fraud or deception on a third person is against public policy. Black agrees to pay White $1000 if White (a prominent nuclear engineer) will recommend Black to Gray for a job as a nuclear engineer. White knows nothing of Black's qualifications for the job. Even though White wrote the recommendation and Black succeeded in getting the job, it is likely that White could not get the $1,000 by court action. The contract would be void as against public policy.

In contrast, payment for a recommendation from an employment agency would be quite enforceable. The employment agency is in the business of recommending people for jobs for a fee. Thus, those who hire employees through an employment agency have knowledge of the usual arrangement; those who hire based on individual recommendations have a right to assume that such recommendations are given freely and without prejudice.

♦ Breach of Trust or Confidence

Contracts that breach a trust or a confidential relationship are against public policy and will not be supported in a court. *Agency* is one such fiduciary relationship—the agent acts for his principal in dealing with others. White, as Black's agent, contracts with Green for the purchase of steel. Unknown to Black, White is to receive a 5 percent kickback from Green for placing the order with Green. Even though the steel purchase occurred, the five percent kickback agreement could not be enforced in court.

♦ Sources of Public Policy

Public policy is found in the federal and state constitutions, in statutes, in judicial decisions, and in the decisions and practices of government agencies. Lacking these, the court must depend on its own sense of moral duty and justice for its decisions.

EFFECT OF ILLEGALITY ON CONTRACTS

Generally, no action in law or in equity can succeed if it is based on an illegal agreement. That is, court actions based on agreements that are illegal, immoral, against public policy, intended to prompt the commission of a crime, or are forbidden by statute probably will be unsuccessful. (Because defining "public policy" is sometimes difficult, it is also difficult to predict whether a given agreement will be held to violate the "public policy.") The defense of illegality is allowed, not as a protective device for the defendant but as a disability to the plaintiff. The public is better protected against dishonest transactions if the court places this stumbling block in the path of those attempting to avoid the law. Suppose a member of a gang of thieves, on being cheated out of a share of the loot, went into court to try to force a split. What should the court do as to the division of the spoils? The court would contend that the agreement to split is void, being based on a crime. The public interest is best served if the court does everything in its power to discourage the crime. Thus, as far as the loot-splitting is concerned, the case would be dismissed.

Public interest, then, is the determining factor in such cases. The court's decision is based on its opinion of what would be the greatest aid to the public. If the public interest would be promoted better by a decision for the plaintiff, the court will so decide. Consider a confidence game in which the victim is placed somewhat afoul of the law by an agreement with the confidence men (as is usually the case). Swallowing pride and guilt alike, the victim asks the court to order the return of the money taken. The court may decide to do just that as a discouragement to continued practice of such con games on the public.

Disaffirmance of an illegal contract while the illegal portion of the agreement is still executory often allows the person disaffirming the agreement to recover the lawful consideration previously given by that person. A person paying money to another to have a crime committed ordinarily may disaffirm the contract so that the crime may be stopped and thus obtain restoration of the money. Here again, the public interest is the determining factor, with the court acting as the representative of the public interest.

VAN HOSEN v. BANKERS TRUST COMPANY
200 N.W.2d 504 (Iowa 1972)

By action in equity plaintiff Hugh Van Hosen seeks declaratory relief from the forfeiture provision of a private pension plan established by defendant Bankers Trust Company and administered by it as trustee. Trial court adjudged the controverted proviso violative of public policy, therefore void and unenforceable. Defendants appeal. We affirm.

Van Hosen worked for defendant bank almost 33 years prior to his resignation in 1967. Starting as a messenger he ultimately became a vice-president and manager of installment loans. While so employed certain pension and profit-sharing plans were initiated and at times amended by the bank. Among them was the instantly involved "Bankers Trust Company Primary and Supplemental Pension Plan and Trust Agreement." Relevant provisions thereof will be later set forth.

Upon termination of his aforesaid employment, plaintiff sold insurance with a Des Moines agency for about 18 months. He was then offered a position in the commercial loan department of another Des Moines bank, admittedly an institution in competition with defendant bank. Thereupon Van Hosen requested a waiver by defendant of the subject pension forfeiture clause. After some apparently unavoidable delay, during which plaintiff accepted proffered employment with the competing institution, his requested forfeiture waiver was refused. Defendant bank has since withheld retirement benefits payable to plaintiff under the plan here involved. Incidentally, no issue regarding trade secrets is instantly presented.

In support of their appeal, defendants contend (1) the forfeiture provision is not a part of the plaintiff's employment compensation, and not in restraint of trade, (2) the court may enforce a forfeiture provision in a pension plan to that extent reasonably necessary to protect the employer's legitimate interest without imposing undue hardship upon the employee so long as public interest is not adversely affected. We shall deal generally with these propositions.

♦♦♦

II. *Murphy* v. *R. J. Reynolds Tobacco Co.* involved a problem akin to that instantly presented. In the cited case we approvingly quoted this from *Cantor* v. *Berkshire Life Ins. Co.*

"'The concept of employees' rights and of the place of the so-called fringe benefits in relationship to employees' remuneration has undergone a substantial change in recent years. . . .

"'There has been . . . in recent years a gradual trend away from the gratuity theory of pensions. The courts, recognizing that a consideration flows to an employer as a result of such pension plans, in the form of a more stable and a more contented labor force, have determined that such arrangements will give rise to contractual rights enforceable by the employee who has complied with all the conditions of the plan, even though he has made no actual monetary contribution to the fund.'"

To the same effect is IA Corbin on Contracts, sec. 153, at 18-20:

"A promise by an employer to pay a bonus or a pension to an employee in case the latter continues to serve for a stated period is not enforceable when made; but the employee can accept the offer by continuing to serve as requested, even though he makes no promise. There is no mutuality of obligation; but there is sufficient consideration in the form of service rendered. Indeed, the employer's offered promise becomes irrevocable by him as soon as the employee has rendered any substantial service in the process of accepting; and this is true in spite of the fact that the employee may be privileged to quit the service at any time."

It is thus evident plaintiff's action stands squarely in contract and this is true whether defendants' pension program be characterized as a contributory or noncontributory plan.

III. These are the relevant portions of the Agreement here involved.

Under Article III "competing institution" is defined as a financial organization doing a correspondent bank business or any financial enterprise in Des Moines, Polk County, Iowa.

Article XI provides, in essence, a participating "member," upon completion of 20 or more years' continuous regular service with defendant bank shall, upon leaving for any reason, have a vested right under the Agreement, subject to the following material part of Article XII: "In the event a Member shall become employed by a competing institution without the consent of the board of directors of the Bank, . . . then in . . . such event all of the rights of such Member, and the rights of any person claiming by, through, or under such Member, to benefits under the terms of this Plan shall be *forfeited and terminated*; and, in the event that any annuity benefits shall have been purchased for such Member, such benefits shall be terminated both as to the rights of such Member thereunder or the rights of any joint annuitant or other beneficiary thereunder and such benefits as may thereafter accrue under the terms of any such annuity shall be paid to the Trustee and applied by the Trustee to the costs of administering the Plan." (Emphasis supplied)

"Forfeited," in the contractual sense, ordinarily means the taking away or loss of rights and interest in property. . . . Forfeiture is also here unquestionably made positive and interminable by conjunctive use of the word "terminate." This terminology unquestionably denotes an expiration, extinction, and cessation of all affected contractual rights. . . .

From the foregoing flows the unavoidable conclusion that under the terms of Article XII, if it be held enforceable, plaintiff forever lost any and all rights held under the Agreement by merely becoming an employee of a competing institution.

IV. Trial court likened Article XII to restraints on post-employment competition, sometimes contained in employer-employee contracts. In so doing the court leaned heavily on *Baker* v. *Starkey,* and thereupon found Article XII, being in restraint of trade, was violative of public policy, therefore unenforceable. . . . In that regard, as heretofore disclosed, various courts have recently espoused diametrically opposing views on the subject at hand. Some hold such a pension forfeiture clause as that now before us, unlike those restraints often embodied in employment contracts, does not preclude a benefitted retiree from engaging in work for a competitor of the pension-paying former employer, therefore is not in restraint of trade. . . . Other courts have held the threat of economic loss to a pension recipient by affiliation with a competitor of the former employer constitutes, in effect, an impermissible restraint on competition. . . .

For reasons later revealed, however, it is neither essential nor do we here adopt trial court's analogical approach, or the divergent concepts enunciated in Rochester and Food Fair Stores, supra.

V. It should be inceptionally understood:

"(A) man is permitted to make a contract that will result in a forfeiture, and when it is clear from the terms of the contract that the parties have so agreed, a court of equity, as well as a court of law, will enforce the forfeiture. . . . If one contracts for forfeiture, while the law will scan it closely and will look upon it with disfavor, yet as a part of the contract it must be enforced. . . . While it is true forfeitures are not favored in the law, they are not outlaws. In *Fairgrave* v. *Illinois Bankers Life Association* we said: 'If the parties have by the terms of their agreement fixed the consequences of a forfeiture on the violation of it, this becomes the law of their contract, and by it they are to be governed.' . . .

On the other hand, since forfeitures are not favored, those claiming them should show the equities are clearly on their side.

In the absence of a statute declaring void any such *in terrorem* provision as that contained in Article XII, and we have none, the enforceability of a contractual divestiture is usually determined by application of the public policy or reasonableness standard. This in turn necessitates a balancing of interests. . . .

So we resort to a weighing of interests in order to determine the enforceability or unenforceability of Article XII. If that proviso be found reasonable then it is enforceable. Conversely, in event such imposes an unconscionable burden on plaintiff it will be deemed unenforceable.

VI. First to be considered is defendant employer's objective in providing for a divestment of pension rights acquired by a former employee who engaged in post-termination competition.

As indicated by the quote in *Murphy* v. *R. J. Reynolds Tobacco Co.,* . . . an employer's basic purpose in creating an employee pension program is to encourage loyal and productive career service on the part of all possible participants. In fairness it must be conceded that most if not all employers, like employees, are participating in a constant struggle for survival. To that end they must make every reasonable effort to gain and hold the good will of all possible customers.

It is thus apparent that in the organization and administration of an employee pension program an employer does have legitimate business interests at stake, including good will, as to which he is entitled to reasonable protection.

But there are counterbalancing factors that must also be evaluated.

Under existing economic conditions a person accepting employment will generally evaluate both take-home pay and fringe benefits. In any event, when a career employee retires, either voluntarily or involuntarily, he or she often experiences a traumatic economic change. Furthermore, many pensioners cannot, at the moment, qualify for social security and must resort to other employment for supplementary income. Usually, in such cases, work openings in the employee's accustomed field of endeavor are not readily available.

And if such a position is obtained, comparatively inconsequential or no attendant marginal benefits are ordinarily provided, or ultimately acquired.

Also, while engaged in working for a second employer, such pension rights as may have been acquired during service with a prior employer are, to the retiree, relatively insignificant. But when retirement from any subsequent employment occurs, the result can be chaotic, absent restoration of any pension rights acquired through extended service with a prior employer. The harshness of such a situation is self-evident.

Moreover, private pension plans have a humanitarian purpose in that, like employment security, they extend to those benefitted some degree of financial independence at a time when earning ability and related income may be impaired or ended. See The Code 1971, Section 96.2. It is in turn evident these programs have become increasingly vital to our socioeconomic community welfare. By the same token society today has a material interest in the orderly development and administration of all pension plans, public or private. Thus public policy comes into play. . . .

It therefore follows, the infinite forfeiture and termination of all pension rights instantly acquired by plaintiff through prior affiliation with defendant bank, merely by accepting employment with a competing institution, imposes an unjust and uncivic penalty on plaintiff at the same time disproportionately benefiting these defendants.

We now hold Article XII of the Agreement, being so unreasonable as to be in violation of public policy, is accordingly unenforceable.

VII. Trial court, in reaching the same result, did so upon a different premise from that here adopted. But such is of no consequence.

Affirmed.

REVIEW QUESTIONS

1. Black, an engineer, agrees to act as an expert witness for White in a court case. In return, Black is to receive a $50,000 fee plus expenses for the expert witness services. Is the contract lawful? Why or why not?
2. In question 1's example, assume that Black agrees to accept $50,000 if White wins, or payment for expenses only if White loses. Is the contract lawful? Why or why not?
3. Gray is a process engineer for Brown. Green has made the low bid on equipment for a manufacturing operation that Gray had begun setting up. The equipment meets the specifications as well as equipment proposed by other bidders. To improve the chance for acceptance of Green's bid, Green offered to give Gray 2 percent of the bid price if Green's equipment is used. If the equipment is bought, can Gray get the two percent? After receiving the two percent offer, what if anything, should Gray have done about it?
4. Explain why an insurance contract is not a wagering contract.
5. What is public policy?
6. Why are contingency fees looked on with disfavor by courts?
7. In the case of *Van Hosen* v. *Bankers Trust Company* the employee's acts clearly violated a provision of the pension section of his employment contract. Outline the legal reasoning by which the court affirmed the judgment for the employee.

Chapter 11

Statute of Frauds

An oral contract is generally just as enforceable as a written contract, but a written one has at least one major advantage—its terms are easier to ascertain. The relationships of the parties are set forth for interpretation by the individuals concerned and by a court of law if the need should arise. Although it is true that in most circumstances an oral contract is as enforceable as a written one, it is sound common sense to put into writing any contract of more than a trivial nature.

ENGLISH STATUTE OF FRAUDS

In England, the courts were faced with many cases concerning oral agreements. Instances of perjured testimony were frequent as the contending parties attempted to prove or disprove the existence or terms of contracts. In 1677, the English Statute of Frauds was passed as "An Act for the Prevention of Frauds and Perjuries" to relieve the courts of the necessity of considering certain types of contracts unless their terms were set forth in writing and signed. The statute did not require a formally drawn instrument. The only writing required is the minimum needed to establish the material provisions of the agreement. Generally, the writing had to fulfill all of the following criteria:

- Reasonably identify the subject of the contract
- Indicate that a contract has been made between the parties
- State the essential terms of the contract
- Be signed by or on behalf of the party to be charged

The writing was not required to be all on one instrument, but if it was not, a connection between the various instruments usually had to be apparent from the documents themselves. The statute did not require a signature from both parties. Only the person sought to be charged with the contractual duties has to have signed. The signature itself could consist of initials, be rubber stamped, in ink or pencil, or anything else intended by the party to constitute identification and assent; it could appear anywhere on the document.

The English Statute of Frauds consisted of several sections, but only the sections numbered 4 and 17 are of major importance to us today. These two sections have become law in all of the states in the United States with only minor modifications.

♦ Fourth Section

According to the fourth section, "no action shall be brought" on certain types of contracts unless the agreement that is the basis for such action is in writing and signed by the defendant or the defendant's agent. The following types of contracts are subject to the requirement of a written and signed instrument:

1. Promises of an executor or an administrator of an estate to pay the debts of the deceased from the executor's or administrator's own estate
2. Promises to act as surety for the debt of another
3. Promises based on marriage as a consideration
4. Promises involving real property
5. Promises that cannot be performed within a year

♦ Seventeenth Section

The seventeenth section states the requirements for an enforceable contract having to do with the sale of goods, wares, or merchandise. It says that unless one of the following three actions is done to secure such a transaction where the consideration is at least ten pounds sterling, the contract will be unenforceable at law:

1. Part of the goods, wares, or merchandise must be accepted by the buyer.
2. The buyer must pay something in earnest toward the cost of the goods.
3. Some note or memorandum of the agreement must be made and signed.

The 10 pounds sterling minimum has been changed to $500 by each of the states with the adoption of the Uniform Commercial Code.

Promises that cannot be performed within a year and contracts for the sale of goods in excess of $500 are the two types of contracts most commonly encountered.

Promises Requiring More Than a Year to Perform

The legal interpretation of a year is important here. Generally, the time starts to run from the time of making the agreement, not from the time when performance is begun. An oral contract, then, to work for another for a year, starting two days from now, would be unenforceable under the English Statute of Frauds. In most jurisdictions it is held that parts of days do not count in the running of time. If a contract is made today, time will start to run on that contract tomorrow.

The statute refers to contracts "not to be performed within the space of one year of the making thereof." The court interpretation of this statement is that it means contracts that, by their terms, cannot be performed within one year. White promises orally to pay Black $80,000 if Black will build a certain house for White. No time limit is set on the construction. The fact that actual construction took place over a two-year period would not bring the contract under the Statute of Frauds. If the contract could be completed within one year no matter how improbable this may be, an oral contract for its performance is binding. By this reasoning, a contract to work for someone "for life" or to support someone "for life" is capable of being performed in one year and need not be in writing to be binding.

If the performance of an oral contract is to take place in less than a year and is extended from time to time by increments of less than a year in such a way that performance continues for more than a year, such a contract is valid. For instance, an oral lease contract for nine months might be continued orally at the end of the period to run for another nine months for a total of eighteen months.

Suppose that an oral contract cannot be fully performed within a year. If one of the parties nonetheless fully performs, is that party unable to enforce the contract against the other? The majority of courts treat full performance by a party as removing the contract from the effect of the one-year requirement of the Statute of Frauds, even though the party took more than a year to perform. Accordingly, most courts would enforce such a contract. Some courts, however, hold that one party's full performance still does not remove the contract from the written requirement of the Statute of Frauds.

In situations where the Statute of Frauds applies but there has been at least a partial performance, most courts allow the party who has partially performed to recover the value of the services rendered. In some situations, one party may be estopped (that is, precluded) from asserting the Statute of Frauds as a defense if the other party has already performed in whole or in part. Courts are usually careful in such situations. They will refuse to allow the Statute of Frauds to be used as a defense by someone who uses it to secure an unfair advantage and thus "defraud" the other party to the agreement. In short, the courts refuse to allow the Statute of Frauds as a defense when doing so helps one party perpetrate a fraud.

Sales of Goods

As noted, a large majority of the states have adopted the Uniform Commercial Code (UCC), which governs the sale of goods within those states. The UCC drafters adopted the rule of the seventeenth section of the English Statute of Frauds in slightly altered wording.

If a transaction involves goods of a value great enough to be governed by the Statute of Frauds in the particular state, an oral agreement will be valid only if (1) some of the goods are accepted by the buyer, or if

(2) the buyer pays part of the purchase price in earnest, or (3) the buyer signs some note or memorandum as to the terms of the sale, or (4) the party to be charged with the contract formally admits that a contract was made, or (5) if the goods are to be specially made, are not suitable for sale in the ordinary course of the seller's business, and the seller has taken substantial steps to begin performance of the contract. Payment of part of the purchase price may be made in money or in anything of value to the parties.

What actually constitutes goods under the UCC has led the courts into some difficulty on occasion. It is apparent that if the goods already exist, a contract for their sale will not involve services. However, if the contract is for the purchase of goods not now in existence but to be made by the seller, there is a question whether this is not a contract for services to be performed. If the contract is primarily for services rather than goods, the UCC (and the accompanying UCC Statute of Frauds) will not apply.

Even when the UCC applies, however, there are still exceptions to the UCC Statute of Frauds. For example, the UCC provides the previously noted exception for an oral contract for the sale of goods that are to be specially made for the buyer and are not suitable for sale by the seller in its ordinary course of business. Such an oral sales contract is enforceable if the seller has begun making the goods or has made commitments for procuring the goods. Black wishes to purchase a gate to match a very old wrought iron fence around Black's house. Black contracts orally with White to have such a gate specially made. The contract will be enforced. This, of course, is a logical rule, because the gate probably would be unsalable to anyone else.

Another exception to the UCC rule is if the party against whom enforcement of the contract is sought admits in pleadings, testimony, or in court that a contract was made. Thus, testimony from a buyer who agrees that a contract was made but denies its enforceability for one reason or another removes the availability of the UCC Statute of Frauds as a defense (although the buyer may indeed have other defenses). If a written confirmation is sent by one merchant to another and the receiving party knows its contents, the confirmation can serve as the writing and satisfy the UCC Statute of Frauds if the receiving party fails to object within 10 days after receiving it.

Promises of Executors or Administrators

In law an executor is a person appointed in a will by the testator[1] to execute, or put in force, the terms of the instrument. An administrator serves a somewhat similar function where the deceased died without a will *(intestate)*. The administrator is appointed by a court to collect the assets of the estate of the deceased, pay the estate's debts, and distribute the remaining estate to the heirs—those persons entitled to it by law.

If the executor of an estate agrees to pay the debts of the deceased out of the executor's own estate, the executor is acting as a surety for the debt of another. In effect, the executor is saying: "If the estate of the deceased is insufficient to pay you, I will pay the debt." This situation is covered under surety contracts, which are described next.

Promises to Act as Surety

A promise to act as a surety for another's debts is quite similar to an executor's promise to pay the debts of the deceased. The requirement that a contract to answer for the debt, default, or miscarriage of another person must be in writing covers all types of guaranty and surety contracts. Lending institutions frequently require either collateral or a responsible cosigner as security for a loan. If White wishes to borrow $1,000 from Black Loan Company, a cosigner may be required for the loan. If Gray is to act as surety, agreeing to pay if White does not pay, Gray's agreement to do so must be in writing to be enforceable against Gray.

It is appropriate here to distinguish between *primary* and *secondary promises*. An oral primary promise (also called an *original promise*) is enforceable but an oral secondary promise (also called an *collateral promise*) is not. Wording of the promise can be quite important, but the apparent intent of the promisor when the promise is made is even more important. If the wording and other facts make clear an intent such that "If White does not pay, I will," the promise to pay is a secondary promise. However, if the circumstances show the intent to be that "I will pay White's loan," the promise is a primary one. The primary promise is enforceable against the promisor if it is made either orally or in writing. Even though a "primary" promise

1. The testator is the person who made the will

may be for someone else's debt, the court views such a promise as an independent obligation that can be enforced, even if made orally. Similarly, when someone agrees to make good any loss sustained by the plaintiff if the plaintiff acts as surety for a third party, the promise is primary in nature and enforceable even though the promise was oral. This comes closer to an undertaking of indemnity or insurance than surety.

Indemnity (insurance) contracts need not be in writing. Under such contracts, no liability is held to exist until an obligation arises between the insured and some third party. In an indemnity agreement, one person agrees to pay another's debts to third persons. For example, your insurance company agrees to pay your debts to persons injured in an automobile accident as a result of your negligence. The difference between a promise to act as a surety and to insure or indemnify you, however, is that the promise of indemnity is made to the *debtor* (that is, the insurer makes its promise to the insured), not to the *creditor*.

♦ Promises in Consideration of Marriage

Marriage is said to be the highest consideration known to law. This provision of the Statute of Frauds applies particularly to situations in which the agreement to marry is based on consideration such as a marriage settlement. If Mr. White agreed to pay Black $10,000 in consideration of Black's marriage to White's daughter, the contract would have to be in writing to be enforceable. The statement that contracts in which marriage is to be a consideration must be in writing does not mean that when a man and woman simply agree to get married, with neither giving up anything but their unmarried status, such an agreement would have to be in writing. Oral promises to marry are actionable under the common law if one of the parties attempts to breach the promise to marry the other.

Real Property Transactions

Transactions involving "lands, tenements, and hereditaments" (that is, real property) must be in writing to be enforceable at law. Real property law is treated more thoroughly in Chapter 19. It is necessary here, however, to distinguish between real and personal property. *Real property* is anciently defined as consisting of land and those things permanently attached to it, or *immovables*. *Personal property*, then, includes the movables, or things not firmly attached to the land. These definitions comprise only a part of the distinction currently applied by the courts.

Minerals in the soil, water rights, and trees on the land, for instance, are usually considered by the courts as real property requiring a writing for their sale. An exception exists, though, in dealing with these natural fruits of the land. When the contract contemplates immediate severance or removal of these things, an oral contract to such effect will be enforceable.

Black orally sells White a stand of timber to be cut and sold by White with Black to receive payment as cutting proceeds. If cutting is to begin immediately, the oral contract is enforceable; if it is to begin 10 months from now, it is unenforceable.

However, things such as cultivated crops, which result from human effort, usually may be the subject of an enforceable oral contract. For instance, a contract to sell the fruit in an orchard as it ripens would be enforceable under the Statute of Frauds, even though made orally. Lease contracts involve an interest in land but, for most purposes, are considered personal property. An oral lease for a year or less is valid, but a lease must be in writing if it is to run longer than a year.

Part performance of an oral contract for the purchase of real property can influence a court to disregard the Statute of Frauds. It is only in an unusual case, however, that the courts will enforce such an oral contract. For example, suppose that the buyer set about building a new house on the land. The extent of improvements made by the buyer is so vast and material as to make it unjust to hold the oral agreement unenforceable.

Effect of the Statute

In some states, the Statute of Frauds is phrased so as to make contracts void unless in compliance with the Statute of Frauds. In most states, though, a contract that fails to comply with the Statute of Frauds is merely voidable. In these latter states, then, the Statute of Frauds operates as a defense to the enforcement of the contract. If the Statute of Frauds is not raised as a defense, the court will enforce the contract.

Even if the Statute of Frauds applies, a party that has performed may still be able to recover something. When one party has performed under such an oral contract, that party has probably enriched the other party. A court of equity may recognize the obligation created by such a performance and enforce payment by the other party under a quasi-contract theory of recovery. Generally, when the result would be inequitable or grossly unfair, the courts will not allow the statute to stand as a defense.

When the oral contract that should have been written is either completely executory or completely executed, the courts often will not consider enforcing it. The parties, generally, are left where they are found.

DAVIS v. CROWN CENTRAL PETROLEUM CORPORATION
483 F.2d 1014 (4th Cir. 1973)

These two cases, combined on hearing because of their similarity of issues, are unfortunate by-products of the current oil shortage. Both plaintiffs and defendant are victims of the shortage in one form or another. They may properly be termed independents in their own particular type of operation. The plaintiffs, both citizens of North Carolina, on the one hand, are small independent oil dealers with their main operations in North Carolina. The defendant, on the other hand, is a refiner, dependent almost entirely on producers for its supply of crude oil. It has for some years been selling its product to the plaintiffs. In anticipation of the oil shortage, which all parties in the industry apparently foresaw, discussions were had among the parties as to future supplies. It is contended by the plaintiffs that the defendant agreed to supply them with certain fixed quantities of gasoline. As the energy crisis deepened, the suppliers of the defendant reduced drastically its supply of crude oil. It accordingly proceeded to allocate on a lower percentage its deliveries to its contract customers and to notify customers such as the plaintiffs, whom it denominated noncontract customers, that it would make no further sales to them. These actions followed that notification. The plaintiff Davis filed his action originally in the Western District of North Carolina, seeking injunctive relief against what he claimed was a breach of contract involving irreparable injury and damages for violation of the Sherman Act. Jurisdiction was based on diversity of citizenship and federal question. The other action was first instituted in the State Court and removed to the District Court for the Middle District of North Carolina. The plaintiff in this action sought similar relief to that demanded in Davis, and federal jurisdiction was predicated on similar grounds.[2] Preliminary injunctive relief was sought in both cases by the plaintiffs and were granted in both instances by the District Court having jurisdiction of the actions. It is from these grants of injunctive relief that the defendant has appealed in each case. Since there is similarity both of facts and issues in both cases, we ordered the appeals argued together and we shall dispose of both together. We reverse in both cases.

In granting temporary injunctive relief, the District Court in each instance made specific Findings of Fact. The defendant contends that these Findings show on their face that the temporary injunctive relief was improvidently granted. Specifically, it attacks the Finding made by the District Court in each case that the plaintiffs were likely to prevail eventually on the merits. Such a finding is necessary for the granting of preliminary injunctive relief. . . . This basic contention presents the issue on appeal.

In support of its conclusion that the plaintiffs were likely to prevail, the District Court in each case found that there was an oral agreement between the plaintiff and the defendant whereby the defendant was to supply the plaintiffs with their gasoline requirements. . . .

In the Davis case, the District Court found there was an agreement between the plaintiff and the defendant "for the purchase of goods vastly exceeding $500.00," which was "not written nor evidenced by any writing" and which, though obligating defendant to supply the plaintiff as much 1,200,000 gallons per month, did not obligate the plaintiff "to purchase that quantity nor

---

2. The state court clearly lacked jurisdiction of the action under the Sherman Act but, whether jurisdiction existed after and as result of removal to Federal Court, *see Freeman v. Bee Machine Co.*

any set quantity of gasoline." Although it recognized that the defendant was suffering hardship too, on account of the oil crisis, the District Court found that "(T)he balance of hardship favors the relief sought by the plaintiff" and "(T)hat the public interest will not be harmed by the issuance of an injunction." It accordingly ordered the defendant to "continue to supply plaintiff with at least 300,000 gallons of gasoline products per month . . . pending further orders of this Court. . . ."

The defendant denied in both cases the existence of any valid agreement. In support of this position, it asserts the alleged agreements are lacking in mutuality and definiteness of duration. Moreover, it argues that if there were any agreement, it was void as violative of the controlling Statute of Frauds. It is conceded by all parties that the North Carolina law is controlling on these several contentions.

We find it necessary to consider only the contention based on the Statute of Frauds. The Findings of Fact in each case make it clear that any agreement between the plaintiffs and the defendant was within the North Carolina Statute of Frauds. That statute, unlike its counterpart in other jurisdictions where its scope is merely remedial, "affects the substance as well as the remedy." . . . On the basis of the present Findings of Fact in each case, the statute would be a complete bar to a ruling that there was a valid contract or agreement between the plaintiffs, on the one hand, and the defendant, on the other. It is true, as the plaintiffs have argued, that in exceptional cases, courts of equity will find an estoppel against the enforcement of the statute, but such an estoppel can arise in North Carolina only "upon grounds of fraud" on the part of him who relies on the statute. . . . The District Court in neither case made a finding of "fraud" on the part of the defendant and, absent such finding, there can be no estoppel. Nor, on the facts in the record before us, would it appear that any such finding would have been in order. In *United Merchants & Mfrs.* v. *South Carolina El. & G. Co.* . . . it was stated:

> "A mere failure or refusal to perform an oral contract, within the statute, is not such fraud, within the meaning of this rule, as will take the case out of the operation of the statute, and this is ordinarily true even though the other party has changed his position to his injury."

After all, as the District Court in one of the cases observed, the plight of the defendant was not substantially different from that of the plaintiffs. It was experiencing hardships which forced it to take action it obviously did not relish.

The claim of the plaintiffs is appealing and our sympathies are with them. As the District Courts indicated, it is equitable in periods of scarcity of basic materials for the Government to inaugurate a program of mandatory allocations of the materials. This, however, is a power to be exercised by the legislative branch of Government. The power of the Court extends only to the enforcement of valid contracts and does not comprehend the power to make mandatory allocations of scarce products on the basis of any consideration of the public interest. Public reports indicate that Congress is cognizant of the problem presented by these actions and is giving active consideration to the establishment of a program of mandatory allocations that would apply from the oil producer down to the oil retailer. It is earnestly hoped that Congress will take the necessary steps to establish such controls. For that control, however, the parties must look to the Congress and not to the Courts.

The District Courts were in clear error in finding on the record before them that there was the reasonable likelihood that the plaintiffs would prevail on the merits. For that reason, the granting of injunctive relief during the pendency of the actions was improper in both cases and the injunctions are hereby vacated.

Reversed.

McCOLLUM v. BENNETT
424 N.E.2d 90 (Ill. App. Ct. 1981)

This is an appeal from a judgment of the circuit court of Peoria County in favor of the defendants, Harvey and Catherine Bennett, and against the plaintiffs, Robert and Mona McCollum. On January 31, 1976, the parties executed an agreement, the pertinent part of which reads: "We, Harvey M. & Catherine A. Bennett agree to sell said Business known as Harvey's Towing and Used Auto Parts & Property located at 1601 S.W. Adams St." The crux of the issue in the instant case is the meaning of the aforementioned passage. Plaintiffs contend that the passage refers not only to the business and personal property located at the address mentioned, but to the real property there as well. Defendants agree that the passage refers to the business and personal property but contend that it does not refer to real property.

In March 1976 the plaintiffs filed suit for specific performance of the agreement. They then filed an amended complaint requesting instead damages for breach of the agreement. The defendants filed an answer to the amended complaint admitting they operated the towing business, that the parties had executed the agreement, and that the agreement states in part that the defendant agreed to sell the business and property located at 1601 S.W. Adams Street. Defendants also asserted their readiness to perform the terms of the agreement. In response to the plaintiffs' request for admission of facts, defendants admitted that the lot at that address matched the legal address listed in the request for admission.

In May 1980 plaintiffs filed a second amended complaint. The second amended complaint alleged that the agreement was for the sale of real property as well as personal property. Defendants filed a motion to dismiss, raising the issues of the Statute of Frauds and election of remedy. The trial judge granted defendants' motion to dismiss the complaint with prejudice for the reason that the agreement sued on "is in violation of the Statute of Frauds in that the alleged real estate to be sold is not described or mentioned in the contract, and further that the contract does not contain the essential terms and conditions of the alleged agreement upon which the claim for damage is based."

On appeal, plaintiffs raise three issues: (1) whether the agreement violated the Statute of Frauds; (2) whether the plaintiffs waived their right to assert the Statute of Frauds; and (3) whether the doctrine of election of remedy precludes plaintiffs from amending their complaint from specific performance to recovery of damages.

We affirm.

We initially deal with whether the agreement violated the Statute of Frauds. We believe the trial court's holding that the Statute of Frauds was violated was correct. There is nothing in the terms of the written agreement that indicates an agreement that the defendants were selling the real property located 1601 S.W. Adams Street. It merely recites that the business and property located at that address are to be sold by the defendants and purchased by the plaintiffs. As the trial court stated, the real property to be sold is not mentioned in the written agreement, nor are any of the terms and conditions mentioned. Therefore, any agreement between the parties to sell the real property was oral and, therefore, unenforceable under the Statute of Frauds, unless this defense was waived by the defendants.

Therefore, we next examine plaintiff's second issue: whether the defendants waived the defense of the Statute of Frauds. We find no such waiver. The plaintiffs correctly state that the Statute of Frauds may be waived by an acknowledgment of the agreement and the subject matter thereof. However, in the instant case, while the defendants admit the existence of an agreement, they have steadfastly denied that the agreement referred to real property. Plaintiffs contend that the defendants' admission that the property 1601 S.W. Adams Street was the same as the legal description contained in plaintiffs' request for admission of facts is evidence of an agreement to sell the real property. We disagree. Simply admitting that the address of real property conforms to its legal description is inadequate to prove the existence of an agreement to sell that real property. In the absence of any evidence showing defendants acknowledged an agreement to sell the real property, we find no waiver of the defense of the Statute of Frauds.

Because we hold that any agreement to sell the real property is unenforceable under the Statute of Frauds, we need not reach plaintiffs' final issue. Accordingly, the judgment of the trial court of Peoria County is affirmed.

Affirmed.

REVIEW QUESTIONS

1. Why is a written contract better than an oral one?
2. Name three ways in which the seventeenth section of the English Statute of Frauds differs from the fourth section.
3. What is the difference between surety and insurance or indemnity?
4. Are the following things real property or personal property?
   a. A gas-operated water heater.
   b. A bird bath.
   c. Flowers and shrubs in a flower bed.
   d. A television aerial mounted on a roof.
   e. A window-mounted air conditioner.
5. Black orally contracts with White to build a brick garage for Black for $5000. When the garage is finished Black refuses to pay the price, claiming that the garage is an addition to real property and, therefore, should have been in writing to be enforceable. Can White collect? Why or why not?
6. Green orally contracts with the Gray Die Shop to build a punch press die for $2,000. The production for which Green was going to use the die is canceled and Green refuses to accept and pay for the die. Gray claims that the contract was for a service and, thus, Gray is entitled to payment. Green points out that dies are the usual product of the shop and, therefore, that the contract had to be in writing to be enforceable. Who is right? Why?
7. In *Davis* v. *Crown Central Petroleum Corporation,* suppose Crown Central had arranged binding contracts with all its outlets to supply them based on their "needs." What recourse would it have if its supply of crude oil were suddenly cut in half?
8. In the case of *McCollum* v. *Bennett,* the statement in question refers to " . . . Business known as Harvey's Towing and Used Auto Parts & Property located at 1601 S.W. Adams St." Would it have been unreasonable for the court to interpret this to mean all property at that address, both personal and real?

Chapter 12

Third-Party Rights

A contract is a voluntary, intentional, and personal relationship. The parties to the contract are the ones who determine the rights and obligations to be exchanged. Ordinarily, only those who are directly involved have rights stemming from the contract. In law, the relationship between parties to a contract is known as *privity of contract;* generally, only a person who is *in privity* with another may enforce any rights under the contract.

A strict interpretation of the privity of contract principle would prevent many common transactions of considerable value and convenience in our economy. Indeed, the early common law that developed in England during the middle ages refused to recognize attempts to assign a contract right. These rules developed when wealth was essentially either land or tangible items (such as horses, food, gold, or the like). In today's economy, wealth is instead embodied in intangibles like bank accounts, stocks and bonds, accounts receivable, and so forth. Therefore, the courts recognize three major exceptions:

1. Rights assigned by a party to a contract to a third party are enforceable by the third party.
2. Rights arising from third-party beneficiary contracts are enforceable by the beneficiary.
3. Rights arising under a manufacturer's warranty of its products are often enforceable by persons besides the one who actually bought the product.

ASSIGNMENT

Probably the most common form of assignment occurs when an indebtedness that is not yet due is assigned to another for value. Suppose White Company sells industrial machinery. In payment for some machinery, the White Company received $5,000 from Black Company on delivery of the machinery and also the promise of Black Company to pay another $10,000, six months after delivery. That promise to pay is often embodied in a separate written instrument called a *note*. If the White Company finds itself in need of funds, it may be able to sell this right to payment from Black to someone else (possibly a local bank). The bank would pay something less than face value (known as *discounting*) and then take over Black's note with the same rights White had.

An assignment involves at least three parties. The *obligor* (or debtor) is the party to the original contract who now finds that because of the assignment, the obligation is now owed to someone not in privity of contract with him. The *obligee* (or creditor, or assignor) is the person to whom the obligor originally owed the duty to perform. Now, because of the assignment, the right to that performance has been assigned to another. The *assignee* is the one to whom the duty has been assigned and is the person to whom the obligor now owes the duty of performance.

Most contract rights are assignable if there is no stipulation to the contrary in the contract. Many contracts state that the rights and obligations involved cannot be assigned by one party without the other party's consent. Generally, an assignment in violation of a contract provision renders the assignment voidable at the option of the obligor. Such antiassignment clauses are not necessarily enforceable. For example, the Uniform Commercial Code (UCC) states that a right to receive damages for breach of a contract for the sale of goods can be assigned despite an agreement to the contrary.

There are situations in which contract rights are not freely assignable (even absent a contract clause that prohibits an assignment). Generally, a contract right cannot be assigned if any of the following applies:

- The assignment would materially change the other party's duty.
- The assignment would materially increase the other party's burden or risk under the contract.
- The assignment would materially impair the other party's chance of obtaining performance of the contract in return.

What amounts to a "material" change or increased burden obviously involves many shades of gray and may vary from situation to situation.

Contract rights to personal services and services based on the skill of an individual are treated exceptionally; and it seems right that they should be. If Black contracts to perform personal services for White, it should not be possible to force Black into a choice of either performing those services for a third person (to whom White has assigned such rights) or breaching the contract. The same reasoning applies where the rights involved are in the nature of a trust or confidence. For example, suppose Black Machinery Co. obtains a license to certain patents and trade secrets of White. If the license agreement requires Black Machinery Co. to maintain all of the trade secrets in confidence (as such agreements usually do), is it fair to allow Black Machinery Co. to assign the license agreement to Green Co.? The answer will be "no" in many situations.

Assignment Parties

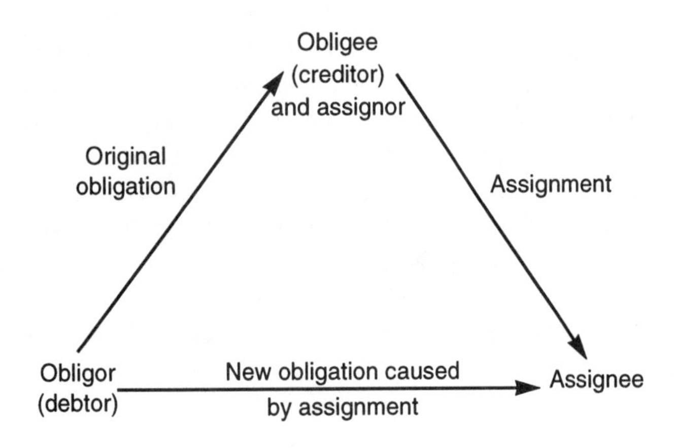

♦ Rights and Duties

If you can transfer your rights under a contract to someone else, can you also transfer your obligations or duties? The terminology used is important. Rights are assigned. Duties, however, are delegated. This distinction is helpful in considering the legal issues involved. (Unfortunately, not all lawyers and courts use these terms correctly.)

In contrast to rights, duties generally cannot be freely delegated (in the absence of the obligee's consent). At least, one cannot relieve himself of liability for nonperformance or poor performance by delegating the duties to someone else. Even if the duties are delegated to another, the person delegating the duties is still responsible for their performance. The delegating person is essentially a *guarantor* of the performance. Black hires White to build a structure. White subcontracts the plumbing to Gray. White's subcontract with Gray does not relieve White of White's responsibility for Black's structure, including the plumbing. But White is only responsible as a guarantor if White was the one to deal with Gray. White's responsibility is relieved if Black makes the plumbing arrangements with Gray.

Assignment and delegation differ from *novation.* In a novation, both rights and duties are effectively transferred. The distinction depends on the number of parties involved. If all three parties agree to the substitution of one party for another, the effect is a new agreement by means of a novation. Assignment of rights requires only an agreement between assignor and assignee. The obligor or debtor must, of course, be informed of the assignment if it is to be binding on the obligor. Also, the obligation must be made no greater or more burdensome as a result of the assignment—the obligor may not be forced to do more than the obligor originally agreed to do.

♦ Assignee's Rights

The assignee (the third party) acquires no better right than the assignor had. The assignee's right to performance is subject to any fault the obligor could have found with the right in the hands of the original obligee. In other words, if there is a defense available to the obligor (the obligee's fraud or misrepresentation, for instance) the obligor may use this defense against the assignee's claim just as the obligor could have against the original obligee. If the obligor exercises such a claim, the assignee is left with only an action against the assignor (who is also the obligee) for whatever consideration the assignor has given up. If the obligor has made partial payment to the assignor or has a counterclaim against the assignor, the right obtained by the assignee may be subject to these claims.

Black has a $6,000 claim against White for installation of a machine. Black assigns this claim to Gray in return for operating cash. If the installation proves to be substantially less than a proper performance, requiring White to spend $1,000 more to put it in operating condition, this amount might be used as a set-off against the $6,000 claim. As a result, Gray would get only $5,000 from White and have to look to Black for the other $1,000.

An assignee may have rights against the assignor for a breach of any express (or implied) warranties made by the assignor. Generally, there are several implied warranties in an assignment. The assignor *warrants* (guarantees, according to common parlance) that there is a valid claim, that the assignor has a right to assign it, that there are no defenses available against it except as noted in the assignment, and that any document delivered is genuine and not a forgery. If any of these is not as warranted, the assignee has a right of action against the assignor. However, the assignor does not warrant that the obligor will be able to perform. If the obligor becomes bankrupt, for instance, the assignee may have to settle for considerably less than the amount bargained for. The assignee takes the same risk here as taken in any claim where the assignee becomes the obligee.

A *gratuitous assignment* creates no rights. If an assignment is made gratuitously (the assignor getting nothing back for it), the assignee gets nothing but a promise of a future gift. Such a promise is unenforceable. Until the assignee actually gets something, the gratuitous assignment may be recalled and avoided at will by the assignor.

♦ Notice

A debtor cannot be charged with nonpayment of the debt if the debtor innocently pays a creditor who has assigned the obligation. To make the assignment legally effective against the debtor, there must be *notification*. When an assignment of rights is made, the assignee obtains, along with the rights, a practical duty to notify the obligor of the assignment. If such notice is not given and the obligor pays the debt to the original creditor, the obligor's debt is discharged. The assignee is then left with an action for recovery against the assignor. If notice has been given to the obligor and the obligor still pays the assignor, such payment does not discharge the obligor's debt.

It is possible, though unlawful, to make subsequent assignments of the same right. The results for the respective assignees are somewhat controversial. Some court holdings are based on the argument that after the first assignment the assignor had nothing more to assign, so the assignee received nothing. Consequently, subsequent assignments are null and void. Another view is taken by other courts. According to them, the first assignee to notify the obligor has priority as to the claim. The argument here is that only proper notification completes the assignment. The losers, in either instance, have only their actions against the assignor (who may have since disappeared). Some assignments of contract rights are subject to statutory filing requirement. For example, the UCC provides that, as between two assignees (each of whom paid value for the assigned rights) of the same rights, the first who files the appropriate notice with the appropriate county clerk or state agency (as provided for by the statute) is the prevailing party.

♦ Wages

The assignment of future earnings for a present debt is governed by statute in most states. These statutes vary considerably, some making such assignments void. Many set a maximum amount (for example, 25 percent) that may be assigned from expected wages. In states where wage assignments are upheld, the right to lawfully assign future wages usually depends on present employment. One must be presently employed to make an effective assignment of expected wages. Generally, if a person is employed, that person can assign his or her wages, even though it might be argued that the employment may be terminated at the option of the employer. However, an unemployed person cannot assign expected wages from hoped for employment, even though the hopes may be well founded.

THIRD-PARTY BENEFICIARY CONTRACTS

If Black and White make a contract that will benefit Gray, what right does Gray have? The nature of the contract, the intent of the parties, and prior indebtedness existing between them all have bearing on Gray's rights. Courts are not in complete agreement in their holdings in such cases. Given the same set of facts a case may be decided quite differently in two different jurisdictions. Only in insurance contracts is the law regarding third-party beneficiaries likely to be applied uniformly.

♦ Insurance

Probably the most common type of third-party beneficiary contract is the life insurance contract. There are two possibilities relating to the beneficiary's right to recover following death of the insured. First, if the insured has purchased insurance from an insurer, the insured has the right to name anyone as beneficiary. When the insured dies, the beneficiary has a right of action to collect the amount of the policy from the insurer. White contracts with the Black Insurance Co. to insure White's own life, agreeing to pay the stipulated annual premiums, and naming Gray as beneficiary. When White dies, Gray can collect.

The second situation differs from the first in that it is the beneficiary who purchases the insurance. Gray contracts with the Black Insurance Company to insure White's life. Gray agrees to pay the annual premiums and is to be beneficiary. For Gray's right as beneficiary to exist, Gray must be able to show an insurable interest in White's life; Gray must risk a loss that would occur on White's death. In other words, there must be some anticipated benefit to Gray resulting from White's continued existence. An *insurable interest* is present where the insured is a member of the beneficiary's immediate family. It does not extend to outsiders unless there is an economic tie involved, as between creditor and debtor or partners in an enterprise. Sometimes a business will obtain "key man" insurance for its president or other key employees. Presumably, if such employees are key to the business' continued existence, then the business should have an insurable interest.

An insurable interest in property extends to those who have an ownership or lien interest in the property or merely possession of it with the attendant risk of loss. Both mortgagor and mortgagee, for instance, have an insurable interest in real estate being purchased under a mortgage contract. There is a limit to the insurable interest created by an economic tie, though; usually the courts consider the justification for insurance to be limited to the amount of economic benefit expected or the risk of loss involved.

The time when the insurable interest must exist differs in the two types of insurance. In life insurance it must be shown to have existed at the time when the insurance contract was made. A later change in the relationship does not serve to terminate the insurance. In property insurance, however, the insurable interest often must exist when the loss is suffered.

♦ Donee Beneficiary

An unsealed promise of a future gift is generally worthless as far as enforceability is concerned. That is, it is worthless if the donee tries to enforce it against the donor—as a contract it is unenforceable for lack of consideration. A donee beneficiary has a right to enforce a contract created to benefit that person. Gray is a third-party beneficiary if Black and White agree that White will pay Gray $500. Gray has an enforceable legal claim against White. However, Gray would not have a right of action against Black, who is Gray's benefactor.

♦ Creditor Beneficiary

Assume that Gray is to receive $500 in payment of an obligation Black owes to Gray. If Black and White make a contract whereby Black's consideration is to be paid for by White's paying $500 to Gray, Gray is a *creditor beneficiary* of that agreement. As such, Gray may enforce the claim against either Black or White. As a creditor beneficiary, Gray is in a stronger position than as a donee beneficiary. Gray may collect from either the promisor or the promisee. Gray can sue the promisee on the previous debt or Gray can elect to force the promisor to perform.

♦ Incidental Beneficiary

Many contracts are made that indirectly benefit third parties. Failure of the promisor to perform properly does not necessarily give an *incidental beneficiary* the right to take legal action. Generally, if there was no intent by the contracting parties to benefit the third party, the third party has no legally enforceable interest in the contract. Brown hires White (a landscape architect) to landscape Brown's estate. Gray, who lives next door, will incidentally benefit as a result of the landscaping. However, Gray has no legally enforceable interest in the completion of the landscaping. Gray was not a party to the contract, and whatever benefit Gray might have received was merely incidental to the contract's primary purpose.

Now suppose Brown is a manufacturer and supplier of parts for the automotive industry. Brown contracts with Green, a builder of automation equipment, for Green to build and install an automatic machine. Gray supplies most of Green's steel. Although Gray would be likely to benefit from the automation contract, the agreement is not for Gray's benefit, and Gray could not enforce it.

Contracts in which the government is a party are sometimes considered to be third-party beneficiary contracts. This is the case when it is reasoned that citizens are to benefit from contracts made by the governing body, and that they have an interest in these contracts because they pay taxes and thereby acquire rights to benefits. If a small segment of a community is to receive benefits and possibly pay a special assessment toward them, the argument for enforcement by a citizen is even stronger. However, the right of a citizen to take action as a third party beneficiary to enforce a government-made contract is largely a matter of local statute.

A more recent rule reformulates the older categories of donee beneficiaries, creditor beneficiaries, and incidental beneficiaries. Under this newer rule, "intended beneficiaries" may enforce the contract. To be an intended beneficiary, the third party must show that recognizing a right to performance is appropriate to give effect to the parties' intentions and either (a) the performance of the promise will satisfy an obligation of the promisee to the beneficiary or (b) the promisee intended to give the third party the benefit of the promised performance.

WARRANTIES

Traditionally, any promises or warranties made by one party to a contract benefitted only the other party and persons in privity with that party. However, the requirement of privity has eroded over time. Today, a seller's warranties about the goods sold often flow to persons besides the purchaser of the goods. The courts recognize that a number of people will use a product like a car. Essentially, the courts have allowed persons besides the buyer to sue for injuries received due to a breach of warranty. These concepts are discussed in greater detail in the chapter on Product Liability.

The Uniform Commercial Code adopted three alternatives that govern the "flowdown" of the warranties provided by sellers of goods:

1. The seller's warranty extends to any person in the family or household of the buyer or who is a guest if it is reasonable to expect that such a person may be affected by the goods and that person suffers personal injuries due to a breach of the warranty.
2. The second alternative expands the first to include all persons who may reasonably be expected to be affected by the goods and who are personally injured due to a breach.
3. The third alternative extends the second to include persons who may reasonably be expected to be affected by the goods and who are "injured" by a breach (that is, the injury need not be a personal injury). Different states have adopted different alternatives, thus leading to a variety of rules.

FRAZIER, INC. v. 20TH CENTURY BUILDERS, INC.
198 N.W.2d 478 (Neb. 1972)

This appeal is from the sustaining of a motion for summary judgment in a third party action. The principal action was a mechanic's lien foreclosure brought by Frazier, Inc., against Majors, Inc., et al., in which Omaha Poured Concrete Company was subsequently included as a defendant. Majors filed a cross-petition against Frazier and Omaha Poured, alleging damages as the result of the negligence of Frazier in the installation of the plumbing system that resulted in the collapse of a cement floor poured by Omaha Poured. Omaha Poured was given leave to file a third party complaint against Transamerica Insurance Company as a third party defendant, to recover under the terms of an insurance policy providing coverage to Omaha Poured for consequential damages. We affirm.

Frazier, as a subcontractor under Omaha Poured, installed the plumbing in a building being erected for Majors. The plumbing was installed with removable sleeves around the floor drain to permit the installation of a poured concrete floor. After the completion of the building, Majors moved its plastic moulding machines onto the concrete floor and started operations. These machines required the discharge of large quantities of water through the drain. After the first day's operation, the concrete floor pulled away from the wall, buckled, and collapsed. It was then discovered that the sleeves had not been properly installed, and after they were removed the water, instead of running off through the drain, went through an unsealed gap into the ground under the floor, undermining it, causing the floor to sink. Notice was immediately given to Transamerica who investigated the damage. Omaha Poured replaced the floor at its own expense. The present third party action involves the amount paid to settle the consequential damages sustained by Majors because of the collapse of the floor, and does not include any part of the cost of replacing the floor.

Omaha Poured was granted leave to file its third party action against Transamerica January 22, 1969. The petition filed January 30, 1969, alleged that Transamerica is obligated to defend Omaha Poured, and to pay any sums of money that may be awarded against Omaha Poured as the result of consequential damages sustained by Majors. Omaha Poured alleged it complied with all of the conditions precedent of said policy, but Transamerica refused to provide coverage for said incident which falls within the terms and conditions of the policy, and has refused to defend Omaha Poured. On June 5, 1969, the issues raised by the third party petition were ordered separated for trial purposes from the issues raised by the principal pleadings.

On July 14, 1969, the day the main action was called for trial, Omaha Poured made a settlement with the other defendants, and judgment was entered against it for Majors' consequential damages in the amount of $14,298.22. The agreement included a covenant that the judgment creditor would not execute against the personal assets of Omaha Poured for more than $1,500 of the judgment, and would thereafter look to the Transamerica policy for all remaining sums thereunder.

Immediately previous, and on the day judgment was entered, Majors amended its cross-petition to include a special allegation of negligence against Omaha Poured. No notice was given Transamerica of this amendment, but a copy of the judgment entry was mailed to it the same day the judgment was entered.

The trial court sustained Omaha Poured's motion for summary judgment, specifically finding that there was no substantial controversy as to the following material facts: (1) Transamerica knew of the collapse of the floor shortly after it occurred and knew that the potential loss and claim by Majors, Inc., would include not only damage to the floor itself but the very consequential damages which Majors, Inc., ultimately sought by the instant litigation. (2) Pursuant to investigation by representatives of Transamerica, it unequivocally and unconditionally denied any coverage on the policy of insurance for the incident involved herein. (3) Transamerica refused to accede to the demand of Omaha Poured that it take over and defend the crosspetition filed against Omaha Poured by Majors, Inc. (4) Omaha Poured, without the knowledge or consent of Transamerica, entered into a compromise settlement and agreement with Majors, Inc., resulting in a consent judgment against Omaha Poured for the consequential damage sustained by Majors, Inc., with an agreement that no more than $1,500 of the judgment would be collected from Omaha Poured personally. (5) The amendments to the crosspetition of Majors, Inc., just prior to the entry of the judgment were made without the knowledge or consent of Transamerica. (6) The evidence is sufficient to support a finding of legal liability by Omaha Poured to Majors, Inc., for the consequential damage loss to Majors, Inc., resulting from the collapse of the floor, in the amount of the judgment.

The trial court made the following specific findings of law: First, the policy in question affords coverage to Omaha Poured for the consequential damage claim asserted by Majors, Inc. Exclusions raised by Transamerica in support of its denial of coverage do not apply. Second, once Transamerica denied coverage to Omaha Poured in the manner in which it did, Transamerica breached its contract with Omaha Poured, relieving Omaha Poured of any obligation under the contract to notify or deal with Transamerica any further regarding the claim.

Third, the allegations contained in the cross-petition of Majors, Inc., prior to the amendments made on July 14, 1969, together with all the pleadings in the case,were sufficient to compel Transamerica to assume the defense of Omaha Poured, and its refusal of Omaha Poured's demand that it do so constituted a breach of its contract with Omaha Poured.

Fourth, Transamerica's breach of its contract with Omaha Poured, by denying coverage and refusing to defend its insured, entitled Omaha Poured to enter into the most favorable settlement and consent judgment possible, which it did.

Fifth, the consent judgment, supported by the evidence, was obtained in good faith and without fraud or collusion. Therefore, having declined to defend the action when called upon to do so, Transamerica may not again litigate the issues that resulted in the judgment.

Only one of the court's findings of fact set out above is questioned in Transamerica's brief, that being No. (2). . . . Dwight Stephens, Transamerica's claim representative who first investigated the claim, stated in his affidavit that after completing his investigation he informed Omaha Poured's representative that no coverage was afforded to the insured for the collapse of the floor for the reason that the aforementioned policy did not provide coverage for completed operations and products' liability, and that he thereafter reported his findings to his superior, Harold Pace. The affidavit of Harold Pace states that coverage was denied under said policy shortly after May 27, 1967, for the reason that coverage was not afforded thereunder to the insured for completed operations and products liability coverage. There can be no question that Transamerica at all times was insisting that the damage was for completed operations and products liability coverage and it was not covered under the policy. Transamerica overlooked the fact that its policy did insure Omaha Poured for consequential damages and that the action brought by Majors, Inc., was for consequential damages resulting from the incident. Nowhere in Transamerica's brief does it challenge, by assignment of error or discussion, the finding of the trial court that the Transamerica policy in question affords coverage to Omaha Poured for the consequential damage claim asserted by Majors, Inc. Nor, incidentally, does Transamerica challenge in any manner the trial court's conclusion that as a matter of law the "exclusions raised by third party defendant in support of its denial of coverage" do not apply.

♦♦♦

As to Transamerica's third assignment of error, when we pierce the allegations of the pleadings there is no question the material issues are those of law and not of fact.

As to Transamerica's fourth assignment of error, there is not the slightest evidence of fraud or collusion. The settlement was consummated on the morning of the scheduled trial of the main action. Omaha Poured, left to its own resources, at the last moment made a settlement to avoid increased liability. Omaha Poured must have realized, as would any reasonable person in its position, that prudence required it to protect its assets as best it could. There was no question about its liability for consequential damages. There is not the slightest merit to this assignment.

Nor is there merit in Transamerica's fifth assignment of error. Transamerica investigated the incident and denied coverage. It further refused to tender a defense for Omaha Poured. Exhibit 84 to the deposition of Transamerica's claim manager, is a letter from attorneys for Omaha Poured to Transamerica under date of January 10, 1969, reminding it of two previous denials of coverage and of a previous letter dated August 31, 1967, to the effect that Omaha Poured was taking steps to protect itself from the effects of the wrongful denial of coverage. The letter of January 10, 1969, gave Transamerica the trial date of the main action, again tendered the defense of Omaha Poured, and advised that in the event it again declined to provide a defense, a third party petition would be filed. To hold that Omaha Poured had any additional obligation to keep Transamerica informed of the developments would be absurd. Transamerica's sixth assignment of error misconstrues the decisions of this court on the point in question. In *National Union Fire Ins. Co.* v. *Bruecks, . . .* we said: "It would seem that the obligation to defend a suit for an insured should be determined on the basis of whether the petition filed against him attempts to allege a liability within the terms of the policy." Clearly, the inference from the crosspetition and the prayer before the amendment included Omaha Poured, and sufficiently inferred liability to require action on the part of Transamerica. If there was the slightest question, certainly the proceedings culminating in the filing of this third party action, which occurred six months before the amendment in question, would have removed any doubt. Transamerica's seventh assignment of error is predicated on its assertion that affidavits offered into evidence, notarized by an attorney of record, are improper and should be excluded. We direct counsels' attention to section 25-1245. R.S. Supp., 1969, which provides: "An affidavit may be made in and out of this state before any person authorized to take depositions, and must be authenticated in the same way. An attorney at law who is attorney for a party in any proceedings in any court of this state shall not be disqualified as the person before whom the affidavit is made by reason of such representation."

In view of previous comment, Transamerica's eighth assignment of error requires no further discussion. The trial court correctly sustained the motion for summary judgment. The judgment is affirmed.

Affirmed.

REVIEW QUESTIONS

1. What is meant by privity of contract?
2. Black is purchasing a house under a real estate mortgage held by the White Mortgage Company. Under the terms of the mortgage, Black is obligated to make a payment to White Mortgage Company each month. Part of the payment is for interest, another part reduces the mortgage balance, and a third part is deposited in an escrow account to pay for taxes and insurance. Black sells his house to Gray, with Gray agreeing to take over the mortgage payments. If Gray defaults, can White Mortgage Co. take action to recover from Black?
3. Why must the obligor be notified of an assignment?
4. What is meant by insurable interest?
5. In the case of *Frazier, Inc.* v. *20th Century Builders, Inc.,* if the insurer reimbursed the owner for the damages caused by collapse of the floor, would the insurer then have a right to take action for reimbursement against the subcontractor, Frazier, Inc., based on Frazier's alleged negligence? Why did Majors amend its cross-petition just before the entry of the agreed judgment to include an allegation of negligence on the part of Omaha Poured?

Chapter 13

Performance, Excuse of Performance, and Breach

Contracts, as you have observed, are made every day. They are quite ordinary agreements. The terms of the agreements are usually carried out with no dispute. When you complete your obligations under a contract (such as by paying the purchase price of this book), you have performed under the contract. This chapter deals with the exceptional situations, where a party has bargained away more than intended, or where performance became more difficult than expected. Public attention is called to the exceptional cases; they are publicized in court cases and newspaper articles. It must be kept in mind, however, that these cases are the exception to the general rule, and that the vast majority of contracts are performed satisfactorily and with benefit to both parties.

Chapter 13 shows how the courts interpret contracts. Many contracts are clearly worded and the meaning of a contract is not disputed. In this chapter, we shall see how the parties are held to their bargain, what constitutes performance, or a real offer to perform. We shall see when the law will discharge a person's obligations or, in effect, excuse that person's performance. Then, in Chapter 14, we shall examine the remedies available if a contract is breached by one of the parties to it.

PERFORMANCE

Theoretically, a party to whom a contractual obligation is owed has a right to precise performance of the obligation. Anything short of that performance constitutes a breach of the agreement. However, performance or nonperformance is not always as self-evident as it might appear. The nature of the obligation undertaken determines the required performance. As noted earlier in Chapter 9 regarding consideration, when a liquidated amount of money is owed, payment of a lesser amount will not satisfy the obligation. Where the contract requires the construction of an office building, though, exact performance may not always occur and cannot seriously be expected. This is particularly true where research and development is to be part of the performance of the contract. Such is the case in many engineering contracts.

♦ Conditions

Frequently contracts are written so that the apparent intent of the parties is that performance by one party will precede performance by the other. For instance, one might contract to have a machine built and installed in a plant with payment to be made when the installation is complete. A condition such as the installation would constitute a *condition precedent* in a contract. If the parties to a contract agree that, on the occurrence of some future event, an obligation will come into existence, the event is a condition precedent. One common type of condition precedent is obtaining an architect's certificate. The passage of time until the completion date is also often a condition precedent.

A frequent source of confusion is the distinction between a condition precedent and a condition subsequent. A *condition subsequent* is a condition that, when it occurs, ends an existing contract. Assume, for example, that White's house is insured by the Black Insurance Company. The insurance policy states that if the house is ever left vacant for a period of 30 days or more, the insurance will no longer be in effect. Such a conditional event is used as a condition subsequent.

♦ Architect's Approval

When an architect's approval is required as a condition precedent to the owner's duty to pay the contractor, the courts usually enforce the contractual requirement. It must be recognized, however, that architects are human too, and they are capable of human failings. The architect may have died or may be insane when the certificate is required, or the architect may unreasonably or fraudulently refuse to issue the certificate. Under such extreme circumstances the court ordinarily dispenses with the requirement of the architect's certificate. If there is any sound reason for the architect's objections, though, the court will probably enforce the requirement.

♦ Satisfactory Performance

Where the purchaser of a certain performance must be satisfied, two possibilities exist. The nature of the contract may be such that the only test of satisfaction is the personal taste of the buyer. Black, an artist, agrees to paint White's portrait to White's satisfaction for $5,000. White may never be satisfied, even though to a third person the portrait appears to be a perfect likeness. White could conceivably state, after each submission of the portrait for approval, that "it just isn't me" and require Black to continue. If the court determined the contract to be binding (the consideration of Black's performance might be termed illusory), Black would have to continue to paint or else breach the contract.

The nature of the contract may be such that, even though satisfaction of the buyer is specified, it provides for mechanical or operational suitability. In this case, if performance is such that it would satisfy a reasonable person, the court will deem the condition satisfied. Black agrees to install an air conditioning system for White, to White's satisfaction. After the installation, if White is not satisfied, Black may still be able to collect by proving that the air conditioning system will do all that a reasonable person could expect it to do.

♦ Completion Date

Time limits are frequently stated in contracts. Where performance is to take place by a certain date and it actually extends beyond that date, any of several situations may result. If the time for performance is not important, there may not even be a breach of the agreement. Generally, time is important. Thus, many contracts include provisions for liquidated damages (sometimes called *penalty clauses*) that state an amount to be paid for each day's delay in performance beyond the date specified in the contract. If the amount specified is a reasonable estimate of the difficult-to-calculate damages that would be caused by delay, the provision should be enforced.

If the specified amount of the liquidated damages is excessive, the court may hold that the clause is really a penalty clause and will refuse to enforce it. If there is no liquidated damages clause and performance runs beyond the time agreed on, the party aggrieved can sue the other party for damages due to the delay. Usually, either actual or liquidated damages merely serve to reduce the price paid for the work. Ordinarily, nothing but an unreasonably late performance gives the buyer sufficient grounds to terminate the contract and pay nothing. Where performance is not unreasonable, the buyer must accept the performance, but at a price reduced by the damages suffered.

In certain circumstances, the time of performance is critical in a contract. In this event, a clause to the effect that "time is of the essence of the contract" is usually included. In such instances the courts generally will allow termination of the contract for late performance.

♦ Substantial Performance

Many building contracts or contracts in which machinery or equipment is to be built and installed are performed in a manner that deviates from the exact terms of what is specified. Some flaw in such a project can be found by looking long enough and thoroughly enough. If variations from specifications were not allowed, few projects would be undertaken and fewer yet would be paid for. Those that were completed would cost so dearly that very few could afford them. As it is, if a building contract calls for a concrete floor of a uniform five-inch thickness, it is not likely that the resulting floor will be precisely five inches thick throughout. However, if the result accomplished approximates closely the result specified, the contractor can still recover the contract price; the performance has been *substantially* the same as that specified.

If gross inaccuracies occur or substitutions are found to have been made that do not reasonably satisfy the specifications, the claim is no longer *substantial performance*. Just where the line is drawn between substantial performance and an outright breach is a question of fact, and is usually submitted to a jury.

Cases involving substantial performance frequently result in the performer receiving the amount for which the parties contracted, less the other party's damages. The extent of damages is usually determined by the value of the performance as rendered compared with the value of the performance as specified, or the cost of additional work to complete the performance properly.

Substantial performance assumes that the performer has performed the contract in good faith and with the intent that the performance would agree with the standards specified. If this is not the case—if the performer willfully abandons the performance or the work is unreasonably poor—then there has been a breach of the contract and not a substantial performance.

♦ Impossibility of Performance

As it deals with impossibility of performance, the law is not entirely clear or settled. In certain cases involving death or illness of the promisor, definite statements can be made. In other cases, particularly where performance has turned out to be much more difficult than was anticipated, general conclusions can be drawn from the majority of decisions, but numerous exceptions may also be cited.

Generally, when a person undertakes a contractual obligation, that person assumes certain risks. One can hedge against such risks in various ways. Contract clauses can be used that state that, if certain things occur, you will be excused from performance. A contractor can add enough money to the price to cover "contingencies." You can purchase insurance against the risks you assume. In fact, almost the only hazard you cannot cover in some way is liability for your performance if the public is injured by it. For example, injuries resulting from the collapse of a public building caused by the contractor's negligence could allow recovery against the contractor by either the injured person or the insurance company. Presumably, the contractor will have its own liability insurance to cover such claims.

Death or Illness

Ordinarily, the death or illness of a party to a contract does not discharge that contract. If a person dies or becomes incapacitated by illness, that person's estate or those appointed to act for that person must take over and complete the obligations. Only where a contract is such that personal services are involved will death or incapacity due to illness serve as a lawful excuse for nonperformance. For instance, death or illness of a free-lance consulting engineer would discharge the engineer's remaining obligations to her clients.

Destruction of an Essential to the Contract

Destruction of an essential to a contract means that something is destroyed without which the contract cannot be performed. Black hires the White Construction Company to build an addition to Black's plant. Before the work is begun, the plant is destroyed by fire without fault of either party. White Construction Company's obligation is terminated. If the White Construction Company had begun work and were, say, half finished, White's obligation to complete the structure would be ended, but White could collect for the work the company had completed in addition to any materials that had been accepted by Black. If, as a third possibility, the contract had been for the building of a structure by itself (not an addition to an existing structure), and if the work again were half finished when the building was destroyed, White Construction Company's obligation would not be ended and the contractor would have to rebuild. One cannot make an addition to a structure that no longer exists, but you can build a separate structure even though the first attempt to do so was destroyed.

Unexpected Hardship

As previously indicated, one who contracts to perform in some way runs the risk that conditions may not remain as they are when the contract is entered into and that conditions may not be as they seem. It is not an uncommon experience to have a materials price increase or a wage increase cut deeply into the profit margin. If the cause of hardship is anything the contractor reasonably could have anticipated, the courts generally will not relieve the contractor of the contractual duties. It is only where the difficulties that have arisen are of a nature such that no one reasonably could have anticipated them that the court may, in some way, either relieve the burden or lighten the load on the contractor. Under such circumstances a subsequent contract with the owner, whereby the contractor is to receive additional money for performance, may be enforceable in court. Or the court may enforce a subsequent contract to give the contractor more time in which to perform. It should be remembered, though, that these rulings are exceptions; if the difficulty was foreseeable the law often gives no relief. A subsequent contract based on a foreseeable difficulty that did arise (for example, a materials price rise) may be void for lack of consideration as to the increase in the contract price.

Commercial Frustration

The doctrine of commercial frustration is often treated in the same manner as the destruction of an essential to a contract. In the United States courts, the result is the same. *Commercial frustration* commonly results when a contract is made to take advantage of some future event not controlled by either party. The event is then called off and, as a result, the purpose of the contract is said to have been frustrated. Black leases a concession stand from White for a certain week during which an athletic event is to be held. The athletic event is called off (or moved to a different location). The courts would allow Black to avoid the lease contract. Similarly, if a law were passed preventing such an event, Black probably would not be held to the lease.

♦ Prevention of Performance

It is implied in every contract that each party will allow the other to perform his or her obligations. If one party prevents the other's performance of the contract, that party discharges the other party's obligation. The interfering party also has breached the contract and may be sued for damages. Black sells White some standing timber, giving White a license to use a private road to the timber. Black prevents White from using the road. White's obligation is terminated, and White may sue Black for damages.

♦ Waiver

A *waiver* consists of voluntarily giving up a right to which one is legally entitled. To waive a right, the party waiving it must first know that he or she is entitled to it. That person must also intend to give up the right. If Black purchases a machine from White according to a detailed specification, Black can expect to receive the described machine. If the actual machine received varies significantly from the description, Black has a right to refuse to accept it. If, with knowledge of the differences involved, Black agrees to accept it for a slightly lower price, then keeps the machine and uses it, Black probably has waived the right to return the machine to White and get one according to the specifications.

♦ Agreement

Renunciation

If two parties have a right to make a contract by an agreement between them, it is only reasonable that they could also agree to disagree. If no rights of a third person are involved, the parties may discharge their contract by mutual agreement without performance in several ways. The parties may agree merely not to be bound by the terms of the original agreement or they may make a new contract agreement involving the same subject matter, thus discharging the old contract. The original agreement itself may specify some event, the occurrence of which will end the contractual relation (that is, a condition subsequent). Both of the parties may ignore their rights under the contract, each going about their respective business in such a manner that a waiver of performance may be implied from their actions. When a contract is discharged by such methods, the release of one party constitutes the consideration for the release of the other.

Accord and Satisfaction

Accord and satisfaction occurs when a party agrees to accept a substitute performance for the one to which that party was entitled. Ordinarily this occurs when there has been a breach of performance by one party, giving the other party a right to sue for damages. In common terminology, this is the "settlement out of court" one frequently hears about in connection with both contract and tort cases.

To be effective as a discharge, both accord and satisfaction must have occurred. *Accord* refers to a separate agreement substituted for the original one. *Satisfaction* occurs when the conditions of the accord have been met; that is, the performance of the "accord." As a practical matter, most disputes never reach court. Most cases that reach court are settled by one side paying some amount to the other.

Novation

A *novation* replaces one of the parties to a contract. For a novation to be legally effective, all the parties to a contract must agree to it. As a simple illustration, assume that Black owes White $100. White owes Gray $100. If the three parties agree that Gray will collect the $100 debt from Black, a novation has occurred that completely relieves White of White's obligation to Gray. Black no longer owes White $100, but Black has a legally enforceable obligation to pay Gray. Common examples of novation occur when a person buys a house or a car from another, substituting himself or herself as a mortgagor and agreeing to make the loan payments to the lending institution (with consent by the lending institution to the substitution).

Arbitration

A court action for damages is sometimes impractical because of the time required for it, the cost involved, or some other reason. In many states and in federal court actions it is possible to substitute a procedure known as *arbitration* for a court action. Arbitration is a procedure in which a dispute is submitted to an impartial umpire or board of umpires whose decision on the matter is final and binding. The legality of the procedure depends upon the statutes of the state in which the suit arises or the Federal Arbitration Act. In common law, arbitration has no standing; even if a decision were rendered by an arbitrator, the case might still be taken to court and the arbitration would have no effect. Many of the states and the federal government have seen in arbitration a means of relieving crowded court dockets and have passed laws setting up the procedure and giving an arbitration decision almost the same force as a court judgment. In these jurisdictions almost any controversy in which damages are requested can be submitted to arbitration—not just contract cases, but tort cases and even property settlements following divorces.

Arbitration has several inherent advantages when it is compared to court proceedings. Probably the main advantage is found in specialization—the disputing parties decide among themselves what person is to act as judge and jury. This allows them to select someone who has a specialized knowledge of the field involved—someone who would not have to be educated on the general technical principles before deciding the case. Often this results in a more equitable decision than a judge and jury might render in court.

A second advantage of arbitration is the speed of the procedure. Court dockets are quite crowded; it is not unusual for a year to pass before a particular case comes up, and delays as long as five years occur in some jurisdictions. In the intervening period witnesses may die or move away, and memories dim. Arbitration affords an immediate solution. Today's dispute may be settled yet today or tomorrow if the parties so desire. It is only necessary to select a disinterested person and submit the dispute to him or her for decision. A minimum of formality is involved.

Cost-saving is a third advantage that may result from arbitration. In addition to the saving of whatever monetary value may be attached to waiting time, the cost of the procedure itself is often less than court costs.

Arbitration is not bound by the evidence rules encountered in a court of law. Whether this is an advantage or a disadvantage is questionable and would depend largely on the case. However, if the arbitrator considers hearsay testimony, for instance, as desirable in determining an issue, such testimony can be taken.

In jurisdictions where arbitration is used, the proceeding is conducted as an extrajudicial action of a court. Questions of law may be submitted to the court for determination, and the final arbitration award is enforced by the court.

There are three main legal requirement for arbitration.

1. The parties must agree to arbitrate—when the agreement to arbitrate took place is of little matter as long as the parties did agree at some time prior to the arbitration.
2. A formal document known as a submission must be prepared by the parties and given to the court.

The submission is roughly a combination of the complaint and reply required in a court case; it presents the issue to be decided.

3. The arbitrator(s) must be impartial and disinterred parties. If these requirements are met, the arbitrator's award will bind the parties.

The popularity of arbitration as a means of settling disputes has increased considerably in recent years. Probably the main reason for this is the efficiency of the procedure. It is conceivable that laws *requiring* arbitration in certain types of civil cases may be passed to further relieve the courts of burdensome cases in civil disputes.

♦ Tender of Performance

Tender of performance, if refused, may discharge the obligations of the party tendering performance. Three conditions, however, generally must be present in a lawful tender of performance:

1. The party offering to perform must be ready, willing, and able to perform the obligation called for.
2. The offer to perform must be made in a reasonable manner at the proper time and place according to the contract.
3. The tender must be unconditional.

Not only is the obligation discharged if such a tender is refused but the party who wrongfully refused the tender has breached the contract and may be sued by the tendering party. There is one notable exception to the general rule by which a tender of performance discharges an obligation. If the tender is an offer to pay a debt that is due and payable in money, the debt is not discharged by refusal of the payment. However, there are at least two rather important effects:

1. The accrual of interest is stopped.
2. Any liens used to secure the debt are discharged.

If the debt is payable in money, an offer to pay with anything other than legal tender may be refused by the creditor; the creditor is under no duty to accept a check, for instance. If the offer to pay is made before maturity of the debt, the creditor also need not accept. In such cases, if the creditor rejects the offer to pay for the reasons indicated, there has been no tender of performance.

♦ *Anticipatory Breach*

A breach of contract usually results from a failure to perform one or more contractual obligations as agreed and at the proper time. However, a contract may be breached before the time of performance has arrived. If the party who is to perform notifies the other party that he cannot or will not perform when the time comes to do so, that party is said to have repudiated the contract. Because the repudiation occurred before the time for performance, that party has committed an *anticipatory repudiation (or anticipatory breach)* of the contract. Such an anticipatory repudiation gives the other party several choices:

- The nonrepudiating party may accept the repudiation and immediately sue the other party for whatever damages may have been caused.
- The nonrepudiating party may treat the repudiation as inoperative, await performance, and (if there is no performance) hold the breaching party responsible for the resulting damages.
- The nonrepudiating party may accept the repudiation and obtain performance from someone else if it is possible to do so.

To constitute an anticipatory repudiation, a party must demonstrate a distinct and unequivocal refusal to perform as promised. Such a refusal may be shown by words or by conduct. Often, there is no express statement by a party to the effect of "I will not perform as promised." Hence, the courts often consider the circumstances to see whether the statements or conduct really amounted to a repudiation. If the nonrepudiating party continues to urge the other to perform, for example, the court may hold that urging performance demonstrates that the other's conduct was not a clear and absolute refusal to perform and thus was not an anticipatory repudiation.

Suppose Black Construction Company agrees, as general contractor, to build a structure for White. Gray is hired as subcontractor to do the electrical work. A month before the electrical work is to be undertaken, Gray informs Black that Gray cannot do it because of other commitments. At this point, if there were no rule as to anticipatory breach, Black would be in quite a dilemma. If Black obtained the electrical work from someone else and Gray subsequently had a change of heart, Gray could demand to be allowed to perform and sue Black if denied the opportunity to do so. If Black hired a second subcontractor and then Gray returned and performed the work, the second subcontractor could sue. If Black waited until Gray's time for performance was past and the contract was breached, thus causing Black to be late with the contract with White, then White could sue Black.

The rules relating to anticipatory repudiation give Black a way out. Black can hire another electrical contractor to do the work without fear that Gray will be able to take successful action against Black. Before Black actually hires another subcontractor, though, Gray has the right to withdraw the repudiation, thus resuming Gray's obligations under the contract. Generally, the nonrepudiating party should be careful to be sure that the other party has clearly expressed a refusal to perform before taking action.

Anticipatory repudiation does not apply to the payment of a debt. Although a debtor may notify a creditor that the debt will not be paid when due, the creditor must wait until the duty to pay has actually been breached before taking action. The creditor, of course, is not likely to be placed in a dilemma similar to that in which Black, in the foregoing example, would be in the absence of the rules relating to anticipatory repudiation.

BY OPERATION OF LAW

Certain laws have been passed and rules developed to provide for contract discharge as a matter of law. Chapter 10, "Lawful Subject Matter," discussed the result of a change in legislation, showing the effect of legislation in discharging contracts. Here we will consider alteration of the contract, the statute of limitations, bankruptcy, and creditor's compositions.

♦ Alteration of the Contract

If a contract is intentionally altered by one of the parties, without the consent of the other, the obligations of the other party under the contract may be discharged. A party cannot be held to the other's later changes in the contract terms if that party never consented to the changes. Black uses White as surety to secure a $500 loan from Gray, dated July 1 and due September 1. During August Black concludes that the debt cannot be paid when due. Black asks Gray for a loan extension to October 1, to which Gray agrees. On October 1, Gray, discovering that Black has moved without leaving a forwarding address, turns to White as surety.

White's surety agreement would not be enforceable under these circumstances unless White had consented to the extension. Generally, when one party makes a material alteration of a written contract without the other's consent, the contract is rendered unenforceable. However, if the nonconsenting party later finds out about the alteration and then continues to treat the contract as in force (such as by continuing to perform under the contract), then the alteration may be said to have been ratified and accepted. If the alteration is ratified, the contract will be enforceable even by the party who made the alteration.

♦ Statute of Limitations

Each of the states has adopted a statute (or statutes) that limits the time during which a lawsuit can be brought for a breach of contract. Many of the states specify a certain length of time for oral contracts, a longer time for written contracts, and a still longer time for contracts under seal. The state of Florida, for instance, specifies four years for oral contracts and five years for written contracts. Texas, however, has a general rule of four years for all contracts, written or oral. A party who has a right of action on a contract must take such action within the time limits stated in the statute of limitations; otherwise, the party will lose the right to take such action. The time is figured from the date the cause of action accrued, which is usually when the contract was breached. However, there is the possibility of renewal whenever the debt is acknowledged, such as by part payment. Under most such statutes, time does not continue to run while the person who has breached is outside the state.

Suppose Black orally hires White to add a roof to a structure. The contract is made March 1, 1995. White finishes performance on March 31, 1995, but is never paid. If the statute of limitations states five years for this type of contract, White has until March 31, 2000, to commence a court action for recovery. In some states, if Black made a part payment or in some other way acknowledged the debt on, say, April 15, 1997, the time for White to take action would not expire until April 15, 2002. Similarly, if Black left the state for a year, White would have until March 31, 2001, to begin the suit. If a suit is not begun by the dates mentioned, however, White loses the right to take action for a recovery from Black.

Although the legal duty to perform a contract may be discharged by the statute of limitations, it may be reinstated by the debtor. An act or promise by the debtor by which the debtor resumes the obligation will revive the debt and give the contract new life under the statute. If Black made a part payment for the roof in 2003, the statute of limitations would start to run again. Much the same is true of bankruptcy, discussed below. Reacknowledgment by the debtor of a debt discharged in bankruptcy serves to reinstate its legal life despite the discharge.

♦ Bankruptcy

When a person owes more than that person can pay, should the law help or leave the person to fend for himself or herself? Should one creditor be allowed to receive payment for that creditor's entire debt at the expense of other creditors? These and similar questions have been debated by legislatures since the problem of bankruptcy was first recognized.

It is not surprising that the Constitution gives Congress the power to establish "uniform laws on the subject of bankruptcies throughout the United States." Debt was the greatest single cause of imprisonment at the time of the American Revolution. Inability to pay a debt was a prison offense. In fact, forgiving a debtor's obligations and allowing the debtor to begin again with a clean slate is a somewhat recent innovation in the law. Bankruptcy proceedings for the purpose of paying off creditors are not new, but it is only recently that the debtor could receive a discharge of the debtor's obligations in such an action.

Three federal bankruptcy laws were passed and repealed after very short lives before the Bankruptcy Act was passed in 1898. Eighty years later it was replaced by the Bankruptcy Reform Act of 1978. Our present act has two main purposes:

- To pay off to the greatest reasonable extent the obligations of the debtor to the debtor's creditors either by liquidation of assets or by continuing the operation of a business or by devoting a portion of wages to the purpose.
- To discharge an honest debtor from future liability on the obligations.

The legal machinery enacted to carry out these functions is quite lengthy and complicated. The purpose of this text will be adequately served if we consider only an abbreviated version of the process and some of the key concepts.

The Bankruptcy Reform Act is a federal law. Bankruptcy petitions are filed in a Bankruptcy Court, which is theoretically a part of a corresponding U.S. district court. Appeals from the bankruptcy court may be taken to the district court, then to a U.S. circuit court of appeals and, finally, to the U.S. Supreme Court.

To Whom the Law Applies

The Bankruptcy Reform Act applies to all entities (individuals, businesses, municipalities, railroads) with one major exception. Financial institutions such as banks, savings and loans, credit unions, and the like generally have separate statutes to govern their insolvency. A special bankruptcy chapter, number 13, is designed primarily to give relief in cases of consumer debt. Two chapters, 7 and 11, are designed primarily to give relief in the form of liquidation and reorganization, respectively. Chapter 9 of the act handles municipalities with debt problems.

The law distinguishes between *voluntary* and *involuntary bankruptcy*. Almost anyone who has the capacity to make a contract may become bankrupt voluntarily. The prospective voluntary bankrupt is not even required to be insolvent, although as a practical matter bankruptcy would seem pointless otherwise.

Involuntary Bankruptcy

Involuntary bankruptcy is confined to Chapters 7 and 11 of the act, and there are restrictions about who may be driven into bankruptcy. Specific exceptions are made for farmers, charitable corporations, insurance companies, and financial institutions indicated earlier. Such debtors are protected from involuntary bankruptcy. Involuntary bankruptcy is usually begun by three or more creditors, holding a total of $5,000 or more of unsecured obligations, jointly filing a petition in a bankruptcy court. If there are fewer than 12 creditors, only one creditor with a secured claim of $5,000 is needed to start the bankruptcy proceeding. If the debtor is a partnership, one or more of the partners may file.

Consider the Black Manufacturing Company, a corporation in which Mr. Black holds most of the stock. One of Black's competitors has originated a new product that has captured many of Black's customers, so Black's sales have diminished to a point well below break-even. With revenues so low, Black has lost the ability to cover current obligations. Payments to several creditors have been postponed. Others have had to be satisfied with minimal partial payments. It has become evident to the creditors that Black Manufacturing is in trouble. Three of Black's creditors decide they have had enough and file a bankruptcy petition in the local U.S. Bankruptcy Court. The result is a "Chapter 11 " bankruptcy.

Chapter 11 Bankruptcy

Generally, under Chapter 11, the debtor is allowed to continue whatever business there is until there is a court order to the contrary. The debtor may continue to use, acquire, or dispose of property in the same manner as though no petition had been filed. Of course, counter measures are available in case of abuse of the creditors' rights by the debtor. For example, the court may appoint an interim trustee to preserve the estate until the creditors' meeting determines what is to be done with it.

Meeting of Interested Parties (Creditors' Meeting)

The idea of bankruptcy is to satisfy the requirements of creditors and others with an interest in the debtor's estate. To accomplish this, one of the first orders of business following the bankruptcy petition in involuntary bankruptcy is the calling of a meeting of interested parties. If the debtor is a partnership or corporation, the rights of the various partners or equity interests (for example, stockholders) are also considered. In the course of the meeting the debtor is examined (under oath) to determine the size of the debtor's estate. The debtor is required to reveal all assets, encumbrances on those assets, and the obligations of the estate. It is essentially a fact-finding mission.

A second objective is to allow legitimate claims against the estate. Those creditors with full security for their obligations, of course, are not really involved here—as, for example, one who holds a first mortgage on any real estate involved. Only if the sale of the property might not yield enough to pay off the remaining debt would the mortgage holder have a right to actively participate in the meeting. So then, both secured and unsecured obligations are considered at the meeting.

Along with the acknowledgment of legitimate preexisting debts there is the necessity to approve the costs of the bankruptcy—the administrative expenses. There are court costs, attorneys' fees, accountants' fees, trustee's fees, and the like to be allowed. In involuntary bankruptcy, a trustee is usually appointed at the meeting to take over the estate, collect the assets, and/or run the business according to the creditors' wishes. The bankruptcy court may or may not have appointed an interim trustee following the filing of the petition. In any case, it is up to the membership of the meeting to appoint someone to continue as the bankruptcy trustee.

The final item of business in the creditors' meeting in a Chapter 11 involuntary bankruptcy is the formulation of a plan. (Chapter 7 contemplates the liquidation of a business.) The debtor is generally given 120 days following the "order for relief" to propose a plan that would pay off creditors and continue the business. Confirmation of the debtor's plan can take some time, but if after 6 months from the start of the action the debtor's plan has not been confirmed, it becomes the committee's business to form a plan. Note that any single creditor or the trustee may file a plan (after the appropriate time). Hence, there may be a number of reorganization plans proposed in a given bankruptcy proceeding.

A Chapter 11 plan must set forth a means of continuing the business. (The most likely alternative if no plan is available is to shift to a Chapter 7 liquidation.) To avoid the shift to liquidation, the plan must provide a result for each creditor that is at least as good as the creditor would receive under a Chapter 7 bankruptcy. The plan may provide for retention or sale of property, satisfaction or modification of liens, curing of defaults, changes of security interests, and/or mergers or consolidations. Generally there is wide latitude in the possible actions to be included in such a plan.

In our Black Manufacturing Company case, Black is questioned under oath and the extent of the company's obligations and assets is clarified. The creditors are rather worried about their prospects of recovery and must decide whether they want to either continue the business or try to liquidate it. Because Black's management may be partially at fault for the problems, Gray is appointed by the creditors' committee and confirmed by the bankruptcy judge as the trustee for the company. The committee's decision about continuation or liquidation of the company now depends largely on Gray's opinion of Black's plan for the future. Black could, for example, present an appealing plan contemplating a product improvement or expansion into a new market. Lacking some innovation like this, the committee may simply decide to liquidate the company, satisfying committee members' debts with whatever the liquidation brings.

The plan must be confirmed. This simply means that the bankruptcy court must hold a confirmation hearing to ascertain whether the plan complies with (a) all applicable laws; (b) agency rules (for example, the Securities and Exchange Commission); and (c) provisions of Chapter 11 of the bankruptcy law. The plan also must have been accepted by at least one "impaired" class of claims (that is, a group of creditors who stand to recover less than the full amount of the obligations of the debtor). The court is not required to confirm a plan unless and until it is convinced of its probable success—that the plan will not lead to eventual liquidation of the debtor.

If Black's plan for the future of the company is unappealing to the committee members, they may well reject it. The logical next step, if that is the case, is to convert the Chapter 11 proceeding into a Chapter 7 proceeding.

Chapter 7 Bankruptcy

The objective of an involuntary Chapter 7 bankruptcy is to liquidate the debtor's available assets to pay off the creditors. The case is begun by a creditors' petition followed by the bankruptcy court's response, the *order for relief.*

At the time of the order for relief or before (if the court decides it is necessary), an interim trustee may be appointed to collect and preserve the assets of the estate, the debtor's property. The interim trustee is later replaced by a trustee (perhaps the same person) whose primary duties are to (a) collect and liquidate the property of the estate in the best interest of all parties concerned, (b) investigate the financial affairs of the debtor, (c) examine the creditors' proofs of claims and object to improper claims, (d) perhaps oppose discharge of the debtor, and (e) file a final report. If a business is involved, the trustee may be authorized to operate it for a limited period.

The main business of the creditors' committee meeting is electing the trustee, arranging to work with the trustee, and determining the allowances of claims. Secured claims, of course, have superior standing. A mortgagee or other lienholder has first call on the asset(s) on which the lien was established. However, to the extent that the value of the asset is insufficient to cover the debt involved, the lien holder is also an unsecured creditor. On the other side of the coin, excess value of the asset—beyond that required to satisfy the debt—is available to those who hold unsecured claims. Payoff of unsecured claims follows a priority schedule in which administrative expenses have top priority, followed by unsecured claims in which timely notice was given, employees' wages up to a set amount, taxes, and other claims.

Now let us assume it is Black, rather than the Black Manufacturing Company, that is in financial difficulty. Certain personal losses have occurred that convince Black that the prospect of paying all the bills, including the amounts in arrears to creditors, is dim indeed. If Black's payments to the creditors were a bit smaller, Black might be able to make it; if they could be made a shade smaller yet, Black could pay off the arrears and at least be on a current payment status. If Black's unsecured indebtedness is less than $100,000 and his secured indebtedness is less than $350,000, it may be that Chapter 13 bankruptcy provides a way out of Black's problems.

Chapter 13 Bankruptcy

The title of Chapter 13 is "Adjustment of Debts of an Individual with Regular Income." The main problem treated by Chapter 13 is consumer indebtedness. It is available to a debtor on a voluntary bankruptcy basis.

A trustee is appointed by the bankruptcy court to perform most of the same duties as those in Chapter 7, except for the collection and selling of the debtor's assets. It is up to the debtor to devise a means of paying off the debts. This, of course, may be done by assigning a part of the debtor's wages to this purpose or by assigning a portion of earnings from selfemployment or a business. Whatever the source of income, courts usually allow the debtor broad freedom to choose an acceptable plan and operate within it as long as it appears the past due debts to creditors will be paid off.

The period of operation of the plan is nominally limited to three years, but it may be extended to five years with the court's permission. Creditors cannot force the debtor into a plan unacceptable to the debtor for the same reasons they can't force a debtor into a Chapter 13 bankruptcy in the first place. (They could, of course, force the debtor into a Chapter 7 bankruptcy.) Only the debtor knows the debtor's family requirements, family plans, and other such personal information. Secured creditors in a Chapter 13 bankruptcy may be paid off in many ways, one of which is for the debtor to return goods on which a lien was placed.

Confirmation of the debtor's plan requires only the affirmation of the parties with an economic interest in the action as well as the court's blessing as to the plan's legality. One of the court's confirmation considerations is that the plan's contemplated payoff to the unsecured creditors is at least as much as they would have realized under a Chapter 7 liquidation.

The Estate

Bankruptcy contemplates rehabilitation of the debtor. Depriving the debtor of all property so that the debtor winds up requiring aid from a government welfare program isn't rehabilitation. Because of this, certain real and personal property cannot be taken. Each state has "homestead laws" that specify certain minimums, but the Bankruptcy Reform Act also lists some exemptions. Various forms of income and benefit exceptions are also found in nonbankruptcy parts of the U.S. Code. For example, social security and certain other retirement benefits are exempt from bankruptcy, as are veteran's benefits.

The property of the debtor's estate for bankruptcy purposes consists of all the debtor's assets that are not exempt. In addition to the obvious types of property mentioned above, this includes patent rights, copyrights, trademarks, liens held by the debtor on property owned by a third party, and community property. Generally, the estate rights are limited to those the debtor had at the beginning of the case.

Discharge

Under Chapter 7 and Chapter 13, discharge of the debtor's obligations is limited to individuals. The reason for the Chapter 7 limitation is that even though a corporation or partnership may go through the Chapter 7 liquidation, what is left after bankruptcy is nothing but a hollow shell. The effect of the debtor's discharge in bankruptcy is to void existing debts and enjoin the beginning or continuation of any action based on discharged debts.

Certain debts, however, are not discharged by bankruptcy, and some acts by the debtor may prevent the discharge of obligations. Generally, the following debts are not dischargeable in bankruptcy:

- Tax obligations
- Debts and continuing obligations for alimony, maintenance, and child support
- Liability for fraud or willful or malicious injury
- Certain debts for fines, penalties, or forfeitures for the benefit of a governmental unit
- Debts resulting from fraud, embezzlement, or larceny while acting in a fiduciary capacity
- Debts for educational loans less than five years past due
- Debts remaining after a previous bankruptcy proceeding
- Debts not listed or listed too late to be included with the bankruptcy

Discharge of the debtor's remaining obligations is generally forthcoming if the debtor has dealt honestly and fairly with the creditors during the bankruptcy action. Most of the reasons for denying discharge are based on actual or reasonably suspected deceit by the debtor. Discharge is denied for the following reasons (Chapter 7):

- The debtor is not an individual.
- The debtor transferred, destroyed, or concealed some of the debtor's property within a year prior to the filing of the bankruptcy petition, intending to thereby defraud the creditors.
- The debtor failed to keep or preserve accounts of transactions.
- The debtor made fraudulent statements in the bankruptcy examination regarding the estate or the debtor's financial affairs.
- The debtor failed to explain satisfactorily any losses of assets.
- The debtor refused to obey a lawful court order during the proceedings.
- The debtor was granted a discharge under Chapter 7 or Chapter 11 (or their predecessors) in a case commenced within six years prior to the date the petition was filed.
- The debtor was granted a discharge under Chapter 13 within the past six years unless the debtor had paid at least 70 percent of the amounts of the unsecured claims in the prior case.

Generally, confirmation of a Chapter 11 plan discharges the debtor. The main exception to this is when the Chapter 11 action essentially amounts to the Chapter 7 liquidation. In such an instance, the Chapter 7 limitations just noted would apply. Under a Chapter 13 bankruptcy, the court is required to discharge the debtor after all payments under the plan have been made. However, there is a noticeable reluctance of courts to confirm plans that do not contemplate the payment of at least 70 percent or so of the amounts of claims by unsecured creditors.

Once the debtor has been through the bankruptcy mill and has received a discharge, the debtor's contractual obligations are ended. New property acquired, and new transactions and business dealings are free from interference by former creditors. The debtor is given new economic life.

♦ Creditors' Compositions

It is often to the advantage of both the creditors and the debtor to avoid bankruptcy proceedings. There are many costs of bankruptcy and each cost reduces the assets to be divided. It is costly, for instance, to pay the trustee to maintain and then dispose of the property involved. There is a much less expensive procedure available. The *creditors' composition* accomplishes almost the same thing as bankruptcy. It discharges the debtor's obligations. Each participating creditor gets some return on the creditor's account receivable. The procedure is informal but binding. It is not necessary for all the creditors to join in the composition; two or more are sufficient—if only one creditor is involved it is not a creditor's composition and the remainder of the debtor's obligation is not discharged, as pointed out in Chapter 9, "Consideration."

For example, Black owes White $1,000, Gray $2,000, and Brown $3,000. Black has $3,000 cash available plus various other assets but cannot pay all the debts and remain solvent. Black meets with White, Gray, and Brown, telling them of the situation. The creditors are faced with the possibility of bankruptcy proceedings where, after the costs are paid and Black's assets sold for whatever price they may bring, the creditors may get $0.25 for each dollar of debt. As an alternative, the creditors may choose to divide Black's cash assets in any way they see fit, $0.50 for each dollar being one such possibility. The creditors may agree, instead of taking a straight percentage, to divide the assets in some other way that is satisfactory to each. They might agree, for instance, that White will receive $600, Gray $1,000, and Brown $1,400. There are several explanations of the consideration involved in a creditor's composition. Perhaps the most common one holds that the consideration received by each creditor for giving up the right to sue for the remainder of the debt is found in the forbearance of the same right by the other creditors. Several states have statutes that specify the conditions and procedures for creditors' compositions. Where such statutes exist, examination of the consideration involved is unnecessary.

COOK ASSOCIATES INC. v. COLONIAL BROACH & MACH. CO.

304 N.E.2d 27 (Ill. App. Ct. 1973)

Defendant, a Delaware corporation doing business in Michigan, hired an employee, Dean Averbeck, who had been referred to defendant a year earlier by plaintiff, an Illinois employment agency. When plaintiff filed suit in Illinois to recover the fee for its services, defendant filed a special appearance and motion to quash service of summons alleging that the court had no jurisdiction over its person. This motion was denied, and a trial on the merits was conducted. Defendant made a motion for judgment at the close of plaintiff's evidence, and, after the jury rendered a verdict in favor of plaintiff, a post-trial motion for judgment notwithstanding the verdict, but these motions, too, were denied. Defendant now appeals from the judgment entered against it and from the denial of the three motions described. Defendant raises two issues on appeal: (1) the Circuit Court of Cook County did not have *in personam* jurisdiction over defendant; and (2) Michigan law, which requires that an employment agency contract be in writing, governed the agreement between the parties and, therefore, the contract involved is not enforceable.

There is no dispute over the essential facts. Plaintiff is a private employment agency incorporated in Illinois with its place of business in Chicago. Plaintiff places men only in executive positions paying $15,000 or more. Personnel placement in these higher-paid positions can be very sensitive, and, at times, prospective employers can supply the employment agency with only the qualifications of the needed employee without being able to name the particular job available. With executive placements, it is not unusual for there to be a lag between the time the applicant is referred to an employer and the time he is hired by that employer. In all instances, the fee for plaintiff's referral services is paid by the prospective employer, not the applicant placed. Although plaintiff's policy was that a referral would be applicable to a placement for a two year period, such policy was not printed on its fee schedule in effect at the time of the referral involved, but was added to the schedule April 1, 1968. Plaintiff's fee schedule is on file with the Illinois Department of Labor. It is not plaintiff's custom to have the applicant sign a contract with the agency.

In December 1967 Dean Averbeck, a resident of Wisconsin, responded to an advertisement placed by plaintiff in *Metal Working News*, a business journal in the field of machine tools. This he did by sending a résumé to Robert Danon, an employment counselor and research man with plaintiff agency, and telling him that he was looking for new employment. A week later Averbeck came to plaintiff's office in Chicago for an interview with Danon. Danon explained to Averbeck that the agency would work with him, and that the agency expected to be advised by Averbeck of all interviews he had with employers to which plaintiff might refer him. Danon made out a résumé which represented Averbeck as a sales manager and which contained Averbeck's personal data and education and employment background. This "flyer" was sent to prospective employers in the machine tool industry, and by way of a large mailing, along with the résumé of 3,000 other applicants, to prospective employers in the metal working industry in general. Both mailings went to companies located throughout the United States.

Defendant is a manufacturing company which, as stated above, is incorporated in Delaware, with its place of business in Michigan. Defendant is not registered as a foreign corporation in Illinois, does not maintain offices in Illinois, is not listed in the Illinois telephone directory, and none of its employees reside in Illinois.

On January 23, 1968, E. H. Jones, Executive Vice president of defendant Company, telephoned Danon and expressed an interest in the man represented by code number RD390 (Averbeck). He asked for the man's name, the name of his employer, and his earnings at that time. Before giving this information to Jones, Danon advised him that plaintiff was licensed to operate its agency only if the employer were to pay the referral fee upon hiring the job applicant. Jones indicated that such an arrangement would be agreeable to him. According to Danon's testimony, he also told Jones that the referral was good for two years, and that the employer would have to pay the applicant's interview and relocation expenses if the applicant were hired. Jones requested a copy of Averbeck's résumé and asked Danon to have Averbeck call him for an interview the next time he was in the Detroit area. In response to Danon's question as to the type of job which defendant had available, Jones said, "Something in sales."

That same day, Danon called Averbeck at his home in Wisconsin and learned that Averbeck was in Lansing, Michigan, looking for employment in that area. The next morning, Danon called Averbeck in Lansing and told him to call defendant and arrange an interview. Danon then called Jones to say that Averbeck was in the Detroit area and would be telephoning him later that day.

On January 25, 1968, Danon sent Averbeck's résumé to Jones and enclosed a copy of plaintiff's fee schedule. On March 12, 1968, Danon called Averbeck to check various referrals with him and was told that he had not yet had an interview with defendant. On March 18, 1968, Danon called Jones, who stated that he had not yet interviewed Averbeck.

However, Averbeck had, in fact, been interviewed by defendant on January 24, 1968, but was not hired at that time. On January 18, 1969, a year later, defendant contacted Averbeck to return for another interview. Averbeck was hired on January 30, 1969, and began to work for defendant on March 3, 1969. He started in the Sales Department at $17,000 per year with the understanding that he would be made a Sales Manager at a later date. His salary was subsequently increased to $18,000 per year. He left the firm in January 1971.

On December 9, 1969, Averbeck called Danon to inform him that he was considering a change in jobs and incidentally advised him that he had started to work for defendant on March 1, 1969, after defendant had offered him a job as a sales manager at $18,000 per year. He was unhappy with the company, however, and wanted to relocate in either Cleveland or Chicago.

Danon sent Averbeck his old résumé to be updated, and when it was returned, it bore the notation in Averbeck's handwriting, "3/l/69 to present, Colonial Broach & Machine Company, Warren, Michigan, sales manager."

On January 6, 1970, Danon called Jones to say that he was glad Averbeck had been hired and that defendant must have overlooked plaintiff's fee. Jones agreed that Averbeck had been contacted, interviewed, and hired due to plaintiff's services, but that Averbeck, when interviewed, had denied being represented by plaintiff. When Danon stated that plaintiff had, indeed, represented Averbeck at the time of the referral for the original interview in January 1968, Jones replied that he would talk to his people and call back the next day. On January 7, 1970, when Jones did not call back, Danon had an invoice prepared and sent it to defendant. Plaintiff's fee was computed on the basis of 1% per each $1000 of Averbeck's starting annual salary. Since Jones had told Danon, in their telephone conversation of January 6, 1970, that Averbeck's starting annual salary had been $16,000, the fee billed 16% of $16,000 or $2560. The bill was never paid, and plaintiff brought suit. (It was later confirmed that Averbeck's starting annual salary had been $17,000, and the complaint was amended to state that fact and to increase the amount paid to 17% of $17,000 or $2890 plus interest.)

Defendant first contends that the trial court did not acquire jurisdiction over its person by way of Section 17 of the Illinois Civil Practice Act, known as the Illinois "long-arm" statute . . . and that, therefore, the court first erred in denying its motion to quash service of process and dismiss the suit, and erred again in denying its subsequent motions for judgment in its favor. The "long-arm" statute provides, in pertinent part, that a non-Illinois resident submits himself to the jurisdiction of the Illinois courts by "the transaction of any business within this State."... Defendant asserts that by merely telephoning plaintiff in response to an unrequested solicitation sent to it in Michigan by plaintiff, its conduct did not amount to a "transaction of any business" within the meaning of the statute, and that therefore it cannot be subjected to the jurisdiction of the Illinois courts without a violation of due process.

As noted by the Supreme Court in *Nelson* v. *Miller,* 11 Ill. 2d 378, 143 N.E. 2D 673, Section 17 of the Civil Practice Act, which that court held to be valid, reflected the legislative intent to assert jurisdiction over nonresident defendants to the extent permitted by the due process clause. Due process requires only that before an Illinois court can acquire *in personam* jurisdiction over a nonresident defendant, that defendant must have had certain minimum contacts with Illinois such that maintenance of the suit in this state does not offend traditional notions of fair play and substantial justice. . . .

Personal jurisdiction over a nonresident does not depend upon the physical presence of the defendant within the state; it is sufficient that the act or transaction itself has a substantial connection with the forum state. . . . It is also sufficient if the nonresident defendant's contact with the state involves only a single business transaction. . . .

Reviewing the facts of this particular case, plaintiff initiated the contact between the parties by sending to defendant a "flyer" regarding available job applicants. However, the mailing of the "flyer" represented only an offer to do business, and defendant was under no obligation to respond. It was defendant, then, who initiated the business transaction in question by telephoning plaintiff and requesting that plaintiff divulge the name of a possible prospective employee whose partial identification had been gleaned from the "flyer." Defendant also asked plaintiff to send defendant a résumé of that applicant, and notify him to contact defendant for an interview. In the same conversation, a contract was created when defendant agreed that, in return for plaintiff's services, it would pay plaintiff's fee if it were subsequently to hire the applicant.

An employment agency's business is to put a prospective employee in touch with a prospective employer, and the agency's services end once the two principals are made aware of each other's identities so that negotiations can be commenced between them. Although defendant's only contact within this state was a telephone call, that call was all that was necessary for defendant to achieve its purpose. Once defendant informed plaintiff that it was interested in a certain person to fill a position and agreed to pay plaintiff's referral fee if it eventually hired that person, defendant knew, or should have known, that it had entered into a contract with an Illinois agency, that the agency would perform its services from its office in Illinois, that the fee, if due, would be paid to plaintiff in Illinois, and if the fee were not paid as promised, defendant might be liable to suit in the Illinois courts. We find that the necessary jurisdictional minimum contacts for the purposes of *in personam* jurisdiction over this defendant resulted from that single business transaction between the parties, and we are of the opinion that this conclusion does not violate notions of fair play and substantial justice in requiring defendant to defend this suit in the Illinois courts.

At the time of oral argument in this case, defendant cited an opinion of a federal judge (Northern District of Illinois) in *U.S. Railway Equipment Co.* v. *Port Huron and Detroit Railroad Co.*, 58 F.R.D. 588, holding that the Illinois "long-arm" statute had not served to bring the defendant into court for *in personam* jurisdiction. We have studied that opinion and believe it is distinguishable on the facts from the case at bar. Defendant contends, however, that even if Illinois courts had properly acquired *in personam* jurisdiction over defendant, the court erred in denying its motions for judgment in its favor because the contract, under Michigan Law, was unenforceable because plaintiff was not licensed in Michigan as an employment agency and the contract between the parties was not in writing.[1]

Although plaintiff questions whether a nonresident employment agency must first obtain a license in Michigan before referring a job applicant to a Michigan employer, we need not reach that issue as we find that Illinois law is applicable to construction of the contract between the parties and resolution of the instant case. Under Illinois law, if a contract is made in one state with the intention that it be performed in another, and the states are governed by different laws, the law of the place where the contract is to be performed will control as to the contract's validity and will prevail over the law where the contract was entered into. . . . The contract in question, providing for payment by defendant in exchange for services rendered by plaintiff, was oral, and the parties did not specify in their telephone conversation as to where the services were to be performed. However, as previously mentioned herein, the nature of the contract was such that both parties knew plaintiff's services would be performed from its office in Illinois. The fact that the interviewing and hiring took place in Michigan is of no consequence since the services which were the basis of the contract and for which defendant would be liable for payment to plaintiff had already been performed in Illinois. Therefore, since Illinois is the state where the contract was to be performed, and the contract is valid and enforceable under Illinois law, the trial court did not err in denying defendant's motions for judgment in its favor.

It should also be noted that we are dealing with a contractual situation and, therefore, with obligations voluntarily undertaken. That fact, in itself, creates a presumption in favor of applying the law of the state which would validate the contract. . . . Since the contract, under defendant's analysis, would have been unenforceable under Michigan law, the trial court, as an additional ground for its decision, could properly have presumed that the parties intended to

1. Mich. Stat. 17.393, M.C.L.A. sect. 408.603. No person shall open, operate, or maintain an employment agency in the state of Michigan without first procuring a license from the state superintendent of private employment bureaus....Mich. Stat. 17.406, M.C.L.A. sect. 408.616. Every employment agent licensed under class 2 (the appropriate class for employment agencies placing executives) shall enter into a written agreement with every employee, employer, or both, for service to be rendered for which a charge is to be made....

enter into a valid and enforceable agreement, necessarily in this case involving the application of Illinois law. The judgment of the circuit court is affirmed.

Affirmed.

DIVERSIFIED ENVIRONMENTS v. OLIVETTI, ETC.

461 F. Supp. 286 (M.D. Pa. 1978)

This is an action for damages brought by a lessee of a computerized accounting system manufactured and sold by the Defendant, Olivetti Corporation of America (Olivetti). Plaintiff, Diversified Environments, Inc., (Diversified) alleges that the Olivetti computer that it leased has never been made operational and that Defendant is liable for damages on theories of breach of express and implied warranties. Olivetti defends on the bases that it has fully performed and alternatively that it was excused from performance because Diversified unreasonably refused to permit it to effectuate its contractual duties. Plaintiff seeks relief for the total payments made under the lease agreement for the computer, the cost of paper products, and other consequential damages. The following are the Court's findings of fact and conclusions of law.

Findings of fact

1. Plaintiff, Diversified Environments, Inc., is a Pennsylvania corporation engaged in the selling of temperature and energy control systems with its principal place of business at Camp Hill, Pennsylvania.
2. Defendant, Olivetti Corporation of America, is a Delaware corporation engaged in the business of selling computer services with local business offices in Harrisburg, Pennsylvania.
3. Jurisdiction is based upon diversity of citizenship and an amount in controversy in excess of ten thousand dollars.
4. In July of 1974, Tim L. Fleegal, a sales representative of Olivetti contacted Diversified's President, Charles E. Andiorio, Jr., for the purpose of inducing him to buy or lease a computerized accounting system.
5. Mr. Andiorio subsequently met with Mr. Fleegal and explained in detail the nature of Diversified's business and particularly noted that the most tedious part of his duties was the preparation of specifications to be used in submitting bids on jobs.
6. Mr. Fleegal also met with Mr. Warren Beck, who was primarily responsible for the Plaintiff's accounting system, during July and Mr. Fleegal was made aware of all of the accounting procedures of Diversified and that Diversified's accounting records had to be compatible with Barber Colman Co., for whom Diversified was a manufacturing representative.
7. Neither Mr. Beck nor Mr. Andiorio was familiar with computer systems and they relied upon Mr. Fleegal's expertise.
8. After Mr. Fleegal became thoroughly familiar with the operations of Plaintiff's business he stated that the Olivetti P-603 Computer System would meet all of the Plaintiff's requirements and perform all of the functions that were discussed.
9. Mr. Fleegal was qualified to sell only the P-603 Computer System, which was an accounting computer, and not the Olivetti word-processing machines.
10. At the time of the discussions, Mr. Fleegal stated to both Mr. Beck and Mr. Andiorio that utilization of the P-603 Computer System would save the Plaintiff both time and expense by reducing manpower and record keeping.

11. Mr. Fleegal represented to Mr. Andiorio that he would only need to push a button and he would have the specifications, that Mr. Andiorio would save half of his time, and that the computer would enable Diversified to do without one of its secretaries.
12. On July 26, 1974, Mr. Fleegal submitted a proposal to Mr. Andiorio and advised that the P-603 accounting computer could effectively meet all of the objectives discussed between the parties.
13. The proposal specifically set forth that the P-603 computer could perform the functions of specification writing, estimating, accounts payable, job cost, prime cost analysis, accounts receivable, and check.
14. The proposal stated that the total cost of the system was $9,590.00 which included all programming, forms design, initial operator training, delivery, and installation of equipment.
15. It further provided: "in dealing with Olivetti you do business with a firm which herein guarantees in writing the exact performance of the system both machine and program. Only after these assurances have been met can we ship the machine and bill you as a customer. In addition, you have my personal assurance and that of the Harrisburg Management that all of our resources will be employed toward your complete satisfaction in the system."
16. During July or August of 1974, parts of the proposed computer package were demonstrated to Mr. Beck in the Defendant's office; however, at no time were either the specification writing or estimating demonstrated.
17. On August 7, 1974, Mr. Beck signed in two places, on behalf of Diversified, a "Customer Software Acceptance" form provided by Mr. Fleegal.
18. The Customer Software Acceptance form contains three places for signatures and Mr. Beck signed his name after the following statements on the form:

    1. "I agree that the system explained to me with regard to this application is correct in all respects and that any alterations after this date could result in additional charges according to the current published program rates.

    2. This application as described in section 1 has been demonstrated to me in its final programmed form and I accept it as being a complete and workable solution."
19. Mr. Beck did not sign after the third line which stated:

    "The program described in section 2 above has now been installed and the relevant personnel have been fully advised of its capabilities. I have received complete program documentation."
20. Mr. Fleegal advised Mr. Beck that Diversified would owe absolutely no financial obligation until the third line of the form was signed, and it was this promise that actually prompted Mr. Beck to sign the first two lines of the form.
21. This "Customer Software Acceptance" form was considered by the defendant as an agreement or contract between the parties.
22. After the signing of the form, the computer and various programs were ordered by Mr. Fleegal.
23. Around this same period of time, in July or August of 1974, Mr. Fleegal also assured Mr. Andiorio that Diversified would not be bound to accept the computer unless and until a third signature was placed on the acceptance form as acceptance and approval of the complete system.
24. Mr. Fleegal promised to personally oversee the installation and implementation of the complete computer system and also promised that Plaintiff's operators would be fully trained and if the training proved unsuccessful that Olivetti girls would be available to run the computer.
25. Mr. Fleegal also promised both Mr. Beck and Mr. Andiorio that either he or other Olivetti staff would transfer all the necessary information onto the computer cards and the Plaintiff's only obligation was to show the Olivetti staff where the information that was needed as a data base was stored.
26. Along with the other representations, Mr. Fleegal told Mr. Andiorio that Plaintiff would not owe one cent until the computer was fully in operation and they were completely satisfied.
27. On these conditions, the computer was placed at Plaintiff's business during September of 1974; however, it was not operational at that time.
28. Mr. Fleegal then contacted an equipment leasing company, Equipment Funding, Inc., (Equipment Funding) and made the arrangements for Equipment Funding's purchase of the computer and software, and the subsequent leasing of it to the Plaintiff.

29. Mr. Fleegal received a lease agreement from Equipment Funding and took it to Mr. Beck for his signature on October 3, 1975.
30. Mr. Beck signed the lease agreement on October 3, 1975, after Mr. Fleegal assured him that it was just a mere formality, an application for a lease, and that Plaintiff would not be obligated to keep the system or to make any payments until it was completely satisfied.
31. Subsequently, Mr. Beck received a payment or coupon book from Equipment Funding and on approaching Mr. Fleegal, he was told to disregard it as no payments were due until the computer was operative and until they signed the third line of the Software Acceptance form as their acceptance.
32. The lease was not to be effective until an "Acceptance Certificate" was signed by the Plaintiff.
33. In mid-November, Mr. O'Brien, from Equipment Funding, and Mr. Fleegal went to Plaintiff's place of business for the purpose of obtaining execution of the "Acceptance Certificate" and Mr. Beck initially refused to sign the certificate.
34. Mr. Fleegal then induced Mr. Beck to sign the acceptance on the promises that the computer would be running by the first of the year and on the guarantee that the computer system would be acceptable to the Plaintiff.
35. Subsequently, Mr. Fleegal and another Olivetti employee, Barbara Slagle, made several visits to Diversified for the purpose of installation.
36. Mrs. Slagle, a customer software representative, met with Plaintiff's employee on December 18, 1974, and together they set up the payroll data base and the employee was taught how to update the data.
37. The payroll function was the only function Mrs. Slagle was supposed to teach, as Mr. Fleegal was to teach all of the other functions listed in the proposal.
38. Mrs. Slagle incurred no problems in either teaching Plaintiff's employee or in getting the base data for the payroll function.
39. Mr. Fleegal made several visits during the month of December 1974 for the purpose of training Plaintiff's employee on the numerous other functions; however, he neither transferred the data as previously promised nor was he successful in training the designated employee on any of the other functions.
40. Mr. Fleegal spent a substantial amount of his time during these visits in discussing his religious beliefs with Plaintiff's employees.
41. After receiving complaints from his employees, Mr. Andiorio refused to allow Mr. Fleegal to return to the premises.
42. During Mr. Fleegal's visits at Plaintiff's business in December of 1975, he never requested any information from Mr. Andiorio with respect to the specification writing or estimating, even though the specification writing was to be the primary use of the computer.
43. By January 7, 1975, the only data that had been transferred was the payroll base data, and the computer was only capable of being utilized for this minor function at that time.
44. Plaintiff, pursuant to prior representations, in early January, stopped further training and requested the Defendant to remove the computer from their premises, as the computer was at that time nearly worthless because data had not been transferred, employees had not been trained, and Plaintiff was completely unsatisfied with the computer.
45. Defendant refused to remove the machine and by letter of January 20, 1975, District Manager, T. F. Meade, Jr., replied, "The relevant personnel have been trained, and "our only remaining obligation is to continue to assist you in the fullest implementation of this system which you requested and for which you contracted."
46. Plaintiff continued making the lease payments after consulting with counsel, as it did not want to place its credit rating in jeopardy, and was advised to institute suit against Olivetti instead.
47. A meeting was subsequently held in February 1975 with Mr. Meade, Mr. O'Brien, Mr. Andiorio, and others in attendance, and Defendant took the position that it had fully performed.
48. No further training was conducted in 1975 and Defendant was not requested by Plaintiff to conduct training during this period.
49. This suit was filed on April 13, 1976.
50. Subsequent to the institution of this action, Mrs. Slagle returned in December of 1976 to train another employee and to attempt to prepare the computer for Plaintiff's utilization.

51. After one day or a day and one-half this installation attempt was rejected by the Plaintiff on the advice of his employee who was trained in computer systems for the reasons that the P-603 was not compatible with the Barber Colman bookkeeping system and because use of the accounting computer for specification writing was cost prohibitive.
52. Use of the P-603 computer for specification writing required that the information be placed on cards at the rate of one paragraph per card, which for Plaintiff's needs would have required likely over a thousand cards at two dollars each, and which would have required someone to manually pick the necessary cards out of a file and insert them into the P-603 computer.
53. The P-603 computer was not designed to perform word processing as that required for specifications writing, which was the Plaintiff's primary concern.
54. Expenses incurred by Plaintiff with respect to this transaction include $15,126.00 for payments due under the lease, $145.54 for paper products, and an accountant's bill of $387.50.

Discussion

While a number of the theories raised by the Plaintiff are indeed relevant and applicable to the facts of this case, it is unnecessary to discuss the theories of warranty and misrepresentation, as Defendant's breach of its contractual duties is sufficient to impose liability. The parol evidence rule is clearly inapplicable here, as there was no integrated written agreement between the parties that fully and completely stated the entire agreement. The only writings involved are the written proposal and the Customer Software Acceptance form, neither of which completely embody the agreement between the parties. We find that the oral agreement to provide certain services formed part of the contractual relationship and that the oral agreement obligated Defendant to perform the transfer of data, the training of employees, and generally to make the computer functional. These obligations were breached as the base data has only been transferred for the relatively minor payroll function and the only training that has been completed is to this payroll function. Plaintiff's employees have not been trained on the remaining functions, the ones most important to the Plaintiff, and the base data has never been transferred onto the computer cards. Defendant asserts that it should be excused from performance because the Plaintiff prevented it from performing its obligation.

While not totally devoid of merit, this argument fails to relieve the Defendant from liability. As noted in the findings, the computer was installed at Plaintiff's place of business in September of 1974 and the Plaintiff was fully receptive to having the computer made operational until sometime in early January of 1975. Plaintiff assigned an employee for the computer training and Mrs. Slagle encountered no problems in either the training or the transfer of data for the payroll function. The problems arose from Mr. Fleegal's failure to succeed in training the employees on the other numerous functions and failure to transfer the base data as was initially promised and which formed a part of the understanding between the parties. This could have been due to the fact that this was Mr. Fleegal's first sale and his first attempt at training individuals in the use of the computer. Regardless of this, it is clear that no one was trained to perform the other functions and that Mr. Fleegal interfered with Plaintiff's employees to some extent with his religious discussions.

It was only after Defendant's attempts at performance were proving unsuccessful that Plaintiff demanded the computer be removed and training halted. This was reasonable under the circumstances as Defendant had advised Plaintiff numerous times that it had no financial or contractual obligation until they signed the third line of the Customer Software Acceptance form as an indication that they were completely satisfied and that the computer was fully operational. As noted above, the Defendant took the position that it had performed and that the employee of Plaintiff had been trained, which was simply not true. The Plaintiff remained receptive to having its employees trained, even after suit was filed, until it determined that the computer was neither economically feasible nor compatible with other bookkeeping requirements. In short, Plaintiff's refusal to permit training at the two times noted was reasonable under the circumstances and was not a material interference with Defendant's obligation. The argument that it was excused because base data was not supplied is similarly unpersuasive as Mrs. Slagle encountered no such problem, Mr. Fleegal never advised Mr. Andiorio of any uncooperative conduct of his employees, and because the letter of Mr. Meade of January 1975 never raised the problem of obtaining base data. Therefore, no excuse exists for Defendant's nonperformance. . . .

Defendant also argues that Plaintiff has failed to join an indispensable party, Equipment Funding, because complete relief cannot be accorded among those already parties. This argument is not valid for at least two reasons. First, complete relief can be accorded between the present parties. Defendant breached its contractual obligation, and defendant has made no showing that Equipment Funding is an indispensable party. Second, the Defendant should be stopped from even raising the issue, as it was only due to Defendant's misrepresentation that the lease was initially entered and the acceptance certificate signed.

The final argument of the Defendant is that Plaintiff failed to mitigate its damages. While Plaintiff was under an obligation to mitigate damages, it was Defendant's burden to prove that Plaintiff failed in this obligation. . . . Defendant did not present any facts at trial that established a means of mitigation for the Plaintiff. Instead, the evidence showed that Plaintiff tried to return the computer and that after this was rejected by the Defendant, Plaintiff gave Defendant other opportunities to perform its contractual duties. The breach was material and the measure of damages is that which was caused by the breach. . . . Under the circumstances of this case, the damages are the cost of the computer and paper products, the interest that Plaintiff was obligated to pay under the lease, and the expense incurred for an accountant on the request of Defendant's agent, totaling $15,659.04.

Conclusions of Law

1. A contractual relationship exists between Diversified and Olivetti, part of which is the oral agreement by Olivetti to perform the transfer of all base data and to train Plaintiff's employees.
2. The contractual duty owed to the Plaintiff was breached by the defendant as Defendant substantially failed to carry out its obligations of transferring the base data, of training Plaintiff's employees, and of making the computer system operational.
3. The injury suffered by the Plaintiff due to Defendant's breach is $15,659.04, and judgment will be entered for the Plaintiff in this amount.

REVIEW QUESTIONS

1. Black Tool and Die Company agreed to make a punch press die for White for $10,000. A one-month delivery time was agreed on. Black was ready to begin work on the die when White called and told Black to hold up until further notice. White then shopped around in an attempt to find a lower price for the die. White could not find a better price and called Black about two weeks later to tell Black to go ahead on the die, but Black refused, saying that its work schedule was now such that the die could not be completed within six months. White claims breach of contract and threatens to sue. What is the likely outcome of the case? Why?
2. Why is it necessary for courts to recognize an anticipatory repudiation in connection with contracts involving a structure that is to be built and installed?
3. Black, under a contract with White, built an automatic assembly machine to assemble drive mechanisms for automobile window regulators. The contract calls for a machine capable of producing 1,500 assemblies per hour. The resulting machine ran at a speed that would easily produce 1,500 assemblies per hour. However, its longest run since installation a month ago has only been about two minutes before it jammed. Frequently it will run only one or two pieces before stopping. The cause of jamming is slight variations in the dimensions of the component parts of the window regulator assemblies. The parts are manufactured by White in the same manner they have been produced for many years. Black had access to unlimited quantities of the parts while the machine was being built. According to the contract, Black's performance was finished when the installation of the machine was completed. White has paid $162,000 of the $180,000 agreed price of the machine. Black demands payment of the remainder. White claims a right to retain part or all of the $18,000 to compensate him for efforts spent in making the machine work. Has Black substantially performed? Can Black recover the remaining $18,000? Why or why not?
4. Green leased a building near two metal working plants for a period of five years. The lease specified no restrictions for use of the building. Green set up a tool and die shop that operated profitably for about a year when the first plant left and relocated in another city. Shortly thereafter the second plant was liquidated in bankruptcy. Green wants to avoid the lease, claiming commercial frustration in that the remaining tool and die work is insufficient to be profitable. Is Green bound by the lease or can Green get out? Why?
5. In regard to the case of *Cook Associates, Inc. v. Colonial Broach and Mach. Co.*, summarize the reasoning by which Illinois obtained jurisdiction over defendant. What is meant by a "long arm" statute?
6. In *Diversified Environments* v. *Olivetti*, how could Diversified have prevented its computer disaster while still accomplishing its desired objectives?
7. Look at finding of fact number 46 in *Diversified Environments* v. *Olivetti*. Why would Diversified's lawyers advise Diversified to keep making the lease payments on the computer equipment? (Hint: Perhaps the advice had little or nothing to do with Diversified's credit rating.)

Chapter 14

Remedies

We have already considered the requirements for the formation of a lawful contract. Chapter 13 examined the discharge of the obligations imposed on the parties by such contracts. We have observed that not all contractual obligations are discharged as the parties originally intended. When a party fails to perform as promised, a breach of contract has occurred. Each party to a contract has a right to obtain proper performance for the rights or performance given up. If this right is not satisfied, the law affords a remedy. The extent and type of remedy afforded is determined by the nature and extent of the breach.

We have defined a contract as an agreement enforceable at law. The enforceability aspect of the definition sets a contract apart from other agreements. Saying that contract rights are enforceable at law implies that there must be remedies available for their breach. An old equity maxim is that "For every right, a remedy." This pertains to contract rights as well as to other rights the law recognizes and protects.

Legal and equitable remedies exist to enforce a right, to prevent the violation of a right, and to compensate for an injury. Chapters 21 and 22 consider "tort" laws. Generally, this body of law defines the circumstances in which an injury due to another's actions or inactions will be compensated. Probably the most common remedy sought and obtained in both contract and tort is monetary damages. However, there are numerous instances in which damages will not afford an adequate or complete remedy. For this reason, other remedies have developed. Such remedies as *restitution, specific performance, injunction, rescission,* and *reformation* are examples of remedies that may be available under particular circumstances. In addition, a court of equity, with its historical origin based on unusual remedies, can combine and select remedies or, if necessary, invent new ones to fit new circumstances.

This chapter considers only the more common remedies of damages, restitution, specific performance, and the injunction. One or a combination of these remedies will be appropriate in nearly any case in which an engineer is likely to be involved.

DAMAGES

Damages are the compensation in money awarded by a court to one who has suffered a loss, detriment, or injury. Most breach of contract cases involve some sort of claim for damages. When the breach is not of a nature such that compensation is really justified, the injured party may still win the case and be awarded *nominal damages* (such as an amount of one dollar). An award of nominal damages merely means the court has recognized that there was an invasion of a technical right of the plaintiff. Of course, as with any other award, the loser will probably be assessed the court costs.

♦ Compensatory Damages

The usual reason for undertaking a damage action is to obtain *compensation* for an injury to one's person, property, or rights. To obtain compensatory damages it is necessary to prove that a right existed and was violated; in addition, the amount of damages suffered must be established with reasonable certainty. If the amount of damages cannot be reasonably established, the result is likely to be either an award of nominal damages or else no award of damages at all. The amount of compensation to which the plaintiff may be entitled usually is a question for juries.

The basic purpose of awarding compensatory damages is to make the plaintiff whole for any losses suffered by the defendant's wrongful conduct. The approach used to determine the compensatory damages awarded may vary depending on whether the dispute involves a contract right or a tort.

The *contract approach* attempts to compensate the plaintiff for his or her lost expectations. This approach does so by compensating for any out-of-pocket costs to the plaintiff and other damages such as profits missed as a result of the contract breach. The objective of this method is to place the plaintiff in the position that the plaintiff would have enjoyed if the contract had not been breached.

Of course, the damages claimed must be directly connected to the contract breached if the plaintiff is to be compensated. Black has a contract to build automation equipment for White. Gray states that if the automation works properly for White, Gray will be interested in a similar installation. Black may even have submitted preliminary plans and drawings to Gray. White then breaches the contract with Black. Black could collect compensation for the costs incurred on White's contract plus the profits Black could reasonably expect from White. However, Black probably could get nothing from White to cover Black's anticipated profit from Gray; Black's claim probably would be considered too speculative.

The *tort approach* attempts to compensate the plaintiff for the plaintiff's actual losses. The plaintiff's out-of-pocket costs are considered, and lost profits (or lost wages for an individual) also may be considered. The results, in terms of money damages, can be quite different depending on which theory is followed. However, the results also can be quite similar; a tort case can look very much like a contract case, and damages for a breach of contract can appear to be the same as those for a tort. It is not unusual to see someone pursue a recovery under both contract and tort theories in the same case. How can this happen? If you're injured in a car accident, you may decide to sue the manufacturer for breach of warranty and negligence, as well as for strict products liability. All of these types of actions are discussed later.

♦ Types of Compensatory Damages.

At this point, it is worth noting that different types of compensatory damages exist. More importantly, however, different rules apply to different types of damages with respect to the question of what damages can be recovered in a given case. "General damages" are damages that logically and naturally flow from the wrongful conduct or breach of contract.

Each kind of legal injury has its own kind of associated general damages. For example, the general damages a buyer of land might recover due to the seller's fraud may be different from the general damages one might recover due to injuries received in a car wreck. Usually, general damages need not be "foreseeable." In other words, it does not matter that the defendant did not, or even could not be expected to, foresee the nature and extent of the plaintiff's injuries. The courts essentially presume such foreseeability with respect to what are classified as general damages. Moreover, a plaintiff usually is not required to "specially plead" (that is, specifically include a claim for) general damages. Thus, when a plaintiff prepares the complaint to be filed against the defendant, the plaintiff's general allegation of damages will be enough.

As opposed to general damages, a plaintiff usually needs to specifically plead "special damages." Special damages consist of those damages resulting from a party's particular circumstances. Courts usually consider lost profits as a type of special damages. To recover such damages, the plaintiff must specifically plead in some detail the facts relating to the damages in the complaint. Moreover, special damages must be foreseeable to be recovered.

Suppose White Manufacturing Co. agrees with Black Inc. to buy a surface mounting system for use by White in manufacturing circuit boards to be included in personal computers. The system fails to work. Before White discovers the problems, the system ruins about $1,000 worth of White's inventory. Because the system fails to work, White also cannot deliver 20,000 circuit boards it promised to Green Computers. As a result, White cannot obtain the $20,000 profit it would have made on the contract with Green. It appears likely that the $1,000 would be general damages, whereas the $20,000 would be considered special damages. Should White be able to recover the $20,000 from Black? Assuming that the amount of lost profits can be shown with reasonable certainty and it is clear that White would have made the profits "but for" Black's breach of contract, the issue becomes one of foreseeability. If White had told Black, prior to the execution of their contract, that White's contract with Green would yield White $20,000 in profits, and had further told Black that White needed the system to perform the contract with Green, then White's lost profits seem likely to be considered "foreseeable" and therefore recoverable.

♦ Exemplary (Punitive) Damages

In certain cases the courts allow a party to recover more damages than the reasonable compensation for the injury or wrong suffered. Where a person's rights were violated or an injury caused under circumstances such as fraud, malice, oppression, or other despicable conduct by the defendant, *exemplary damages* (also called *punitive damages*) may be awarded. Such damages have a primary purpose of punishing the defendant for the reprehensible conduct and making the defendant an example to deter others from similar conduct. Secondarily, exemplary damages are added compensation for the shame, degradation, or mental anguish suffered by the plaintiff.

Exemplary damages are much more likely to be allowed in tort cases than in contract cases. At common law, exemplary damages could not be recovered for a breach of contract, no matter how willful the breach was. Exemplary damages are governed generally by the state laws on damages and the relevant principles vary from state to state. Large awards of punitive damages often grab headlines and generate a considerable amount of debate. Are they necessary or even useful to deter wrongful conduct? Do they deter outrageous corporate conduct and make the world safer? Recent efforts to persuade the U.S. Supreme Court that punitive damages are unconstitutional as "cruel and unusual punishment" have been unsuccessful.

♦ Liquidated Damages

It has become almost standard practice to provide some form of remedy such as "liquidated damages" in the wording of engineering contracts.

Liquidated damages are damages that are liquidated in the sense that they are a known, predetermined amount or measure of damages, such as damages of $100 per day for each day a contractor is late in completing the contract. Such provisions are enforceable in court only if they are made in good faith by the parties and are a reasonable estimation of the uncertain amount of the damages caused by delay.

If a court views the contract provision as one intended to secure performance rather than provide for just compensation, the court is likely to view the clause as a "penalty." If considered a penalty, the contract clause will not be enforced.

♦ Duty to Mitigate Damages

If there were no recourse to the courts to recover damages when your rights were invaded, you would certainly make every effort to keep your damages as small as possible. The law imposes a duty by requiring the plaintiff to *mitigate* (that is, minimize) the damages suffered by taking reasonable steps to minimize or avoid additional damages. If one is injured, every reasonable effort to keep the injury to a minimum should be taken. For instance, if an employment contract (to run for a certain period of time) is breached by the employer, the employee must actively seek work elsewhere. If the employee is successful, the difference between the two salaries (assuming the original job paid more) will be awarded; if the employee is unsuccessful after a reasonable effort to find subsequent employment, the total lost pay may be recovered. If the employee does not make a reasonable effort to find subsequent employment or refuses suitable work, the employee's damages may be considerably diminished. However, the employee probably need not take work for which he or she was not suited (for instance, an experienced engineer would be unlikely to be criticized for refusing employment as a farmhand); neither would the employee be required to move a great distance.

To further illustrate the concept of mitigation, suppose Black is a manufacturer of appliance parts, particularly chrome-plated ones. Black has a contract with White whereby White is to supply Black with nickel, at a stated price, for use in the copper-nickel-chromium plating process. During the life of the agreement, White raises the price of the nickel supplied, thus breaching the contract. Black could try to find an alternate source of nickel or pay White's increased price. Black might even use the increased price as an excuse to cease manufacturing appliance parts for Black's customers, relying on White's breach to cover any losses Black might sustain.

Either of the first two alternatives might be considered reasonable as an attempt to mitigate damages. Black probably cannot simply breach Black's contracts with the appliance manufacturers and pass along to White the damages assessed against Black. Neither could Black maintain an action for lost profits if Black ceased manufacturing parts on this basis. Black's damages suit probably will get only the difference between the contract price and the price actually paid for the nickel.

It might appear in the mitigating situation just described that the equitable remedy of specific performance would be fair. Such is usually not the case, however, unless a statute exists to make it available. Without a statutory provision, specific performance probably would be denied on the basis that monetary damages would be a sufficient remedy. Black could conceivably obtain the same quality and quantity of nickel from other suppliers, with the difference in price being the only source of damages.

RESTITUTION

Restitution is generally similar to an award of money damages. Awards based on restitution are usually made in money. Such awards are based on the plaintiff's having parted in good faith with consideration and the defendant's having breached a duty owed the plaintiff. The difference between restitution and damages lies in the purpose and amount of the award. *Restitution* only restores what is lost or the value of the thing given. This remedy focuses on the value received by the defendant that would be unjustly retained unless returned to the plaintiff. There is no attempt, as in a damages action, to compensate the plaintiff for what the plaintiff expected to gain (such as the plaintiff's lost profits). In effect, the defendant is required to return whatever consideration was received from the plaintiff. The courts, however, do not adhere strictly to this rule. They will not apply it where, by so doing, they offer a shield to the defendant for the defendant's wrongdoing. For example, if the defendant has made a profit through wrongful conduct, the profits of the defendant may be awarded to the plaintiff. The injured party generally is required to return what was received where it is reasonably possible to do so.

EQUITABLE REMEDIES

Specific performance and the injunction are the principal equitable remedies. Neither remedy may be used where an adequate remedy at law (for example, damages, restitution, or a statutory remedy) is available. However, either remedy may be used in conjunction with damages where damages alone would be an insufficient remedy. When either an injunction or specific performance would require extensive supervision of the court for enforcement, an attempt should be made to find a different remedy. When, for instance, specific performance of a contract to perform computer maintenance services is requested, it is likely that the court would deny the request. Such a remedy seems likely to require a fair amount of court supervision; hence, the court would look for a more appropriate remedy.

♦ Specific Performance

Simply put, the remedy of *specific performance* takes the form of a court ordering a party to perform the specific action required by a contract. The most common, though not exclusive, use of the remedy of specific performance occurs when a unique piece of property is involved. A piece of land, such as a city lot or a farm, is unique. So is an original painting by an old master or a tailor-made piece of automation equipment.

Courts, since ancient times, have considered land as unique (the extension of this concept to other types of property is of more recent origin). If a contract to sell a particular piece of land is breached by the seller, sufficient money damages might be awarded to allow neighboring property to be purchased. However, no two pieces of land have the same location, and it is likely that there would be other tangible and intangible differences; for instance, the view from one location may give a sense of security or be more appealing than the view from another location. Also, consider the uniqueness of the neighbors adjoining different lots.

Courts usually will not require specific performance of personal service contracts. However, a court might be persuaded to prohibit a person who has contracted to perform services to another from performing the same services for anyone else.

♦ Injunction

Originally, injunctions were only *prohibitive* in nature ("thou shalt not"). Now, in most jurisdictions, an injunction may be either prohibitive or *mandatory* ("thou shalt"). An order for specific performance can be viewed as a type of mandatory injunction. Even where injunctions must be prohibitive, it is possible to write what is, in effect, a mandatory injunction. In one case a tenant, enraged at his landlord, piled garbage on the front lawn of the tenant house before leaving. Though a mandatory injunction ordering the tenant to remove the garbage could not be issued in that state, a prohibitive injunction did the job just as well. The tenant was prohibited from allowing the garbage to remain on the lawn at the former residence.

The injunction is often used where irreparable injury to property is imminent—not just "possible," but very probable. The probable damage must also be a type of damage that could not be satisfactorily repaired. Damage to one's good will or reputation, for example, is generally considered irreparable.

In addition to a "permanent" injunction entered after a trial as part of the court's final judgment, courts can enter interim orders that amount to injunctions controlling a person's conduct pending a trial or final resolution of the case. In emergency situations, courts can issue "temporary restraining orders." Such orders are a type of injunction and are usually limited in duration to a matter of days or weeks. Pending a final trial on the merits, a court can also issue a preliminary injunction. Usually, preliminary injunctions are entered only after a hearing; unlike a temporary restraining order, a preliminary injunction usually remains effective up to the entry of a court's final judgment. The purpose of a temporary restraining order or preliminary injunction is usually to preserve the "status quo" pending a full trial.

The speed and convenience of the injunction as a means of enforcing a law often has appeal to the litigants (the parties to a lawsuit). If a particular law can be made to call for an injunction or a "cease and desist" order to be issued when the law is violated, the time and expense of a jury trial are often avoided. It is necessary only that the order be issued and probable violators informed. Any further violation would be a contempt of court possibly resulting in a jail sentence or a fine. The speed and simplicity of the injunction is often appealing to the party filing a lawsuit. Generally, however, the courts view injunctions as extraordinary remedies that are available only in limited circumstances.

ENFORCEMENT OF REMEDIES

A remedy without enforcement would be meaningless. The law must "have teeth" if it is to be effective. There are three common means of enforcing court awards against the loser in a suit at law; execution, garnishment, and attachment. Enforcement of equitable remedies usually takes the form of contempt of court.

After a judgment is rendered, the loser is expected to comply with that judgment. When security has been posted, the award may be deducted from it. When no security has been pledged and the loser does not comply with the court's order (by paying the amount of damages awarded), the other party may return to court for an order to confiscate property to satisfy the judgment.

Such an order is a *writ of execution*. The writ is addressed to the sheriff or other law enforcement officer and gives the officer the right to seize as much of the loser's property (both real and personal) as may be necessary to satisfy the judgment. The property so obtained is then sold at an execution sale, the proceeds being used to satisfy the award of damages, with any remainder going back to the loser.

Garnishment is the legal mechanism used to obtain the loser's property that is held by or under the control of a third party. Notice is given to the third party to turn over the judgment debtor's property in satisfaction of the obligation and to not pay a debt to the judgment debtor until the judgment is paid.

Attachment is a legal mechanism used when the defendant is not within the jurisdiction of the court, but some of the defendant's property is available. The attachment process usually takes place before the court proceedings to ensure that the plaintiff, if successful in court, will be paid the amount of the judgment. Because the seizure of the property occurs before a trial is held, the plaintiff is usually required to post a bond to protect the defendant. An attachment prevents the defendant from removing property from the court's jurisdiction prior to the court's judgment. If the defendant wishes to remove the attachment, the defendant can post a counter bond in an amount sufficient to cover the plaintiff's claim.

Execution, garnishment, and attachment are all limited by statutes in the various states. The homestead laws that limit creditors' rights also protect the loser in a damages action. Also, many state laws severely restrict garnishment, particularly when the wages of the head of a household are concerned.

Another device that should be considered as security or enforcement is the *mechanic's lien*. It is a lien against not only the structure built, improved, or worked on, but also generally against the land on which the structure rests. Generally, such a lien may be established by any unpaid laborers, contractors, subcontractors, materialmen, or others having a hand in the work involved. The mechanic's lien does not exist under common law or equity; instead, it is a creation of state statutes. The statutes tend to be similar and tend to emphasize benefits to laborers, materialmen, and subcontractors rather than those to contractors, engineers, or architects. The reason for this is that the latter group is much more likely to be in position to pursue a remedy for nonpayment by means of breach of contract.

Contempt of court is the principal means of enforcing equitable remedies such as injunctions or an order requiring specific performance. The extent of the enforcement is pretty much within the court's discretion. Generally, an order duly issued by a court that has been properly served on the persons to whom the order is directed must be obeyed by those persons. Any violation may amount to contempt. The court has considerable discretion, though, in assessing an appropriate fine or penalty. A court may order the offending party to serve time in jail, pay damages caused by the contemptuous (or contumacious) conduct, or set a daily fine for each day the contempt (violation of the order) continues. The remedy chosen may be coercive (that is, to try to force compliance with court's earlier order) or punitive (to punish for the violation of the order). In addition, the contempt proceedings may be criminal in nature or civil in nature. Contempt procedures can vary widely.

KROEGER v. FRANCHISE EQUITIES, INCORPORATED
212 N.W.2d 348 (Neb. 1973)

This was an action for damages for breach of contract. The defendant appeals from a judgment for the plaintiff in the amount of $4,000.

On November 9,1970, the plaintiff entered into a subcontract with the defendant whereby the plaintiff agreed to furnish labor, materials, and equipment and perform all carpentry work required to construct a service station and restaurant in Omaha, Nebraska. The contract provided the plaintiff would begin and terminate his work according to a "Critical Path Schedule" which was a detailed construction schedule set out in the contract. It provided the carpentry work would be performed between December 3, 1970, and January 28, 1971.

The plaintiff commenced work on November 20, 1970. In December the plaintiff received a telephone call from the defendant's job superintendent, Burt Smith, who said he "was shutting the job down." The plaintiff heard nothing further from the defendant until April 1971. The plaintiff then did 8 hours' additional work on the job but refused to complete the contract because carpenters' wages were about to increase and the job had been figured originally on the basis of being winter work.

The entire evidence at the trial consisted of the written contract and the plaintiff's testimony. The defendant's answer alleged the plaintiff had agreed to the closing down of the job because of the weather. The evidence was to the contrary. The plaintiff testified he did not agree to closing down the job; that there was temporary shielding and a portable heating unit on the job; and the weather did not prevent construction from proceeding.

The defendant also claimed it had a right to reschedule the work under the contract. The contract provided that if the critical path schedule was revised, the plaintiff would receive 3 days' notice when the job was to be ready for him and claims for damage for delays were to be made promptly. We believe that a delay in all construction from December to April was not a "revision" of the critical path schedule within the meaning of the contract and the contemplation of the parties.

The petition alleged the plaintiff had performed work and services in the amount of $800 and was further damaged in the amount of $3,500. The plaintiff testified he and his employees had spent 139 hours on the job; the job had been estimated on the basis of 1,000 hours' work at $8 per hour plus 15 percent profit; and he had been paid $241 by the defendant.

The plaintiff was entitled to recover the value of the work performed plus the profit he would have received if he had been allowed to complete the job within the time provided in the contract. . . . The plaintiff was required to furnish appropriate data to enable the trier of fact to find the damages with reasonable certainty. . . . Damages for breach of contract that are susceptible of definite proof are recoverable only to the extent the evidence affords a basis of ascertaining their amount in money with reasonable certainty. . . .

There was no satisfactory evidence of the reasonable value of the work performed and no evidence as to the profit the plaintiff would have realized if he had been permitted to finish the job as the contract originally provided. The contract required the plaintiff to furnish both labor and materials for "carpentry per plans and specifications" but there was no evidence concerning the plans and specifications and no evidence concerning the labor cost to plaintiff and his overhead expenses.

The evidence was not sufficient to sustain the finding that the plaintiff was damaged in the amount of $4,000. The judgment is reversed and the cause remanded for a new trial on the issue of damages only.

WIEBE CONST. CO v. SCHOOL DIST. OF MILLARD

255 N.W.2d 413 (Neb. 1977)

This is an action by Wiebe Construction Company against the School District of Millard, arising out of a contract entered into by the parties on July 22, 1969, under the terms of which Wiebe agreed to construct for the district a project identified as the Millard High School stadium at a cost price of $556,365. In accordance with provisions for changes in the contract contained therein, the contract was modified by the parties by change orders Nos. 1, 2, 3, and 4, calling for additional work and payments in the amount of $2,172.18, $2,067, and $8,553. The original contract required completion of the work within "350 consecutive calendar days" after " 'Notice to Proceed.' " The contract also provided that the contractor pay liquidated damages in the sum of $100 per day for each day of delay in performance.

Wiebe's petition contained three causes of action as follows: (1) $56,915.72 for balance unpaid on the contract price; (2) $42,042 for increased costs by reason of delay caused when problems not anticipated by the parties arose and the district and its engineers delayed in making decisions as to necessary specification changes; and (3) $1,361.21 for extra work performed in reconstruction of a sidewalk.

The district filed an answer and counterclaim. In its answer the district alleged that Wiebe failed to complete the work properly. It denied that unanticipated conditions were encountered and alleged that in any event they should have been anticipated by the contractor. In its counterclaim the district made factual allegations, and prayed for liquidated damages at $100 per day for 224 days' delay in completion of the project and for damages for certain defects in performance, including, among others, the claim that planks for certain stadium seats were not in accordance with contract specifications. Its total prayer for damages was in the sum of $62,112.65. In its reply Wiebe alleged, among other things, that the delay in construction was caused by indecision of the district's engineers and agents as to whether specification changes were required and what these changes would be.

The trial of the case began before a jury and after 5 days before the jury the parties waived a jury trial, presenting the remainder of the evidence and submitting the case to the trial judge.

The trial court entered a judgment in part as follows: "...the court finds generally in favor of the plaintiff and that there is due to the plaintiff from the defendant on the causes of action set forth in plaintiff's petition the sum of $44,880.68." The court also allowed interest at 6 percent from October 1, 1971, until April 30, 1976, the date of judgment, in the amount of $12,342.18. The court dismissed with prejudice the counterclaim of the district.

We will first discuss the claim of the district that it was entitled to liquidated damages for delay in performance. The entire project consisted of a football field, stadium, appurtenances, and a running track. The delay in performance arose only in connection with the running track portion of the contract. The evidence would permit the trial court to find the following. After construction was commenced a part of the site was found to be underlain with ground water and to be in a very spongy condition. When this matter was called to the attention of the engineers in November of 1969, it appeared that the condition would require a change in contract specifications and not merely more difficult work on the part of the contractor. As a consequence of this situation the engineers, in a letter from Wiebe, stated: "It would seem to us that a more reasonable approach to resolving this matter would be to wait until spring of 1970 to determine the change in the work." In a letter dated June 1, 1970, the engineers wrote to Wiebe: "We will provide you with additional details on the extra work to be performed for stabilization of the running track subgrade in the very near future." A letter from the engineers to Wiebe dated June 15, 1970, contained the following item: "6. The track stabilization detail will be made available, for construction purposes, upon the completion of all items listed above and no work should be performed on the track curb, subgrade base, etc., until these items have been corrected." As one consequence of the water condition, change order No. 4, dated September 1, 1970, and executed by the parties a few days later, was entered into. It called for an extra in the form of a drainage system which was in fact constructed. The system was not, however, completely effective. After that there were extended discussions between the parties as to what, if any, further specification changes were required. The engineers hired a soil expert who made recommendations. Under date of September 3, 1970, a fifth change order, which was a revision of several previously proposed change orders, was offered by the district's engineers. This change order made some additional specification changes in the preparation of the subgrade of the track and also provided for a 120-day additional time extension on the contract period in addition to a 28-day extension which had been granted by one of the earlier change orders. The proposed changes in change order No. 5 were not agreed to by Wiebe because of, among other things, a claim that it could not, on the basis of the proposed specification changes, make any reasonable estimate of quantities of certain materials involved.

It is now necessary to note portions of the four change orders which were in fact adopted as these portions pertain to the "contract period." In change order No. 1 the contract period was designated as August 18, 1969, to August 3, 1970 (350 days). The second change order provided for the same contract period. The third change order stated the contract period as 378 days and contained no substantive contract changes otherwise. Change order No. 4, dated September 1, 1970, in addition to the drainage system change, provided: "Contract period to be determined." Following change order No. 4 there were no further determinations as to what would be the contract period, but on September 15 or 16, 1970, Wiebe was instructed to proceed under the original track specifications. On October 28, 1970, the engineers performed a test of soil conditions and it was found the ground was too wet to proceed at that time. Wiebe, however, indicated that he would proceed nonetheless if the district would accept the risk of frost damage to the track. No agreement was reached at that point. The track was completed on October 25, 1971.

The foregoing evidence would be sufficient to support a finding that the district had waived the provision for time of performance, or that the parties had modified the time for performance of the contract. A contractual provision providing for an award of liquidated damages for delay in performance may be waived.... Thus dismissal of the part of the counterclaim concerning liquidated damages for delay in performance was supported by the evidence and cannot be said to be clearly wrong.

The evidence as to whether some of the work done was not in accordance with contractual specifications is to some limited extent in conflict. It presented simply a factual issue for the trial judge to determine. The court's dismissal of the counterclaim for defective work is therefore also supported by the evidence. We now turn to the issue of the propriety of the award of prejudgment interest. The district argues that a reasonable controversy existed as to Wiebe's right to recover the contract balance and the amount was a matter of dispute, therefore the claim was unliquidated and under the previous holdings of this court no prejudgment interest could be allowed.... In the cited case, although the contract price was a fixed amount, we upheld the trial court's denial of prejudgment interest because the contract was ambiguous as to the amount of work to be done for the stated contract price and there was a factual question as to whether all the required work had been done. This, we said, made the plaintiff's right to recovery, as well as the amount, a subject of dispute, therefore the claim was unliquidated even though the jury awarded the plaintiff the exact amount of his claim.

In none of the Nebraska cases cited by the district do we have the situation which confronts us here. Wiebe, in its first cause of action, sought a sum certain fixed by the terms of the contract. The district sought to offset the amount owed by a claim for liquidated damages for delay in performance as well as by amounts for claimed defects in the performance of the work. Wiebe relies upon the proposition that where the plaintiff's claim is liquidated, the existence of an unliquidated set-off does not prevent the recovery of prejudgment interest.... It is apparent that Wiebe's claim on its first cause of action for the balance owed on the contract is, by itself, a liquidated claim. It is the amount owed, computed in accordance with the terms of the contract. The amount was disputed only because of the claims the district made in its counterclaim. Because of the limited assignment of error, the only part of the counterclaim before us on this appeal is the claim for the contractually stipulated daily damages for delay in performance. The trial judge found against the district on that part, as well as on those for defective performance.

♦♦♦

We hold that in an action for a liquidated sum which represents a balance owing on a contract, the amount claimed does not become an unliquidated claim merely because of the assertion of an offset, and that if the trier of fact finds against the defendant on the offset, prejudgment interest should be awarded on the plaintiff's claim. In accord is *Raymond International, Inc.* v. *Bookcliff Constr., Inc.,* supra, a diversity case in which the trier of fact was the judge, a jury being waived, and where the court was required to apply Nebraska law. There the trier of fact found against the party asserting the offset and, after pointing out that where the contract if clear and unambiguous and the defense untenable, the Nebraska Supreme Court held that prejudgment interest is to be awarded. The federal court then went on to say: "The existence of an unliquidated set-off or counterclaim does not in this case bar interest prior to the entry of judgment.... In *Hansen* v. *Covell,* supra, cited by Wiebe, the California court has gone a bit further. It treats the offset, if allowed, as a payment and awards interest on the contract balance after deducting the offset. Whether we wish to go that far can be decided when a pertinent case reaches us. The trial court did not err in awarding prejudgment interest.

We now turn to Wiebe's cross-appeal in which it asserts that it should be awarded the full balance of the contract price in the amount of $56,915.72 instead of the $44,880.68 awarded by the court, a difference of $12,035.04. The only basis for reducing the contract balance would have been by allowing some portion of the counterclaim. This the trier of fact expressly disallowed. The general finding for the plaintiff and the disallowance and dismissal of the counterclaim cannot be reconciled with the award to the plaintiff of an amount less than the contract balance. "In a contract action, if the plaintiff has fully performed his contract he is entitled to the contract price."... Wiebe has not assigned as error disallowance of the counterclaim except that part on liquidated damages. This we have already treated.

The judgment is reversed and the cause remanded for further proceedings consonant with this opinion.

Reversed and Remanded.

REVIEW QUESTIONS

1. Black hired White Automation to build a special machine to be used by Black in the manufacture of automobile door handles. The door handles were to be sold to Gray Motor Company. The price of the special machine was to be $200,000. Shortly before work was to begin on the special machine, Gray canceled the order for door handles and Black immediately canceled the contract for the special machine. Does White have a right to resort to legal action? If so, against whom and for how much?
2. What is the purpose of (a) nominal damages, (b) compensatory damages, (c) exemplary damages?
3. Distinguish between general damages and special damages.
4. Green bought a new car from Brown Motor Sales for $30,000. A few days later, Green attempted to pass another vehicle, and the steering linkage locked when Green turned the wheels to the left. The resulting crash destroyed the car and sent Green to the hospital. Green's hospital bill amounted to $8,800; during the first month away from Green's job, Green received full pay from her employer ($3000 per month), but in the next two months she did not. When she returned to work with a 20 percent disability, she was asked to take a job paying $2,000 per month because of her inability to perform her former job. Green is 45 years old. Examination of the wrecked automobile showed that one joint in the steering linkage appeared too tight and that there was no grease fitting at the joint and, apparently, there never had been one even though a hole had been drilled and tapped for the fitting. Does Green have a right of action against anyone? If so, for how much? Based on what theory?
5. Why should punitive damages be awarded to a plaintiff in a fraud action but denied when the action is based on a breach of contract?
6. In *Kroeger* v. *Franchise Equities, Incorporated,* what additional evidence will be required to establish the extent of damages?
7. In the case of *Wiebe Const. Co.* v. *School Dist. of Millard,* calculate the total award of damages Wiebe may expect based on the court's opinion.

Chapter 15

Sales and Warranties

The building of any structure, whether a productive machine, a building to house it, a road, or a bridge, requires the purchase of many things. In a large organization, a purchasing department buys what the engineer specifies; in a small firm, the engineer may need to do some of the buying. In either case, the engineer's job includes at least effective recommendation of the goods and services to be purchased and a vital concern with their adequacy and delivery. A manufacturing engineer or process engineer may not be responsible for making the actual purchase; nevertheless, the engineer remains very much concerned with the items to be bought, because they will be components of the final structure or process. In addition, a target date for completion must be met, and delay in receipt and installation of components may be quite costly. From this standpoint, the engineer must be concerned also with the transportation of the items purchased, and even with the financial arrangements involved.

When things go wrong—when a vendor claims that a product will perform according to the engineer's specifications, but the product fails to do so, for example—does the engineer have any recourse? Such problems do arise, and engineers will benefit from some understanding of their rights and responsibilities regarding sales and warranties.

The term *sale* can refer to a transfer of real property, personal property, or even intangible property. As it is used in this chapter, however, it refers only to the transfer of ownership of tangible personal property, for a price usually stated in money. In other words, we are here concerned only with contracts for the sale of goods. The law that governs such transactions is set forth in practically every state by the Uniform Commercial Code (UCC).

OWNERSHIP

Two concepts basic to the sale of goods are *ownership* and *risk of loss*. Because a sale involves the transfer of ownership, it can be important to know who owns the goods at a particular time. Generally, only the party who owns the goods can lose them, but this is where the risk-of-loss concept comes in. When goods are stolen or destroyed by fire, flood, or other catastrophe, the one who loses (at least, initially) is the party who had the risk of loss.

The UCC allows the parties a great deal of freedom to determine when title (or ownership) passes from seller to buyer and when the risk of loss passes from seller to buyer. The parties can have risk of loss pass at the same time as title passes. However, there is no requirement that they do so. In most situations, a buyer or seller can obtain insurance to cover possible losses or damage to the goods. Thus, buyer and seller may negotiate about when and how risk of loss and title are transferred and about which party pays for insurance (or, if no insurance is purchased, who bears the risks). Consequently, the time when title and risk of loss passes from seller to buyer and the shifting of ownership risks are important.

♦ Title

For title to pass from seller to buyer, the parties must identify the goods as those to which the contract refers. This identification can be made at any time and in any manner explicitly agreed to by the parties. The goods do not have to be in a deliverable state before they can be identified. Also, despite the identification of goods (and the passing of title), the risk of loss remains on the seller until the risk is passed, either as agreed by the parties or pursuant to the terms of the UCC (some of which are discussed below).

Generally, title to goods passes in any manner and on any conditions explicitly agreed to by the parties. In the absence of an explicit agreement, title passes to the buyer at the time and place at which the seller completes performance with reference to the physical delivery of the goods.

For example, Black, in need of 20 cooling fans, visits White's warehouse, where White has about 100 such fans stored. After examining a few sample fans, Black agrees to take 20 of them at an agreed price per fan. At this point there is no sale: White still owns the entire 100 fans, and should anything happen to them, the entire loss is White's. Now assume that Black asks White to separate out 20 fans. Black intends to go get a truck to pick up the fans. If no other material thing remains to be done, Black has title to the 20 separated fans. Suppose now, however, that the parties agreed to have White deliver the fans to Black. Since it appears to be their intent to have title pass on delivery, title would not pass until delivery had occurred. Of course, if the fans were not yet in existence and Black purchased them merely by a description or a sample, title could not pass until the fans were made or at least started and tagged or appropriated for (or somehow "identified" as relating to) Black's contract.

♦ Risk of Loss

F.O.B., F.A.S., C.I.F., and C. & F. are abbreviations commonly used in contracts to describe the terms concerning shipment of goods from seller to buyer. Among other considerations, such terms specify the point at which the risk of loss passes from seller to buyer. For instance, *F.O.B. San Francisco* (which stands for *free on board*) indicates that when the seller has delivered goods into a carrier's possession in San Francisco with instructions that they be delivered to the buyer, the risk of loss passes to the buyer. Until the goods reach this point, though, the seller suffers if the goods are damaged or lost. *F.A.S.* generally means the same thing with respect to delivery by ship. The term *F.A.S. San Francisco*, then, would mean that the seller is required to deliver the goods alongside the vessel in the manner usual in that port or on a dock designated by the buyer.

C.I.F. and *C. & F.* are similar to the concepts of F.O.B. and F.A.S., but they require the seller to do more than just hand the goods over to the common carrier. In a C. & F. (cost and freight) sale, the price includes the cost of the goods and freight to the named destination; moreover, the seller must obtain a negotiable bill of lading for the goods, load the goods, and send all the documents plus an invoice to the buyer. A C.I.F. (cost, insurance, and freight) contract requires the seller to do everything required in a C. & F. sale in addition to insuring the goods. The buyer must pay the price of the goods (which, of course, includes shipping costs and any other extras) when the documents arrive. In addition, the buyer has no right to wait for the goods or inspect them prior to making payment; instead, the buyer must make payment against tender of the required documents. If the goods are lost or damaged in transit, the buyer has no action available against the seller. The buyer, however, would probably recover under the insurance policy and, failing that, the buyer can file a legal action against the carrier.

Despite the standard definitions given above, the buyer and seller can use these abbreviations in their contracts and still negotiate their own terms. That is, if the wording of their agreement makes it clear that they intended another meaning (for instance, giving the buyer the right to approve of the goods before paying for them), such will be the court's interpretation. The point is that where these terms are used and not modified by some other agreement, the UCC definitions dictate their meaning.

♦ Security Interest

In our economy, buying on credit accounts for a large number of sales. People (and businesses) buy goods and services now and agree to pay later. But the simple promise to pay later leaves something to be desired. When "later" comes, the buyer may have many debts and very few assets. Without some added security, the seller may be reduced to the status of creditor in bankruptcy and receive only a few cents for each dollar of the debt. In response to these concerns, arrangements have been devised to improve the seller's position (and willingness to sell on credit). Under the UCC the arrangement is known as a *security interest*. When a seller has a security interest in the goods, the seller retains certain rights to the goods sold until the buyer has paid for them.

Consider Black, who is in need of a refrigerator and lacks funds sufficient to pay the entire cash price. Having located the refrigerator of choice, Black arranges to purchase the refrigerator by paying $90 immediately and agreeing to make a series of 10 monthly payments of $90 each to the Brown Appliance Company. To "secure" Black's promise, Brown wants to retain a security interest. To fortify Brown's rights, Brown may want to file a *financing statement* regarding the security interest. Among other things, a financing statement should include the names and addresses of the debtor and the secured party, and a description of the property covered by the security interest. Most often, a seller of goods or a lender usually prepares the financing statement by completing a blank form. The place of filing differs from state to state; Brown may want to file it in the county where Black lives and in the secretary of state's office. If Brown were to keep the refrigerator pending full payment, the filing of the financing statement would not be required. This is because Brown's interest is a special type of security interest. It is a "purchase money security interest" (that is, Brown loaned Black the "purchase money" for the collateral) in a consumer good.

Default

Still considering Black's refrigerator purchase, suppose that Black makes eight of the ten required payments but cannot make the last two. In other words, Black defaults. The possibility of this happening was, of course, Brown's reason for obtaining the security interest in the first place. So now what can Brown do about it?

When a debtor defaults, the secured party (in this case, Brown) can regain possession of the goods. In other words, Brown can *repossess* Black's refrigerator. However, Brown has certain obligations, and Black has certain rights even after default and repossession. If the secured party elects to repossess (other methods of enforcing the security interest may be appropriate), that party must proceed in a commercially reasonable manner. For example, if the goods are resold, the secured party must use the proceeds to pay the debtor's obligation and the reasonable costs of repossession and then turn any remainder back to the debtor.

However, repossession and resale constitute only one of several possible remedies available to the secured party. The remedy the secured party will elect to use depends to a considerable extent, of course, on the nature of the goods involved. One alternative is simply to obtain a court judgment against the debtor in default, using the security agreement as proof of the debt. The debtor, then, retains possession of the goods, and the secured party can use the powers of the court to collect the debt. Another alternative is to repossess the goods and lease them, using the rental payments to satisfy the balance due and costs. The secured party may sometimes even repossess the goods and keep them to satisfy the debt, but the debtor must be notified that the secured party intends to do this. The UCC attempts to balance the rights of both parties. Brown, as secured party, has a variety of remedies if Black defaults on the refrigerator payments. On the other hand, Black has a right to expect fair treatment from Brown if, for some reason, it becomes impossible for Black to make the payments.

♦ Sale on Approval and Sale or Return

A device frequently used to sell goods is to place them in the prospective buyer's hands for a period of time. Of course, this method can be highly effective because the consumer often finds it difficult to surrender possession at the end of the trial period. This technique can also be highly effective with merchants: they can take goods to sell and then return them if they do not sell. The UCC addresses two distinct types of such sales activity.

In a *sale on approval*, the buyer is given possession of the goods to use or inspect for a period of time. While the goods are in the prospective buyer's hands, the title to them rests with the seller. Thus, any loss of the goods is the seller's loss, with, of course, the right to recover if the prospective buyer caused the loss. *Approval* may be either express or implied. Expression by the buyer of a willingness to take title to the goods constitutes approval. If a time limit is stated and the goods are held beyond the time limit, the buyer's approval may be implied. If no time limit was stated but the goods are held beyond a reasonable period, approval also may be implied. If the buyer uses the goods as his or her own, exceeding what would be considered a reasonable trial, the buyer's acceptance also may be implied. Until the buyer has registered approval, though, the prospective buyer in a sale on approval arrangement is merely a bailee[1] of the seller's goods.

---

1. A *bailee* is one who possesses property belonging to another (the *bailor*) with that person's acquiescence.

A sale in which the seller gives the merchant or dealer the right to return the goods at the merchant's option is a *sale or return* transaction. In such sales, the risks of ownership of the property pass to the merchant. Frequently, such sales take the form of sales to a merchant "on consignment" or "on memorandum." Return of the goods revests title and ownership risks in the seller. If a time limit is stated and the goods are not returned in that time, the sale to the merchant becomes final. If no time limit is expressly agreed to by the parties, the merchant has a reasonable time in which to return the goods.

SALES CONTRACTS

The law of contracts sets forth certain rules for the formation, interpretation, and discharge of contracts. If the parties to a contract agree on a specific provision, that provision controls the contract. But if the parties do not agree to a specific provision, the UCC's rules probably will supply the applicable rule of law. In short, the UCC fills any gaps in the parties' agreement.

According to the law of contracts, the offeror determines the terms of the acceptance (even, perhaps, the means of acceptance). Any alteration of the offer by the offeree in the intended acceptance constitutes a counteroffer. The formation of contracts, both at common law and under the UCC, was considered in Chapter 7.

Many contracts that would have been considered unenforceable for a lack of certainty are enforceable under the UCC. For example, the parties may choose to determine the quantity of product to be delivered later. Or they may choose to base the price on a future market quotation or a later arrangement by the parties. Such contracts would probably be considered questionable as "gentlemen's agreements" (or agreements to later agree) under the common law of contracts. But under the UCC, such agreements probably have the legal status of fully enforceable contracts.

The main purpose of the UCC is to reflect the intent of the parties more accurately. If it is apparent that the parties intended to make a contract, the law attempts to enforce the essence of the agreement even if some terms are missing or indefinite. Generally, the parties retain a great deal of freedom to structure their relationship. Where the contract terms are unclear (either because of ambiguities in the language used or the parties' failure to consider certain issues), the UCC attempts to provide rules that supply or define the contract's terms.

WARRANTIES

A *warranty* (or guarantee) provides added assurance to a buyer. The common-law courts early on adopted the concept of privity of contract. Under this view, only a party to the contract could sue on the warranty. Thus, if you bought a car and a family member was injured while driving it, your family member could not sue for a breach of warranty, such as that the car's steering system was defective. (As discussed in the chapter on products liability, that concept has eroded in terms of who may receive the benefits of a warranty.) In this section, however, we will discuss the two major types of warranties: express warranties and implied warranties.

♦ Express Warranties

A warranty is a promissory statement. In making a warranty, the vendor makes a contract somewhat similar to insurance. The vendor essentially agrees to assume a risk that would normally be borne by the buyer. Here, it is important to distinguish between statements that may be fraudulent and warranty statements. You recall that *fraud* involves a false representation of a fact—something in the past or present. *Warranty*, on the other hand, has to do with either the present or the future.

When Black sells White an automatic screw machine, for example, and tells White that it has just been overhauled, when in truth it has not been, such a statement is fraudulent. If Black, on the other hand, promises to repair the machine if it should break down in the first year White uses it, this statement constitutes a warranty. Black is promising to take over White's risk of repairing the machine during the year after White's purchase.

Statements having to do with the present, though, are a little harder to distinguish. A statement by Black that all the collets and pushers on the machine have been replaced and are new would be a statement regarding fact; if untrue, the statement could be the basis of fraud. In contrast, a statement that the machine is in such a condition that it will be useful in the manufacture of White's product would constitute a warranty. Another distinction between a warranty and a fraudulent representation is that a warranty becomes part of a contract. In cases where a false representation has been made to induce a sale, the buyer may elect to pursue either a remedy for fraud or a remedy for breach of warranty.

Opinions

A warranty is a statement of fact, not one of opinion or judgment. With few exceptions, a statement made as the seller's opinion cannot constitute a warranty. A statement by the vendor that the merchandise is "first rate," "the best," or "superior quality" is usually construed as sales talk, or "puffing." The courts have long adhered to the idea of *caveat emptor* ("let the buyer beware"). According to this logic, buyers are free to inspect goods before buying them and are free not to buy if the seller does not allow them to inspect or if they find something wrong.

The recent tendency in the law, however, is to place more responsibility upon the seller. It might even be called *caveat vendor*. This trend is based upon the notion that many statements a vendor makes, in effect, relieve the buyer of the duty to ascertain the value of the goods. An affirmation of value having this effect has been held to be a warranty in some recent decisions. Also, many states have adopted statutes aimed at certain deceptive trade practices. Under such statutes, representations may be actionable even though the representations fail to rise to the level of a warranty.

Description or Sample

Contracts for the sale of goods often involve samples of the goods the buyer is to receive or descriptions of them. The use of such descriptions or samples generally constitutes an *express warranty* of what the buyer is to get. Because the buyer normally does not have an opportunity to examine the goods when a description or sample is used, the seller would have a chance to substitute inferior goods. The warranty is meant to prevent this. Accordingly, the UCC provides that any description of the goods (or any sample or model) that is made part of the "basis of the bargain" creates an express warranty that the goods will conform to the description (or the sample or model, as the case may be).

♦ Implied Warranties

In addition to the seller's express warranties, there may be *implied warranties*. These implied warranties automatically exist in all sales contracts unless they are expressly disclaimed by the parties.

Good Title

In a contract to sell goods, there is an implied warranty that the buyer will receive *good title*. This guarantees that there are no rights or liens on the goods other than those of which the buyer is made aware. In other words, no other person has valid claim to the merchandise in question. Specific exceptions to the warranty of good title exist in the form of sheriff's sales, certain auctions, mortgagee's sales, and the like. Such sales are often authorized by law, but they include no assurances regarding prior owners or lien holders.

With a related warranty, the seller guarantees that the goods are free of any rightful claim of a third person by way of infringement or the like. Such a warranty applies to sellers who regularly deal in the goods sold. Thus, a seller of an automated materials-handling system impliedly warrants that the system does not infringe someone else's patents. However, if the seller builds goods in accordance with the specifications furnished by the buyer, the buyer indemnifies the seller against any claim against the seller that might arise out of such specifications.

Fitness for Described Purpose

The seller is assumed to be more familiar with the goods than the purchaser. For example, a buyer makes a purchase by describing to the seller what the goods are to do, with the seller selecting and then supplying the goods for that purpose. Because the buyer relies on the seller's skill or judgment to select appropriate goods, the goods come with an implied warranty of fitness for the buyer's purpose (as he or she described it to the seller). This warranty is not applicable, however, if the buyer orders goods according to a trade name or trade specification. When the buyer orders in such a way, it makes no difference if the seller knows of the intended purpose and believes the goods will not fit the purpose.

For example, Black requires a punch press for a blanking operation on a production line being set up. Black's calculations (in error) show a requirement of a 40-ton press. White, with full knowledge of the purpose intended, is called upon to supply the press and does so. On the first day of operation, the punch-press crank breaks, and Black brings an action on the implied warranty of fitness for the purpose of the press.

Since Black specified the press, Black cannot recover for any breach of such a warranty. If, on the other hand, Black had asked White to supply a press for the blanking operation, allowing White to determine the required press size, the fitness of the press for Black's purpose would have been impliedly warranted.

Merchantability

All goods sold by a merchant are subject to an implied warranty of merchantability. To be *merchantable*, the goods must at least (1) pass without objection in the trade under the contract description; (2) be of fair to average quality within the description (in the case of fungible goods); (3) be fit for the ordinary purposes for which such goods are used; (4) be of even kind, quality, and quantity within each unit and among all units involved; (5) be adequately packaged and labeled as may be required by the agreement; and (6) conform to the promises or statements of fact made on the container or label. In the above example, if it could be shown that the punchpress crankshaft had not been properly heat treated and, therefore, could not withstand a 40-ton force, Black might recover for breach of the implied warranty of merchantability.

♦ Disclaimers

Parties to a sales contract are free to contract in any lawful manner they choose. Frequently, sales contracts contain *disclaimers*—clauses that, in effect, state that there are no warranties, express or implied, and that the buyer takes the goods at his or her own risk. As long as public policy is not seriously involved, such clauses are usually lawful and binding. A contract provision that the buyer takes the goods "as is" or "with all faults" will be held to relieve the seller of all implied warranties. Of course, even in an "as is" sale, the merchandise sold must be what it is purported to be. That is, sale of a vertical milling machine indicates that what is bought will constitute a vertical milling machine even though the term "as is" is used in the sale.

Generally, any such disclaimer of the implied warranty of merchantability must mention merchantability and, if in writing, it must be "conspicuous." This means that the disclaimer must stand out from the rest of the writing and attract the attention of the reader for it to be effective as a disclaimer. In other words, if one wishes to exclude implied warranties, the seller cannot hide such wishes from the buyer at the time of the sale.

Moreover, warranties are construed to be cumulative. Thus, the buyer's rights under any implied warranties are added to the buyer's rights under any express warranties. Generally, the courts try to construe disclaimers to be reasonably consistent with any express warranties; in cases of doubt, however, courts often construe disclaimers of warranties narrowly (i.e., in favor of the buyer).

PERFORMANCE AND BREACH

The parties' obligations in a contract are for the seller to transfer and deliver the goods and for the buyer to accept and pay for the goods in accordance with the contract. In other words, the seller is to package and ship goods that conform to the contract (i.e., meet or exceed the warranties applicable) so that the goods arrive at the time and place specified by the agreement. Assuming that the seller meets these obligations, the buyer is obligated to accept the goods and to pay the contract price. In the vast majority of such sales contracts, the parties perform without a problem. The rest of this chapter considers the occasions when a party does not.

The UCC requires that the seller "tender delivery" by placing conforming goods at the buyer's disposal. The manner, time, and place for tender are to be determined by the parties' agreement and the relevant provisions of the UCC. Unless otherwise agreed, the buyer has a right to inspect the goods in a reasonable manner before accepting or paying for them.

If the goods conform to the contract, the buyer is obligated to accept them and pay for them. If the goods received by the buyer are nonconforming, however, the buyer has several options. The buyer may (1) accept the goods sent and pay for them at the contract price, subject to the buyer's right to sue for the cost of the nonconformance of the goods; (2) reject the entire shipment of goods; or (3) accept any conforming commercial unit or units and reject the rest.

If the buyer rejects the goods, he or she must inform the seller within a "seasonable" time after rejection and state the reason for rejection. The seller may then instruct the buyer as to disposition of the goods. If no instructions as to the disposition of the goods are forthcoming, the buyer is simply required to act reasonably according to the nature of goods and the circumstances. If the buyer rejects the goods but the time allowed for the seller's performance has not yet passed, the seller may notify the buyer of the seller's intention to "cure."

A seller may then cure by making a delivery of conforming goods within the contract time for performance.

A buyer may accept the goods by expressly telling the seller that he or she accepts them. Probably more often, though, the buyer accepts the goods by continuing to retain possession of them and to use them after a reasonable opportunity to inspect them has come and gone. If the buyer accepts nonconforming goods, the buyer must notify the seller of the nonconformity within a reasonable time after the buyer discovered (or should have discovered) the nonconformity. A failure to give such notice usually will bar the buyer from any remedy.

In some situations, a buyer may accept nonconforming goods because of the seller's representations about fixing any problems. (Refer back to Mr. Fleegal's statements to Diversified Environments in the case of *Diversified Environments* v. *Olivetti*, at the end of Chapter 13.) A buyer may revoke an acceptance of a lot or commercial unit whose nonconformity "substantially" impairs its value if the acceptance was made either (1) on the reasonable assumption that its nonconformity would be cured, but the nonconformity has not been "seasonably" cured or (2) without discovery of the nonconformity, and such acceptance was reasonably induced either by the difficulty of discovery or by the seller's assurances. Thus, if a seller assured that a nonconformity would be fixed (but did not fix the nonconformity) or assured that a defect or problem was really not a nonconformity, then the buyer may revoke the acceptance. If the defect is latent and cannot be reasonably detected by normal inspection techniques, a buyer may also have a right to revoke an earlier acceptance.

Suppose Green ships the Brown Company 10,000 zubit components known as ZIP-2s. Brown Company's receiving inspection shows the lot of ZIP-2s to consist of about 20 percent defective units and, therefore, to be unacceptable. Brown Company rejects the entire shipment of 10,000 ZIP-2s and notifies Green. Green may require the entire lot of 10,000 to be returned or may ask Brown Company to sort the 10,000. If the nonconformance could shut down a production line, the idea of mitigation of damages would probably require Brown Company to sort and use the good ones. In any event, Green should be liable for any added costs caused by the nonconformance of the shipment. If the ZIP-2s are nonconforming, Green probably will be liable for Brown Company's inspection costs. On the other hand, Green would not be liable for such inspection costs if the goods entirely conformed to the contract specification.

BUYER'S REMEDIES

If the seller breaches the contract, the buyer has a right to take legal action. The parties may anticipate the buyer's damages, however, in a "liquidated damages" clause in the contract. As previously noted, such a clause may be enforced. If considered a penalty, however, the clause will not be enforced.

If the seller breaches the contract by failing to deliver the goods or by repudiating the contract, the buyer has several options available. First, the buyer may procure substitute goods from another source. The buyer's damages, then, would reflect the difference between the costs of cover (i.e., the price for the cover goods) less the contract price, together with any costs resulting from the delay involved or the costs of obtaining the goods from another source. Second, the buyer may seek damages based on the difference between the market price for the goods at the time the buyer learned of the seller's breach less the contract price for such goods. Additional damages, such as lost profits, also may be recoverable.

The UCC also provides the buyer with the right to replevin (or recover possession of) the goods if, after reasonable effort, the buyer cannot procure substitute goods or if the circumstances indicate that such effort will be unavailing. Finally, if the buyer can establish that the goods are unique, the buyer can resort to the equitable remedy of specific performance. In effect, the buyer can seek an order requiring the seller to deliver the goods to the buyer.

If the buyer rightfully rejects or revokes acceptance, the buyer has a security interest in the goods to the extent of any part payments or expenses reasonably incurred in their inspection, receipt, transportation, or the like. At that point, then, the buyer's rights are those of the secured party.

What if the seller delivers the goods as required, but the goods are nonconforming? Damages for breach of contract may take the form of a setoff or deduction from the price to be paid for the goods. If making a partial payment may leave the buyer in breach for failure to pay the contract price, the buyer must notify the seller of his or her intention to do so. The buyer's right to take legal action based on the seller's breach eventually ends if the buyer does nothing about it. The statute of limitations for such claims specifies this length of time as no longer than four years. Similarly, the seller's time to sue a buyer for breach is four years. In many states, statutes allow the parties to shorten this time limit in their agreement; such statutes, however, usually set a minimum time, such as one year, to which the parties can agree as a shortened statute of limitations.

SELLER'S REMEDIES

If the seller delivers goods to the buyer and those goods conform to the buyer's specification, the seller expects to be paid for them. The buyer, however, might refuse to accept the goods, might repudiate the contract, or might not pay the price. The UCC provides the seller with a set of remedies in such cases. One rather obvious remedy is to sue the buyer for damages. Of course, the seller also may simply cancel the contract if the buyer breaches. The seller has other alternatives, too, and they are discussed below.

♦ Withholding Delivery

If the buyer breaches the whole contract, the seller has a right to retain the goods and withhold delivery. If the goods are in the hands of a common carrier or other bailee when the seller has the right to withhold delivery, the buyer's breach allows the seller to stop delivery of the goods by notifying the carrier or bailee. However, the seller's right to stop delivery automatically ceases when the goods are received by the buyer, when a bailee of the goods (except a carrier) acknowledges to the buyer that the bailee holds the goods for the buyer, or when any negotiable document of title is negotiated to the buyer.

A problem arises when the goods are being manufactured for the buyer and are to be shipped in a series of shipments. When the buyer's breach occurs, the goods may be in various stages of completion. In such situations, the seller must decide whether to finish any work in process. Depending on the nature of the goods, they might conceivably be finished and sold to another buyer. If the first buyer is the only customer for them, however, further work on the inventory would seem pointless.

♦ Seller's Resale

If the buyer breaches the contract, the seller has a right to resell the goods to another. The seller may do this either by public or private sale. The UCC imposes certain conditions on such sales to protect the buyer. For example, the terms, method, manner, time, and place of resale must be commercially reasonable. If the sale is public, even the seller may purchase the goods. Any buyer at the resale takes the goods free of any claim by the original buyer.

If the resale yields a lower price than the original buyer was to pay (i.e., the contract price), the seller has a right to recover from the buyer the difference between the resale price and the contract price, plus incidental costs. However, if resale yields a profit for the seller, the original buyer has no claim on that profit.

Unless the parties have agreed otherwise, when one party breaches a contract, the other's right to compensatory damages usually includes lost profits. In other words, the goal is to place the non-breaching party in the position he or she would have enjoyed had no breach occurred. Since most commercial transactions are undertaken for a profit motive, claims for lost profits are fairly common.

PLEASANT v. WARRICK

590 So2d 214, 25 ALR5th 922 (Ala. 1991)

Ingram, Justice.

The plaintiff, E.L. Pleasant, sued Rodney Warrick, John Deere Industrial Equipment Company, and Deere Credit Services, Inc. ("Deere Credit"), for conversion, negligence, and wantonness for destruction of a logging skidder, which he contends had been wrongfully repossessed. All of the defendants filed timely motions for summary judgment supported by affidavits and deposition excerpts. Contending that Pleasant was in default at the time of the repossession and that the logging skidder had been lawfully repossessed. The trial court entered summary judgment for the defendants.

The dispositive issue on appeal is whether the trial court erred in granting the defendants' motions for summary judgment.

The record, in pertinent part, reveals the following: Pleasant, a timber subcontractor, purchased a John Deere 440C logging skidder from Warrior Tractor and Equipment Company, Inc., an independent John Deere dealer. Pleasant traded a used John Deere 440B skidder and received a credit of $7,117.21 against the total purchase price of $25,268.24, leaving a balance due of $18,151.03. The skidder was used as collateral for a security agreement related to the $18,151.03 balance. Pleasant agreed to make 36 consecutive monthly payments of $721.66 each. The security agreement was transferred or assigned to defendant Deere Credit for administration and collection.

The security agreement, which was signed by Pleasant, contained the following pertinent language:

"This contract shall be in default if I (we) shall fail to pay any installment when due.... In any such event Lender may take possession of any Goods in which Lender has a Security Interest and exercise any other remedies provided by law, and may immediately and without notice declare the entire balance of this contract due and payable....

"...Waiver or condonation of any breach or default shall not constitute a waiver of any other or subsequent breach or default."

The evidence is not in dispute that Pleasant was behind with his payments almost from the beginning. In fact, Pleasant admitted that his account was in default and that Deere Credit had sent him past-due notices of his default. At the time of repossession, which was only 20 months after Pleasant had purchased the skidder, Pleasant was approximately 6 months past due in making the payments pursuant to the security agreement. The record shows that Deere Credit, through its agent, Rodney Warrick, made every reasonable effort toward working out Pleasant's default in order to bring his account current. There was evidence, although disputed, that, instead of a full payment every 30 days, Pleasant was permitted to make partial payments at more frequent intervals. Pleasant contends that this constituted a modification of the original agreement and that, therefore, there was no default. Although we find that Deere Credit may have tried to accommodate Pleasant in an effort to bring Pleasants' account current, we do not find that this accommodation constituted a modification. Nevertheless, even if we concluded that there was a modification, Pleasant admitted that he did not make all of the partial payments. Further, the record reveals that two of the partial payments were made with checks that were dishonored for insufficient funds. The record does not reveal that Pleasant brought his account current at any time.

The record further reveals that on September 1, 1989, Warrick met with Pleasant concerning his account, which was then over $4,000 past due. Warrick had spoken with Pleasant's wife the previous night and had told her that he was coming to see Pleasant about his past-due account. Warrick testified that he obtained directions to the skidder and that he then drove to its location and verified its identity. Pleasant testified that Warrick did not tell him that he was going to repossess the skidder, only that he was going to check the condition of it, and that he would then come to Pleasant's home later that afternoon to get some money Pleasant would have for him.

When Warrick found the skidder, he testified that he attempted to obtain a truck to haul the skidder. When no truck was available, Warrick asked for and received permission to park the skidder at a local dealership over the weekend. When Warrick returned to the skidder, he testified that he checked the oil, the water in the radiator, and the fuel. He then repossessed the skidder; he started its engine and drove it down the highway toward the dealership where he would leave it. No one else was present when he drove the skidder away.

On his way to the dealership, the unexpected occurred. Warrick testified that he heard the engine backfire and saw steam coming from the engine. Assuming that the skidder had simply overheated, he parked the skidder, locked the brakes, and removed the ignition key. He walked back to his vehicle, and as he drove past the skidder, he saw smoke and found the skidder to be on fire. He testified that he believed the fire was beyond his control, so he called the fire department.

The origin of the fire is unknown, and Pleasant did not introduce any evidence concerning the cause of the fire. The investigation report from the fire department lists the cause as "unknown." Warrick testified that the skidder was steaming, but that it was not on fire, when he parked it to return to his car.

Warrick reported the fire loss to the insurance company, and Pleasant filed and insurance claim and proof-of-loss statement. The claim was approved, and proceeds form the insurance policy were forwarded to Deere Credit. Deere Credit retained the amount equal to Pleasant's debt ($13,589.51) and forwarded a check for the balance of $3,598.26 to Pleasant, which Pleasant cashed.

In reviewing a summary judgment, this Court uses the same standard as that of the trial court to test the sufficiency of the evidence.... The party moving for a summary judgment must make a prima facia showing that there is no genuine issue of material fact....The burden then shifts to the nonmoving party to show by substantial evidence the existence of a genuine issue of material fact....

The applicable law regarding secured transactions and repossessions is well settled in Alabama. Ala. Code 1975, sect. 7-9-503 provides:

"Unless otherwise agreed a secured party has on default the right to take possession of the collateral. In taking possession a secured party may proceed without judicial process if this can be done without breach of the peace...."

This section allows the secured party, after default, to take possession of collateral without judicial process if possession can be accomplished without risk of injury to the secured party or to any innocent bystanders....The secured party may repossess collateral at his own convenience and is not required to make demand for possession or have the debtor's consent prior to taking possession....

Here, we find the evidence to be undisputed that Pleasant was in default on his account, The record is clear that his payment history over the 20-month period prior to the repossession was irregular and that his account was never brought up to date. The record is undisputed that Pleasant failed to meet his regular payment schedule, failed to make promised payments on his arrearage, and remitted checks that were not honored because of insufficient funds. The record is further undisputed that at the time of repossession, Pleasant's account was over $4,000 in arrears. Indeed, the record is clear that Pleasant was in default and that pursuant to sect. 7-9-503, the defendants were entitled to take possession of the skidder.

The record is also undisputed that the defendants did not commit any breach of peace in obtaining the skidder. There was no evidence of any actual or constructive force used when the skidder was repossessed. In fact, no one other than Warrick was present when the repossession took place. Nor do we find that Warrick repossessed the skidder through fraud or trickery.... Pleasant contends that Warrick "tricked" him into informing Warrick of the location of the skidder. However, in view of the payment history of this case, we do not agree. Pleasant was well aware of his $4,000 arrearage, as well as the fact that there was a strong possibility that the skidder would be repossessed.

In view of the above, Warrick's repossession of the skidder was lawful; i.e., Pleasant was in default and no breach of the peace occurred while taking possession of the collateral. Therefore, we find that Pleasant's claims of conversion, negligence, and wantonness ar without merit....

The judgment in this case is due to be affirmed.

AFFIRMED.

CITY OF MARSHALL, TEX. v. BRYANT AIR CONDITIONING

650 F.2d 724 (5th Cir. 1981)

One short year after the now legendary heat wave of 1980 burned its way through the South, we are forced to consider the paradoxically chilling thought of a Texas summer without air-conditioning. Plaintiff-appellees claim that they were forced to simmer through several summers due to defendant-appellant's deceptive trade practices and breach of warranty with regard to the sale and maintenance of air conditioning equipment. While we can certainly sympathize with appellees' frustration and perspiration, we are required by Texas law to reverse the judgment in their favor and remand the case to the trial court.

I. "Summer in the City"

In 1973 defendant-appellant Carrier Corporation ("Carrier"), through its Bryant Air Conditioning Company division, manufactured six air conditioner units which were sold for use in the City of Marshall, Texas ("Marshall") public library. In 1974 Carrier, again through its Bryant division, manufactured ten air conditioner units which were sold for use in the Wiley College men's dormitory.

Each of these units was sold by Carrier pursuant to an express warranty which provided that (1) Carrier warranted the components and parts of the air-conditioning equipment to be free from defects in material or workmanship for a period of one year after installation, (2) Carrier agreed to "repair or replace, at its option" certain components or parts which were found to be defective during the one-year period, and (3) for an additional four-year period, Carrier agreed to "repair and replace, at its option, certain parts or components in the 'Refrigeration System' found to be defective." Record at 49.

According to plaintiff-appellees, the equipment never worked properly and required considerable maintenance and repair by Entex Corporation ("Entex"), a gas utility company. Plaintiff-appellees claim that the problems were due to the unavailability of replacement parts and hence the failure of the Carrier warranty. In 1977 the sweltering situation allegedly reached near-crisis proportions, and in October of that year Carrier sent the late Ralph Kemp ("Kemp") to look into the matter. During his visits to Marshall and Wiley College, Kemp apparently stated that the air-conditioning equipment could be repaired and adjusted to give satisfactory performance, that Carrier would supply necessary replacement parts, and that Carrier would provide factory-trained personnel to work on the units. However, shortly after Kemp examined the allegedly defective air-conditioning equipment, Wiley College and Marshall entered into an agreement with Entex which provided that Entex would replace the Carrier air conditioners with new equipment at its own cost in return for Marshall's and Wiley College's legal rights against Carrier. Pursuant to this agreement, new air-conditioning equipment was installed at the Marshall library and Wiley College dormitory.

Both Marshall and Wiley College brought state court suits against Carrier seeking damages for breach of warranty and violations of the Texas Deceptive Trade Practices—Consumer Protection Act ("Deceptive Trade Practices Act"). Carrier removed the actions to federal court based on diversity of citizenship, and the federal district court granted Entex's motions to intervene as assignee of Marshall's and Wiley College's claims. The actions were then consolidated and tried before a jury. After a four-day trial, the jury returned a verdict in favor of plaintiffs on both the breach of warranty and deceptive trade practices issues. Because the two claims represented alternative grounds of recovery, the trial court entered judgment pursuant only to the greater of the two verdict amounts, awarding treble damages and attorney's fees under the Texas Deceptive Trade Practices Act. Carrier appeals from the trial court's judgment.

II. "We CAN Work It Out": The Deceptive Trade Practices Act Claim

It is undisputed that the basis of plaintiff-appellees' Deceptive Trade Practices Act claim in this case centers around statements made by Carrier's representative Ralph Kemp during his visits to Wiley College and the Marshall public library in the autumn of 1977. As noted above, after examining the allegedly defective air-conditioning equipment, the late Mr. Kemp apparently stated that the units could be repaired and adjusted to give satisfactory performance, that Carrier would provide factory-trained personnel to work on the units, and that Carrier would supply necessary replacement parts. Plaintiff-appellees argued at trial that had these representations been fulfilled, there would have been no need to replace the allegedly defective air-conditioning equipment. Plaintiff-appellees then reasoned that they were entitled to the cost of replacing the Carrier air conditioners with new units. The jury apparently agreed, and the trial court entered judgment trebling the $77,000 replacement cost verdict for plaintiffs.

On appeal, Carrier raises a number of issues which it claims preclude a Deceptive Trade Practices Act judgment in the case at bar. However, we need go no further than a consideration of one of the basic elements of a Texas Deceptive Trade Practices Act claim: the requirement that a plaintiff recover only for damages actually caused by the allegedly deceptive trade practice. Because there is no proof in the record of this case of any damages caused by an allegedly deceptive trade practice, the trial court should have directed a verdict in favor of defendant on the Deceptive Trade Practices Act claim.[2] We therefore must reverse the trial court's judgment and order dismissal of the Deceptive Trade Practices Act claim in this case.

It is clear that under the Texas Deceptive Trade Practices Act, the allegedly deceptive trade practices must cause the plaintiff to be actually damaged before he can recover.... In the case at bar, plaintiff-appellees' damages, if any, were caused solely by the failure of the air-conditioning equipment and alleged breach of warranty. There is no evidence that Kemp's allegedly deceptive trade practices caused any actual damages whatsoever. While it is true that if Kemp's alleged representations were fulfilled by Carrier, plaintiff-appellees would not have incurred the replacement cost of new air-conditioning equipment, the statements themselves in no way caused this cost or additional costs. The allegedly defective equipment had been purchased years before Kemp's statements were made, and so this is not a case in which an allegedly deceptive trade practice caused a plaintiff to buy defective equipment. Moreover, shortly after Kemp's visit, plaintiff-appellees purchased new air-conditioning equipment, and so this is not a case in which an allegedly deceptive trade practice caused a plaintiff to enter into a detrimental transaction, or to refrain from entering into a beneficial transaction. Rather, it is a case in which the allegedly deceptive trade pactice did not cause plaintiffs to alter their plans or behavior, and did not cause them any harm whatsoever. The only damage in the case at bar—if any—was the result of an equipment failure beginning some time before the allegedly deceptive statements were made. Plaintiff-appellees may have suffered through several summers because of the air-conditioning equipment failure and alleged breach of warranty, but they did not suffer any more due to Kemp's statements made in the autumn of 1977.[3]

---

2. "A directed verdict should be granted if, considering all the evidence in the light most favorable to the party against whom the verdict is directed, reasonable men could not reach a contrary verdict."... In the case at bar, the defendant moved for a directed verdict but the trial court refused to grant the motion....
3. Shortly after Kemp's autumn visit to Wiley College and the Marshall public library, and before the start of the next sweltering summer, Entex—the company handling virtually all air-conditioning matters for Wiley College and the Marshall public library—replaced the allegedly defective equipment with new air conditioners. However, the decision to replace the Carrier air conditioners with new equipment appears to have been made before Kemp's visit. ... Hence, Kemp's visit was viewed by plaintiff-appellees as nothing more than as an opportunity to bargain for a cash settlement with Carrier. ... and the alleged representations made by Kemp could not have caused plaintiff-appellees to do anything differently or to suffer any harm.

Because "reasonable men could not reach a contrary verdict,"...the trial court should have directed a verdict in defendant's favor. We therefore reverse the judgment of the trial court and order dismissal of plaintiff-appellees' Deceptive Trade Practices Act claim.[4]

III. "Promises, Promises": The Breach of Warranty Issues

As an alternative ground of recovery, the jury awarded plaintiffs breach of warranty damages in the amount of $49,000. Appellants argue that this verdict cannot be sustained due to a number of errors below. We agree with appellants that the jury was not properly instructed on the issue of reasonable notice under Texas warranty law, and we therefore reverse the jury verdict and remand for a new trial on this issue.[5]

Texas law "requires notification by the buyer to the seller that a breach of warranty has occurred," so that the seller has an opportunity to cure the breach..... In the case at bar—despite timely objection by defendant-appellant—the jury was not instructed that such notice was required.[6] Rather, the jury was told that the essential elements of a breach of warranty claim in the case at the bar were simply that the air-conditioning equipment was defective in material and workmanship, and that defendant failed to repair or replace defective components or parts in accordance with the warranty.... Because the issue of reasonable notice was improperly omitted from the trial court's jury charge, we must remand for a new trial on the breach of warranty issue....

Plaintiff-appellees note that the jury was asked—via special interrogatory—whether Carrier was given a reasonable opportunity to cure the equipment defects or malfunctions. Since the jury answered that Carrier was given such an opportunity, plaintiff-appellees reason that the notice issue has been resolved in their favor.... While this argument looks persuasive at first glance, upon closer scrutiny it is seen to lack merit. The special interrogatory made it clear that the jury was to consider the notice issue only if the *Deceptive Trade Practices Act* claim was resolved in favor of plaintiffs. Hence, the jury was asked to resolve the notice issue only with regard to claims arising from statements made by Carrier representative Kemp in the autumn of 1977. Since the allegedly defective equipment was installed in 1973 and 1974, since plaintiff-appellees complained that the equipment never worked properly, and since the primary warranty was for only one year, adequate notice with regard to claims arising from allegedly deceptive trade practices in 1977 is irrelevant with regard to claims arising from an alleged breach of warranty.[7]

---

4. Because we reverse the Deceptive Trade Practices Act judgment on this ground, we do not reach the remaining Deceptive Trade Practices Act issues raised by appellant.
5. We reject appellant's contentions that, as a matter of law, Carrier fulfilled its warranty, and that, again as a matter of law, plaintiff-appellees failed to give Carrier adequate notice of the breach. There was ample evidence both of the breach and of notice to Carrier to submit these issues to the jury.... Likewise, we reject appellees' contention that adequate notice was proven as a matter of law.... Considering the record as a whole, the issue of reasonable notice was clearly a jury question and the court's erroneous instruction on this issue therefore requires reversal.

   Because we reverse and remand with regard to the breach of warranty claim based on the trial court's erroneous instruction regarding notice, we do not reach appellant's other breach of warranty claims which—if correct—would require the same result.
6. Contrary to plaintiff-appellees' contention, the issue "whether Carrier was given a reasonable opportunity to cure the alleged defects or malfunctions" was presented in the pre-trial order by defendant-appellant and hence the issue of notice was adequately raised. . . . Moreover, defendant-appellant raised a timely objection to the court's jury charge on breach of warranty, arguing that the trial court erroneously failed to instruct the jury that notice is a necessary element of a breach of warranty claim. . . .
7. For example, the jury's answer to this special interrogatory may simply mean that the jury found that plaintiffs gave Carrier reasonable notice of Carrier's failure to fulfill the alleged representations made by Kemp. This notice is a far cry from the reasonable notice required with regard to an alleged breach of warranty for equipment installed several years earlier.

Moreover, as noted above, the court's instruction on the essential elements of a breach of warranty claim clearly omitted the notice issue. The jury was not asked to consider whether Carrier was given a reasonable opportunity to cure equipment defects with regard to the alleged breach of warranty, and the instruction below was therefore erroneous.[8]

IV. Conclusion: "The Second Time Around"...

Because there is no evidence that plaintiff-appellees were actually damaged by the allegedly deceptive trade practices in the case at bar, we reverse the judgment of the trial court and order that the Deceptive Trade Practices Act claim be dismissed. Because the jury was improperly instructed on breach of warranty, we remand for a new trial on this issue.

REVERSED and REMANDED.

8. Appellant also suggests that the plaintiffs' claim is time barred by the applicable statute of limitations. Because the record below is not sufficiently developed concerning when plaintiffs' cause of action accrued for statute of limitations purposes, we do not reach this issue. Nothing in this opinion should be read as in any way precluding defendant-appellant from raising the statute of limitations issue on remand.

REVIEW QUESTIONS

1. Brown, in Seattle, Washington, ordered a machine from the White Company in Detroit, Michigan. The contract stated that the machine was to be shipped F.O.B. Detroit. At the bottom of the sheet the note "we will deliver the machine to you in Seattle at our cost" was written in longhand and signed by White. En route between Detroit and Seattle the machine was destroyed in a cyclone. There was no insurance. Whose machine was lost? Why? Who pays for or bears the costs of the loss of the machine? Why?
2. Black bought an automatic machine from White to replace a machine that had previously required two employees to operate. In selling the machine to Black, White pointed out the savings in the cost of the employees' time. White also stated that "Productive capacity will be doubled" with the new machine. Hourly production increased even more than double — 1,100 pieces per hour against 500 pieces per hour previously. However, there was a problem involved. The machine seldom ran an hour without breaking down. Whenever it broke down, either one or two machine service personnel would be called upon to fix it. After a month or so, one person was ordered to stand by for breakdown whenever the machine was scheduled to run. Weekly production on the automatic machine is only slightly greater than previous weekly production on the hand-operated machine. Black has charged White with fraud. Can Black recover on this basis? Why or why not? Is Black stuck with the machine? What defense does White have?
3. Distinguish between *express* and *implied warranties.*
4. How does the existence of a security interest protect a creditor from debtor default?
5. What rights does the debtor have in a security interest arrangement?
6. Distinguish between *sale on approval* and *sale or return.*

7. What is required of the seller to sell something *as is*?
8. What remedies are available to a seller if the buyer refuses to pay for goods the buyer has accepted?
9. Referring to the case of *Pleasant v. Warrick*: a. What would be both the legal and practical implications of repossessing the logging skidder if Pleasant had paid half of the outstanding balance and then was three weeks behind with a payment? b. Suppose the skidder had not been destroyed by fire and Deere Credit had been able to realize only $10,000 upon the resale of the used skidder. How much of the $10,000 would be paid to Pleasant?
10. What must the plaintiff in *City of Marshall, Texas v. Bryant Air Conditioning* prove to recover under the Carrier Corporation warranty when the case comes up for retrial?

Part Three

ENGINEERING CONTRACTS

The general public is well aware of construction contracts that determine how buildings are to be erected, bridges built, and highways created. The media often report the stories of such construction projects and refer to the underlying contracts. Legal problems and entanglements concerning contracts in the civil engineering, architectural engineering, and construction engineering fields occur frequently enough that special courses in contract writing are commonly included in those curricula. The need for such education in mechanical, industrial, electrical, chemical, and other engineering curricula may be less apparent. However, all engineers seem to be involved in writing specifications and interpreting contract documents at some time during their careers. Some, especially those who rise in management, spend their entire working lives dealing with contracts. Thus, all engineers need to be aware of laws regarding ownership, independent contracting, and agency, and of contract documents that spell out work to be done and the responsibilities of the parties to contracts.

Chapter 16

Contracting Procedure

Before delving into the contracting procedure, several terms and relationships between contracting parties need to be discussed. Although the following discussion focuses on construction contracts, most (if not all) of the discussion applies equally to other large projects, such as systems or software development or the like.

PARTIES TO A CONTRACT

The parties to an engineering contract generally are the owner, the engineer or architect, and the contractor. The owner is the party for whom the work is to be done. He or she is the one to whom the others look for payment for services. The owner has the final authority in questions as to what is to be included or left out of a project. The owner may be a private individual, president of a corporation, chair of a board of directors, or a public official charged with the responsibility for the project.

As the terms will be used here, *engineer* and *architect* are virtually interchangeable. At present, considerable controversy (including court cases) exists over the meanings of the terms and the work to be considered the proper field of each. No attempt will be made to add to that controversy here. It will be assumed that the person is properly employed, whether an architect or engineer. Such a person acts as an *agent* of the owner. Specifically, that person furnishes the technical and professional skill necessary in the planning and administration of the project to accomplish the owner's purpose. That person is the owner's designer, supervisor, investigator, and adviser. Most state laws require such a person to be registered as a professional. This person may be a consultant or an employee of the owner. Generally, this is the person with whom the contractor deals directly.

The third party to the contract, the contractor, is the one who undertakes the actual construction of the project. The contractor furnishes the labor, materials, and equipment with which to complete the job. The term *contractor* is frequently further broken down into *general contractor* and *subcontractor*.

The general contractor agrees to accept the responsibility for the complete project and frequently undertakes the major portion of the project. Subcontractors are hired for particular specialties by the general contractor and, in effect, work for the general contractor while completing the portions of the project for which they have been hired. The general contractor retains responsibility for the entire project, even when the wiring, the plumbing, and the roofing are done by various subcontractors.

TYPES OF CONSTRUCTION

Independent contractors construct much of the capital wealth of our country. Most of the buildings, machines, bridges, utilities, production facilities, and so forth responsible for our standard of living were built under contract. There are two major types of such construction—public and private. The two types differ significantly in the motives and relationships involved with each.

♦ Private Works

Private projects are those that are undertaken for an individual or a company. The restrictions imposed upon the parties are those of contract law in general. The agreement may be achieved by advertising and then choosing the best bid, or by direct negotiation without advertising. The acceptance may be oral or in writing; in fact, the entire contract could be oral if the parties so choose.

The profit motive is the usual reason for construction of private works. The owners believe the project will give them a desirable return on investment. However, this need not necessarily be the reason for the project. In private works, the reason could be nothing more than a personal whim of the owner. The desirability of the project is not open to question by the public as long as no one is harmed by it.

♦ Public Works

The motive for construction of public works is public demand or need. Financial return sufficient to justify the investment is frequently of less than primary importance. Money for private works comes either from direct payments from available funds, such as tax receipts, or from loans, such as bonds.

Voluminous statutes usually govern the letting of public-works contracts. These laws apply certain restrictions to the contracting procedure. Generally, formal advertising and bidding are required to ensure competition among the bidders. The "lowest responsible bidder" gets the job. Usually the means of advertising and the length of time the advertisement is to run are specified. Changes in plans or specifications often cannot be made after the award of the contract, even though such changes might be beneficial. Acceptance of a bid may not be effective until it has been ratified by a legislature or a legislative committee.

Once a contractor's bid on a public project has been accepted, the amount the contractor will get is fixed. The official in charge of the project cannot agree to pay more, regardless of the apparent justice of the contractor's claim.

The purpose of the restrictions in letting public projects is, of course, to prevent dishonesty, collusion, and fraud among bidders and between bidders and public officials. Sometimes the restrictions seem cumbersome and unfair; sometimes the results are less than might have been accomplished without them. At times well-meaning contractors fail to survive their ignorance of the law. The results in general, though, seem beneficial to the public.

CONSTRUCTION AND MANUFACTURING

A construction project and the manufacture of goods for a market are basically similar. Each is concerned with the use of labor and equipment to turn raw materials into a finished product of some sort. Each must deal with economic considerations in acquiring raw material and labor. Each aims for efficient management in an effort to show a profit. Both are concerned with production schedules that must be met if penalties of one kind or another are to be avoided. Quality must be maintained, and costs must be minimized.

The main distinction between *construction* and *manufacturing* stems from the location of the product. In most construction, whether it is a building or a piece of productive equipment, the place where the product is built is the place where it will stay. A construction product is usually custom-made, built to specific requirements, and not to be reproduced. Such a product does not lend itself to the economies of standardization of method, as do the products of most manufacturing companies.

DIRECT EMPLOYMENT OR CONTRACT

After the decision has been made by the owner to undertake a construction project, the question frequently arises as to whether the work should be done by the available staff, by contract, or by a combination of both. The decision involves many factors; two very important ones are the relative size of the job and the skill of the staff. Many jobs are too large for the present staff or too small to be submitted to an outside contractor, so the decision is clear. But many projects of intermediate size could be completed either way. It is with these intermediate-size jobs that the present discussion is concerned. The owner (or engineer) should consider a number of benefits and disadvantages in choosing the best procedure.

♦ Cost

Certain savings in cost are apparent when direct employment is used: (1) The owner does not have to pay the contractor's profit margin if the owner does the work. (2) The amount added in by the contractor for *contingencies* is saved if no contingencies arise. The owner who expects to save costs in this area is really anticipating winnings from a gamble. If the owner's staff is inexperienced, the odds of winning are not good. On the other hand, an experienced contractor can often see and avert incidents that would otherwise constitute contingencies. (3) The cost of making *multiple estimates* is avoided if the owner's staff undertakes the job.

♦ Flexibility

A project undertaken by direct employment has more flexibility. If it becomes necessary to make changes while the job is under way, the owner has only to order the changes made. When a contractor is hired to build according to plans and specifications, and work is under way, change proposals will usually meet with resistance. Contractors are interested in completing the project as soon as possible so that they can get paid and go on to the next job. They will be tempted to charge heavily for the delay caused by a change.

♦ Subsequent Maintenance

A machine or other structure is more readily repaired by the original builder than by maintenance personnel who have had no experience with it. Therefore, the issue of subsequent maintenance might induce the owner to use as many of his or her own people as possible.

♦ Grievances

Labor problems are usually somewhat reduced by using present crews. The possibility of disputes about what a given set of workers can or cannot do under a labor contract exists, but the likelihood of these and other disputes (such as grievances) arising can be minimized by using direct employees. The individuals forming the nucleus of the crew, at least, have probably learned to live with each other and with others employed by the owner in different jobs.

♦ Specialization

The great and sometimes decisive advantage of hiring a contractor for a project is that the contractor is a specialist. The contractor's specialty is in labor, supervision, and procurement of materials. It is probably this factor, more than any other, that gives the contractor an advantage.

♦ Public Relations

The public relations programs of many large concerns tip the scales in favor of hiring contractors to undertake jobs for them. This factor plays a big part when a company sets up an operation in a new community. To establish itself in a favorable light, the company may hire local people to set up its facilities, even though it has a staff capable of doing the job.

PAYMENT ARRANGEMENTS

An owner may pay for work undertaken by a contractor in four basic ways. In addition to these basic payment arrangements, a nearly infinite number of variations and combinations of them exists, each adjusted to a given project. Here, the features of each basic arrangement will be considered primarily from the owner's standpoint.

♦ Lump-Sum Contract

If an individual purchases a product manufactured by a company, that person usually knows the price before making the agreement to purchase. The price of an automobile, for example, is arranged between the dealer and the buyer before the sale is actually made. The *lump-sum* contract arrangement gives the owner the same assurances as to the price to be paid for the job. This aspect of the lump-sum contract appeals to people, and it is the main advantage of the arrangement from the owner's standpoint.

The fixed-price, or lump-sum, contract has several inherent disadvantages, however. Probably the main one is the antagonistic interests of the owner and the contractor. Once the contract is signed, the contractor's main interest is in making a profit on the job and in doing it as quickly as possible. The contractor has an incentive to do no more than the minimum requirements set forth in the plans and specifications. With an unscrupulous contractor, the results may be shoddy work and a poor structure. Even a responsible contractor may be tempted to cut corners if contingencies start eating into the profit margin or if the contractor is already losing on the job. The owner's interest, on the other hand, is to obtain the best possible structure for the agreed price.

Under a lump-sum contract, changes can be quite costly. The change represents an impediment to the contractor's speedy completion of the job, and the contractor is in the position to dictate the cost of changes. The owner has hired the contractor to do the job, and the contractor is on the premises with equipment to render the service called for in the contract. The owner's alternative to paying the contractor's price for a change is to wait for completion of the job and then hire someone else to make the change. This is often very costly.

A lump-sum contract requires that considerable time and money be spent by the contractor in examining the site, estimating, and drawing up and submitting a bid before the job can start. The delays may run to months or even years, and the cost of delay can also be considerable. If the job is to be started as quickly as possible, the lump-sum contract is not recommended. In addition, if the work required is at all indefinite or uncertain, the lump-sum contract should not be used. The contractor will have to add in a sufficient amount to cover the uncertainty to show a profit.

The greater the degree of uncertainty, the greater the probable spread in bidders' proposals. Prices on a 500-ton press, for example, might vary as much as 6 or 7 percent between high and low bids among five or six press manufacturers. If the equipment were a piece of automation with only the raw material and the end product known, the highest bid might be four or five times that of the lowest. Each contractor would try to hedge as well as possible against a large number of unknowns and gamble that the amount submitted would result in a successful bid.

Lump-sum contracts are used commonly and successfully, but their successful use is generally restricted to situations in which unknowns are at a minimum. They are not appropriate (1) where uncertainties exist; (2) where a speedy start is necessary—emergency work, for instance; or (3) where ongoing plant operations will interfere with the contractor's work. Where such circumstances exist, one of the two basic "cost-plus" types of contract can be used.

♦ Cost-Plus-Percentage Contract

The most rapid means of starting a project is through the use of the *cost-plus-percentage* pricing arrangement. In an emergency, an owner may have to start work immediately, while the plans for the completed structure are still being drawn up. Using a cost-plus-percentage arrangement, today's phone call can result in action today. It is this feature that prompted the widespread use of such contracts by the federal government during World War II. Valuable time would have been lost in estimating and bidding if they had not been used.

With the cost-plus-percentage arrangement, the owner usually pays all the contractor's costs plus an added percentage (often 15 percent) of these costs as the contractor's profit. *Cost* refers to the cost of materials and services directly connected with the owner's structure. Such items as the contractor's overhead, staff salaries, and the like are usually excluded unless direct connection to the project can be shown.

A major advantage of the cost-plus-percentage contract is its flexibility. Changes may be made readily when the owner desires them. Because the cost to the owner for making changes includes profit to the contractor, the contractor has little reason to object to them.

One disadvantage of this approach is that the risks involved in construction are assumed by the owner. In fact, the owner does not know with any certainty what the project will cost until it is completed. On the other hand, if no adverse conditions (contingencies) arise, the benefit goes to the owner.

Another disadvantage is that the contractor's profit is tied to costs, so the costs may be quite high. Gold-plated doorknobs may show up, for example. The contractor has what amounts to an incentive for dishonesty—a financial reward for running up costs. The most honest of contractors (or people in any profession, for that matter) would be tempted to be inefficient under the circumstances. Even if the contractor pursues the completion of the project in the best interests of the owner, efficiency is not assured. Supervisors and workers are not likely to put forth outstanding effort for their employer if they will gain nothing as a result. The need for the owner to police the project is apparent.

It is because of this disadvantage that the federal government looks with disfavor on the cost-plus-percentage contract. Renegotiation of contracts and the recapture of excess profits on contracts following

World War II left the government with a rather bitter attitude toward the arrangement.

♦ Cost-Plus-Fixed-Fee Contract

The remedy for the main undesirable feature of the cost-plus-percentage contract is found in the *cost-plus-fixed-fee* arrangement. Here, the contractor is paid a fixed fee as the profit, and the owner picks up the tab for all the contractor's costs of undertaking the project. The amount of the fixed fee is subject to negotiation between owner and contractor or bid by the contractor based upon estimates of the cost of the completed project. The advance settlement of the amount of the fixed fee precludes an immediate start on the project, but the time required is not nearly as great as that necessary for a lump-sum contract.

The cost-plus-fixed-fee contract is probably the best basic arrangement from the standpoint of all parties. The contractor has no incentive to run up costs or work inefficiently. In fact, with a view toward maximizing profit in a given period of time, the contractor has an incentive to hasten the completion of the project to obtain the fee.

However, less flexibility exists with this arrangement due to the contractor's desire to complete the job rapidly. A proposal by the owner for a major, timeconsuming change in the project is likely to be met with objections and a request for more money. But if the contractor is required to spend a longer time in pursuit of the fee, it is fair to expect that fee to be increased.

As in the cost-plus-percentage contract, the owner runs the risk of contingencies, but here the owner is in a somewhat better position. Under the cost-plus-percentage arrangement, the occurrence of contingencies tends to increase the contractor's ultimate profit. Such contracts, then, have built-in incentives to bring about contingencies or, at least, not to actively avoid them. In contrast, in the cost-plus-fixed-fee arrangement, the desire for early completion is an incentive for the contractor to avoid contingencies if possible.

♦ Unit-Price Contract

Certain kinds of structures may be conveniently built under the *unit-price* type of arrangement. This scheme applies best where a large amount of the same kind of work must be done. In the building of a road, for instance, the main elements are excavating, filling, and pouring concrete. The contract would specify so much per cubic yard of excavation (plus an extra amount if rock formation is encountered), so much per cubic yard of fill, and so much per cubic yard of concrete. The price per unit includes the contractor's cost per unit plus an amount for contingencies, overhead, and profit. The engineer or architect usually estimates the quantities required ahead of time for the benefit of both owner and contractor. The actual quantities may be considerably different from the estimate, but the contractor is paid according to the actual quantities required.

Usually the unit-price arrangement is used in conjunction with a lump-sum contract. The combination is usually quite beneficial, largely because it is flexible, reflecting the realities of most projects. In nearly every job, for example, there are some elements (such as clearing, grading, and cleaning up the site) that do not lend themselves to unit pricing.

With a unit-price contract, much of the contractor's uncertainty is relieved, and this is reflected in a lower estimate for contingencies. The risks not assumed by the contractor, though, must be borne by the owner, and the owner has no precise knowledge of the cost of the job until its termination.

Just as is true with a lump-sum contract, the job cannot begin immediately under a unit-price arrangement. The contractor requires time to examine the premises and prepare an estimate for the owner. Normal bidding procedure is usually used in the award of such contracts.

♦ Variations

The types of payment arrangements described above are the four basic forms of construction contracts. In addition to these, many variations, or hybrids, exist—many of them tailored to particular needs of projects. Probably the most common variation involves the addition of an incentive system of some sort. Profit-sharing and percentage-of-cost saving are examples. Another kind of variation involves the addition of a kind of reverse liquidated damages clause. That is, the contractor is offered a fixed amount per day additionally if the contractor finishes ahead of schedule.

Another variant form of contract is the *management contract*. In such a contract, the owner hires a contractor, not necessarily to undertake the work with the contractor's own organization, but to oversee the job, often hiring others to do the work. Frequently the contractor's duties include managing work to be done with the owner's labor force. Work undertaken by the owner's people is called work under a *force account*.

The contractor under a management contract still holds independent contractor status. The contractor agrees to produce a result, but the contractor's actions do not bind the owner under a management contract as an agent would bind the principal. The management contract is, therefore, distinct from the agency relationship, which exists between an owner and the engineer or architect. Usually the services of an engineer or architect are not used.

STAGES OF A PROJECT

A project starts as an idea or a dream for the future. If it is soundly conceived, planned, and developed, the dream may be realized. Failure in any of these areas can turn a dream into a nightmare. In discussing the development of a project in the following sections, let us consider the lump-sum or the unit-price type of contract arrangement, since each requires extensive preliminary work.

♦ Feasibility Studies

Regardless of how beneficial a project may seem to its originator, it is nearly always advisable to conduct a preliminary investigation. Although wars and other emergencies may preclude such an investigation, a state of emergency is somewhat unusual and presumably won't prevent most preliminary investigations. The purpose of the *feasibility study* is to answer several preliminary questions:

1. What is the cost of undertaking the project? This cost is usually determined on the basis of either annual cost or present worth (or capitalized cost if perpetual service is contemplated).
2. Can the objective be attained better (or cheaper) in some other way? The comparison is made on the basis of annual cost or present worth.
3. What benefits will result from the use of the various alternatives? In most private projects and some public projects, the ratio of economic benefit to cost must be sufficient to justify the project. In many public works, a crying need may push consideration of economic benefit into the background.

The preliminary investigations and reports are usually not intensive. Some schemes are shown to be obviously impractical after only minimal data are obtained. Where the scheme appears to be profitable, though, the expenditure of considerable time and money on the feasibility study may be justified. Surveys and investigations of such elements as markets, sources of raw material and labor, costs of transportation, and applicable laws and ordinances may become necessary.

Even when a tentative decision has been made to go ahead with a project based on reports that show it will pay off, it is usually worthwhile to investigate cheaper ways of achieving the same goal. Most such studies can be undertaken at a reasonable cost. However, the more closely balanced the evidence, the more detailed must be the study. Investigations about alternatives can save money in the long run, but wisdom must dictate the point at which a study should be terminated and a decision about the project made. Feasibility studies do not complete projects. Thus, when the benefits of a course of action (or inaction, as the case may be) are clear, the feasibility study has usually served its purpose.

♦ Design

If the feasibility study indicates the project is desirable, and the decision is made to go ahead, the next step is to design the structure. Normally, all designs are complete in some detail before work is started. A complete feasibility study usually includes sketches of possible layouts. The design, however, includes functional requirements, layouts, and dimensions. Preparation of the drawings is a part of the engineer's or architect's task. When the drawings and specifications are completed, they combine to give the basic information on the project.

The design drawings provide the owner and engineer with a picture to use in talking to others about the project. A copy of the design drawings of the proposed construction ordinarily must be filed with the proper authorities to obtain a building permit.

♦ Legal Arrangements

If such arrangements have not already been made, land must be obtained for the project by purchase or by exercise of eminent domain if the project is a public one. Access to railroad sidings, highway connections, and utility services must all be considered. If a building is being constructed, zoning is likely to be a consideration, and building codes may need to be satisfied.

Another legal matter is the engineer's or architect's right to practice in the state. State licensing laws generally require that a registered professional engineer sign construction plans or issue them under his or her seal. All members of the engineering staff need not be registered though. If a member of the staff who is registered takes responsibility for the plans, the law is satisfied.

If the project is to be undertaken for the public, strict adherence to laws governing such projects is necessary. The order and appropriation for the project must be passed before the work is undertaken. In many types of public construction, minimum standards must be met.

♦ Preparation of Contract Documents

During the preliminary stages of a project, it is necessary to prepare contract documents that will guide the remainder of the project. One of the main purposes of these documents is to set forth the relationships between the parties. If this purpose is to be accomplished adequately, the documents must be prepared with great care and skill. The engineer's task of preparing these documents often seems dull and routine—an obstacle to be overcome to get to something more interesting. Still, careful attention to this task prevents many future controversies, and it may make controversies that do arise easier to settle.

The pieces that make up a construction contract generally include (1) the advertisement, (2) the instructions to bidders, (3) the proposal, (4) the agreement, (5) the bonds, (6) the general conditions, and (7) the specifications and drawings. Each component of the documents that constitute a construction contract has a name or heading. The parts do not bear the same names in all contracts, and the material covered under the headings is not always the same. For example, what is covered in " general provisions" in a federal government contract might be covered in "general conditions" or "information for bidders" in a state or private contract. Certain instances of overlap may be noted. Ideally, a subject should be thoroughly treated in only one section, with reference made to this section whenever the occasion arises again in the documents. In larger projects, though, duplication of coverage is not unusual.

The American Institute of Architects has attempted to standardize contract documents. It has drawn up and copyrighted documents for use in building construction contracts. The standards will be the focus of our discussion because they are national standards that are widely used in connection with construction projects. The federal government and most state governments also have standardized contract components; the same is true of most large companies. Engineers will need a short indoctrination to become familiar with the particular terms and usages of a new employer.

♦ Advertisement

The *advertisement* is usually the last contract document prepared, but it is the first one seen by the contractor. The primary purpose of the advertisement is to obtain competitive bidding on the project. In public contracts, a minimum of three bidders on any project may be required, and a special authorization may be needed if an award is to be made otherwise. In private work, there is no legal requirement to advertise. Contract awards are often made to contractors with whom the owner has successfully dealt on previous occasions. Many companies consider competitive bidding to be desirable, however, and they have established formal procedures to require it.

A second purpose of the advertisement is to attract appropriate bidders—those interested in the type of work involved. To accomplish this end, the wording of the advertisement should give a clear, general picture of what is to be done. The advertisement should not attract those who have insufficient capacity or who are not interested in the type or location of the work.

The title of the advertisement is important, too. It must attract attention, and it should state very generally what is to be built. The phrase "Notice to Contractors" or "Call for Bids" together with a phrase generally describing the type of structure, such as "Elementary School," serves these functions well. A glance attracts the qualified contractor and prompts the contractor to read further.

The information to be included in the advertisement varies with the type of project contemplated, but the following facts should usually appear:

1. The kind of job.
2. Location of the project—construction equipment is rather costly to move.
3. The owner and the engineer—previous contracts with an owner or engineer may influence a contractor to submit a bid.
4. Approximate size of the project.
5. The date and place for receipt of bids, and the date and place of opening and reading bids.
6. How an award is to be made (for example, "to lowest responsible bidder"), whether the owner

reserves the right to reject any and all bids, and when the contract is to be signed.
7. Notice of any deposits, bid bonds, or performance bonds to be required.
8. Procedure for withdrawing bids if withdrawal is to be allowed.
9. Where copies of the contract documents may be obtained.
10. Any special conditions.

Finally, the advertising medium should be carefully selected. Public contracts must be advertised in local newspapers. For private contracts, local newspapers and trade journals make good advertising media. Whatever the medium, however, brevity best serves the purposes of advertisement.

♦ Instructions for Bidders

Contractors who remain interested in the project after reading the advertisement are offered the opportunity to obtain more information, usually from the engineer. It is customary to require a deposit (usually from $50 to $500) from the contractor when the contractor obtains a set of the contract documents.

The "Instructions for Bidders" or "Information for Bidders" may be compiled as an independent document. Probably just as frequently, however, this information section is the first portion of the specifications. Variations do occur though. For example, items included in the instructions for bidders in one project may be found in the general conditions document or in the general provisions section of the specifications in another project.

Much of the information contained in the instructions is the same as that given in the advertisement, but it appears in expanded form—the "what," "where," "when," "how," and "for whom" are given in more detail. Additional inclusions in the instructions vary considerably from one project to another, but the most common are listed below:

1. A requirement that the bidders follow the proposal form.
2. A statement as to whether alternative proposals will be considered.
3. The proper signing of bids (e.g., the use of a power of attorney).
4. Qualification or prequalification of bidders. This is a requirement to satisfy the owner that the contractor has the necessary skill, financial means, staff, and equipment to do the job.
5. Discrepancies. Provision is usually made for review with the engineer of any discrepancies appearing in the contract documents.
6. Examination of the site. An invitation is normally extended to bidders to examine the site with a warning that no additional compensation will be allowed as a result of conditions of which the bidders could have informed themselves.
7. Provision for return of required bid deposits to unsuccessful bidders.

♦ The Proposal

The *proposal* is a formal offer by the contractor to do the required work of the project. Acceptance of the proposal often completes the formation of the contract, so considerable care must be exercised by the engineer in drawing it up.

Proposal forms are usually standardized for all bidders on a job. If the proposal is to be broken down into components (as in a unit-price contract), all bidders must break it down in the same way so that their proposals can be compared. Even if only one lump-sum bid is requested, all bidders must use the same proposal form so that bids can be considered on an equivalent basis. The following elements may be contained in a proposal form:

1. The price or prices for which the contractor agrees to perform the work involved.
2. The time when work is to start and when it is to be finished.
3. A statement that no fraud or collusion exists. Of particular concern is the possibility of fraud or collusion among bidders or between any bidder and a representative or employee of the owner.
4. A statement that the bidder accepts responsibility for having examined the site.
5. A proffer of any required bid bond or guarantee and agreement to furnish whatever other bonds may be required as well as an agreement to forfeit these bonds or portions of them under the conditions stipulated.
6. Acknowledgement that the various other documents are to become parts of the contract.
7. A listing of subcontractors if required.
8. Signatures of contractor and witnesses.

♦ Agreement

To a lawyer, the agreement probably would consist of all the various documents. However, the parties often use the term *agreement* for the short "outline" of the transaction. The term *contract* is preferred when one is referring to the complete array of contract documents.

Usually the agreement is quite short—one page or possibly two—and it describes the work to be done, largely by reference to the other contract documents. It also states the owner's consideration (the price to be paid for the work) and the means and time of payment. Although they are included in the other contract documents incorporated by reference, a few terms are usually repeated in the agreement. If time of completion is an essential element of the project, that date and amount of liquidated damages for a failure to complete on time are mentioned. If there is to be a warranty of the work by the contractor for, say, a year after completion, that is usually included. Any amounts to be held back from payments to the contractor are also set forth.

♦ Bonds

A bond gives the owner financial protection in case of default by the contractor. Three types of bonds are commonly used in contracting—the *bid bond*, the *performance bond*, and the *labor and material payment bond*. Each protects the owner by assuming a risk the owner would otherwise normally run in hiring a contractor to undertake the work.

The Bid Bond

A *bid bond* assures the owner that the bidder will sign the agreement to do the work if that bid is the one accepted. In theory, the amount of the bid bond covers the owner's loss if the lowest responsible bidder fails to sign the agreement and the owner must then turn to the next higher bidder. Usually, however, the owner does not require a bid bond as such. Instead, the ower requires a deposit that may be satisfied by a bid bond, a certified check, or some other security posted by the contractor to assure that the owner will be reimbursed in case of a failure to sign. The chosen bidder forfeits the bid security only if he or she fails to agree to the project. It is returned to the bidder after the agreement is made. The bid securities of unsuccessful bidders are, of course, returned to them.

The Performance Bond

Through the *performance bond*, the owner's risk that the contractor may fail to complete the project is passed on to a surety company. The risk involved may be very slight—most contractors are in business to stay—but there are still risks. The contractor may meet with insurmountable obstacles. Unforeseen price rises, strikes, fires, floods, storms, or unsuspected subsoil conditions, for example, can financially ruin even the best-backed contractors.

If the contractor is faced with such disaster, insolvency may result, and if there is no performance bond, the owner may be left trying to obtain blood from a stone. However, if a performance bond has been required, completion of the structure is guaranteed by the surety. Many surety companies maintain facilities that can be pressed into service on such occasions; more frequently, the surety company hires another contractor to complete the job.

The cost of a performance bond will, of course, be passed on to the owner in the price to be paid for the work. This price varies, but in many situations will be around 1 percent of the bid price for the contract. For this price, the owner not only gets risk protection but a preselection of contractors. Surety companies are quite choosy about the contractors they agree to back. The irresponsible and the "fly-by-nights" are poor risks, and a reputable surety company will not back them. The requirement of a performance bond, then, allows the owner to select from the best.

The Labor and Material Payment Bond

The *labor and material payment bond* offers protection against labor and material liens, known generally as *mechanics' liens*. The lien laws vary considerably from state to state. In any state, though, if a contractor fails to pay for the labor or material, a lien may be obtained against the property the labor or material was used to improve. In addition to the labor and material supplier's liens, the contractor, subcontractor, engineer, or architect may secure payment by recording a lien. Based upon the assumption of the owner's honesty and integrity in dealing with the contractor and engineer or architect, discussion here will focus on labor liens and material suppliers' liens.

Mechanics' liens secure payment for anything connected with the improvement of real estate. The security is an encumbrance upon the property that makes it more difficult to sell or mortgage. Technical procedures and delays, specified in lien laws, must be followed to remove the encumbrances. Owners may protect themselves from such encumbrances in two ways: (1) by withholding sufficient funds from payments to the contractor to pay for labor and materials if the contractor fails to do so and (2) by requiring the contractor to obtain a labor and material payment bond.

To withhold part payment, the owner must know the amount that he or she may need later. This usually requires the owner to get a sworn statement from the contractor as to the outstanding bills before making payment to the contractor. Some state lien laws require such a sworn statement if a payment bond has not been provided in the contract. The procedure can become somewhat cumbersome, however. Requirement of a labor and material payment bond from the contractor is much simpler for the owner. According to the bond, the surety company assumes liability for the contractor's unpaid bills.

When the federal government is owner in a project, a performance bond and a labor and material payment bond are automatically required. The Miller Act (passed in 1935) requires such bonds for federal contracts exceeding $25,000.

♦ General Conditions

In a set of construction contract documents, the document that sets out the general relationships between the parties is usually referred to as the *General Conditions*. This document deals with rights, obligations, authority, and responsibility of the parties. Topics covered differ from contract to contract, but the following are almost invariably present.

- A statement near the beginning attests to the *unity* of the contract documents—meaning that a requirement in one document is just as binding as if it appeared in all of them. Such unity is necessary since the specifications for even a small project would reach tremendous size if each specification had to be followed by all conditions pertaining to it.
- The right of inspection (of the site) by the owner or the owner's representative at any time is usually reserved. The documents usually are written so that the owner "may" inspect the site but is not required to do so. It is common also to allow inspection by public officials as a general condition.
- Conditions for terminating the contract by the owner and by the contractor are set forth. Bankruptcy of the contractor and failure by the owner to make scheduled payments to the contractor when due are usually made conditions for termination.
- The insurance program for the project is outlined. Provision is often made that worker's compensation will cover medical costs and partial wage payments in case of injury to a worker as a result of the worker's employment by the contractor. The owner's protective liability insurance protects the owner from contingent liability if the contractor's operations should injure anyone. Fire insurance and vehicle liability are among other types of insurance commonly required.
- The extent or limits of authority should be spelled out clearly. For example, the engineer or architect is to make decisions on the work involved in the project. That person usually must then be given the right to stop the work if it appears necessary to do so.
- Provision is often made for arbitration as a final step in any dispute between the owner and the contractor.
- A requirement is usually included that the contractor keep the site clean as work progresses and that the contractor clean up the site thoroughly when finished.

♦ Final Considerations

Almost invariably the owner is under considerable pressure to get things started soon after the decision has been made to undertake a project. Advertising and bidding procedures require time, however, and to shorten the time allowed for the initial stages is usually an act of folly. Prospective bidders need time to investigate the work site and to plan their work. If they do not have enough time for the investigation and planning stages, many unknowns will remain. Generally, the less the bidder knows, the higher will be the bid price. There are few fears greater than fear of the unknown, and fear of unknowns in a proposed project drives the price up. Owners need time also, to carefully consider the quality of the bidders and their bids.

Collusion

Most owners, engineers, and contractors are honest people. There are some of each, though, who will eagerly forget ethics and the law to seize an unfair advantage. Unfair advantages take many forms, and there are many "shades of gray" between fair, honest practices and unethical or illegal ones. The time for receiving proposals, for example, should be fixed and inflexible. It is unfair to give one contractor a longer time to prepare a proposal. Similarly, the opening of the proposals should take place at the time specified. Also, if an owner or engineer gives preference to one contractor over another in other ways, either by giving the contractor added vital information privately or by writing out specifications, that preference constitutes an unfair advantage. If the owner unfairly helps a contractor, it will almost certainly be resented by the others, even if it's not illegal.

Collusion occurs occasionally when a limited number of responsible bidders bid on a project. The bidders get together and decide who will take the job and for what price; the other bidders then submit prices higher than that submitted by the "successful" bidder. The result of these and other forms of collusion is higher cost to the owner.

The existence of such practices is the main reason for governmental restrictions in letting public contracts. Collusion is, of course, against public policy and the Sherman Anti-Trust Act. It is punishable by both fines and imprisonment. In addition, up to triple damages can be recovered in a civil action by those injured.

Lowest Responsible Bidder

A contract award for a public project normally goes to the "lowest responsible bidder." Many factors enter into the determination of the contractor who is to be awarded the contract. If *prequalification* has been required, the decision is based upon price alone. The nonmonetary factors used to determine the lowest responsible bidder are those that indicate a reasonable likelihood of completing the contract. The chief factors considered include the following:

- The contractor's *reputation*. Such things as shoddy work, constant bickering for extra payments, and an uncooperative or lackadaisical attitude are causes for caution. Rejection of a bidder on the basis of reputation often leads to arguments, but each person builds a reputation and has to live by it.
- The contractor's *finances*. Inadequate finances or credit can be a source of trouble in a project. Even though a bond is required of the contractor, finding a substitute would cause delay, and the project may suffer.
- The contractor's *experience*. If the project is a new field for the contractor, the contractor may have to experiment, whereas an experienced contractor would automatically know what to do. Of course, the work of an inexperienced contractor might be better because of a new approach or new ideas. The best guide is the contractor's past performance. The size of the project is also an important consideration. A contractor's success on a $40,000 job does not ensure success on a million-dollar project.
- The contractor's *equipment*. Most contracting work requires expensive, often specialized equipment. Without access to the proper equipment, the contractor may be doomed to failure despite good intentions. A trip by the engineer to examine the contractor's machinery and equipment may be recommended.
- The contractor's *staff*. The skill of the contractor's staff members in their areas of specialization can mean the success or failure of an enterprise. Purchasing of materials and supplies, for instance, can add to profit or be an excessive addition to cost. The same or similar items are often sold for widely divergent prices by different suppliers. A highly skilled purchasing agent will know how to get the best prices.

Disqualification of a bidder is a rather serious step, and the engineer should be sure of the reasons before taking such action. Charges of favoritism or debates over qualifications compel the engineer to overlook any personal feelings about a articular bidder. With all bids, the owner's right to reject any and all proposals should be reserved for use if needed.

CONTRACTOR'S COSTS

What is the major reason why contractors lose money on jobs? Probably the best answer is that they bid too low. And why would a contractor bid too low? Often, the contractor has overlooked one or more of the elements of a proposal. The contractor must consider five basic components in his or her proposal to "make out" on the job. These are (1) direct labor, (2) materials, (3) overhead, (4) contingencies, and (5) profit.

♦ Direct Labor

In putting together the elements necessary for any kind of structure, be it a special machine or a building, direct labor is an important item. The cost-per-hour for toolmakers, carpenters, brick layers, and other specialists adds up to a significant amount. The contractor can usually anticipate and estimate this direct payment cost with reasonable accuracy. Other costs tied to direct labor, however, are often overlooked.

Fringes

Fringe benefits include payments (other than direct wages) by an employer caused by the presence of the worker on the payroll and those imposed on the employer by a governmental requirement or by collective bargaining. Many payments are fringes according to this definition. Some examples follow:

- Worker's compensation. To conform with a state's worker's compensation law, the employer usually pays an insurance company or a state agency an amount set in accordance with the amount of the employer's payroll. The amount paid by the employer is determined either from a rate manual or according to the employer's injury experience. These costs can run from a few cents per $100 on office employees to nearly $20 per $100 of payroll on certain types of construction workers. Such costs can vary from state to state.
- Social security. Employer's payments to social security are equal to those deducted from the worker's wage. When the Social Security Act was first passed, these payments were little more than a trivial annoyance; now they constitute a significant cost to the employer.
- Holiday pay and vacation pay. A recent collective bargaining goal has been to add each employee's birthday as a holiday for that employee. Although these payments are made for time when no work is performed, they are still costs of direct labor.
- Medical insurance and life insurance. It is the policy of many employers to share the cost of these benefits with their employees.
- Overtime premiums and shift premiums.
- Retirement funds. Such funds will later supplement an employee's social security benefits.
- Stock-buying plans and profit-sharing plans. In a stock-buying plan, the employer usually helps the employee become a part owner in the company. Frequently, deductions are taken from the employee's wages and then the employer adds to the employee's investment a few years later. Turnover is reduced if the employee must wait for the employer's addition to the investment. Profit-sharing plans work in many ways—one way being the division of company profits after the first, say, $100,000 of profit in a particular year. The increased incentive and company loyalty that result from such plans cannot be denied—but neither can the added cost.

Availability and Transportation of Personnel

With increased automation, it has become more and more possible—and even desirable—to locate plants without reference to the location of the labor supply. Structures are being built in the frozen regions near the Arctic Circle and the Antarctic and in desert regions, for example.The contractor who undertakes such installations has the problem of getting people to go to such places and work on the projects. Often this requires that relocation expenses and housing be furnished. In less spectacular instances, a contractor often overlooks these costs. For example, in many locations there may be an abundance of people but a shortage of specialists. Getting specialists there may entail payment by the contractor for daily time in transit or for relocation.

♦ Materials

One of the largest costs of a project is the cost of the materials that go into it. Purchasing can make or break a contractor. There is, in effect, an added profit for the contractor who, because of a knowledge of sources for materials and supplies, is able to make fortunate procurements throughout the project.

The cost of materials and, thus, profits can also be affected by prompt payment of invoices. It is common practice among suppliers to offer discounts for prompt payment—for example, 2 percent off for payment within 10 days, net in 30 days. Although 2 percent on any one bill may not amount to much, the accumulation can be significant, especially over the life of a long and costly project.

Availability and Transportation of Materials

Local availability of materials and supplies is an important consideration. *Apparent* local availability may not be sufficient though, because contractors can run into very high local prices. Usually, it pays a contractor to investigate and make sure of local supplies before bidding on work in unfamiliar settings.

One story runs something like this: The Black Construction Company, of a neighboring state, was the successful bidder on a contract, bidding in competition with local contractors. The bid was made on the basis of cement, mortar, sand, and gravel being locally available (suppliers in the area were quite plentiful). Black set up to begin operations on the site, but when Black Company tried to buy concrete ingredients, it found that they were "earmarked" for its former bidding competitors. As a result, Black had to transport cement, mortar sand, and gravel several hundred miles, resulting in a loss on the contract.

Storage

The owner often allows the contractor to use the owner's storage facilities if such facilities are available. In many contracts, though, such facilities are not available or are inadequate. The successful bidder should include storage costs in planning the bid.

♦ Overhead

These are the costs that many small contractors or those new in the field may tend to overlook; as a result, they seem to "make out" on their contracts but lose money in the long run.

The contractor's overhead consists of many things: maintenance of an office, equipment maintenance, taxes, supervision, depreciation on equipment, utilities, and others. A discussion of all the items that normally constitute overhead would be beyond the scope and purpose of this book. Only one, the cost of estimating, will be discussed here because it is so often overlooked.

The costs of investigation, planning, and estimating usually run somewhere around 2 percent of the bid price on a project. Two percent does not seem too much, but very few contractors get every job on which they bid. The average is probably about one successful bid out of every six or so submitted. In that case, the cost of making a successful bid is substantial. It is worth noting that this cost constitutes one of the wastes of competitive bidding practices. There have been instances in which owners solicited bids with no real intention of going through with their projects.

♦ Contingencies

Very few projects run smoothly from start to finish. It is rare that every day is a weather-working day or that delays do not occur for one reason or another. To cover the unexpected events, the contractor usually adds an item known as *contingencies* into the bid. The cost of this item runs anywhere from 5 percent to 15 percent of the total, depending upon the likelihood that hazards will occur. If liquidated damages are a prospect, contingencies may be quite high. If all work to be done is clearly and completely shown, and the probability of hazards is small, the amount for contingencies will be reduced.

♦ Profit

People who work for others as independent contractors do so in an endeavor to make a profit. Much of the growth of our economy can be attributed to the intent to make such a profit. Contractors are entitled, as is anyone else, to be paid a reasonable return on the projects they undertake. The percentage of profit varies with such things as the size of the job, competition, and other factors, but there must be profit if the contractor is to survive.

HAYNES v. COLUMBIA PRODUCERS, INC.
344 P.2d 1032 (Wash. 1959)

E. R. Haynes, a general contractor, brought this action to recover money allegedly due him from the defendant, Columbia Producers, Inc., under four contracts entered into between the parties. At the trial the work done pursuant to each of the four contracts was referred to by a separate job number. For convenience, we will refer to the contracts in the same manner. The first contract (Job 5–14) was the only one of the four in writing. It provided for the construction of a grain elevator and was a "fixed-price" contract. The second contract (Job 5–16) was also for a fixed amount. It was for the construction of a foundation and concrete slabs for a steel warehouse to be erected adjacent to the elevator. The third and fourth agreements were both "cost-plus" contracts. One of them (Job 5–19) provided for the labor and material for the construction of an undertunnel and foundation, and a slab, for a storage warehouse. The fourth contract (Job 5–26) was for the installation of machinery to be used in the storage warehouse building.

One of the principal causes of the difficulties between plaintiff and defendant was that all four contracts were being performed simultaneously, and the work was being done at the same location. During the trial, the corporation contended, *inter alia,* that the contractor had billed many items to the cost-plus contracts, whereas, in fact, these items should have been charged to the "fixed-price" contracts.

Plaintiff Haynes performed all work required under the contracts. The sole dispute at the trial concerned the amount, if any, which defendant corporation owed the plaintiff. The trial court entered judgment for the plaintiff in the sum of $28,052.99, plus interest. Both parties have appealed, the plaintiff claiming a greater amount than was awarded to him, and the defendant claiming that the amount of the judgment was too large. For convenience, we shall hereinafter refer to the appellant as Columbia Producers and to the respondent-cross-appellant as Haynes.

Under the machinery installation contract (Job 5–26), Haynes agreed to install certain machinery in Columbia Producers' elevator. Columbia Producers purchased this machinery from a third party, the Carter-Miller Company, thereby receiving a discount of $1,948. Columbia Producers directed Carter-Miller to credit the $1,948 discount to Haynes, who had an account with Carter-Miller for other purchases completely independent from the work he was then doing for Columbia Producers. The obvious effect was, of course, that Haynes owed Carter-Miller $1,948 less than he had owed before the discount was credited to his account.

Columbia Producers contended that the "plus" amount agreed to as the contractor's fee on Job 5–26 was the discount that the corporation caused to be credited to Haynes's account with Carter-Miller. The trial court found that the parties had agreed that the "plus" amount would be 10 percent of cost, and made no finding relative to the Carter-Miller discount.

The appellant, Columbia Producers, does not assign error to the finding of the trial court that the parties agreed to a contractor's fee of 10 percent. Its assignment of error is directed to the failure of the trial court to make any finding relative to the discount. It is urged by Columbia Producers that the effect of the trial court's decision on this point is to give a double allowance to Haynes. With this we agree.

Columbia Producers directed CarterMiller to credit Haynes with this discount, under the mistaken impression that this was what the parties had agreed to as the contractor's fee on Job 5–26. The fact that Columbia Producers was mistaken on this point does not entitle Haynes to a 10 percent fee and the discount. Haynes's contention that Columbia Producers gave him this discount merely as a gratuity is not supported by the record, and is not consistent with the business relationship existing between these parties. In short, Columbia Producers' first assignment of error is well taken, and it should be allowed $1,948 as an offset against the judgment awarded Haynes. As part of Haynes's costs on the machinery installation contract (Job 5–26), the trial court allowed him $1,061.30 for labor used to install certain bin dividers. Columbia Producers contend that the obligation to install these dividers existed under Job 5–14, and that, consequently, it was error to allow these costs as a part of Job 5–26. This is not merely a dispute without substance, as it will be remembered that Job 5–14 is a "fixed-price" contract, whereas Job 5–26 is a "cost-plus" contract.

The contract entered into between these parties for the construction of the elevator (Job 5–14) obligated Haynes to furnish labor and all other costs necessary to complete a reinforced concrete elevator as shown on certain plans. These plans were introduced into evidence, and it is readily apparent that they provided for the construction of bin dividers.

The oral contract made for the installation of machinery (Job 5–26) was subdivided as follows: (a) hopper bin bottoms; (b) spoutings; (c) belt conveyors; and (d) general machinery installation. It will be noticed that there was no obligation under this contract for installation of any bin dividers. Rather, this obligation clearly existed under the contract entered into for the construction of the elevator. Accordingly, the judgment awarded Haynes should be reduced in the further amount of $1,061.30 (independent of any interest calculation).

Columbia Producers next contends that the trial court erred in refusing to credit it with $2,467.18, representing the amount charged by Haynes for certain sheet steel, and the transportation thereof, for use in the construction of bin bottoms and hoppers on Job 5–26. The total bill for this item was $4,164.18, representing 38,352 pounds of steel, plus transportation. Columbia Producers claims that only 20,000 pounds were actually used on Job 5–26. The trial court, faced with conflicting testimony, found for Haynes on this issue. We have carefully examined the record and conclude that it supports this finding.

Finally, Columbia Producers urges that the trial court erred in allowing Haynes interest on the amount found due him. In this connection it asserts that a trial of this case was necessary in order to determine how much was owing from it to Haynes, and that, consequently, this amount was not liquidated in the sense that it could have been determined prior to trial. Columbia Producers bases this assertion on the fact that Haynes allegedly failed to furnish adequate or proper billings, in that many items were being charged to the cost plus contracts which should have been charged to the fixed-price contracts. Columbia Producers asserts that, in addition, many other items on the bill were overcharges. Even assuming these facts to be substantially true, we conclude that the trial court was correct in the allowance of interest.

♦♦♦

We turn now to a discussion of the assignments of error raised by Haynes on his cross-appeal, only one of which we need to discuss in detail.

The written contract entered into for the construction of the elevator (Job 5–14) provided that Haynes should be allowed extra compensation if it became necessary for him to excavate any "rock." The trial court found that Haynes had expended $1,814.11 for extra excavation within the terms of the contract. However, the trial court also found that, under the original plans, Haynes was obligated to construct a basement, size 40′ by 40′. It is admitted that the basement as it was finally constructed was only 20′ by 40′. The trial court determined that there was a saving to Haynes of $1,690 in the construction of the smaller basement; consequently, the $1,814.11 allowed for the rock excavation was reduced by the amount of $1,690.

Haynes contends that there is no substantial evidence in the record to support the trial court's finding that, under the contract, he was originally obligated to construct a basement, 40′ by 40′. Haynes argues that the agreement between the parties was originally for a basement 20′ by 40′. We believe there is merit to this contention.

The written contract did not specify what the dimensions of the basement were to be. The contract merely required Haynes to construct the basement in accordance with certain plans. From the plans it is difficult to ascertain the actual size the basement was to be. For example, one of Columbia Producers' witnesses, an engineer, testified that the plans called for the basement to be 40′ by 40′; on cross-examination, however, he admitted that the plans showed only a one-sided view of the basement. The only probative evidence concerning the agreement between the parties which we have found in the record is the testimony of Haynes. He testified, that, originally, it was agreed the basement would be 40′ by 20′. As there appears to be no evidence to contradict this testimony, it follows that the trial court erred in finding that the original plans specified the basement should be 40′ by 40′, and in awarding the amount of $1,690 as a credit to Columbia Producers.

The remaining four assignments of error raised by Haynes are all directed to the trial court's findings of fact. By his pleadings, as modified by a pretrial order, Haynes claimed (1) that he was entitled to $3,874 for extra work done on Job 5–16 for the reason that he was required to level certain ground before commencing construction of the warehouse required by that job, (2) that he was entitled to $980.27 for cement delivered, (3) that he was entitled to $982.25 for excavation done on Job 5–19 over and above the amount allowed by the trial court, and (4) that he had filed a claim of lien against Columbia Producers within 90 days after the termination of work as required by statute. The trial court found adversely to Haynes on each of these claims. Our examination of the record discloses that the evidence amply supports these findings.

The net effect of our decision is to allow Haynes $1,319.30 (independent of any recalculation of interest) less than was allowed him by the trial court. Under the circumstances, neither party shall recover costs on this appeal.

The cause is remanded with directions to the trial court to enter judgment in accordance with the views expressed herein.

HOUGH v. ZEHRNER

302 N.E.2d 881
(Ind. Ct. App. 1973)

I. STATEMENT ON THE APPEAL

Hough delivered 1,944.6 tons of crushed stone to Zehrner's junk and salvage yard where a new commercial garage for trucks was being constructed. The crushed stone was used for the driveway and parking area around the commercial garage where the muddy condition of the ground would not support commercial truck travel. When Hough was unable to obtain payment for the delivered crushed stone, he filed a mechanic's lien under I.C. 1971, 32-8-3-1....His foreclosure suit resulted in a judgment against him. The trial court concluded that Hough's lien was not within the scope of the statute. Hough filed his motion to correct errors, which presents this question on appeal:

> Does the scope of the mechanic's lien statute encompass a materialman who delivers crushed stone for a driveway and parking area which is to be used in conjunction with a commercial garage being constructed?

Our opinion construes the above statute and concludes that such a materialman is entitled to a mechanic's lien. We reverse the trial court's judgment.

II. STATEMENT OF THE FACTS

Hough entered into an oral contract with Caprio and Phebus in October 1968 to supply crushed stone which would be used in the construction of a driveway and parking facility at Zehrner's junk and salvage yard where a new commercial garage was being built. The first delivery of stone to Zehrner was on October 16, 1968, and the last load of crushed stone arrived on November 1, 1968. The total deliveries amounted to 1,944.6 tons of crushed stone for an agreed price of $2,916.90. Caprio and Phebus had a commitment to Zehrner for the delivery of the crushed stone. All parties knew of the agreement with Hough and the intended use of the crushed stone. When Hough was unable to receive payment from Caprio and Phebus, he filed his notice of claim for a mechanic's lien on December 30, 1968. The foreclosure action which commenced on June 17, 1968, resulted in a judgment against Hough. The trial court entered the following judgment:

> The Court having had this matter under advisement and being duly advised in the premises, now finds:
> "That the crushed stone delivered by plaintiff to premises of defendant, John Zehrner, a/k/a John Zehner, was used in the filling of holes on the land and in covering certain lands to make a parking and driving area for trucks. Said crushed stone was not used in the erection of any building or structure.
>
> The Indiana Supreme Court has often said that the mechanic's lien statutes are in derogation of the common law and must be strictly construed. This Court can find no authority in statute giving rise to a mechanic's lien in this cause.
>
> It is THEREFORE, ORDERED, ADJUDGED, AND DECREED by the Court that the plaintiff take nothing by way of his complaint from the defendant, John Zehrner, a/k/a John Zehner, and that said named defendants recover their costs herein."

Hough timely filed his motion to correct errors, which raises the question set forth below for our consideration on appeal.

III. STATEMENT OF THE ISSUE

The issue is one of statutory construction....We will examine the statute to determine:

> Does the scope of the mechanic's lien statute encompass a materialman who delivers crushed stone for a driveway and parking area which is to be used in conjunction with a commercial garage being constructed?

We conclude in our "Statement on the Law" below that the scope of the statute does encompass such a materialman.

IV. STATEMENT ON THE LAW

The statute here under consideration, contains the following language:

> That contractors, subcontractors, mechanics, journeymen, laborers, and all persons performing labor or furnishing materials or machinery for the erection, altering, repair-

> ing, or removing of any house, mill, manufactory, or other building, bridge, reservoir, systems of waterworks, or other structures, or for construction, altering, repairing, or removing any walk or sidewalk, whether such walk or sidewalk be on the land or bordering thereon, stile, well, drain, drainage ditch, sewer, or cistern may have a lien separately or jointly upon the house, mill, manufactory, or other building, bridge, reservoir, system of waterworks, or other structure, sidewalk, walk, stile, well, drain, drainage ditch, sewer, or cistern which they may have erected, altered, repaired, or removed or for which they may have furnished materials or machinery of any description, and, on the interest of the owner of the lot or parcel of land on which it stands or with which it is connected to extent of the value of any labor done, material furnished or either....

This statute is in derogation of the common law and should be strictly construed as to its scope....Any lien claimant under this statute has the burden of proof to establish that his claim is within the scope of the statute. Once this has been successfully accomplished, a liberal construction will be given the statute so that its purpose can be accomplished....The purpose of the statute was expressed by our Supreme Court in *Moore-Mansfield Construction Co.* v. *Indianapolis, New Castle, & Toledo Railway Co....*

> The mechanics' lien laws of America, in general, reveal the underlying motive of justice and equity in dedicating, primarily, buildings and the land on which they are erected to the payment of the labor and materials incorporated, and which have given to them an increased value. The purpose is to promote justice and honesty, and to prevent the inequity of an owner enjoying the fruits of the labor and materials furnished by others, without recompense....

Looking to the language in the statute which would be indicative of its scope, we extract the following:

> ...persons...furnishing materials...for...other structures, or for construction (of)...any walk or sidewalk, whether such walk or sidewalk be on the land or bordering thereon, description, and, on the interest of the owner of the lot or parcel of land on which it stands or with which it is connected to the extent of the value of any...material furnished...

It would be absurd to limit the scope of the statute to those items specifically and expressly mentioned....Descriptive materials are blatantly absent; therefore, materials are inescapably bound with the nature of "... other structures." These words should be given their ordinary and literal significance first....

Webster's Third New International Dictionary (1970) defines "other" as " ...that which is remaining or additional; being the ones distinct from the one or those first-mentioned or understood." "Structure" is defined as the "...example from T. W. Arnold: "demolish any building, highway, road, railroad, excavation or other structure." The Random House Dictionary of the English Language (unabridged ed., 1969) defines "other" as " ...additional or further; different or distinct from the one or ones mentioned or implied; building, construction, or organization; arrangement of parts, elements, or constituents; a pyramidal structure...."It further defines "structure" as "... anything composed of parts arranged together in some way; ..." In *McCormack v. Bertschinger*...where it was contended"...that the labor and material which were furnished in the construction of the garage, driveway, walks, and retaining wall, and not in the construction of the house itself, were not lienable, ..." since they were not structures, the Oregon Supreme Court held that a driveway was a structure even though it had not been specifically or expressly set forth in the statute. The Oregon Supreme Court applied the maxim of *noscitur a sociis*.

Zehrner urges that the doctrine of *ejusdem generis* would exclude the inclusion of a driveway as a structure under the statute. *Ejusdem generis* is only an illustration of the broader maxim of noscitur a sociis that general and specific words being associated together take color from each other.[1] But, when it is clear, as it is in the present case, that the less general words were inserted for a distinct object to carry out the purpose of the statute, then the general word ought to govern. It would be a mistake to permit *ejusdem generis* to pervert the construction of the statute and defeat the intention of the legislature....The cardinal rule of statutory construction mandates that this Court give effect to the intent of the legislature....

The drilling of a water well was held to be a structure.... Architectural services are within the scope of a mechanic's lien statute....In other jurisdictions where the statutory language is similar, grading was held to be reasonably necessary for the proper construction of a house and a part of its "erection." Even landscaping in Illinois has been included as part of the "erection" of a structure thereby bringing it within the lien statute....

The rationale of our Supreme Court in *Wells v. Christian*...might well be applied to the present case. In *Wells v. Christian*, supra, our Supreme Court stated:

> The mains and pipes laid down...are clearly a part of the apparatus necessary to accomplish the objects for which such heat plant was erected. They constitute a part of the machinery by means of which the business of supplying heat to others must be carried on. The laying of these connecting pipes, essential to the exercise of the franchise held by the appellees, was a part of the erection of the manufacturing plant....The hauling of materials to be used in the performance of the work, and hauling away the surplus earth excavated, were incidental matters inseparably connected with the principal undertaking, and constituted items of labor for which a lien may be acquired...
>
> The work which appellant performed being directly and necessarily connected with the erection of the appellees' heating system....The claim sued upon and found to be due and unpaid is within the protection of the statute, and appellant, having taken the steps necessary to perfect a lien upon the real estate described, was entitled to a decree of foreclosure.

Without the crushed stone, heavy trucks attempting to use the commercial garage would become mired down in the soft, moist earth. Moving heavy trucks about the parking area would be cumbersome and difficult. Crushed stone for the driveway and parking area was a necessary and essential part of the commercial garage structure and its subsequent functional use. Hough's claim of lien is well within the scope of the statute. We do not deem it necessary to draw obvious analogies with walks or sidewalks or the obvious value imparted to Zehrner's commercial garage by having a usable driveway.

---

1. The rule of statutory construction known as *noscitur a sociis*, which means that the meaning of a doubtful word may be ascertained by reference to the meaning of words associated with it, is the underlying authority for the application of the *ejusdem generis* rule.... The rule of *ejusdem generis* is that general words following an enumeration of particular cases apply only to cases of the same kind as expressly mentioned ....

V. DECISION OF THE COURT

Crushed stone delivered for a driveway and parking area to be used in conjunction with a commercial garage is a structure within the scope of the mechanic's lien statute.... The judgment of the trial court should be and the same hereby is reversed.

YALE DEVELOPMENT CO. v. AURORA PIZZA HUT, INC.
420 N.E.2d 823 (Ill. App. Ct. 1981)

This case involves an appeal and cross-appeal from a judgment against defendant, Aurora Pizza Hut, Inc., finding it breached a written contract but limiting plaintiff's recovery to $1,000 under the terms of a liquidated damages clause. The sole contention of plaintiff, Yale Development Co., on appeal is that the liquidated damages clause should not be enforced and that it should be able to recover compensatory damages in excess of the amount stipulated. Defendant, in its cross-appeal, argues that its termination of the contract did not constitute an anticipatory breach and, in the alternative, that if such action did amount to a breach, the plaintiff is not entitled to recover because of its subsequent inability to perform under the terms of the contract. The parties executed a contract on September 19,1975, in which defendant agreed to pay $90,000 in exchange for property located at Route 53 and Butterfield Road in unincorporated Du Page County.

The sale was specifically made contingent upon plaintiff being able to obtain a zoning change and liquor license which would allow the construction and operation of a Pizza Hut restaurant. Although time was made of the essence, no specific time was mentioned in the contract.

Plaintiff filed an application for rezoning with the Du Page County Zoning Board of Appeals on December 14, 1975. After a public hearing, the Zoning Board of Appeals recommended denial of plaintiff's petition, and on February 23, 1976, the County Board of Du Page County concurred in the denial. Plaintiff then filed a complaint in the circuit court seeking a declaratory judgment that the existing zoning was void and that the intended use of the premises be permitted. A bench trial was conducted on October 7, 1976, and the case was taken under advisement on January 5, 1977, after trial briefs were submitted. On February 24, 1977, before the court issued a decision, defendant sent plaintiff a letter purporting to terminate the contract "inasmuch as the city (*sic*) has denied our right to rezone the property." Subsequently, the court denied rezoning of the property, of which the court below took judicial notice in its written reasons for decision. Plaintiff then filed the present action alleging that defendant was in breach for repudiating the contract prior to the time performance was due. The trial court held for the plaintiff but entered an order limiting plaintiff's recovery to $1,000 under the terms of the liquidated damages clause. This appeal and cross-appeal followed, and we reverse on the basis of the defendant's cross-appeal. We find the second issue raised by defendant's cross-appeal to be dispositive of the case at bar but turn first to the question of breach.

Defendant's first contention on cross-appeal is that its good faith termination of the contract did not constitute an anticipatory breach under these facts and circumstances since plaintiff was unable to obtain the rezoning of the subject property within a reasonable period of time. We disagree. Defendant's notice of termination came at a time when the plaintiff had an action for declaratory relief to rezone the property pending before the circuit court. Defendant reasons that the rezoning had not been accomplished in the 17-month period between the time of execution and the notice of termination and that there was no reason to believe the pending action would be successful when proceedings before the Zoning Board of Appeals and the County Board were not. Although the contract did not specify the time for performance, the law will imply a reasonable time....The intention of the parties controls what time is reasonable, and the court must look to the surrounding circumstances to discover the intention of the parties....A reasonable time for performance is such time as is necessary to do conveniently what the contract requires....Under the facts and circumstances of the present case we do not view a 17-month period as being unreasonable when the necessity of rezoning was contemplated by both parties. The record also indicates that the parties had had previous similar dealings which required rezoning actions and that two of such dealings required periods in excess of two years to be resolved. In light of these previous contracts and the period of time typically involved in obtaining a rezoning, we cannot say that a 17-month period was manifestly unreasonable and not within the intention of the parties. The trial court, therefore, correctly viewed the February 24 letter as an anticipatory breach.

Under the law of anticipatory repudiation, when one party to a contract repudiates his obligations before the time performance is due, the other party may elect to treat the repudiation as a breach and sue immediately for damages....This leads us to defendant's second contention and the one we find dispositive of the case at bar. Defendant contends that, assuming the letter of February 24 is regarded as an anticipatory breach, the plaintiff cannot maintain an action on the contract due to his subsequent inability to fulfill the condition precedent of obtaining the rezoning. We agree. Although it was not necessary for plaintiff to actually tender performance since an anticipatory breach excuses the non-repudiating party from further performance, ...we hold that a plaintiff suing on an anticipatory breach must still show a willingness and ability to perform had not the breach occurred. While it appears that no Illinois court has yet had the opportunity to consider this precise question, Illinois courts have consistently held that where both parties to a contract are in default, there can be no recovery by either against the other....It has also been held that where one party has put it beyond his power to perform, although no demand or tender is necessary to allow the other to recover for breach, the plaintiff must show that he was ready and willing to perform the contract on his part. Similarly, in *Nation Oil Co.* v. *R.C. Davoust Co., Inc.*, ...the court held that a party seeking to recover for breach of contract must show that he has performed or offered to perform his own obligation under the contract or that such performance was excused. Although none of these cases involve anticipatory breach, we consider their basic principles to be applicable to the case at bar.

As these cases demonstrate, if plaintiff had elected to wait until the time for defendant's performance to sue, instead of bringing suit immediately on the anticipatory breach, it would have been required to show a tender or offer of performance on its part, or that its performance was excused. Due to plaintiff's inability to procure a rezoning, it could not have tendered its performance and thus could not have maintained an action for breach at the time when defendant's performance was due. It would be anomalous to allow plaintiff to put itself in a better position by suing immediately on an anticipatory breach and thus avoid the necessity of proving its tender of performance. We therefore hold that where a plaintiff sues on an anticipatory breach prior to the time when defendant's performance is due, he must show an ability and willingness to perform his part of the contract within the time specified in the contract.

Indeed, this seems to be the law in those jurisdictions which have considered the question. For example, in *Ufitec, S.A.* v. *Trade Bank and Trust Co.*, ...the question before the court was whether a holder of a draft drawn against a letter of credit issued by the defendant bank could sue the bank for an anticipatory repudiation due to the bank's act of revoking the letter of credit before it had expired. The court concluded that since the holder never complied with the letter of credit, nor had it shown that it could ever comply during the term of the letter of credit, it could not maintain an action for anticipatory breach. The court stated, "(a)n anticipatory breach, in a proper case, may excuse one from performing a useless act, but it does not excuse one from the obligation of proving readiness, willingness, and the ability to have performed the conditions precedent."

The problem is also discussed in the third edition of Williston's Treatise on Contracts:

> The difficulty is not peculiar to cases of supervening illegality, but is involved in every other case where, after an anticipatory breach, supervening impossibility occurs which would in any event prevent and excuse performance of the contract. The situation differs, also, in only a slight particular, hereafter referred to, from one where it appears after the anticipatory breach that, although there was no legal excuse, the return performance could not or would not have been rendered to the repudiator.
>
> In each of these cases, if all the facts could have been known or foreseen at the time of the repudiation, no cause of action on the contract would have arisen.
>
> It is a practical disadvantage of the doctrine of anticipatory breach that an action may be brought and perhaps judgment obtained before the facts occur which prove that there should have been no recovery. Fortunately, in most cases, this evidence, though not available at the time of the repudiation, becomes available before judgment can be obtained.
>
> It seems clear that if the evidence thus becomes available the plaintiff can recover no substantial damages, and in the case of supervening illegality which is not due to the defendant's fault, there seems no reason to allow even nominal damages.
>
> The loss should rest where chance has placed it; and the same should be true in case of any supervening excusable impossibility. Still more clearly, there should be this result if the facts show that the plaintiff could not or would not have performed for reasons which would not be a legal excuse. The fact that a right of action has already arisen should not preclude the defense. Failure of consideration after a cause of action has arisen bars recovery....

Similarly, comment a of section 277 of the *Restatement of Contracts* makes the following rule applicable to anticipatory breach:

> Where in promises for an agreed exchange a promisor commits a breach, and a right of action for the breach arises, the right of action is extinguished if it appears after the breach that there would have been a total failure to perform the return promise, even if the promisor had not justified the failure by his own breach or otherwise....

Since the plaintiff has failed to obtain rezoning of the subject parcel it cannot demonstrate an ability to perform and, hence, is not entitled to recovery under the contract. The judgment of the trial court is accordingly reversed.

Reversed.

REVIEW QUESTIONS

1. Describe the relationships between the parties to a construction contract.
2. How do public projects and private projects differ?
3. In what ways are contracting and manufacturing similar? How do they differ?
4. If you planned to build a two-car garage on your lot, what factors would you consider in deciding whether to attempt to build the garage yourself or hire a contractor to do it? How does this differ from an industrial situation in which a company is trying to decide whether to build a special machine or hire someone else to build it?
5. Summarize briefly the advantages and disadvantages of: (1) a lump-sum contract; (2) a cost-plus-percentage contract; (3) a cost-plus-fixed-fee contract; and (4) a combination unit-price and lump-sum contract.
6. In *Haynes v. Columbia Producers, Inc.*, what caused the court action? How could it have been prevented in the original arrangement?
7. What purposes are served by a feasibility study? When should it end?
8. Why do governments strictly regulate competitive bidding practices? What formal procedures are required in your state?
9. Locate in a local newspaper an "advertisement" or "call for bids" for a construction project. What details does the advertisement give its reader?
10. What are mechanic's lien laws? How do your state's statutes call for such liens to be established and removed?
11. What are the purposes of each of the three types of bonds commonly required of contractors?
12. What nonmonetary factors determine the qualifications of a bidder?
13. Your employer is about to bid on the building and installation of a shuttle mechanism to automatically move 2"-diameter hollow cylinders 1" high from a lathe to a punch press, to locate them in the press die and actuate the press to extrude the cylinders. What cost elements are likely to be encountered in the work, and which, therefore, should be included in the price bid?
14. Review the statement of Indiana law in *Hough v. Zehrner*. What, if any, work or materials used on a piece of property would not be a sufficient basis for a mechanic's lien?
15. In *Yale v. Pizza Hut*, the defendant was held to have committed an anticipatory breach of the contract. Why, then, did the court refuse damages to the plaintiff?

Chapter 17

Specifications

As the term is used here, *specification* refers to the description of work to be done or things to be purchased. Specifications are most frequently heard of in building construction, but they exist in every area of engineering. The definition above is broad enough to include many things we encounter daily in our dealings with others. If, for example, you hire painters to paint your house and tell them the house is to be white with gray trim, you have made a specification. If you take your car to an auto mechanic and tell the mechanic to do whatever is necessary to remove the "grind" from the transmission, you have made a specification. A person ordering four 1/4" stove bolts, 3" long, from a hardware store clerk is using a specification. The specifications mentioned so far have been oral, but they are just as much specifications as the voluminous written ones for such a construction project as the Hoover Dam.

Building construction is seldom undertaken without specifications; process engineers, project engineers, accountants, and purchasing agents toil over specifications for machinery and equipment. Very few government purchases are made without reference to written specifications. In fact, the federal government only proceeds without such specifications in isolated instances of purchases of services. Generally, if the purchase involved is of sufficient monetary value so that a breach of the specified conditions could cause the purchaser substantial injury, the specifications should be written. Two main reasons exist for putting specifications in writing rather than giving them orally: (1) They create a permanent record for resolving disputes, and (2) they assure planning.

The written record of specifications defines the duties of the parties. The specification then becomes part of the contract. Later, it may be interpreted in court or by an arbitration board if a dispute arises between the parties. Lawsuits on some construction projects have been undertaken years after completion of the projects. Without a permanent record of the rights and obligations of the parties, the result can be utter confusion. Written specifications can also frequently *prevent* costly lawsuits, because they act as a ready reference in controversies between the parties.

The second reason for putting specifications into writing, the planning reason, is based in both practicality and efficiency. It is easy to specify something orally without giving it much thought. When it becomes necessary to reduce the specification to written form, however, the writer becomes more cautious, examining his or her reasons for making a particular specification. The result, in specification writing, is usually a more efficient purchase. The likelihood that the thing specified will do the job required at a lower overall cost is improved greatly by the writing.

About to write 30 strokes per minute as the specification for a punch press, for example, a process engineer may well begin to question this speed. Under certain conditions, a variable-speed press (say 20 to 40 strokes per minute) might be more appropriate, and the writing of the specification calls attention to the question.

SPECIFICATION WRITING

A specification must communicate to the reader what is required by the writer. Its job is essentially quite simple. As such, it can be best accomplished in simple language. Complex sentence structure and complicated wording can often be interpreted more than one way. With simple structure and simple wording, ambiguity is much less likely. By the same token, it is best to avoid jargon, abbreviations, and symbols unless they are of a recognized standard and fully understood in the trade.

Specifications, together with the drawings, show the contractor what is to be done. Each supplements the other. In addition, the specifications indicate the relation between the parties in greater detail than does the agreement. The working drawings show the work to be accomplished, but, frequently, not the quality standards required. The quality standards are usually more conveniently stated in the specifications.

♦ Style

Specifications should be written with great clarity. A specification should not contain flowery language or complicated legal terms. Although they must be brief and terse, the specifications still must be complete. The writer should strive to be brief up to the point where further brevity may be accomplished only by sacrificing some of the meaning in the specification.

Literature written to be read for entertainment frequently refers to a subject in terms of "he...hers," and other such pronouns. The use of such words adds considerably to the ease and enjoyment of reading such material. The vast majority of readers will understand which subject the author is referring to. Those who do not will usually pass over the sentence without hesitation to maintain interest in the story.

The writer of anything that is likely to be interpreted in a court of law, however, cannot afford to risk this kind of openendedness. Where a misinterpretation is at all likely, the subject should be repeated. Consider the sentence: "In case of controversy between the contractor and the inspector, he shall immediately refer the question to the engineer." Who is to refer the question? The contractor could say, "I thought the inspector was to contact the engineer," and vice versa. Replacement of the word "he" with "the contractor" or "the inspector" clears up any doubt.

♦ Precise Wording

Our language consists of many words with similar meanings. It includes general words and specific words. The word *property*, for instance, has a very broad meaning, including all items of personal possessions as well as real estate. Many other terms have broad meanings and should be used very carefully in the wording of specifications. The words used should indicate what is to be accomplished so precisely that no doubt can exist. Commonly used words are often not exact enough in meaning. For instance, it is quite common to hear someone say that something should have an addition on *either* side of it, as in "light on either side of the gateway," when what is intended is that the addition must be present on *each side*, as in "a light on each side of the gateway."

Many other words are commonly used interchangeably, such as *any and all*, *amount* and *quantity*, *malleable* and *ductile*, and *hardness* and *rigidity*, for example. The inexactness of meaning in oral conversation usually causes no problems; but in specification writing the result can be trouble, measured in dollars. There are few instances in the English language where one word means exactly the same thing as another. Care in wording is definitely called for where the writing may have to be reviewed in court.

♦ Reference Specifications

With the wealth of old specifications, texts on specification writing, and standards published by various organizations, it is a rare specification today that is entirely original. As a practical matter, this is desirable. Previous specifications have been tested, errors removed, and the language improved. If the rewriting for the new specification is carefully done, other potential problems may be discovered and eliminated. In addition, the writer is less likely to overlook something in the present specifications if he or she can follow a pattern. However, a word of caution is necessary. Wholesale clipping of paragraphs and requirements from previous specifications can result in problems. While two specifications may be similar in many ways, it is unlikely that any two will be exactly alike. If the writer does not exercise caution in adapting paragraphs and clauses from old to new, certain features of the new job are likely to be different from what was intended. For example, features of the new job could be completely omitted from the new specification.

Many organizations have written standards for various items of equipment or elemental components, making it unnecessary for the specification writer to do more than incorporate these standards by reference. The United States government, for one, has published a vast array of standard specifications. For instance, when the government buys women's slacks for use in the armed forces, the specifications consist mainly of references to military standards and other standardized specifications. The buttons are specified by federal specification number V-B-781, the fasteners by V-F-106, the shipping boxes by NN-B-631, and so on for the cloth, the stitches, the labels, and all other elements of the purchase. The writing of a specification where such standards exist consists largely of determining the appropriate elements and listing the standards in logical order.

The advantage of such standardization is quite obvious. A specialist in one element of the entire assembly has specified the appropriate quality to be used. No person can be a specialist in everything. The specification resulting from the combined efforts of several specialists is quite likely to be better than it would be if the total specification were to be written by any one of them.

ARRANGEMENT

The specification usually consists of at least two parts, one general and the other specific and detailed. The first part is known by different names (such as general conditions, general provisions, or special conditions) in various specifications, and the information included is not well standardized. Most contracts also include a document, previously discussed, known as the "General Conditions," which deals with the basic rights and responsibilities of the parties. To avoid confusion with the more basic document, we will here use the term *general provisions* as the name for the general portion of the specifications.

♦ General Provisions

In the relationships between the parties to a construction contract, there is a vast middle ground between the provisions in the agreement and the general conditions, and the detailed provisions in the specifications. The general conditions are broad enough to apply to any contract work the owner may want and to any contractor he or she may hire.

In contrast, the general provisions part of the specifications pertains to a particular contract or type of contract. It sets the owner's policy as to control of the work, the scope and quality of the work, and any special requirements or precautions. It is here that answers are found to many of the day-to-day questions that arise between the parties. To cover these questions, the general provisions must consider many topics. A few of the topics appear frequently enough to justify consideration here, these being representative of the contents of a general provisions section.

Schedule of Work

A topic that is almost certain to be covered (usually by a requirement for consultation) is the sequence of work. It is poor policy, from both a legal and a cost standpoint, to rigidly control the details of the work to be accomplished. However, there are occasions when the owner must exercise some control to protect the public interest (as in highway construction) or to dovetail the project with the work undertaken by other contractors.

Usually a work schedule is called for in the specifications to help the engineer and contractor plan the work to the best advantage of the owner. Such a work schedule is made by showing graphically the starting and completion dates for the component parts of the project. The work schedule resembles the Gantt Charts so commonly used in production scheduling in industry. The figure that follows shows a simple work schedule chart for multistation, strip-feed punch-press installation. This schedule provides the engineer with the means of obtaining a weekly (or daily or monthly) check against actual performance.

The schedule also contains percentages assigned to each component of the job. These percentages represent estimated portions of the total job for each component and are usually agreed upon by the engineer and the contractor prior to starting work. Such schedules are frequently used in a progressive payment scheme (so much per week or month according to accomplishments during the period). The schedule allows the parties to make a realistic estimate of progress. For instance, assume that the contractor installing the presses shown in the figure is on schedule on February 4 (the end of the week beginning January 30). At this point, he would be entitled to payments totaling 50 percent of the total contract price, less whatever amount may be withheld until final completion and acceptance. The amount to be held back may also

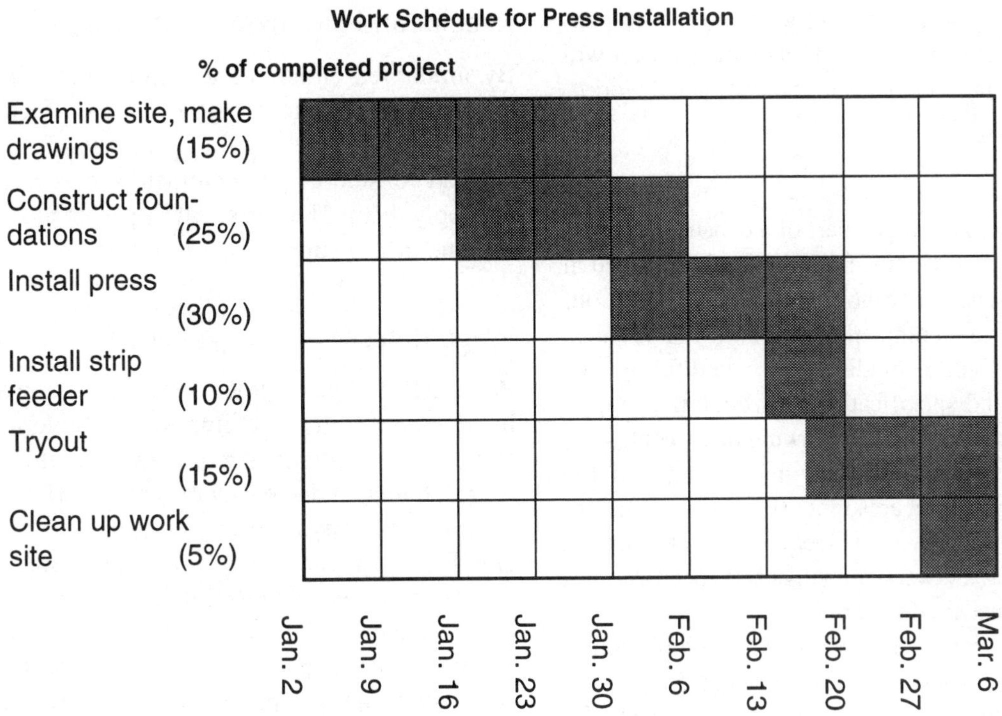

Work Schedule for Press Installation

% of completed project

Examine site, make drawings (15%)
Construct foundations (25%)
Install press (30%)
Install strip feeder (10%)
Tryout (15%)
Clean up work site (5%)

Jan. 2 Jan. 9 Jan. 16 Jan. 23 Jan. 30 Feb. 6 Feb. 13 Feb. 20 Feb. 27 Mar. 6

Figure 17.1 Work schedule for press installation.

be specified in the general provisions. Frequently this amount is set at 10 percent. In the foregoing case, assuming a 10 percent holdback, the contractor should receive 45 percent of the total contract price.

Changes and Extra Work

Very few large contracts are completed without changes being made during the course of the work. Questions of payment for such changes can cause severe disagreement if no provision has been made to anticipate them. A provision for changes and extra work is usually made in either the general conditions or the general provisions portion of the specification. Just as there are four basic means of paying for a contract, there are also four means of paying for extra work:

- Lump sum
- Cost plus percentage
- Cost plus fixed fee
- Unit price

The cost-plus-percentage and unit-price types of contracts present little or no problem as to compensation for changes or added work. The extra work is compensated at the fixed rate in the unit-price contract; the open-endedness of the cost-plus-percentage contract also accommodates changes easily. In the lump-sum and cost-plus-fixed-fee contracts, though, the parties must decide in advance how extra work will be compensated. The contractor is already on the job; it is usually inconvenient and costly to get someone else to take on the added work. If the contractor is unscrupulous and the engineer has not provided adequately for changes, the contractor can make these changes costly. The engineer who gives careful attention to the contract provision regarding payment for changes and extra work may find that time well spent.

Other Suppliers

Work or materials to be supplied by others (either by the owner or other contractors) should be shown in the specification. Frequently this information takes the form of one or more "right of supply" clauses, in which the owner reserves the right to supply motors or other components (often salvaged from worn-out machines). When more than one contractor is to work on a project, the areas of responsibility must be delineated clearly in the specification. This precaution is necessary for avoiding conflict and assuring that someone will be responsible for the details of each item; in other words, it will avoid both overlaps and gaps in the contracted work.

Drawings

The drawings are an integral part of a construction contract. The following statement is commonly included in the general provisions of the specification: "The Contract Documents are complementary, and what is called for by any one shall be binding as if called for by all." The drawings and specifications, particularly, supplement each other as everyday working documents. Occasionally a conflict between a drawing and the specifications becomes apparent. To resolve such conflicts, the general provisions frequently state that "in case of difference between drawings and specifications, the specifications shall govern."

Some provision is usually made for submission of working drawings and sketches for the engineer's approval during the course of the construction. Such details as the number of copies of drawings, even size and type of paper to be used, are sometimes specified.

Information Given

Usually, considerable information pertinent to the proposed project is passed along to the bidders and thus, eventually, to the contractor. Some information may be such that the owner and the engineer do not want to warrant its completeness or the indications given by it. This is commonly true of test borings, for instance, where a defect in the subsoil might not show up in the samples that are taken. To protect the owner in such circumstances, the general provisions usually state that neither the owner nor the engineer will warrant that the information given shows the entire picture. Generally, contractors are held responsible for having made their own examinations of the site.

Services Furnished

When an addition to an existing structure is made, or when machinery or equipment is installed, certain of the services necessary are commonly supplied by the owner. Such services as water, compressed air, electricity, and crane service to unload equipment are frequently furnished by the owner. It is in the owner's interest to mention the availability of such services in the specification, to give contractors a truer picture of the costs they will *not* have to bear.

Receipt and Storage of Materials

By similar reasoning, if the contractor is to be allowed to use the owner's receiving and storage facilities, this should be made clear in the specification. Such facilities add considerably to contractors' costs if they have to supply them. The costs will be passed along to owners who do not supply such facilities.

Wage Rates

Particularly in government contracts, the wages paid by the contractor are important. Several federal statutes (and usually similar state statutes) regulate wages, hours, and even sources of materials on public contracts. By way of example, many federal contracts cite the Davis-Bacon Act (wages), the Buy American Act (purchases of materials), and the Eight-Hour Laws (wages and hours) in their specifications. But the wages paid to workers are of importance even when the owner is a private party. Wages and working conditions poorer than those to which the area is accustomed can cause strikes and other labor problems. Projects are sometimes delayed and even destroyed as a result of labor strife. To avoid trouble of this nature, many specifications for private contracts provide that wage rates and working conditions must be equal to or better than those prevailing in the trade or locality.

Safety

A requirement such as the following, as to safety and accident prevention, usually appears in the specifications: "The contractor shall, at all times, exercise reasonable precautions for the safety of employees in the performance of this contract, and shall comply with all applicable provisions of federal, state, and municipal safety laws and building construction codes."

♦ Detail Specifications

The second major division of the specifications gives the details of the work to be undertaken. These detail specifications, together with the engineer's drawings, state how the job is to be done and what is to be accomplished. In writing the general provisions portion, the engineer has access to guideposts and instructions. When writing the detail specifications, though, the engineer has fewer guideposts and must rely on his or her writing skills and knowledge of what is to be accomplished. Detail specifications are concerned with the materials and work quality in the finished project.

Materials

When someone buys a lathe or a punch press, he or she is really purchasing the ability to turn, bend, blank, or perform some other operation on materials. All a possession can do is provide a service of some kind. Much the same is true in the purchase of materials for a construction project. The materials purchased must render a service. But it is up to the engineer to obtain the best service available for the owner at the most favorable cost. Thus, two factors normally oppose each other in the selection of material: cost and service. In many instances, the same or similar service can be rendered by two quite different materials (or pieces of equipment or machines). Transportation, storage, and inspection costs join other costs to be weighed against the service offered by a particular material. Part of an engineer's "stock in trade" is a knowledge of various means of obtaining services for the employer.

Materials are commonly called for in specifications according to established standards. SAE 1090 steel, for instance, indicates a steel with 90 points of carbon that hardens with proper heat treatment and has some unique physical properties. The various physical properties of materials, such as strength, elasticity, conductivity, and appearance, are important in determining the service they will render in a specific application. Very often, specifications are written with "or equal" clauses; for example, "Electrical controls to be XYZ or equal." The *or equal* means "not literally exact equality." A more precise way of stating the meaning might be *or equivalent*. With this, as with other contract language, the courts interpret wording according to trade usage.

Work Quality

A person who takes an automobile to a mechanic for a valve job usually isn't interested in the order of removal and replacement of the screws, nuts, and bolts. Similarly, an owner is rarely interested in how a particular result is accomplished, providing the result is satisfactory. It is usually far better to specify the results to be accomplished rather than the process by which the results are to be accomplished. For instance, it is much better to specify the compressive strength of concrete a week after pouring than to specify the quantities of cement, sand, and broken stone or gravel and the method of mixing.

There are exceptions, of course, to the principle of specifying results only. A particular method may interfere with the rights of others or with other contractors (e.g., using blasting in place of air hammers). Usually, however, it is sufficient in specifications to require that the work quality be equal to the best available without going into the details of the method. Language like the following might be used: "All sheet-metal work shall be performed and completed in accordance with the best modern sheet-metal practice, and no detail necessary therefor shall be omitted, although specific mention thereof may not be made either in these specifications or on the drawings."

♦ Arbitrary Specifications

Generally, there are two possible relationships between the owner and the contractor. The relationship of owner-independent contractor indicates that the contractor has been hired to produce a particular result. The employer-employee, or master-servant, relationship prevails where the owner (employer) or the owner's agent, the engineer, supervises the work of the contractor too closely.

The owner has two main advantages in retaining the owner-independent contractor relationship. Probably the primary one is that the contractor retains liability for his or her acts. There are many cases on record where, because the supervision was too close or the specifications made the contractor a mere employee of the owner, the owner was held directly liable for the acts of the contractor. Statements like the following make the contractor sound like an employee of the engineer (and, hence, an employee of the owner): "The contractor shall begin and continue work on whatever parts of the project the engineer shall direct, at whatever time the engineer shall direct." Such statements make it easy for a court to find against the owner and hold the owner liable for the actions of the contractor. The second advantage to maintaining the owner-independent contractor relationship (and specifying results rather than means of attaining them) involves liability as well. A contractor who follows a specified procedure cannot be held responsible if the result proves faulty. On the other hand, if the contractor agrees to produce a specified result, with the means of accomplishment left to the contractor, he or she must produce the result. A contractor who fails to produce the specified results may become the target of a damage action or may be required to assure repairs.

The following specification was written and used in the purchase of the described automation. The process engineer who wrote these specifications helped to plan the impact-bar manufacturing process. Later, he supervised the installation of the automation and worked out production problems concerned with it.

X COMPANY SPECIFICATIONS INSTRUCTIONS

♦ Information

Contractor shall consult with the Manufacturing Engineering Department of the X Company, Y Plant, in regards to working procedure, production line operations, reference drawings, plant layout drawings, and these specifications.

♦ Responsibility

Contractor shall be responsible for all field dimensions including interference with adjoining machines, building structure clearances, and installation location. Dimensions, speeds, machine sizes, and locations described in these specifications are only approximate. Reliable data must be ascertained by contractors and be verified by the person or persons concerned before proceeding with a questioned phase of automation and machine construction.

♦ Testing

Contractors shall completely design, fabricate, machine, deliver, and install special machines and automation described in these specifications. This equipment shall be tested for performance after installation and shall simulate actual production rates, as set forth in these specifications. The contractor shall be responsible for the proper functioning of all contracted equipment. The validity of such functioning shall be borne out in the prescribed test phases herein explained. Test phases shall be conducted under the supervision of suitable representatives of the X Company, and the approval of all equipment will be forthcoming from these representatives.

♦ Drawings

After contract is awarded, five (5) sets of preliminary prints representing proposal drawings of the equipment contracted shall be submitted to the Manufacturing Engineering Department for consideration, subject to approval. One set of submitted proposals will be approved by the X Company and returned to the contractor, together with proper contract authorization. The remaining prints shall become the property of the X Company. Errors, omissions, and/or changes affecting these drawings that are discovered or made as the equipment is manufactured shall be corrected by the contractor and submitted to the Manufacturing Engineering Department for approval before altering contracted construction.

The Manufacturing Engineering Department will issue the necessary drawing numbers and titles as required for the various components making up the complete unit. All working drawings shall be supplied by the contractor and shall be made on tracing cloth. Such drawings shall be made on sheets conforming to X Company Standard sizes. They shall be intelligibly drawn and cross-indexed for reference purposes. Tracings, or tracing reproductions, comprising all drawings of the complete project shall be delivered to and become the property of the X Company on or before filing completion notice. These shall include piping diagrams, electrical diagrams, and detail and assembly drawings of all mechanical and structural components of the contracted equipment. All drawings must conform to the X Company Standards and to A.I.S.C. and J.I.C. Electrical, Pneumatic, and Hydraulic Standards.

♦ Rights of Supply

X Company reserves the right to supply any items on the proposed automation and equipment or motors that may be available from their source. Therefore, an itemized list of all standard commercially manufactured equipment shall be submitted to the X Company for approval before purchases are made by contractor.

♦ Consultation

Immediately after the contract is awarded, the successful bidder shall confer with the Manufacturing Engineering Department, X Company, Y Plant, and discuss details of time scheduling and working procedure. Contractor shall then prepare a progress schedule of the

X COMPANY

MANUFACTURING ENGINEERING

DEPARTMENT SPECIFICATION NO. ____________________

P. N. NO. ____________________

SUBJECT ______ AUTOMATIC SHEET LOADER FOR IMPACT BARS. ______

PLANT __Y PLANT__ PROJECT ____________ ITEM NO. ____________

INCLUDED DRAWINGS ____________________

APPROVALS ______ /s/ ______ /s/ ______

______ /s/ ______ /s/ ______

______ /s/ ______

______ /s/ ______

______ /s/ ______

______ /s/ ______

______ /s/ ______

—— REVISIONS ——

DWG. NO.	SHT. NO.	LOCATION	DATE	REMARKS

ISSUED BY ___/s/___

Y PLANT

SHEET NO. ______ OF ___7___ SHEETS

work, complete with starting and completion dates for each phase of the machine setup and equipment he expects to manufacture, and submit it to the Manufacturing Engineering Department, Y Plant, for approval.

♦ Procedure for Delivery

All equipment and machines, either assembled or knocked down, shall be plainly marked with the vendor's name, a description of the article, and X Company Purchase and Item Number. Miscellaneous or loose items shall be crated and marked in a similar manner with a list of contents enclosed in each package. All items or packages shall be listed on the Bill of Lading. The X Company will receive items and furnish crane service, if available, to unload. Contractor shall move equipment to job site. X Company assumes no responsibility for any loss or damage in transit. X Company "general conditions" shall become part of the contract.

GENERAL

♦ Purpose

The purpose of these two (2) sheet loaders is to load sheets singly onto a press feed shuttle from a conveyor on which 14" high stacks of 125 sheets each are placed.

♦ Work

All work shall be performed in accordance with X Company standards, and all workmanship shall be equal to or better than the best modern practices. Job site must be completely cleared of all debris caused by the contractor before completion notice can be accepted. Installation and field work shall be done under the direction of the Y Plant Engineering Department.

♦ Power

Contractor shall furnish all wiring from the X Company distribution panel to the machine control panels. Contractor shall be responsible for all wiring and control devices within his contracted equipment, including mounting, labeling, and operation of all switches, relays, and solenoids. All electrical, pneumatic, and hydraulic installation shall conform to X-J.I.C. Standards. A minimum number of air and water supply lines to any and all equipment shall be furnished by X Company when necessary. 440/220 V., 3 Ph., 60 Cy., A.C. power.

♦ Materials

Machines and allied equipment shall be constructed mainly of casting, rolled structural shapes, gears, chain, and fixtures of good quality, free of rust. Fabrication shall be accomplished by bolting or welding. All holes must be drilled or punched. Burning of holes will not be permitted. All supports shall be properly spaced for uniform transfer of load to floor. Means of securement shall be fixed anchor bolts set in lead or sulphur. All parts shall be designed and detailed with due regard to allowable stresses in materials used. Contractor shall shop paint all materials with one (1) coat of M-426 Red Lead and field paint with one (1) coat of M-340 Grey. Inside of all removable drive guards shall be painted with one (1) coat of M-314 Alert Orange.

Machines and automatic equipment shall be suitably equipped with alemite or equal hydraulic lubrication fittings wherever necessary. Manifold lubrication lines shall be provided on all equipment wherever feasible, employing alemite or equal lubrication manifolds. All air or hydraulic cylinders, valves, motors, speed reducers, relays, starters, and other commercial items shall be of approved X Company selection. All exposed drives, couplings, belts, sprockets, chains, and/or other moving parts shall be completely guarded with enclosed-type easily removable metal guards. All guards to be designed and fabricated in accordance with X Company Sheet Metal Standards. All guards to have the approval of the X Company Local Safety Engineer.

AUTOMATION DESCRIPTION

A. Machines and automation equipment shall be fully automatic wherever possible and interlocked electrically or mechanically.
B. Automation equipment shall be designed in such a manner that conveyed parts can be removed from it at any phase of its operation without difficulty. Exceptions to this rule must have individual approval.
C. Machines and automation equipment shall be provided with guards wherever necessary to prevent loose parts from falling or bouncing from proper locations.

D. All wipers, diverters, and other moving equipment shall be designed and built in a manner that will prevent jamming or damaging of machine parts.
E. All metering devices and automatic equipment must be designed and constructed in a manner that will cause an even and regular distribution and flow of parts.
F. All air cylinders supplied by contractor shall be cushioned at both ends and be provided with suitable speed valves wherever necessary.

DETAILED DESCRIPTION

1. Two (2) magnetic (pneumatic) sheet loaders will be required. They shall be located relative to the 2,300-ton presses and four-chain stack conveyors as shown in the attached sketch. Contractor shall furnish department 240, Manufacturing Engineering, with the approximate overall dimensions of these machines and their allied equipment as soon as possible for Plant Layout purposes.
2. The machines shall be designed to load the 2,300-ton press dies with single sheets removed from stacks of 125 sheets previously placed on a four-chain conveyor. The machines shall be interlocked mechanically or electrically with the 2,300-ton presses and the four-chain conveyors to provide a continuous flow of production. The machines shall be capable of handling single sheets up to 144" in length, 48" in width, and.125" in thickness. Sheet sizes to be handled are shown:

Part	Sheet Size
"A" Rear	.110" × 36" × 95"
"A" Front	.110" × 42" × 118"
"B" Rear	.110" × 28" × 103"
"B" Front	.110" × 46" × 110"

3. Operator push-button station shall be designed and built in such a manner that the machine may be operated as follows:
   a. Fully automatic—with machine loaded
   b. Manual—individual push button for each station
   c. One-cycle operation—automatic with machine not loaded

4. Each machine control panel shall contain sufficient excess room to accommodate the installations of six (6) additional relays.

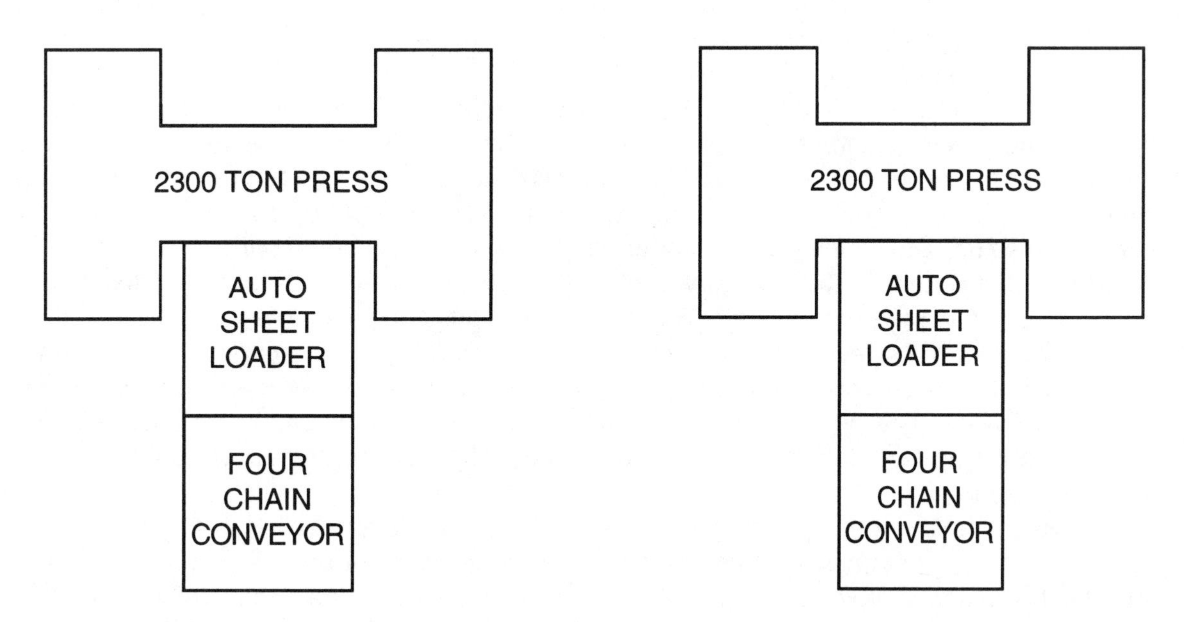

Figure 17.2. Sketch showing relative location of automatic sheet loaders.

5. Outside contractor to furnish the following:
   a. Necessary labor and materials to design, fabricate, and install two (2) complete sheet loading mechanisms with control panels and operator control stations as described above
   b. All necessary safety equipment
   c. All starters, Square "D" type or equal
   d. All field painting
6. X Company to furnish the following:
   a. Crane service, if available, to unload
   b. Plant layout location of all items
   c. Electric, air, and other plant services to within 40' of contracted equipment
7. Assigned "Z" number is 121-ZP-124. Drawing title is "Automatic Sheet Loader for Impact Bars."

SWITZER v. BOZEMAN

106 So.2d 762 (La. Ct. App. 1958)

Defendant-appellant appeals from a judgment costing him in the sum of $269.22 in connection with his contract with subcontractor A. Switzer, plaintiff-appellee, herein. This is a suit in which the only issue is the interpretation of the provision of the plans and specifications. And that is whether certain electrical work was called for in the plans and specifications.

Louisiana State University, as owner, entered into a contract with defendant-appellant, Robert L. Bozeman, a contractor, to remodel the third floor of Nicholson Hall, which remodeling work consisted only in subdividing some of the large rooms and hallways into small offices. The remodeling work was to be in accordance with plans and specifications prepared by architects.

Bozeman then entered into a subcontract with plaintiff-appellee, A. Switzer, under the terms of which Switzer was to do all of the electrical work required under the plans and specifications for $1,183.

During the course of the job, Switzer claimed that the plans and specifications only called for one new electrical ceiling fixture, whereas, in order for the job to be complete, it was necessary to furnish and install many additional new electrical ceiling fixtures. Switzer then requested the architects to grant him an extra allowance from L.S.U., to enable him to install these "claimed" additional new electrical ceiling fixtures. The architect ruled that the plans and specifications were clear and that all new electrical ceiling fixtures required to complete the job were specified in the plans and specifications.

Switzer, despite repeated demands from Bozeman, refused to complete the job, so Bozeman hired another electrical company to finish Switzer's contract. This latter company completed the job at a cost to Bozeman of $687.59. Bozeman then deducted this amount from the electrical subcontract price and paid the difference to Switzer. This suit followed, in which Switzer claimed this $687.59, which was the amount deducted from his contract price.

The district court held that Switzer was wrong and that the new electrical ceiling fixtures required to complete the job were all specified in the plans and specifications, and therefore, covered by Switzer's sub-contract; this item included $102.16 paid the new electrical contractor to install the fixtures, and $238.87 paid for the fixtures themselves. The district court also found that Switzer had not "grounded" the outlets to which the fixtures were attached, and recognized the amount paid by Bozeman to this new electrical contractor of $80.32 to have such work done.

However, the district court held that the $269.22 which Bozeman paid this new electrical contractor to install the electrical work for the telephones was not specified in the plans and specifications, and, therefore, not included in the electrical subcontract that Switzer had with Bozeman. Accordingly, the lower court granted judgment to Switzer against Bozeman for $269.22. We are not favored with written reasons by the trial court.

The only issue involves the interpretation of a contract. Since the plans and specifications by the architects are made part of the contract, we must determine whether they require that the telephone outlets be connected by conduit with the existing telephone panel in the building.

The plans have a legend thereon where various symbols are listed, with the meaning of each symbol shown opposite thereto. For example, there was a small circle with radiating lines, spoken of during the trial as "rising sun," and opposite which were the words "Elect. Clng. Fixture," meaning electrical ceiling fixture. And there were symbols for bookcase, new partition, electrical base outlet, etc. Additionally, there was a symbol of an inverted triangle, opposite which was the word "Telephone." This inverted triangle, meaning telephone, was shown in six places on the plans, indicating that there were to be six telephone outlets installed.

The question involved is whether, by showing such symbol of an inverted triangle that only the outlet itself was called for, or whether that outlet was also to be connected with the telephone panel already in use in the building. Now, if only the outlets were called for and it was not required to connect such outlets up with the panel so that service could be had, it is obvious that another contract would have to be let to connect the outlets up with the panel in order to use the telephones. It would appear to be an absurd consequence to say that the contract called for the outlets, but did not require that they be connected with the panel, in view of the specifications.

The specifications made it clear that this was to be a completed job "including tying utilities all into existing service," and was to be "ready for immediate use." The plans and specifications called for "work evidently necessary within the general intent (thereof)...for the...thorough completion of the work." We quote these provisions of the specifications, viz.:

Scope of work and general items: "Extent of work: It is the intent of these specifications to cover all required labor and materials for the remodeling interior of Nicholson Hall, Third Floor, LSU, Baton Rouge, La.; The contractor shall perform all work required for the completion of this work in accordance with these specifications and accompanying drawings ready for occupancy; including tying utilities all into existing services.

Scope: These specifications, together with the drawings, are intended to cover all labor, materials, and appliances of every kind required to provide all necessary electrical work and electrical fixtures for the remodeling job. All the work shall be ready for immediate use before the same will be accepted.

Extra work: No additional compensation will be allowed for work evidently necessary within the general intent of these specifications and accompanying plans for the proper construction and thorough completion of the work.

Arthur G. McLavy, the electrical contractor who completed the electrical work for Bozeman, testified about this conduit to connect the telephone outlets with the existing telephone panel in the building, viz.:

Q. Now what is your other item?
A. That was to furnish the necessary labor and materials to put in a conduit telephone system as required by the telephone company.
Q. To put in what?
A. A telephone system for existing telephone panels to outlets as shown on the plans according to directions by the telephone company and the University.
Q. Was that the type of work customarily done by the telephone company?
A. No, sir.
Q. Was that work called for in these plans and specifications?
A. You are asking for an opinion now.
Q. Well, you read the plans and specifications, didn't you?
A. Yes, sir.

Q. Was that work called for? Can you point out on the plans and specifications where it is called for?
A. It is not called for on the plans and specifications in so many words, but it's the general practice in any commercial building or any building used as an office building to have a continuous conduit system for the telephone company to install their wires.
Q. What requirement is shown on the plans with reference to telephones?
A. Just the outlets are indicated.
Q. Just the outlets are indicated?
A. Yes.
Q. And from that you draw the conclusion that you are supposed to connect those up with the telephone wires from outside of the building?
A. No.
Q. What is that?
A. You would draw the conclusion from that that you would have to extend from the nearest, I would say, panel or junction box in the building in each and every outlet, a continuous run of conduit.
Q. Is that customary and standard practice in the electrical business in this section in a building such as this?
A. Yes.
Q. Any question about this, any room for debate on that?
A. Of course, there would be room for debate on it.
Q. Well, what is the room for debate?
A. Well, I mean in my mind there is.
Q. Say what is in your mind, Mr. McLavy.
A. Nothing.
Q. The plans indicate here on page A-l the three-cornered white symbol and marked opposite it is shown telephone, and how many of those do they have on this plan? Will you count them?
A. Six.
Q. When you went on the job what work on the telephone installation had been performed before you got there?
A. The outlet box had been installed. The outlet box had been installed in the partition, and the conduit run out at ceiling level.
Q. Then what did you do?
A. I requested a meeting with representatives from Louisiana State University and the telephone company to meet out there, and we laid out how they wanted it done.
Q. What was it that you did?
A. There is an existing panel—I forget which end of the hall it is on—east end, yes. We ran a one-inch conduit from there out into the corridor and down. We could describe it as down close to the west end, and from there branched out to the smaller conduits of the existing outlets, or close to them as directed by the telephone company.
Q. I ask you again, in a building such as this, is that work customarily done by the electrical contractor and not by the telephone people?
A. The telephone company does not run any conduit themselves.

The electrical work in connecting the telephone outlets to the telephone panel by conduit was obviously necessary in accordance with the plans and specifications.

It is our opinion that this work was clearly called for in the plans and specifications. It necessarily follows that it was included in Switzer's bid, which is as follows:

Confirming phoned price this date...Nicholson Hall Elec. per plans and specs., $1183.00.

It is significant that the record reveals no effort by Switzer with respect to any claim with Bozeman or anyone else that the electrical work necessary to connect the telephone outlets to the existing telephone panels in the building was called for in the plans and specifications. Switzer testified that he ran a 1/2-inch conduit from the telephone boxes as shown on the plans by the symbols to points at, in most cases, ceiling height, where he could be instructed to bring them to other places. He further testified that he had met Mr. Wilson on the job once or twice to discuss fixtures and other arrangements. He stated that he had written a letter, but upon a close perusal of the record no letter in evidence disclosed any reference to telephone connections; we therefore conclude that there was no controversy whatsoever with respect to the telephone connections prior to the date of trial.

The record disclosed that Mr. Switzer took the position all along by the letters he wrote and by his testimony in the record that he should have been permitted to charge extra for new light fixtures to be installed on the job, which request was denied by the architect. To clarify this point, we take the liberty of quoting from the correspondence in the record.

Bozeman's letter of August 29, 1956, to Switzer, in part reads:

> ...we take the position, however, that the job must be complete in accordance with the plans, specifications, and job requirements as interpreted by the architect. We will complete our portion of this job by August 31st, 1956, and unless notified to the contrary, in writing, we shall expect the electrical work to be complete also.

Switzer's answer to Bozeman, dated August 30, 1956, in part says:

> Your letter of the 29th received this date. We appreciate your position in this matter...but feel also that the fixtures in question...are not specified...as part of our work. It is not possible for us to complete this work...other than the installation per plans and specifications...which is installed as of this date.

Bozeman's letter to Switzer dated September 4, 1956, by registered mail, in part, reads:

> In your letter of August 30, 1956, you have refused to complete the above mentioned job unless you receive from the architect a change order covering five different items. Since the University officials and the Architects have advised us that this building must be ready for occupancy by Monday, September 10, 1956, and in view of the position that you have taken in this matter, it appears to us that we are forced to act as follows:
>
> We hereby notify you that all of your work must be completed by Saturday, September 8, 1956. Further, if you have not shown sufficient effort by 8 a.m. Thursday, September 6, 1956, to complete this job within the time set forth, we will take over and complete that portion of the electrical work as is now incomplete, and deduct from your contract the cost to us...

It is in evidence that Switzer did not examine the site so that he could be in a position to say where the telephone panel was to which the telephone outlets were to be connected. Nor could he know the conditions that were present so that the type and size of conduit could be determined. It is important to quote from the specifications at this point:

> Examination of Site:...each bidder will be held to have examined the site and satisfied himself as to the existing conditions...that will in any manner affect the work under this contract.

The specifications did include tying utilities all into existing services and require the remodeled part to be ready for immediate use after acceptance by L.S.U. and called for all work evidently necessary to complete the job. Thus, it appears to us that if Switzer had looked, he would have seen the telephone panel on the third floor, and it is obvious that the outlets were to be connected with this panel, and by so doing he would have known the type and size of the panel. Aside from that, he could have called in the architect, the telephone company, and the electrical representative of the University to assist him. Accordingly, for these reasons, the trial court judgment in favor of plaintiff-appellee is hereby reversed and his suit dismissed at his cost.

Reversed.

REVIEW QUESTIONS

1. Why should specifications be written rather than oral?
2. Distinguish between the following:
   a. Malleable and ductile
   b. Strength and rigidity
   c. Force and pressure
   d. Structure and building
   e. Project and operation
   f. Tool and die
   g. Machine and automation
   h. Precise and accurate
3. What are the inherent advantages of standardized specifications?
4. What danger is there in requiring a contractor to perform according to a particular method?
5. Why is the owner interested in the wages the contractor pays to employees?
6. What different machines could be used to obtain a 2″ × 4″ rectangular piece from a sheet of 1/16″-thick aluminum? In what ways could you fasten two pieces of metal together?
7. Write a simple specification for the purchase and installation of a 1 / 3-HP pedestal grinder.
8. What are the inherent advantages and disadvantages in copying portions of old specifications into new ones?
9. Interpreting from *Switzer v. Bozeman*, to what extent do trade or area practices have a bearing on meanings or omissions in specifications?

Chapter 18

Agency

When a person's duties and desired objectives become too numerous to handle, he or she can usually delegate some of those duties to others. This delegation of duties may take one of three forms: (1) the employer-employee (sometimes called master-servant) relationship, (2) the owner-independent contractor relationship, or (3) the agency relationship. In satisfying the normal requirements of a job, an engineer (whether a consultant or an employee) must act as an agent at least part of the time for the person who hired him or her. The engineer's rights and liabilities while so engaged are dictated by the law of *agency*. Agency can be defined as a consensual fiduciary relationship between two persons by which one (the *principal*) has the right to control the conduct of the other (the *agent*), who in turn has the authority to affect the principal's legal relationships with others.

An agent represents a principal in dealing with other persons. Herein lies the distinction between the agency relationship and the employer-employee relationship. An agent is often an employee for many purposes, but special rights and duties are involved in an agency relationship. A lathe operator in a plant, for example, is an employee. If the operator deals with others as a representative of the employer, the operator becomes an agent.

The intent of the parties and the degree of control exercised by the owner determine whether a relationship is one of owner-independent contractor or employer-employee. As indicated, the parties must agree to have one act as an agent for another. If one person decides to negotiate a price for a piece of land prior to receiving any authority to do so, there is no agreement and no agency (although, as discussed later, the parties may subsequently decide to create the agency relationship). The degree of liability of the employer (or owner) for another's conduct is, in turn, determined by the relationship between the employer and the other person whose conduct is at issue. For example, an employer may be held liable for tortious injuries caused by employees acting within the scope of their employment and for harm to them from any injuries that arise from their employment. Although a principal may be liable for an agent's actions or for harm done to an agent while he or she was working for the principal, a principal's liability for injury to, or injury caused by, an independent contractor is usually much more limited. In many instances, particularly worker's compensation cases, a supposed owner-independent contractor relationship was held to be an employer-employee relationship because of the degree of control exercised by the employer (owner).

Consider this example: Black, a manufacturer, hires White as a time-study engineer in Black's plant. In setting production standards and making methods changes, White is Black's employee. White is soon promoted to process engineer and is charged with installing an automatic machine being purchased from Gray Automation. When White deals with Gray Automation and other outsiders, White acts as Black's agent. White fills a dual role, then, for while White acts as an agent in dealing with others, White remains Black's employee. In installing the machine, Gray Automation is an independent contractor unless either

Black or White exercise an excessive degree of control over Gray Automation.

Agency involves three people. First, the *principal*, the person who is represented by the agent and who is the source of the agent's authority. Second, the *agent*, the person who represents the principal. And finally, the *third parties*, the persons with whom the agent deals in the name of the principal.

CREATION OF AGENCY

The agency relationship may be created in any number of ways. Such relationships are commonly created in four situations: by agreement, by ratification, by estoppel, or by necessity. The responsibility and authority of the agent differs slightly in each.

♦ Agreement

In an *agency by agreement*, both parties must intend to create a relationship that amounts to agency. The parties may not consider the relationship to be agency when they enter into it, but if the result amounts to an agency, it will be so construed. The means the parties use to express their intentions to form an agency are ordinarily unimportant, as long as the ideas are exchanged. An agency contract is much the same as any other contract in this respect. The intent may be either expressed or implied; the agreement may be written or oral. The agency agreement is not necessarily a contract, but it usually is. A contract requires consideration, but a simple agreement may create a valid gratuitous agency.

Consider the following example: Black owns a truck; White does not. Black, without being promised any compensation, agrees to transport a machine for White from a freight depot to a machine shop across town. In doing this, Black must deal with others. Even though Black is to be paid nothing, Black's actions are controlled by White when Black picks up, transports, and unloads the machine. Thus, an agency was formed. Now suppose that Black goes alone, but White telephoned ahead to tell the freight depot that Black was to pick up the machine. Again, the parties have formed an agency.

As noted, an agreement to act as another's agent ordinarily may be either oral or written. However, for certain purposes, a written or a sealed instrument may be required. Where the instrument that binds the principal requires a seal from a notary public, the agent's authority usually must be in writing and notarized also. Probably the most common example of this is the *power of attorney*, which establishes the agent as an attorney-in-fact for the principal. Where a transaction requires a public recording, such as in the sale of real estate, any powers of attorney involved are also recorded. Another form of written agency is the corporate *proxy*. By the use of a proxy, a stockholder appoints some particular person to vote the shareholder's shares of stock in a particular way.

♦ Ratification

If a person purports to act on another's behalf and contracts with a third person, but had no authority to do so, the person for whom the "agent" purports to act is not bound to the contract. Much the same is true if an actual agent exceeds the authority given by the principal. In either of these circumstances, the principal may agree to be bound by the terms of the contract. Agreeing to do so is referred to as *agency by ratification*, and it cures any defect in a lack of authority. Such a ratification by the principal is *retroactive*—that is, the ratification goes back to the time the contract was made. The effect of ratification is the same as if the principal had previously retained an agent to act in that particular manner.

For example, suppose Black hires White as a salesman to sell the company's products. White finds a buyer, Gray, for a used milling machine that Black has wanted to sell for some time. White, without delay, contracts in Black's name to sell the machine to Gray. If Black no longer wishes to part with the machine or, possibly, has contracted to sell it to another, Black will not be bound to the agreement with Gray. If Black does not ratify White's agreement with Gray, White will be personally liable for any harm to Gray resulting from a breach of the contract. If Black ratifies the agreement, Black will be held to the contract just as though White had been given specific orders to sell the machine. However, the fact that White lacked the authority to sell the machine does not allow Black to ratify only a portion of the contract. Black must ratify all or nothing. Meanwhile, Gray is not bound to the agreement until Black ratifies. If Gray finds that White acted without authority, he may withdraw before Black's ratification.

There are at least four requirements for a valid ratification:

1. A principal must exist when the supposed agent acts. A corporation, for instance, cannot make a

binding ratification of contracts made in the corporation's name before it was formed. New contracts will be required if the corporation is to be bound.

2. The person acting without authority must act as an agent. If the person acts on his or her own behalf, subsequent ratification by another will not create enforceable obligations between the third party (the supposed principal) and an outsider.
3. The principal must be aware of the facts when the principal ratifies the contract. A principal's actions that would imply ratification have little effect unless the principal knows the facts. Of course, if the principal does not investigate the details when the principal has a duty to do so, the principal's negligence may be interpreted to support the conclusion that the principal did know or should have known the facts.
4. The principal must intend to ratify. Intent and ratification may be interpreted from the principal's actions after the principal obtains knowledge of the transaction and the relevant facts. If the principal does nothing after being informed of the transaction, for example, ratification could be implied from the principal's inaction. In such a situation, the principal is said to have ratified the transaction through acquiescence.

♦ Estoppel

Agency by estoppel arises where one person appears to have the authority to act for another and, despite a lack of real authority, does act in the name of the other. *Apparent* authority is one key to this concept. If the principal acts in such a way that another person appears to be the principal's agent (thus, in effect, deceiving the third party), the principal is then estopped (i.e., prevented) from denying that the other person is an agent. The second key for an agency by estoppel is that the third party must have acted in reliance upon the supposed agent's apparent authority. The third party must, of course, have dealt with the supposed agent to prove agency by estoppel.

♦ Necessity

An *agency by necessity* may occur as a result of an emergency. This occurs rather infrequently, however. If, to save the principal from some disaster, an employee must deal with others without an opportunity to obtain authorization, an agency by necessity is created. Generally, such an agency has some relationship to what the agent's normal duties involve.

Where a wife binds her husband to pay for necessaries (or where a dependent binds a guardian), agency is sometimes said to exist. The most common holding, though, is that the husband (or guardian) had a duty to support, and binding him to such contracts is a result of this duty rather than agency.

♦ Competency of Parties

Since, in agency, the principal is the party to be bound to a third party, the principal must be *competent* to contract. The agent may be a gray-haired man of 60, but if his principal is a minor, the contract is voidable at the minor's option. From this, there arises a practical desirability of investigating the principal before contracting with him or her. Not only could an incompetent principal avoid a contract made by his or her agent; the principal might also avoid the contract with the agent.

Though competency of the principal is of major concern to third parties and agents, competency of the agent is not. Almost anyone may be an agent; it is the principal who is bound. A minor agent contracting with a third party would bind the adult principal to the contract.

AGENT'S AUTHORITY

The agent's authority comes from the principal. It consists of *express* orders or directions the principal gives in addition to authority that may reasonably be implied. *Implied authority* is based on previous dealings between the parties, or local or trade customs. If none of these control the situation, the extent of implied authority is the extent necessary to accomplish the purpose of the agency.

If the agent contracts beyond the scope of his or her authority, the principal is not bound. From this, an obvious burden is placed upon the third party to determine whether the agent has authority to enter into a given contract. The third party is safe in dealing with the agent if he or she can obtain evidence of the agent's mission. Implied authority necessary to accomplish the purpose is included despite a principal's instructions to an agent to the contrary.

For example, Black is a buyer for the White Manufacturing Company. Black's present assignment is to buy a six-station automatic indexing table, drive, and base, to be used in machining small die castings. Black has been specifically told not to buy any tooling with the machine. Nevertheless, Black contracts with Gray for a machine with tooling that, he is told, may be reworked for the die castings. The contract for the machine and the tooling probably will be binding because Black's specific orders to buy the machine might well be taken to imply that he also had a right to purchase tooling for it. If Gray knew of Black's assignment, he would not have a duty to go further and determine any unusual restrictions that might have been imposed on Black.

A third type of authority is *apparent authority*. The reasoning of apparent authority runs so close to that of agency by estoppel as to make the two nearly indistinguishable. Apparent authority exists when, by some act or negligence on the part of the principal, an agent either appears to be clothed with more authority than he or she really has, or a person who is not an agent is made to appear as though he or she were one. A person usually creates apparent authority by conduct or statements that give the impression that he or she is authorized.

Agents are chosen for their particular capabilities and fidelity. The agent is the one the principal trusts. Therefore, the agent generally cannot freely delegate his or her authority to another. If the agent hires a sub-agent, the principal has no liability to the sub-agent for wages or other benefits and is not liable for the sub-agent's acts. The rule is not without exception; if part of the agent's express or implied task is hiring others for the principal, those so hired work for the principal. Obviously, when a principal hires a corporation as an agent, that corporation must hire individuals to perform its duties.

AGENT'S DUTIES

Agency is a *fiduciary* relationship—that is, the principal's trust and the agent's loyalty are implied. The term *fiduciary* comes from Roman law and is used to describe a person who has a duty to act primarily for another's benefit. A fiduciary generally has duties of trust and confidence that are owed to the person for whom the fiduciary acts. The law enforces these qualities in the relationship; a breach can give rise to a cause of action. Furthermore, the agent is personally liable for the results of any disloyalty.

♦ Obedience

An agent owes the principal a duty of strict obedience in all ordinary circumstances. Disobedience constitutes a breach of the agency agreement. It is not the agent's function to question or judge the wisdom of the principal's orders; it is the agent's function to do everything in his or her power to obey them. Of course, when the principal outlines a general purpose to be accomplished, the agent may be required to use judgment and discretion in working out the details. Still, the purpose to be accomplished is not open to question. Direction by the principal is implicit, even in a gratuitous agency (that is, one in which the agent acts without compensation), once the agent has begun to perform as an agent.

Of course, strict obedience is limited to lawful and reasonable acts. The agent need not, for example, follow instructions of an unlawful nature. Neither would an agent be expected to accomplish an impossibility. In an emergency situation, an agent may have the right to fail to obey instructions strictly. The reasoning here is the same as that in authority of necessity: If the agent's failure to follow instructions will save the principal from disaster, the agent's right to disobey is apparent. In such situations, the agent is to act as the principal would direct if the principal knew of the emergency.

♦ Care and Skill

The agency relationship normally implies that the agent will use ordinary care and skill in carrying out his or her duties. The test of whether or not the agent has done so is the test of the *reasonably prudent person*. That is, has the agent acted as a reasonably prudent person would be expected to act in like circumstances? Negligence in following the principal's orders may make the agent liable for payments to the principal; in addition, the agent may have to pay for losses suffered by the principal that could be reasonably anticipated from the agent's failure to follow instructions properly.

If the agent professes to be a specialist (for example, a consultant in some professional field), the standard of skill expected is that which is normally attributed to such a person. The standard is the same whether the person really is such a specialist or not; if the agent fails to perform as a specialist would be expected to perform, the agent is liable. For example, Black, a manufacturer, hires White—an engineering consultant on conveyors—to design, recommend, and oversee the installation of a monorail conveyor system in Black's plant. If, because of very poor planning and design, the conveyor must be removed shortly after its installation, Black may have an action against White. If Black's action is to be successful, he must prove (usually by expert testimony) that anyone possessing the knowledge and skill normally possessed by an engineering consultant specializing in conveyors would have made a more effective design. Damages could include a refund of the consultant's fee plus the cost of the improper installation and the cost of its removal.

♦ To Act for One Principal

An agent has a duty to act for and accept compensation from only one principal. An agent could not, for instance, reasonably represent both buyer and seller in a sales contract. The buyer's interest and the seller's interest are at opposite poles; each desires to get the best possible deal. If the agent represents more than one party to a contract, the transaction is voidable at the option of either principal. The agent is relieved of the responsibility to act for only one party if the parties are told of the multiple relationships and acquiesce. An agent's interest that is adverse to a principal's interests is allowable if the principal knows of the interest, is fully informed of all the relevant facts, and continues the relationship in spite of it.

The agent's compensation should come only from his or her principal. Hence, an agent is not allowed to secretly profit from his or her actions on behalf of the principal. If the agent acquires property in his or her own name that should go to the principal, the agent will be deemed to be holding the property as bailee. Similarly, the agent cannot contract with himself or herself as the third party without fully disclosing the facts to the principal and gaining the principal's consent. Without such disclosure and consent, the principal has the right to avoid the transaction. Kickbacks or secret commissions from third parties are, of course, against public policy.

♦ Accounting

The agent has a duty to account for all money or property involved in agency transactions. Further, the principal's property must be kept separate from the agent's property. If commingled money or property belonging to both principal and agent is lost, the agent must make good the principal's loss. If the property were kept separate and the loss occurred through no fault of the agent, only the principal would lose.

The agent has no right to use the principal's property as his or her own without consent to do so. Using the principal's property without the principal's consent is the same in agency as elsewhere—it is *conversion*.

Closely akin to the agent's duty to account to the principal for money or property is the agent's duty to report information to the principal. This often involves relating offers, financial information, or other details encountered in the course of business negotiations or transactions. Generally, notification to an agent has the same effect as notice to a principal. In either case, the principal is charged with possession of the knowledge.

Loyalty

The essence of agency is the identity of the agent with his or her principal's purpose. Generally, an agent has a duty to act solely for the interests of the principal within the agent's authority. The agent is often in a position to gain personally from information he or she acquires. To use the information to add to the agent's personal fortune is an act of disloyalty. Recovery by the principal for such misuse of information is possible.

The duty of loyalty requires the agent not to use or disclose confidential information, except for the principal's benefit. Indeed, this duty is so strong that it continues even after the termination of the agency. Thus, although the agent may compete with a former principal after termination of the agency, the agent may not compete by using any of the principal's trade secrets, such as secret customer lists, blueprints, or the like.

PRINCIPAL'S DUTIES

Most agency agreements are contractual in nature, and each party has a duty to live up to the agreement. The principal does not, however, have to pay for disloyal service. Neither will the principal have to pay if payment is contingent upon the agent's success (as is the case with commissions for sales) and the agent's efforts do not meet with success.

♦ Payment

In the usual agency contract, the means and amount of payment are stipulated. If such a stipulation is not made, though, the agent is entitled to reasonable payment for his or her services. If principal and agent are not close friends or relatives, and there is no other reason to suggest that the agent acted for nothing, an unliquidated obligation of payment by the principal to the agent exists. A gratuitous agency is, of course, an exception to this.

Payment of an agent on a *commission* basis presents some special problems. When, for example, has the agent earned the commission? What happens if the principal accepts an order through the agent to sell to a third party, but the principal and third party cannot reach an agreement on the terms? Does the principal have to pay if he or she deals directly with the third party? Such problems arise not only in agencies to sell a company's products; they are also common in real estate and other agencies.

Generally, and barring an agreement to the contrary, the agent has performed and is entitled to a commission when the agent has achieved the desired result, such as when the agent has found a buyer and has contracted with the buyer for the principal. A real estate agent (or broker) ordinarily does not have the power to contract for the principal. Therefore, the role of the real estate agent is only to find a buyer who is ready, willing, and able to buy the property at the owner's price. With these conditions satisfied, the principal has an obligation to pay the agent the agreed commission.

If, after the agent has performed the obligation, the principal and the third party do not complete the transaction, the principal is still bound to pay the agent. It matters not that the principal will no longer profit by the transaction.

The agency agreement often states whether a commission is to be paid to the agent if principal and third party deal directly. An *exclusive agency* for the sale of real estate, for instance, requires that the principal pay the real estate agent's commission regardless of who sells the realty. Without the exclusive agency feature, the owner can sell the realty to another and be free of the obligation to pay the agent. The same is generally true of an agent who sells a product. If the sale is made without his or her services, the agent has done nothing to earn a commission and is not entitled to it.

♦ Expenses

The principal is legally bound to pay an agent's expenses. To constitute an obligation of the principal, the expenses must, of course, be connected to the purpose of the agency. Thus, the cost of travel, meals, and overnight hotel or motel accommodations connected with an agent's trip to sell a principal's products should be paid by the principal. Similar costs incurred by the agent on a pleasure trip with his or her family ordinarily would not be covered by the principal.

♦ Indemnity

While the agent has a duty to follow the principal's orders, the principal also has a duty to *indemnify* the agent if the result injures someone and the agent has to pay for the injury. Of course, the agent is not required to perform unlawful acts; he or she is prohibited in the same manner as anyone else from committing a crime. If, though, a tort or crime is committed by the agent innocently following a principal's instructions, the principal is liable for the result. Both principal and agent are liable to the third party for acts committed out of and in the course of the agent's employment; but if the agent has to pay, he or she can recover from the principal. The agent alone is responsible for acts not connected with his or her employment.

Black, for example, is an agent of the White Machinery Company. His function is to answer customers' complaints and thus make the sales of the machines permanent. He is a troubleshooter. As such, he is entitled to compensation according to his agreement with the White Machinery Company. He is also entitled to payment for legitimate expenses in connection with his job. If, while instructing someone in the proper use of a machine, an injury should occur, White Machinery Company would be liable.

THIRD-PARTY RIGHTS AND DUTIES

So far, we have considered the rights and liabilities of two of the agency parties—the principal and the agent. But the third party also has a stake in the relationship. Questions frequently arise concerning the extent to which an agent's transactions bind the principal and a third party, and the extent to which the third party can rely on the agent's claims. These and other questions are considered below.

♦ Duty to Question Agent's Authority

We have noted that the agent has the authority given by the principal. The agent also has authority common to other agencies of a similar nature. Third parties are safe in relying upon the agent's authority to this extent once they have established that they are dealing with an agent of an existing principal. If the third party knows nothing of either the principal or the agent from previous contacts, he or she should determine by some objective means whether the principal and the agency relationship actually exists.

Obviously, if the principal is nonexistent or the principal exists but there is no agency, the third party may part with something of value in good faith and get nothing for it; a person cannot be bound as a principal simply because someone claimed to be his or her agent. Under such circumstances, the third party would be left with an action against the agent only, and it is likely that he or she would be hard to find.

♦ Determination of Who Should Collect

Does the agent have the right to collect from a third party? Usually not, barring specific authority, or trade or local practice to the contrary. Ordinarily, the third party must give the consideration directly to the principal.

Black, for example, represents the White Manufacturing Company. He obtains an order from Gray for a quantity of his principal's product and receives part payment for the goods. Black is not seen again. Who bears the loss of the part payment? It depends upon whether Black had express or implied authority to receive payment. If no authority to receive payment can be found, Gray is the loser to the extent of the payment. The situation would be different if Black had brought the goods along with him; under this circumstance, the right to collect can be implied. If Gray has doubts about Black's authority to collect, Gray ignores them at his or her risk.

Somewhat akin to the agent's authority to receive payment is the authority to sign a negotiable instrument in a principal's name. Authority to do this must be expressly given to have binding effect.

♦ Transactions Binding Agent

In the normal course of affairs, the third party has no action available against the agent in a transaction. If the agent has acted within the scope of his or her express and implied authority, it is the principal and the third party who are bound. It is possible, though, for an agent to act as surety for the principal or to contract with the third party in such a way that it is the agent rather than the principal who is bound. Ordinarily an agent identifies the principal and discloses the fact that he or she is agent for the principal. If the agent however, merely agrees to something in his or her own name or signs as "Black, agent," indicating no principal, he or she may be held personally liable by the third party.

♦ Tort

If, while in the course of a principal's business, the agent commits a tort against a third party, the third party may charge either the agent or the principal with the act. Successful action against the agent then gives the agent a right to recover from the principal, since the principal has a duty to indemnify the agent. An exception to this rule appears to exist where the principal is a minor and the agent an adult. Here the agent must bear the loss.

The third party can engage in tortious conduct as well. For example, if the third party, without cause, brings about the principal's discharge of the agent, the third party has committed a tort. In fact, the rule is more general than this. Anyone who maliciously causes another to lose an employment relationship (without any legal privilege to do so) has probably committed a tort, and an action will lie against him or her.

♦ Undisclosed Principal

Normally third parties are aware that they are dealing with an agent of a known principal. Such is not always the case, though. The agent may not reveal that he or she is working for any principal, thus allowing the third party to assume that the agent is the party to be bound. Or the agent may reveal that he or she represents another without naming the principal (a partially disclosed principal).

In either case, where the agent has not disclosed the principal to the third party, the third party may elect to hold either principal or agent to the contract. If the third party elects to hold the agent, and the agent has acted within the scope of his or her authority, the agent has a right to be indemnified by the principal. Either the principal or the agent may hold the third party to the transaction, but the principal's right to do so is superior to the agent's.

The enforceability of undisclosed-principal transactions appears to be counter to the concept that a contract must be entered into voluntarily and intentionally by the parties. However, the legality and enforceability of such contracts are well established. It is, in effect, an exception to the general rule of contracts. An undisclosed-principal contract will not be enforced, however, where the third party, either expressly or by implication, makes clear his or her intent to deal exclusively with the agent.

Consider this example. White Manufacturing Company wishes to expand its operation into another section of the country. It retains Black to purchase land for the expansion without revealing the company's name. (Such purchases are sometimes undertaken to keep local land prices from soaring.) Black contracts with Gray to buy 200 acres of suitable land. Gray can elect to hold Black to the contract or, when the principal is revealed, hold the White Manufacturing Company. Gray is bound to the contract unless he has either expressed or implied his intent to deal exclusively with Black. If Black, under questioning by Gray, were to deny the existence of a principal, it would be grounds for fraud, making the contract voidable at Gray's election.

TERMINATION

The rules for winding up an agency agreement are about the same as those for winding up any employment agreement, except that the third party must be considered. The usual contract of employment of a so-called white-collar employee is oral and terminable at the option of either employer or employee. If an engineer, a sales agent, or an accountant engaged in such employment decides to leave, it is only necessary to so inform the employer, settle the accounts, and leave. The employer's right to terminate such an agreement is similar. Not all agency contracts are of this simple form, though, and they are not always terminated in this fashion.

♦ By Law

Death, insanity, or bankruptcy of either principal or agent automatically terminates an agency relationship. Death or insanity of the principal is effective even if the agent is not aware of the event. That is, if an agent deals with another after the principal dies or becomes insane, but before the agent is informed of it, the transaction is not binding. If the agency has been created for a specific purpose, destruction of an element essential to the accomplishment of the purpose ends the agency. Similarly, passage of a law that makes the purpose of the agency unlawful, terminates the agency.

♦ By Acts of the Parties

An agency created to accomplish a specific purpose or to last for a stated time is generally not terminable at the option of the parties without possible repercussions. If an agency is created to accomplish a particular purpose, it ends when the purpose is accomplished. If a time limit for the agency is set, it ends when the time runs out.

Where an agency has been created for a purpose or to last a certain time, neither the principal nor the agent may unilaterally terminate the agency without the other's agreement. Both parties may, of course, agree to disagree before the contract is finished. Revocation of the agency by the principal terminates the agency; but if it is done without just cause or agreement, the principal is likely to be held liable for payment to the former agent for the remainder of the term for which the agency was to run. Similarly, renunciation of the principal by the agent ends the agency, but the principal may be allowed recovery. As indicated earlier, disloyalty by the agent would be just cause for early termination by the principal. In any case, termination by a unilateral act of either principal or agent does not become effective until the other party is informed of it.

♦ Agency with an Interest

If the agent has an interest in the subject matter of the agency, the relationship cannot be terminated by an act of the principal. The term *interest* refers to more than just the agreed compensation for the agent's services. Essentially, this term implies part ownership or an equity in the subject matter. This issue might come up in a partnership venture, for example; in most instances, one partner could not very well fire another.

♦ Notice

When an agency is terminated, third parties should be notified. If the agency is terminated by law, notification is considered to have taken place, since death, insanity, or bankruptcy would be a matter of public record. When termination takes place by acts of the parties, though, there is a particular necessity to inform those who have dealt with the agent. If, in ignorance of the dissolution of the agency, a third party deals with an agent as he or she had dealt with the agent before, the principal will be bound. The reason for this is the agent's apparent authority to act for the principal. Because it is the principal who runs the risk of being bound by the ex-agent's actions, a careful principal will make sure that third parties who have dealt with the former agent receive notice that the agency relationship has ended.

Notification to those who have previously dealt with the agent prevents this, but notification is not effective until the third parties receive it. Thus, a notice in a newspaper or trade journal may not be effective notification to third parties.

PARTNERSHIPS AND CORPORATIONS

Each partner in a partnership is an agent for the partnership. Generally, the rules of agency apply to partnerships. As long as a transaction takes place in the normal course of business of the partnership, an agreement by one partner binds all partners to the contract. Transactions beyond the normal scope of the partnership generally require approval of all the partners. Just as notice given to an agent is the virtual equivalent of notice given to the principal, notice given to any partner is also notice to the partnership.

A distinctive feature of a partnership is the agreement of the partners to share the profits. As you might guess, the partners also share the liabilities. Thus, each partner remains liable for the debts of the other partners resulting from their activities on behalf of the partnership. For this reason, many businesses are organized as corporations. In a corporation, the business is owned by the shareholders. The shareholders' liability is limited to their investment in the corporation. If White and Black are partners, for example, and White commits a tort while acting on behalf of the partnership, then White, Black, and the partnership can be held liable. If, however, Black is simply a shareholder in White Corporation, then Black's personal liability is usually limited to the loss of his investment in White Corporation due to a suit against White and the White Corporation.

Another distinctive feature of corporations is that they can "live" forever. A partnership ends when a partner leaves by death or agreement. A corporation's directors, officers, and employees may come and go, but the corporation remains a distinct and unchanging legal entity.

VAUX v. HAMILTON
103 N.W.2d 291 (N. D. 1960)

The two actions involved in this litigation arose out of a collision of an automobile driven by plaintiff Vaux, in which the plaintiff Nixon was a passenger, and a Cadillac automobile driven by the defendant Dorothy Hamilton, which carried the dealer's license of the defendant Day's Auto Brokers, Inc. The accident occurred just west of Jamestown, North Dakota, on U.S. Highway No. 10.

Both actions are against the same defendants and involve the same facts. The cases were consolidated for trial in the district court of Stutsman County, and both cases were argued together on appeal. Both appeals will be considered in one opinion.

The defendant Day's Auto Brokers, Inc., is a foreign corporation engaged in the business of selling used cars in the city of Seattle, Washington. Through its agent, DeLain Belch, the defendant purchased the Cadillac automobile involved in this litigation in the Detroit area. The employee purchasing the car left it with the Midwest Auto Delivery for delivery to the defendant's place of business in Seattle. The delivery service advertised for a driver to deliver the automobile to Seattle, and the advertisement was answered by the defendant Ruby Cuthbert. The agreement under which she was to drive the car to the defendant's place of business in Seattle provided that all gas and oil and other expenses be paid by the driver. While the car was being driven to Seattle, it was involved in a collision just west of Jamestown resulting in the litigation now before the court.

The plaintiffs alleged in their respective complaints that defendant Cuthbert was the agent of defendant Day's Auto Brokers, Inc., in delivering the car. This was denied by the defendant

Day's Auto Brokers, Inc. Verdicts were returned by the jury in favor of both of the plaintiffs and against the defendant Day's Auto Brokers, Inc., and the defendant Dorothy Hamilton, who was driving the car at the time of the collision. The defendant Day's Auto Brokers, Inc., has appealed from the judgments and from orders denying its motion for judgment notwithstanding the verdict or, in the alternative, for a new trial.

The burden of establishing agency rests on the party alleging it, as respects the master's liability for negligence of the alleged servant. . . .

Thus, where existence of agency is denied, the burden of proving agency is on the party asserting its existence.

The evidence as to the existence of agency in this case, relied on by the plaintiffs, consists of the deposition of Henry Freymueller, the president of defendant Day's Auto Brokers, Inc. He testified, on cross-examination by the plaintiffs, that the defendant company obtained the automobile in question by having its employee, DeLain Belch, purchase it and then deliver it to the drive-away firm for shipment after attaching defendant's "intransit" dealer's license; that the defendant was to pay to Midwest Auto Delivery a flat fee to deliver the said automobile. Freymueller further testified:

> Well, we pay them to get the car out here. The discretion of how they deliver it is up to them.

There was the further evidence of the defendant Ruby Cuthbert, the girl who answered an ad of Midwest Auto Delivery, advertising for a driver to deliver a car to Seattle. She stated that she had seen the advertisement of Midwest Auto Delivery in her local paper; that she answered the advertisement and agreed to drive the car in question to Seattle; that she did not know who owned the automobile but that it had a Washington dealer's license on it. She then said:

> It is quite the thing. They advertise in newspapers in Ontario—anyone who wants to go to the West Coast. It is not employment, it is a case if you go to the West Coast, it is a cheap way to go, and they—....

She further stated that she and her two companions, the defendants Dorothy Hamilton and Margaret Jack, were to pay the oil and gas and their own expenses incurred while taking the automobile to Seattle.

On this evidence the jury found the defendant Hamilton to be the agent of defendant Day's Auto Brokers, Inc.

In reviewing the sufficiency of the evidence on appeal from the judgment and from an order denying motion for judgment notwithstanding the verdict or for a new trial, this court will view the evidence in the light most favorable to the verdict. . . .

The record is silent as to whether this was the first occasion on which the defendant Day's Auto Brokers, Inc., had had the Midwest Auto Delivery deliver an automobile for it. While there is evidence that the defendant Day's Auto Broker's, Inc., was in fact the owner of the automobile involved in the collision, ownership of the automobile alone does not establish or prove agency. Neither is ownership alone sufficient to impose liability upon the owner of a car because of the negligence of another who is permitted to use it....

It is true that defendant Cuthbert was performing an act in the interest of defendant Day's Auto Brokers, Inc., but acts of the alleged agent cannot establish agency without evidence showing that the alleged master had knowledge thereof or assented thereto. . . . In this case, the testimony was undisputed that the defendant Day's Auto Brokers, Inc., did not even know of the existence of the three girls who were delivering the automobile until after the accident had occurred.

The limited evidence presented to the jury on this question is more indicative of a relationship of independent contractor than of master and servant. Here, Midwest Auto Delivery was hired for one purpose, namely, to deliver the car to Seattle, and was responsible only for that result; it could accomplish that result in its own way. The record fails to disclose any right of control by defendant Day's Auto Brokers, Inc., over details as to how that result was to be accomplished. Such evidence may be available, however, on a new trial. While, under the present state of the record, on appeal from the judgment and from the order denying the motion for judgment notwithstanding the verdict or for a new trial, there is a failure to sustain the allegations of the complaints as to agency, this court need not order judgment for the defendant but will order a new trial when it appears that the defects may be remedied upon a new trial....

Other specifications of error alleged by the defendant relate largely to instructions given by the court, or instructions refused by the court, relating to matters of agency, independent contractors, and liability of the owner of a vehicle for negligence of one permitted to use it for the user's own purposes. Since a new trial in these cases must be granted and, in view of what we have said above, it is unlikely that these questions will arise upon a new trial, we do not now consider them.

However, one specification of error deals with a matter which may well arise on the retrial of these actions. The trial court overruled an objection by the defendant to the following question put by the plaintiffs to a medical expert, testifying on behalf of the plaintiffs, as to the future pain and suffering of one of the plaintiffs:

> "Doctor, can you state with a reasonable degree of medical certainty that there is a distinct possibility that this might happen?"

An objection that the question was leading, suggestive, speculative, and conjectural was overruled.

The question in the form in which it was asked was clearly objectionable. While there are exceptions to the general rule that an opinion of a witness may not be received in evidence and although, under certain circumstances, the opinion of an expert is admissible, testimony which consists of no more than a mere guess of the witness is not admissible. Such testimony must be as to a definite probability and must not involve, to an excessive degree, the element of speculation or conjecture. The question directed to the medical expert in this case was calling for a mere guess on the part of the doctor "that there is a distinct possibility that this might happen."

Webster defines *possibility* as "the character, state, or fact of being possible, or that which may be conceivable." Thus, even if an event might occur only once in 10,000 times, it still is within the realm of possibility, though very improbable. A medical expert is qualified to express an opinion to a medical certainty, or based on medical probabilities only, but not an opinion based on mere possibilities....

For the reasons stated, the orders denying motion for new trial in the above actions are reversed, and new trials are granted.

MOUNDSVIEW IND. S.D. NO. 621 v. BUETOW & ASSOC.
253 N.W.2d 836 (Minn. 1977)

Buetow & Associates, Inc., (Buetow) entered into an agreement with Moundsview Independent School District No. 621 (Moundsview) to perform architectural services. Buetow agreed to prepare plans and specifications for an addition to a school as well as to provide general supervision of the construction operation. After the completion of construction, a windstorm ripped a portion of the roof off the school, allegedly due to the failure of a contractor to adequately fasten the roof to the building. The trial court granted Buetow's motion for summary judgment based on Buetow's contract with Moundsview, which provided that Buetow was not responsible for the failure of a contractor to follow the plans and specifications. We affirm.

In August 1968, Moundsview retained Buetow to prepare plans and specifications for an addition to an elementary school. At the time of the execution of the agreement, Moundsview had the option of requiring Buetow to provide (1) no supervision, (2) general supervision, or (3) continuous onsite inspection of the construction project by a full-time project representative referred to as a "clerk of the works." Moundsview elected to have Buetow provide only a general supervisory function, the specific language of the contract enumerating the requirements as follows:

> The Architect shall make periodic visits to the site to familiarize himself generally with the progress and quality of the Work and to determine in general if the Work is proceeding in accordance with the Contract Documents. On the basis of his on-site observations as an Architect, he shall endeavor to guard the Owner against defects and deficiencies in the Work of the Contractor. The Architect shall not be required to make exhaustive or continuous on-site inspections to check the quality or quantity of the Work. The Architect shall not be responsible for construction means, methods, techniques, sequences, or procedures, or for safety precautions and programs in connection with the Work, and he shall not be responsible for the Contractor's failure to carry out the Work in accordance with the Contract Documents.
>
> The contract further provides:
>
> The Architect shall not be responsible for the acts or omissions of the Contractor, or any Subcontractors, or any of the Contractor's or Subcontractor's agents or employees, or any other person performing any of the Work.

Buetow prepared plans and specifications requiring the placement of wooden plates upon the concrete walls of the building. The plates were to be fastened to the walls by attaching washers and nuts to one-half inch studs secured in cement. During the 79-week construction period, the president of Buetow made 90 visits to the construction site in performance of Buetow's general supervisory obligation.

On May 19, 1975, a severe windstorm blew a portion of the roof off the building causing damage to the addition and to other portions of the school. It was discovered that the roof had not been secured by washers and nuts to the south wall of the school as required by the plans and specifications.

Moundsview brought an action for damages caused by the roof mishap against Buetow, the general contractor, and the roofing subcontractor. In response to an interrogatory from Buetow requesting Moundsview to state all facts upon which it relied to support its allegations against Buetow, Moundsview replied:

> Defendant Buetow failed to properly supervise the roof construction, failed to supervise and discover the missing nuts and studs and take proper corrective action.

Thereafter, Buetow made a motion for summary judgment, basing its motion upon the affidavit of one of its officers which stated that Buetow did not observe during any of its construction site visits that the washers and nuts had not been fastened to the studs on the south wall. The motion was also accompanied by the architect's contract which the parties entered into and Buetow's interrogatories and Moundsview's answers thereto. Moundsview did not file a responsive affidavit to oppose the motion.

The trial court granted Buetow summary judgment, accompanying its decision by memorandum which states:

> Since the contract of the architect did not require detailed supervision by the architect of the construction project, and since the architect was not contractually liable, as a matter of law, for the acts and omissions of the general contractor or any subcontractor, the architect is entitled to summary judgment under the principle enunciated in *J & J Electric, Inc.* v. *Moen Company....*

Moundsview appeals from the judgment entered pursuant to the order for summary judgment.

The issue presented for consideration is whether there exists a genuine issue of fact in this case that will preclude the entry of summary judgment dismissing the complaint against Buetow.

1. Initially, we note that the rule in Minnesota is that a party cannot rely upon general statements of fact to oppose a motion for summary judgment. Instead, the nonmoving party must demonstrate at the time the motion is made that specific facts are in existence which create a genuine issue for trial....

   The general statements included within Moundsview's complaint and the equally general answers to Buetow's interrogatories are insufficient to create a genuine issue of fact to successfully oppose a motion for summary judgment. Thus, since Moundsview failed to present any specific averments of fact in opposition to the motion for summary judgment, our review of the case is limited to a consideration of the contract between the parties.
2. Moundsview argues that Buetow breached its duty of architectural supervision by failing to discover that a contractor had failed to fasten one side of the roof to the building with washers and nuts as required by the plans and specifications. It is the general rule that the employment of an architect is a matter of contract, and consequently, he is responsible for all the duties enumerated within the contract of employment. . . . An architect, as a professional, is required to perform his services with reasonable care and competence and will be liable in damages for any failure to do so....

Thus, consideration of whether Buetow breached a duty of supervision requires an initial examination of the contract between the parties to determine the parameters of its supervisory obligation. The argument that Buetow breached its duty to supervise would be more persuasive had Moundsview contracted for full-time project representation rather than mere general supervision. An architect's duty to inspect and supervise the construction site pursuant to a contract requiring only general supervision is not as broad as its duty when a "clerk of the works" is required. The mere fact that Buetow received additional compensation for performing the general supervisory service does not serve to expand its responsibilities to an extent equivalent to the duties of a full-time project representative. Moundsview cannot be allowed to gain the benefit of the more detailed "clerk-of-the-works" inspection service while in fact contracting and paying for only a general supervisory service.

Thus, the question of whether Buetow breached its duty to supervise the construction project is to be determined with reference to the general supervisory obligation enumerated in the contract. The contract provided that the architect"...shall not be responsible for the Contractor's failure to carry out the Work in accordance with the Contract Documents." When this section is read in conjunction with the section which provides that "(t)he Architect shall not be responsible for the acts or omissions of the Contractor, or any Subcontractors, or any of the Contractor's or Subcontractors' agents or employees or any other persons performing any of the Work," it is apparent that by the plain language of the contract an architect is exculpated from any liability occasioned by the acts or omissions of a contract. The language of the contract is unambiguous. The failure of a contractor to follow the plans and specifications caused the roof mishap. By virtue of the aforementioned contractual provisions, Buetow is absolved from any liability, as a matter of law, for a contractor's failure to fasten the roof to the building with washers and nuts.

Thus, based upon the language of the architect's contract, Buetow was entitled to summary judgment....

Affirmed.

McCURNIN v. KOHLMEYER & COMPANY

477 F.2d 113 (5th Cir. 1973)

Federal jurisdiction over this case initially was grounded upon a joinder of claims arising under the Commodities Exchange Act, the Securities Act of 1933, and the Securities Exchange Act of 1934 with a diversity claim arising under the Louisiana law of agency. The trial court below determined that none of the customer's (Appellee) federal claims had merit,[1] but, it asserted pendent jurisdiction over the state claim[2] and proceeded to find that the customer was entitled to recovery from the broker (Appellant) under prevailing Louisiana law. We affirm.

The dispute between the parties arose out of trading in the commodity market by McCurnin, the customer. The broker, Kohlmeyer and Company, through its employee Drake, also a co-Appellant, purchased cotton futures for customer at a price in excess of that authorized. He suffered a net loss in the transaction of $26,725.[3] Upon learning of the unauthorized purchase, McCurnin did not immediately and affirmatively repudiate the transaction, at least not by clear and unambiguous conduct. A period of three days elapsed before his market position was liquidated and thus before the full extent of the loss was realized. The broker contends that the customer's delay and his accompanying conduct subsequent to learning of the unauthorized purchase clearly manifested his intention to ratify the purchase and, furthermore, this delay was violative of his duty to mitigate damages.

---

1. As a discussion of the basis alleged for these federal claims and the reasons for the trial court's rejection of them would add nothing to the disposition of this appeal, we abstain analysis.
2. In deciding to assume pendent jurisdiction of the state claim, the trial court correctly made the following determinations: (i) that the federal question raised was not "unsubstantial and frivolous"; (ii) that the state claims arose from identical facts on which the federal remedies were sought; and (iii) that since the case had been fully tried, it was in the interest of justice as well as judicial economy that the issue be decided on what was already a complete record.
3. The suit below was for $15,286.45. This was the amount of his credit balance at the time of the loss. The broker counterclaimed for $11,438.55, the amount remaining due if the customer had to bear the loss.

The trial court, in a thorough and well-reasoned opinion...rejected these arguments. The Judge found that the broker's conduct was violative of two codal Articles of the Louisiana Law of Mandate, La. Civil Code Articles 3010 and 3003.[4]

In response to the contention that the customer ratified, the trial Judge wrote:

> The conclusion that McCurnin failed to repudiate is mistaken. McCurnin had manifested his displeasure to Drake. He had been informed—misinformed—by Drake that there was nothing he could do but complete the transaction. It is true that Drake's optimism about the market had made both McCurnin and Drake sanguine that all might turn out well, but his false hope was never transmuted by McCurnin into approval of Drake's actions.

The Court held that the burden of proving ratification was on the broker and an intention to ratify an unauthorized act cannot be inferred when the conduct can be otherwise explained. For ratification to be implied, it must be shown that the principal actually had knowledge of the material and pertinent facts. Here the judge was entitled to conclude that the customer's error, whether error of law or fact, was clearly induced by the broker. The court held that the customer's effort to repudiate the unauthorized transaction was defeated by the broker.

Though the trial court agreed that the customer owed a duty to minimize his damages, it found that he had acted with reasonable promptness under the circumstances. Immediately upon learning that the broker required that he liquidate his position before they would consider making any adjustment, McCurnin ordered the cotton futures sold.

These were essentially all factual questions. The Judge found the facts. There it ends.

Affirmed.

---

4. Article 3010 of the Louisiana Civil Code provides:

"Art. 3010. The attorney cannot go beyond the limits of his procuration; whatever he does exceeding his power is null and void with regard to the principal, unless ratified by the latter, and the attorney is alone bound by it in his individual capacity."

Article 3003 of the Louisiana Civil Code provides in part:

"Art. 3003. The attorney is responsible, not only for unfaithfulness in his management, but also for his fault or neglect."

REVIEW QUESTIONS

1. Identify the following persons as agents, employees, or independent contractors according to the usual duties involved in their work.
   a. Research chemist working for chemical company
   b. Free-lance consulting engineer in the labor relations field
   c. TV repairman for local department store—on house call
   d. Troubleshooter for steel company (keeps steel sold to customers by recommending proper treatment of a particular heat of steel)
   e. Engineering vice president for local company
   f. Dentist
2. How may the agency relationship be created?
3. Brown is a process engineer for White Manufacturing Company and is about to recommend the purchase of certain machinery and equipment. The Green Equipment Company is one prospective supplier. On a recent trip to the Green Company, Green offered Brown a new station wagon if Green was chosen as the equipment supplier. Brown has always considered herself to be quite ethical, but she is also human, and the station wagon sounds tempting. Neglecting the ethical aspects of the situation, Brown is still faced with certain legal and practical problems. What are Brown's rights if, after recommending Green as supplier, Green fails to produce the station wagon? What can happen to Brown if White Manufacturing finds out about the deal?
4. Gray, engineer for Black, White, and Company, was sent to observe an automation installation at a plant some 50 miles away. On the return trip, he approached an intersection and applied his brakes. His car hit a patch of ice. As a result, he hit another car, injuring its occupants, both cars, and himself. Who is liable for injuries to the other car and its occupants? Who is liable for injury to Gray and Gray's car?
5. In *Vaux* v. *Hamilton* what is the relationship between Hamilton and Midwest Auto Delivery? What is the relationship between Midwest Auto Delivery and Day's Auto Brokers, Inc.? Could Midwest Auto Delivery be held liable for the injuries? Explain your reasoning.
6. Consider the case of *Moundsview Ind. S.D. No.621* v. *Buetow & Assoc.* and respond to the following.
   a. Does the summary judgment in favor of Buetow mean that Moundsview cannot recover for the windstorm damage to its school roof?
   b. Rewrite the contract requirements in such a manner that Buetow is responsible for such disasters in the factual situation described in the case.
7. In *McCurnin* v. *Kohlmeyer & Company*, what minimum additional acts or statements by McCurnin would amount to ratification? Does McCurnin have any right of action against Drake?

Part Four

PROPERTY

Most engineers work for other people—either for a private enterprise or for the public (by working for a local, state, or federal governnment). Engineers use other people's property in their work; hence, they must observe other people's rights. Many situations require the engineer to deal simultaneously with an employer's property and that of others, possibly including his or her own. Effectively handling property and property rights in these circumstances requires some knowledge of property law. Property is either tangible or intangible. Obviously, an automobile is tangible. An example of inangible property would be a patent. Because of the obvious differences between, say, an automobile and the patent on a new plastic, the rights and responsibilities relating to each must be treated differently. The same general thread of ownership rights pertains to both, but the legal rights differ.

Chapter 19

Tangible Property

All of us have things that we consider to be ours: our clothing, books, writing instruments, watches, perhaps a home in the suburbs. These things are our *tangible property*. However, the word *property* can be used in two senses. In its usual sense, it denotes *things* owned by a person. In a broader sense, though, it also refers to the *rights* involved in ownership. These rights, known as *property rights*, signify dominion over the things owned. That is, the right to use and to exclude others from using the things we own, the right of control over and enjoyment of them, and the right to dispose of them.

Tangible property may be further classified as real or personal. In the discussion of the statute of frauds in chapter 11, we distinguished between real and personal property. Generally, *real property* has been defined as land and anything firmly attached to it. *Personal property* is all property other than real property, such as goods, chattels (defined momentarily), choses in action (also defined below), money, and accounts receivable or other evidences of debt. Sometimes, however, personal property is treated as real property; this is called *mixed property*.

PERSONAL PROPERTY

The term *chattels* is often used synonymously with personal property. *Chattels personal in possession* are the tangible items (e.g., a watch, a truck, or a machine in a factory). *Chattels personal in action*, commonly known as *choses in action*, are intangible rights arising from a tort or contract—the right to goods contracted for or the right to recover for injuries suffered in an automobile collision, for example. A *chattel real* is an interest in real property, such as a 10-year lease.

Personal property also includes other types of intangible rights. For example, patents, copyrights, and trademarks are considered personal property. These types of property interests are considered in chapter 20, which deals with "intellectual property."

♦ Acquisition

A person may lawfully obtain ownership of personal property by (1) original acquisition, (2) a procedure of law, or (3) acts of other persons.

Original Acquisition

Unowned things in their natural state become the property of the first person to obtain possession of them. Most things are owned by someone today, but, for example, the possibility of reducing wild animals to personal property still exists. Obtaining ownership by taking possession of such an animal with the intention of becoming its owner is known as acquiring title by *occupancy*.

Generally, property that one *creates* by mental or physical efforts belongs to the creator unless there is an agreement to transfer it to another. Books, inventions, trade names, and other such creations are of this nature. Even where no formal agreement exists, though, the results of such mental or physical efforts may belong to the creator's employer.

Property may be acquired by *accession*—by adding to other property. Generally, an owner of property owns what the property produces and what is added to it. A new windshield of a car or a gear in the transmission, for example, becomes the property of the owner of the automobile. This is particularly true where the addition becomes an integral, built-in part of the whole in such a way that it is not readily detachable. Even where the innocent purchaser of stolen property adds value to it, he or she is merely adding value to property belonging to another.

For example, White buys a car from Gray who, unknown to White, has stolen the car from Black. White adds a new motor, transmission, and paint job to the car. Later, Black locates the car. Black is entitled to regain possession of the car in its improved state. Probably there is no other place where the law adheres so strictly to the principle of *caveat emptor* (the buyer beware) than in the purchase of stolen property.

Accession also applies to a natural increase of purchased property. Here, for example, Black sells White a mare. Shortly after the sale, a foal is born. White is the owner of both the mare and the foal.

Procedure of Law

Property may be distributed according to certain legal procedures. Examples of such procedures include distribution via an intestate death, mortgage foreclosure, judicial sale, and bankruptcy.

- When a person dies and has not left a will, his or her death is termed *intestate*. The various states have, by statute, declared how property shall be distributed in case of such intestate death. Such statutes are called the laws of *descent*. These statutes vary considerably as to who will inherit the estate. If no relatives of the deceased can be found, the property will go to the state—in legal terms, it *escheats* to the state. Even when a person leaves a will, that person is limited somewhat in the way he or she may leave an estate to heirs. A person may not, according to most state laws, leave a spouse or minor children destitute by willing an entire estate to strangers.
- Other property-related statutes provide for mortgage foreclosure when a mortgagor defaults. Although chattel mortgages commonly specify that the mortgagee will take and sell the mortgaged property in case of default, a court procedure is usually possible if such provision has not been made.
- Sales of property may also be undertaken to satisfy a judgment of a court. If the loser fails to voluntarily pay the judgment, the winner can have certain of the loser's property seized (within limits stated in the laws of each state) and then sold to satisfy the judgment.
- When abandoned property is found by the police or sheriff's department in a community, it is kept for a statutory period of time and then sold at public auction. Such sales usually must be advertised and public, with the property going to the highest bidder.
- Under bankruptcy procedures, a trustee may be appointed by the court with the duty to convert the assets of the bankrupt into money. The trustee may take over the property with the right and duty to sell it. Generally the buyer of such property does not get any better title than the seller had.

In judicial sales, sales of abandoned property, and bankruptcy sales, title is usually not warranted by the seller. The seller sells by virtue of a legal right or duty to do so; the buyer assumes the risk that title may not be good.

Consider this example. Black steals White's car and abandons it in a neighboring town. The local police department holds it for the required period of time and then sells it to Gray at a public auction. Later, White finds his car in Gray's possession. White can claim and get his car.

Acts of Other Persons

A person may lawfully acquire title to personal property from others by will, gift, contract, confusion, or abandonment. A person may also acquire possession of property if it is lost or mislaid by another.

The subject of wills was discussed briefly above and will be discussed further under real property. It is sufficient to note here that testators who comply with the law may leave their property to whomever they wish.

If a person acquires property by gift, title to the property follows possession. For example, Black promises White a gift of $500. At this point, White has nothing. The promise of a future gift, either oral or in writing, is unenforceable, since it is unsupported by consideration. Of course, if the written promise of a gift were signed and sealed, a consideration would be imputed under common law, and the promise would be enforceable. As soon as Black actually gives the $500 to White, though, it becomes White's property, and Black loses any claim to it. Under a few exceptional circumstances, the gift may be recoverable, particularly when a third party has rights in the gift.[1]

The subject of acquisition of personal property by contract is covered in chapter 15, "Sales and Warranties." Good title to personal property is warranted in any sale unless there is a disclaimer of the warranty in the contract.

Property may be acquired by what is known as *confusion*, primarily when fungible goods are involved. *Fungible goods* are goods any unit of which is replaceable by any other unit—for example, grain of a particular type, crude oil, or screws in a bin. Such goods are usually sold by weight or measure. If fungible property of two or more owners is mixed together so that the identity of each owner's property is lost, each owner owns an undivided share of the confused mass. For example, after harvest, Black stores 500 bushels of wheat with 700 bushels of wheat belonging to White in a common granary. Each owns an undivided share of the 1,200 bushels of wheat. Destruction of a part of the mass will be shared by each party on the basis of that party's contribution to the total.

If confusion of goods results from the tortious act of one of the parties, the innocent party will be protected. If ownership by the tortfeasor (that is, the person who committed the tort) cannot be determined, the innocent party becomes owner of the total.

Abandoned property is unowned. It becomes the property of the first person to take possession of it. Taking possession of abandoned property is about the same as taking possession of something that has never been owned.

Lost and mislaid properties give rise to some legal problems. In each case, the owner has unintentionally parted with possession of his or her property. In each case the owner still owns the property even though it is no longer in his or her possession. The finder of *lost* property has a right to the property against all persons except the true owner. By contrast, the holder of *mislaid* property has possession of it as a bailee—in other words, he or she is holding it to give to the owner.

The distinction between lost and mislaid property is derived largely from the circumstances in which it is found. If the property is found in such a location that it is apparent that the owner intentionally placed it there and then inadvertently left it, it is mislaid. A purse left on a store counter would be mislaid; if it were found on the floor, it would have been lost.

In many states, the problems involved in lost and mislaid property have been cleared up by statute. The requirements of the statutes usually are met by advertising the property in a local newspaper. If no one claims the property within a certain time after the publication of the advertisement, the finder obtains title to the property.

♦ Bailment

A relationship that closely resembles property ownership is that of *bailment*. The bailment relationship occurs when personal property is left by the bailor (the owner of the property) with a second person, the bailee.

Requirements

Distinguishing bailment from similar relationships requires careful definition. Generally, a bailment is made up of three elements:

1. Title to the property remains with the bailor.
2. Possession of the property is completely surrendered by the bailor to the bailee.
3. The parties intend the return of the bailor's property at the end of the bailment.

Notice the similarity between a bailment and a sale or a trade with a slight delay in it, as in this example. Black stores a spare conveyor at White's warehouse for an agreed period and a fee. Black still has title to the conveyor. White has possession, and the same conveyor is to be returned to Black. Therefore, it is an instance of bailment. With certain other types of property, though, an inherent difficulty exists. If, instead of a conveyor, Black were to store grain in a common granary, he might not expect to get back the identical grain that he stored. The same might be true of animals in a herd and a few other instances where the owner does not expect that the identical property will be returned. Courts are not uniform in all jurisdictions in their holdings under such circumstances. Generally, though, if the owner is not to receive back the identical thing given, it is not bailment. It is held that title passed with

1. This could occur where the property was stolen from another, or where the donor anticipated impending bankruptcy, or where the donor was dying and diminished the property that would go to his or her heirs.

possession and that title to other similar goods will be passed back later. In other words, such a transaction represents a *sale*.

It becomes necessary to find out who owns what when one party or the other goes bankrupt, or a writ is issued pursuant to a judgment against someone's property. Property being held for a bailor by a bailee generally cannot be successfully taken for the bailee's debt. It is possible, though, if a judgment were to be issued against the bailor, to obtain the property from the bailee.

Duty of Care

The person entrusted with the property of another has a duty to care for it. Under a particular set of circumstances, the degree of care may be great, ordinary, or slight. Two primary considerations determine the necessary degree of care: (a) the nature of the property involved and (b) the purpose of the bailment. As to the nature of the property, it is obvious that a person should take greater care of a new computer than of a used anvil. The bailment relationship benefits someone—the bailor, the bailee, or both. If the bailment is to benefit the bailor only, the bailee need exercise only slight care in protecting the bailed property. The bailee is liable only if he or she has been grossly negligent. Such a bailment might occur as a result of the owner requesting a friend to care for his or her property gratuitously.

If property is borrowed for the benefit of the bailee (as one would borrow a neighbor's lawn mower), great care is required of the bailee. The property must be returned in the form in which it was borrowed. About the only damage for which the bailee would not be responsible would be that resulting from an act of God—such as destruction by a cyclone.

Probably the most common form of bailment occurs as a benefit to both bailor and bailee. Whenever the bailor pays the bailee to take goods and alter them in some way, or pays the bailee just to store them and then return them, both parties benefit. Thus, mutual-benefit bailments would occur in the following situations: one party leaves a car at a garage with orders to fix the transmission; one party asks another to transfer a machine from Cleveland, Ohio, to Fort Worth, Texas; and one party stores an unused machine at another's warehouse during a slack period.

When the bailment is to benefit both parties, the bailee is required to use at least ordinary care. *Ordinary care* means the care that a person would be likely to use in preserving his or her own property. The bailee is liable for damage resulting from his or her negligence. If the bailee has used the requisite amount of care in preserving the bailor's property, the bailee will not be held liable for damages. Loss of the property or damage to it, then, will follow title and be borne by the bailor.

Bailee's Right and Duties

The bailee's right to possession of the bailed property is second only to the bailor's right. The bailee may sue a third party to recover the property if necessary. This reflects a general presumption of property law: That is, the person in possession of the property is its owner.

If the bailee gives the bailed property to someone other than the owner or the owner's agent, and it is thereby lost, the bailee is liable. However, if the bailee retrieves the property and returns the property to the person who gave it to him or her originally, assuming that that person is still the owner, the bailee should not be liable, even though the property left the bailee's hands.

The bailor has a duty to disclose any known defects in the bailed property that might harm the bailee or the bailee's employees. If harm results from a failure to disclose such defects, the bailor may be held liable for tort. For example, consider what a bailor of steel drums containing toxic chemicals or radioactive waste should disclose to the warehouse acting as a bailee.

Negligence Liability

Most bailments are contracts. It is possible, therefore, for the bailee, by contract clauses, to remove any or all liability for negligence, but only if he or she is a private bailee. For a public or quasi-public bailee—a hotel or trucking company that offers its services to the public at large, for instance—to make such a contract stipulation probably would be unlawful, especially if the stipulation conflicted with statutes governing such relationships. Such a bailee will be liable for negligence when serving the public regardless of contract clauses to the contrary. If a private bailee insists on eliminating liability, the bailor has only two choices—to do business with the bailee on those terms or do business elsewhere.

♦ Fixtures

A *fixture* is personal property that becomes attached to real property. By the attachment, the personal property becomes a part of the real property. The concrete and other building materials that are worked into a plant become real property. Similarly, a heating unit or a television aerial becomes a fixture when it is attached to a house.

Generally, ownership of a fixture goes to the owner of the real estate to which it is attached. There are so many exceptions to this generality, though, that the principle might be restated this way: *unless something appears to the contrary,* the owner of the real property also owns the fixture. The main condition to the contrary is the intention of the parties when the fixture was attached. If it appears that both parties intended that ownership of the fixture should not go to the owner of the real property, the original owner will have a right to remove it. For example, Black Construction Company undertakes the building of a structure for the White Company. A small building, complete with plumbing and lighting, is erected on the premises as a superintendent's office. Although the superintendent's office is firmly fixed to the ground, it may be removed at the end of the project. The relationship of landlord and tenant often involves the determination of ownership of fixtures. When real property is leased or rented, the tenant normally may install personal property (e.g., machines or conveyors) and then take them when he or she leaves. If the tenant fails to remove and take a fixture, ownership goes to the owner of the land. For example, Black rented a house from White. Requiring hot water, which White's house did not have, Black bought a suitable water heater from Gray Appliance Store, paying 10 percent down. Black installed the water heater, hooked it up to the plumbing, used it for a month, and then moved to another state. Gray Appliance tried to get the water heater back, but could not, since it was now part of White's real property. Gray's only available action is against Black, and he may be hard to reach.

REAL PROPERTY

Real property has been defined as land and anything firmly affixed to it. When an engineer becomes a party to building a road or renovating a manufacturing plant, he or she is concerned with real property. As a citizen in a community, the engineer will either own real property or lease it. As an investor, an engineer may speculate in real estate, since such speculation has become nearly as popular a sport as speculation in the stock market. Because engineers deal with real property in both their personal and professional lives, we will consider here the transfer (or conveyance) of real property and some of the rights and duties created.

We may buy real property, use it pretty much as we please, and transfer it to others with very few restrictions involved. The right to "own one's own home" is almost a part of our heritage. Only when title is threatened or, perhaps, when prescriptive rights (that is, the right to continue using another's property that arises out of longstanding use of that property) are exercised against real property do we become aware that any limitations exist.

♦ Evolution of Property Rights

Our property laws involving real property developed over the centuries in England. When William the Bastard won the Battle of Hastings in 1066, he set about establishing control over England. He accomplished this feat partly by asserting absolute claim to all English land, then parceling it out to those who had helped and supported him. Tied to the land granted by the king was the obligation to serve him.

Under this *feudal system* then, the land belonged to the Crown. The right to hold realty depended upon military service and fealty to the king, and was theoretically terminable at his option. Land was parceled out to the gentry who, in turn, divided it up among their servants. Originally, when a landholder died, another person was appointed to take his place. Very early this practice was replaced by provisions that tended to ensure that the property would remain in the family of the grantee or tenant provided the heir was able to meet the military obligations entailed. The system of estates developed in England is essentially the system that generally remains in place there today.

♦ Kinds of Estates

Estates in real property can be classed as freehold, less than freehold, and future estates.

Freehold

A *freehold estate* is an estate of undetermined duration in real property. It may be an estate in fee simple or it may be an estate for life. An *estate in fee simple* is the highest real property estate known to law. The holder of an estate in fee simple has the right to complete use and enjoyment as long as this use does not harm another and the right to transfer the estate to anyone he or she chooses. Upon death, the holder's estate will be distributed according to his or her will or according to law if the holder dies intestate.

A *life estate* is an estate of undetermined duration and, therefore, a freehold estate. The life upon which the term of the estate depends is usually that of the holder of the estate, but it could be that of anyone else. An estate that is terminable upon some contingency other than death but is, in some way, dependent upon the duration of a person's life is treated the same as a life estate. For example, Black gives an estate to a young widow, terminable when she remarries. She might die without remarrying, and this would also terminate the estate. It is, therefore, treated in law the same as a life estate, but with marriage as an added contingency.

The rights in a life estate are not so complete as they are in a fee simple estate. The holder of a life estate may not sell it to another. Though the holder is allowed to use and enjoy the real property, he or she may not destroy its value. For instance, the holder may not sell the topsoil or remove ornamental trees, although he or she could cut and sell ripe timber.

Less than Freehold

An estate in real property that is to run for a fixed or determinable time is *less than freehold*. In law it is considered as personal property. Thus, a 10-year lease or a grant of property to run "as long as the property is used for educational purposes" would be less than a freehold estate. By contrast, a lease for "99 years, renewable forever" would be a freehold estate, since the duration is undetermined.

Future Estate

A *future estate* is an estate that someone will have when a future event occurs. According to Gray's will, for example, White obtains Grayacre (Gray's estate) for the duration of White's life and, upon White's death, Grayacre will pass to Black. Black thus gets the estate left to White when White dies. Black has a future estate in the property involved.

TRANSFER OF REAL PROPERTY

Real property and real property rights are transferred in four major ways: (1) by will or inheritance, (2) by sale, (3) by gift, and (4) by legal action. We will consider the documents required in these transfers and the principles that govern them.

♦ Will

Originally, the word *will* indicated a disposal of real property only; another document, a *testament*, disposed of the testator's personal property. Thus the use of the phrase *last will and testament* came to be popular when the testator wished to combine the two functions in one document. By common and legally accepted usage, the term *will* today indicates a document that provides for the disposition of both real and personal property. Below, we will consider some aspects of will making.

Age

Under common law, anyone of sound mind and 21 years of age can make a valid will in the United States. Most states, however, have passed laws that reduce the age requirement.

Mental Capacity

A person without his or her proper mental faculties (that is, someone who is insane) cannot make a valid will. The law does not require a towering intellect as a testator, however. The law requires only that the testator have (a) sufficient mental capacity to comprehend his or her property, (b) capacity to consider all persons to whom he or she might desire to leave property, and (c) understanding that he or she is making a will. In these requirements, nothing prohibits an eccentric person from making a will. A person physically or mentally ill may make a will. Even an insane person could, in his or her rational moments, make a valid will.

Who May Inherit

Inheritance is not limited to relatives of the deceased under a valid will. Almost anyone may inherit. Municipalities, universities, and charitable organizations often have benefited from the terms of wills. The law, however, usually prevents inheritance by the murderer of the testator.

Similarly, the testator has a right to disinherit as he or she chooses within the limits of the state statutes. Still, complete disinheritance of the testator's husband or wife cannot be done successfully in most states, and provision for any minor children may be required.

Types of Wills

In addition to the ordinary written will with witnesses, two other types of wills are recognized in some states. The first, the *holographic will*, is one that is written entirely in longhand by the testator. Usually no witnesses are required. Where the holographic will is recognized—in about half the states—it has the same standing as any other will.

The second type, a *nuncupative will*, is an oral will. Such a will usually must be made before a certain number of witnesses, and they, in turn, must reduce it to writing shortly thereafter. The testator cannot will real property to another orally. In fact, in the states where nuncupative wills have legal standing, a limit is usually placed on the amount of personal property that may be willed orally.

Witnesses

A witness to a will should be a disinterested party capable of being a witness in any judicial proceeding. Generally, a minor can act as a witness. A person who stands to gain or lose by the will, though, may be incompetent as a witness. For this reason, it is usually preferred that a witness have no stake in the property being disposed of by the will.

Essentials of a Valid Will

Excluding holographic and nuncupative wills, there are four common requirements for a valid will:

- To pass the testator's real and personal property along to others, the will must be in *writing*. The law does not require any special kind of writing, such as typing or longhand, so long as the will is written. Neither is there a requirement as to the material on which the writing appears. A will chiseled in stone or etched on glass could be as legally binding as one drawn up on a form prepared by an attorney.
- A valid will must be *signed* by the testator and *sealed* (in those states requiring a seal). The signature normally appears at the end of a will, and its validity may be open to question if it appears elsewhere.
- Wills (except holographic) must be witnessed. State laws usually require either two or three persons to attest the signing of a will. The will must be signed in the presence of the witnesses. Depending upon the state's law, the witnesses may or may not need to sign in the presence and sight of each other and of the testator.
- A will must be published. In connection with wills, the word *publication* means something different from its ordinary sense. Publication of a will occurs when the testator declares that this is his or her last will and testament. The witnesses must know that it is a will being signed; they need not necessarily know the terms of it, only that it is a will.

Codicils

A *codicil* is a supplement to a will, such as an amendment. It is used to explain, modify, add to, or revoke a part of an existing will. If there is any question as to the date of the making of the will, the time when the last codicil was drawn is the effective date of the will. The making of a codicil to a will requires the same formality as is required in making a will.

Probate

After the death of the testator, the will is presented for probate to a probate court. State statutes determine the next steps, but the state laws follow a general pattern. Opportunity is given to question the validity of a will. It may be held invalid for fraud, undue influence, improper execution, forgery, mistake, or incapacity of the testator. If the validity is unchallenged or any challenges attempted are unsuccessful, it is *admitted to probate* (i.e., received by the court as a valid statement of the testator's intent). If the will is successfully challenged, a previous will may be reinstated or, if no previous will exists, the result is the same as intestate death.

In drawing up a will, it is customary to name someone as executor or executrix. Upon the death of the testator and probate of the will, the executor is called upon to carry out the terms of the will. This usually must be done *under bond* unless the testator has specifically exempted the executor from bond. If no will was left or if no executor was named in the will, the court will appoint an administrator or administratrix. The functions of an administrator are similar to those of an executor. However, if the decedent left no will, the administrator must follow state laws for distributing property following intestate death. In many situations, family members are named as executor or executrix. This does not prevent them from also being a beneficiary.

♦ Deed

Conveying real property by sale or gift requires a formal transfer. Each state has jurisdiction over the real property within its boundaries. Each has set forth the formalities required to convey ownership from one person to another. Although the statutes vary from state to state, there is a general uniformity in the requirements.

Kinds of Deeds

Two kinds of deeds are in common use in the United States today: *warranty* and *quit-claim*. On certain occasions, either may be used, but the better title is obtained in a warranty deed.

A *warranty deed* warrants that the title obtained by the grantee is good. In any deed, the grantee gets only the title that the grantor has to give; but if a grantee's title under a warranty deed is ever successfully attacked, the grantee may recover any damage suffered from the grantor. By granting a warranty deed, the grantor usually warrants three things: (1) that he or she has good title and the right to convey it; (2) that the property has no encumbrances other than those mentioned in the deed; and (3) that the grantee and his or her heirs or assigns will have quiet, peaceful enjoyment of the property conveyed.

A *quit-claim deed* transfers title but does not warrant it. In effect, it is the conveyance by the grantor of whatever title he or she may have to the property. Such a deed might be used where inheritance of the property by several members of a family sometime in the past has left a clouded title.

Essentials of a Deed

For a deed to be valid, it must usually include several essential elements. The deed must name the grantor and grantee and the consideration (if any) involved; the property must be described or otherwise identified; some words of conveyance must be used; and it must be signed, sealed (in many states), witnessed, delivered, and accepted by the grantee.

Description of property within a city is likely to be by lot number and plat. Rural property may be described according to metes and bounds, or by the Torrens System of sections and fractions. In interpreting a deed, the court will endeavor to carry out intent of the parties even when there is an error in the description. Corner markers or monuments and natural landmarks may show this intent better than descriptions, since these can be seen by the parties. Thus, the presence of such a marker may cause the court to disregard the technical description.

Recording

The law requires any proceeding that affects real estate to be recorded as a notice to the public. Thus, to have full force or standing at law, a deed, mortgage, lien, attachment, or other encumbrance must be filed at the local recorder's office or registry of deeds.

The recording of a deed does not pass title to the property; the making of the deed takes care of that. Still, it is essential to record the deed because the states have various "recording" statutes that adopt and give effect to the idea that a subsequent buyer may obtain good title. Thus, if the grantor were to make a second deed to a second grantee fraudulently, the first grantee may lose the property if the second grantee purchased the property in good faith and without knowledge of the first grant. By recording the deed, the first grantee essentially tells the world of the first grant. Because the grantee is at risk, it is the grantee who makes sure the deed is recorded.

Title Search and Title Insurance

A *title search* involves following the changes in title to a piece of property from the initial grant from the state to the present. The result of the search should show an unbroken chain; a break is cause for suspicion and further search. For example, a will leaving the property to more than one person may be questioned. Any encumbrance on the property is also questioned, to determine whether it has been cleared up. Of course, a search may have to stop at some point. If, for example, because of destruction of records or some other reason, ancient title cannot be cleared, it will usually be certified despite the void.

In most communities *title insurance* may be purchased to warrant title to real property. The title insurance company will search the title and issue insurance for a one-shot fee based on the outcome of the search.

♦ Mortgages

A large proportion of the buyers of real property today do not have sufficient assets to pay cash for their real estate. Buyers must borrow the money from someone. They might borrow the money on a personal loan or a note, but the problem of securing the loan exists. For example, Black borrows $100,000 from White on a note, the money to be used to buy a house and lot. If Black, at some future time, cannot pay an installment on the note, White may obtain a judgment in court for the remainder of the note. However, under most state statutes, Black's homestead and much of his personal property would be protected from execution or attachment. Thus, White would have little real security.

In contrast, a mortgage offers the lender substantially greater security. A real estate *mortgage* is a contract between the mortgagor (the borrower) and the mortgagee (the lender). The mortgagor borrows funds from the mortgagee, perhaps for the purchase of real estate, promising to return the money with interest, and offering the real estate as security.

Mortgage Theories

Mortgages began under common law as *defeasible conveyances*. That is, the mortgage took the form of a deed from mortgagor to mortgagee. The mortgagor could *defeat* the deed (get it back) by paying the loan on which the mortgage was based in the time specified. The mortgagor who missed a payment, though, had nothing. Any default gave the mortgagee absolute right of ownership. The results seemed rather harsh, and the treatment has since become more lenient.

In equity jurisdiction, where mortgage foreclosures are normally handled, certain mortgagor's rights have come to be recognized. Where prior common law considered the mortgagee as owner, allowing the mortgagee to collect rents and profits from the property, the mortgagor now usually has the rights of ownership. A missed payment no longer terminates forever the mortgagor's right; usually the mortgagor has a certain time in which to redeem the property under a right known as *equity of redemption*.

In equity, the view is taken that the mortgage is security for a loan—that it is nothing more than a lien. This *lien theory* now constitutes the accepted reasoning on mortgages in the majority of the states. Some states, though, still hold to the older common law ideas in modified form, under the *title theory* of mortgages.

Formality

Since a real estate mortgage represents an interest in real property, it must be in writing. The form of a mortgage, even in *lien theory* states, is similar to that of a deed, and the same formal requirements usually pertain to both. The mortgage usually must be signed, acknowledged, witnessed, and recorded. When the mortgage is satisfied, this too must be recorded. Recording of the mortgage and its discharge serves as notice to the public of this type of property encumbrance.

Mortgagor's Duties

Security is the reason for a mortgage. While the mortgagor is owner of the mortgaged property, his or her right of ownership must be somewhat restricted. The mortgagor cannot, for example, tear down all buildings, sell off the trees and topsoil, and then allow the mortgagee to take over the worthless remainder. To do so would diminish the mortgagee's security and constitute *waste*. If the mortgagee's security is so threatened, he or she probably has reason to institute foreclosure proceedings.

As the loan balance declines, the mortgagor's right to unlimited use and disposal of the property increases, since less security is required to protect the mortgagee's interest.

The mortgagor usually must pay all taxes, assessments, and insurance on the mortgaged property. Unpaid taxes and assessments become liens and endanger the mortgagee's security. Insurance on the property protects mortgagee and mortgagor alike. In case of near total destruction, the mortgage balance is usually paid first, and any remainder goes to the mortgagor.

The mortgagor must, of course, make the mortgage payments when they are due. Nonpayment constitutes default and gives the mortgagee the right to institute foreclosure. Generally, the mortgagee need not accept early payment of the balance due; he or she has a right to the interest contracted for. The standard FHA and Veteran's Administration mortgages have provisions for early payment, but many other mortgages do not.

Mortgagee's Rights

The mortgagee generally may assign the mortgage note together with the mortgage to a third person, who then has the same rights as the mortgagee, or "stands in the mortgagee's shoes." The mortgagee must, of course, notify the mortgagor of the assignment if the mortgagor is required to pay the assignee. If the mortgagor, lacking knowledge of the assignment, pays the original mortgagee, the mortgagor will diminish the amount of the note by the amount of the payment.

In case of default in payment or a diminishing of the security, the mortgagee may foreclose. Generally, foreclosure proceedings begin with the filing of a *bill* in equity. The bill outlines the mortgagee's rights and the mortgagor's breach of the agreement. If foreclosure is allowed, the court will appoint a master to sell the mortgaged property. The purchaser gets a deed to the property. Court costs are paid first from the proceeds, then from the mortgage balance. If anything from the sale remains, it is returned to the mortgagor.

If the foreclosure sale of the property does not return enough to pay off court costs and the mortgage, the court may issue a *deficiency decree*. Such a decree holds the mortgagor personally liable for the unpaid balance of the obligation. Enforcement by execution or attachment will be likely to follow. The courts lately have exhibited some reluctance to issue deficiency decrees, however. This is especially true where the mortgagee loaned the money for the purchase of real property. The reasoning goes this way: The mortgagee loaned the money on the security of the realty and not on the mortgagor's personal credit. In other words, the mortgagee must have evaluated the risks involved and the mortgagor's security when the loan was made. Recovery, then, should be limited to the price this property will bring; part of the value of the interest charged is payment for risk.

Mortgagor's Rights

The presence of a mortgage on real property does not, of course, prevent reasonable use of the property by the mortgagor or disposal of it subject to the mortgagee's rights. The mortgagor may use and enjoy the mortgaged property in whatever way he or she wishes, as long as the use does no harm to another, including the mortgagee.

The mortgaged property may be willed to another, sold, or given away. If the grantee takes the property merely *subject* to the mortgage, he or she will not be held to have assumed personal liability. In other words, in case of default, a deficiency judgment might still be obtained against the original mortgagor. On the other hand, if the grantee *assumes* the mortgage, he or she should be held to replace the mortgagor in all respects. Hence, a seller of property is usually careful to be sure that he or she has no further obligation if the buyer later defaults.

♦ Land Contracts

An arrangement sometimes used in the purchase of real property is the *land contract*. It resembles a mortgage quite closely under the title theory. According to a land contract, the purchaser, in addition to making a down payment (if any), agrees to make a series of payments. When the buyer reduces the balance to some agreed amount, frequently half the purchase price, the seller will deed the property to the buyer and take a first mortgage.

Land contracts typically include the right of the seller to declare all payments due immediately if the purchaser misses a payment or makes a late one. This, of course, has the effect of forcing forfeiture by the buyer if he or she misses one payment, since the buyer can rarely pay the entire balance in such a situation. If the buyer gives back the land upon default, the seller takes it back with no problem of foreclosure proceedings and public resale. The seller is still owner and merely takes over the property.

This removal of the buyer and repossession of property by the seller with no balancing of the equities involved is known as *strict foreclosure*. Strict foreclosure is allowed in connection with both mortgages and land contracts where the prospective buyer has acted very improperly or is insolvent. State laws govern the handling of land contracts as well as mortgages. Where the buyer has acted in good faith under a land contract and has made only a slight default, the equities will usually be balanced in some manner—but the buyer must ask for such relief. If the buyer merely returns the land to the seller, the buyer gives up any chance to recoup a part of the loss; in other words, the buyer's equity is cut off.

♦ Eminent Domain

A private party has a lesser right to possession and use of property than does a public body. The government's right to take private property for public use is known as *eminent domain* (or condemnation). The right of eminent domain may be exercised by the state, a municipality, or another public entity, but the right is not limited to these bodies. Quasi-public enterprises or private businesses whose functions serve the public at large (e.g., railroads or power companies) also may have this right.

However, the United States Constitution prohibits a taking of private property without just compensation.

Compensation is usually made according to an assessment of the market value of the property taken.

♦ Dedication

When land is required for public use such as a road or school playground, many owners will donate land for the purpose. Such donations are known as *dedications*. Although dedications are usually made expressly by the owner to the public officials involved, they may also be implied. For example, if public use of private property is made continuously for a period of 20 years or longer, dedication may be conclusively presumed.

No formality is required in the offer to dedicate a piece of property to public use. Similarly, no formality is required for acceptance. To complete a dedication, though, there must be some kind of acceptance. When the offer is expressly made, it is usually answered by expressed acceptance. Acceptance may be implied, though, from public activities, such as maintaining the dedicated property. Public maintenance of a privately owned but publicly used road, for instance, would indicate acceptance of the road.

It may seem unnecessary to determine whether a particular piece of property has been dedicated to the public use. The question of tort liability, though, makes title determination important. If someone is injured because of a large hole in the road, the question arises as to who owned the road and who, therefore, had the duty to maintain it.

♦ Adverse Possession

Title to real property may be acquired by *adverse possession*. Legal requirements generally make it quite difficult to acquire title to land in this manner; however, if the requirements are met, a new title to the land is issued to its possessor. That is, adverse possession results not in a transfer of present title, but in an entirely new title being issued.

The right to take title by adverse possession results from the theory in law that doubt and uncertainty as to title to anything should be removed. Reasoning from this, owners should be reasonably diligent in defending their rights.

There are four general requirements to take title by adverse possession:

- Possession must be open, and it must be notorious, actual occupation. The possessor must occupy the realty in the same manner as one might expect of the true owner.
- The possession and occupation must be adverse to the interests of the owner. Thus, a tenant or lease-holder could not obtain title to the realty by adverse possession.
- The adverse possession must be continuous over the statutory period required. For example, if the state statute requires 20 years (as a large number do), two 10-year periods separated by a period when the property was occupied only by the true owner would not suffice.
- There must be either a *claim of right* or *color of title* by the possessor. *Claim of right* is interpreted from acts of the possessor such as improving the land or fencing it. *Color of title* is some symbol of claim of ownership that is in some way defective. Suppose that a former owner deeded the property to two different purchasers. Each would have color of title due to the deeds. Payment of taxes on the property possessed is required by some statutes to show claim of right or color of title.

♦ Prescription

Gaining prescriptive rights is just about the same as gaining title from adverse possession. The only major difference in most states is that prescriptive rights do not give a person ownership of real property. Instead, prescription deals in rights *involved with* real property, particularly easements. As with adverse possession, the use of the property must be open and notoriously adverse to the owner. For example, Black, for many years (a sufficient number according to the state statute), has crossed White's land to get to his own. Black has established an easement by *prescription*. He does not own the path across White's property, but he has the right (called an *easement*) to continue to cross it in the same manner in which he is accustomed to crossing it.

FORMS OF OWNERSHIP

Just as there are many types of property, there are many forms of legal ownership of property. Because engineers may become involved, either personally or professionally, in transactions involving property (whether a house or a patent), the following discussion outlines different types of property ownership.

♦ Trusts

Titles to both real and personal property may be involved in a trust relationship. Trusts involve at least two people—the *trustee*, who either holds or sells property, and the *beneficiary*, who is to benefit from the trust. There are two titles to trust property. The trustee has *legal title*, with the right to sell or otherwise use the property involved. The beneficiary has *equitable title* to the property, since the trustee is required to handle the property involved for the benefit of the beneficiary. Equitable title is regarded by equity as the real ownership, even though legal title is vested in someone else.

♦ Rights in Common

Ownership of property by several persons can arise under at least five situations: partnership, joint tenancy, tenancy in common, tenancy by the entireties, and community property. In each situation, the rights of more than one person are involved in any property dealings.

Partnership

A *partnership* is an association of two or more persons who agree to carry on a business as co-owners for profit. People often join their assets to more effectively carry on an enterprise. Each partner has rights in the partnership property and in other property acquired by the partnership, and each has attendant liabilities. Every partner has a right to act as agent in the business of the firm, thus adding to or disposing of assets in the transactions for which the partnership exists.

As noted in chapter 18, partners in an enterprise have what is known as *unlimited liability* for the debts of the partnership; each stands to lose some or nearly all of his or her personal fortune if the enterprise folds. However, for many purposes, the property of individual partners is separate from that of the partnership. When a solvent partnership is dissolved—possibly because of the death of a partner or agreement to dissolve—each partner has a claim to a share of the partnership assets, but no claim upon the property individually owned by other partners. If a partnership becomes bankrupt, the firm's creditors have first claim upon the partnership assets; the creditors of the bankrupt partner have first claim against his or her individual property.

Joint Tenancy

A *joint tenancy* is created by a will, deed, or other instrument naming two or more parties as joint tenants. Under joint tenancy, each tenant has an equal, undivided interest in the property and the right to use the property. A joint tenant may not exclude the other joint tenants from it. The right of survivorship is a main feature of joint tenancies. In a joint tenancy with rights of survivorship, a joint tenant's interest automatically passes upon death to the surviving joint tenants.

A joint tenant may sell his or her share in the property to another, but the buyer then becomes a tenant in common (defined below) with the remaining joint tenants. The buyer is a tenant in common, but the remaining joint tenants are still joint tenants. Since a right of survivorship acts before a will does, a joint tenant cannot successfully leave an interest in the property to his or her heirs.

Tenancy in Common

A *tenancy in common* is about the same thing as a joint tenancy, except that it includes no right of survivorship. A tenant in common may leave an interest in the property to heirs or sell it to someone. The result is merely a substitution of one or more tenants in common.

Each tenant in common has an undivided part interest in the whole property. Each is entitled to a proportionate share in possession, use of, and profits from the property. If one tenant pays property costs, say taxes, that tenant has the right to contribution from the others.

Tenancy by the Entireties

Tenancy by the entireties might be thought of as a special case of joint tenancy. The relation is created by conveying to husband and wife in a conveyance that states "by the entireties." Neither husband nor wife can destroy the relationship without consent of the other. This type of ownership applies only to spouses.

Community Property

Some nine of our states[2] have a somewhat exceptional treatment of property owned by husband and wife. In those states, property acquired by a couple after their

2. Arizona, California, Idaho, Louisiana, Nevada, New Mexico, Texas, Washington, Wisconsin, and, to a lesser degree, Oklahoma.

marriage is known as *community property*. Each has an equal share in it. Upon the death of either party, the other is entitled to at least half the community property or to the entire amount if there is no will to the contrary.

As noted, community property is property acquired after marriage. It is possible, though, for either husband or wife to have and acquire separate property either before or after marriage. The property that each has when they marry remains "separate" property. Also, if one of the two acquires property after marriage by gift, will, bequest, or descent, it is that spouse's separate property. Property obtained through a trade of separate property remains separate from community property.

♦ Real Property Leases

A *lease* is a contract by which a tenant acquires something less than complete rights in real property owned by another, the landlord. The lease itself is a *chattel real*, that is, personal property. Although a lease is a contract, it involves an interest in land, so it is treated somewhat exceptionally.

Creation and Characteristics

The original statute of frauds considered leases as real property transactions and required them to be in writing. The statute of frauds has been changed in various states, however, so that an oral lease contract to run for a year or less (three years or less in some states) is binding.

A lease creates the relationship of *lessor* and *lessee* or, more commonly, *landlord* and *tenant*. The relationship created is not the same as that between a roomer and proprietor of a rooming house, or an innkeeper and a guest. It involves more than these. Tenants are placed in possession of the property to use it as they please (within the limits of the lease agreement), and as long as they abide by law and the lease, their rights to use and enjoyment are about the same as that of ownership.

The provisions of the lease contract bind the parties. This, of course, is fundamental—but it should be noted that if the lease is written, any oral provisions not reduced to writing are probably worth nothing. For example, Black leases a building from White for the manufacture of boat trailers. White orally promises to rewire the building, but the written lease is silent about the wiring. If White does not rewire the building, Black is likely to be in the market for some extension cords. Black might be able to use the oral promise to show fraud in the creation of the contract, but this would be about the only value of the oral promise.

Those covenants expressed in the lease agreement will be adhered to strictly in a court interpretation. If the Black and White lease above contained a statement that Black "agreed to return the building in as good condition as when received, save for normal wear and tear and natural decay," such clause would hardly sound ominous. However, if the building burned down, Black might find himself replacing it or paying for it under this clause in the lease.

A lease runs for a definite period of time or is terminable at will by either party. It is this feature that makes it something less than a freehold estate. There are four general types of leases: (1) a lease for a definite period of time; whether 1 week or for 99 years, it is still a lease; (2) a lease from year to year, month to month, or week to week; (3) a tenancy at will; and (4) a tenancy at sufferance.

A lease for a definite period of time, say two years, needs little explanation. The tenant's rights end with the passage of time. A *lease from month to month*, might arise from the expiration of a lease that was taken for a definite period of time. A tenant who continues to hold the property with the consent of the landlord after the original lease expires has a lease of this nature. The rent periods in the original lease dictate how long the tenant's new lease right will last (e.g., if the expired lease was to run a year at a fixed amount per month, the new lease runs from month to month).

When the tenant is given possession in such a way that a lease would be presumed and yet no term is called out, the tenant is said to have a *tenancy at will*. It may be terminated at any time by either party.

A *tenancy at sufferance* occurs when the tenant remains in possession of the property without the landlord's consent after expiration of the lease. The landlord may terminate a tenancy at sufferance at any time.

Unless law or the lease prohibits it, a tenant's rights under a lease contract usually may be assigned to another. The result is known either as *assignment* or *sublease*, depending upon how much of the lease contract was assigned. If the entire remainder of the lease is assigned to another, it constitutes an assignment; if the lessee faits to assign all rights under the lease—he or she assigns only part of them—it is a sublease. In a sublease, the original lessee still has property rights in the lease.

Consider the Black-White lease again. Black, finding the manufacture of boat trailers quite seasonal and rather unprofitable, assigns the remaining term of the lease to Gray. If Black has reserved nothing to himself and has assigned the full remainder of the term, Gray is now bound by the terms of the original lease. If Black has assigned to Gray only a part of the building, or has assigned to him only two of the remaining, say, four years under it, this would constitute a sublease. In such a situation, Gray would not be bound by the terms of the original lease, but by the terms of the new one between himself and Black.

Landlord's Rights

The landlord is, of course, entitled to the agreed compensation for the use of the premises. The landlord also has the right to come peaceably upon the premises for purposes of collecting the rent when it is due. In certain states, he or she has the statutory right to exercise a lien against the tenant's personal property if other efforts to collect the rent fail. As a last resort, the landlord may obtain an eviction order against the tenant, thus removing the tenant, if the rent is not paid when due. If the lease only designates a patch of ground, as many do, the destruction of a building there will not reduce the amount of rent, even though the leased property may become untenantable as a result. At the end of the term, the leased premises must be returned to the landlord in substantially the same condition as when leased. Though the landlord is not allowed to interfere with the tenant's enjoyment of the property, the landlord may, after notice, inspect the premises for waste. Also the landlord has a right to come onto the premises for purposes of repairing damage.

The landlord has available an action for waste against the tenant if waste can be shown. But the landlord's available action does not end with the tenant. The landlord still owns the land; she or he is *remainderman*, meaning that she or he is said to have a reversionary interest in the property. The landlord, therefore, has a right to prevent third persons from injuring the property or obtaining easements upon it. The law will support an action by the landlord for recovery or preventive relief as the case may be.

Tenant's Rights

The tenant has a right to the property she or he has leased, free from interference by the landlord. The tenant also has a right to the appurtenances on the property, such as buildings, if such was the intent of the lease. Whether the leasehold will serve the tenant's purposes or not is beside the point, unless fraud or concealment can be proved. Here the rule of *caveat emptor* applies: If the property would not serve the tenant's purposes, she or he should not have leased it. The tenant, of course, is at liberty to use the property for any purpose she or he wishes as long as the use does not violate the law or a provision of the lease.

A tenant often must improve the premises to suit them to her or his purpose. When the term of the lease expires, who owns the improvements? The answer depends upon the extent of improvements and the intent of the parties when the improvements were made. The extent of improvements depends upon two factors: the purpose of the lease and the length of its term. Generally, the landlord is entitled to the return of the property in substantially the same form it was when she or he leased it. A lease to run 100 years, for example, would allow a great deal more alteration than a 1-year lease. If the stated use of the premises or restrictions in the lease make changes obviously necessary, agreement to those changes will be implied. Many leases provide that the landlord owns all improvements. This would show the parties' intent and the landlord would indeed own the improvements.

Taxes and assessments are normally paid by the landlord. If the tenant must pay real estate taxes or assessments to retain the leased property, she or he has a choice of two remedies: (1) pay the agreed rent and maintain a damage action against the landlord or (2) set off the payments against the rent. Of course, the landlord could shift responsibility for taxes and assessments to a tenant as part of the lease agreement.

Liability

The tenant is liable for injuries to her or his employees, guests, or invitees to nearly the same extent as an owner of the property. The tenant has a duty to such persons to keep the premises in a reasonably safe condition.

Unless the law or a covenant in the lease requires it, the landlord is generally under no duty to repair the premises. It follows, then, that the landlord was usually not held liable to outsiders for injuries sustained by them. In the past, it was only rarely that the landlord was held liable for injury to a tenant. When a defective condition of the premises was known to the landlord, and the landlord did not reveal it to the tenant, the landlord was held liable. However, there seems to be a recent trend in which landlords are being found liable more often.

♦ Easement

An *easement* is an interest in land. It gives a person a right to do something with the real property of another or a right to have another avoid doing something with her or his property. An easement is heritable, assignable, and irrevocable. These features distinguish an easement from a *license*, which is the revocable and unassignable permission or authority to use the property of another. A valid license may be given orally.

Consider this example. Black and White own adjoining property. Black has secured written permission to cross White's land. The writing states that the right to cross White's land pertains to "Black, his heirs or assigns." Such a grant would be held an easement, since it is capable of being assigned to another. If Black sells his land to Gray, the easement may be transferred to Gray in the sale. A license, on the other hand, could not be transferred.

There are several types of *natural easements*. An owner of a building, for example, owns a right (in the form of an easement) to prevent a neighbor from excavating in such a way as to cause her or his building to tend to fall into the hole. This is known as the right of *lateral support*. For further example, if Black sells White a piece of property that is completely surrounded by Black's property (there is no other means of access except by air travel), White has an implied natural easement across Black's property. This is also known as a *way of necessity*. Although the idea did not become popular in the United States, a natural easement to light and air developed in English law: No structure could be built that interfered greatly with the natural light available to a neighbor.

Easements, other than natural, are created by grant or prescription. The grant may be in the form of a deed of the easement right itself, or of a covenant in the deed that transfers the real property. All manner of easement rights are created by grant or prescription—roads, power lines, gas lines, or sewers may cross land under such easements. Even raising the water level and inundating part of someone's land by damming a stream may involve an easement.

WATER RIGHTS

A person who owns a piece of real property ordinarily owns the things on it and under it as well. From this, the conclusion might logically be drawn that water on and beneath the land belongs to the owner of that land. Actually, the rights may or may not extend to ownership of the water, depending upon the jurisdiction. Water rights are handled in several ways, ranging from individual ownership to state ownership.

The question of water ownership has lately become more and more pressing, as our population has increased considerably, bringing with it an increased demand by each individual for water. In some areas, such as southern California, water rights are hot political issues. As public demand for water increases still further, we are likely to see continued legislation in this field. Although laws and court cases lack uniformity regarding water rights, some generalities may be stated.

♦ Boundaries

The extent of real property is often limited by a watercourse or a body of water. Under common law, the defining of property limits in such a manner depends upon the nature of the body of water.

If, for example, a nonnavigable stream separates the property of two *riparian owners*—owners of property bordering on a stream or other body of water—each owns to the center of the stream channel, or to the "thread of the stream," as it is known. Shifting the stream to a new channel does not change the rights of the two owners. The property line remains as before if the channel change comes about suddenly or in such a way that the old channel may continue to be identified.

In contrast, riparian owners *are* entitled to additions to their property that come about gradually as a stream adds to one shore or another. If, as a result of these natural accretions, the channel gradually changes, the dividing line of the properties will also change; the property of one riparian owner will be extended at the expense of the neighbor across the stream.

The owner of property bordering on water affected by the tides generally owns only to the high-water mark. The foreshore (the land between the high-water mark and low-water mark) is public property, belonging to the state. Land bordering upon navigable lakes or streams is generally owned to the low-water mark. In addition, the owner usually has the right to build a pier extending to the line of navigability. The difficulty of establishing a fixed property line by the concepts of high-water or low-water marks has led some of the states to establish riparian property lines by other means. Lines so established are not so apt to fluctuate with droughts or floods.

When a stream divides two states, state ownership does not follow any general pattern. One state may own all of the stream or none of it; in other cases, the center of the stream or the channel may be the dividing line.

♦ Riparian Rights and Duties

A riparian owner generally has the right to *reasonable use* of water bordering his or her property. *Reasonable use* is a little difficult to define, however, and each controversy over the right to use water must be decided on its own merits. Generally, it means that the owner may use the water as long as the use has no significant effect upon the quantity, quality, and velocity of the stream. The owner has a duty not to pollute the water that his or her property borders. For instance, the owner cannot dump garbage or sewage into a stream bordering the property. This dumping would constitute an unreasonable use, since it would change the quality of the water with respect to the downstream riparian owners. Domestic use of water (for drinking or bathing) is held to be more important than either agricultural or industrial uses.

♦ Underground Water

Issues concerning water and the rights to its use are not limited to the rivers and lakes on the surface of land. Who, for example, owns water in the soil (percolating water) or underground rivers? Generally, the right to use and the duty not to pollute extend to underground streams as well as surface streams.

A large quantity of water is present beneath the soil as water with no appreciable direction of flow. The results of tapping this source of water are often quite unpredictable. The owner of land generally has the right to drill a well and capture a quantity of this water for his or her own use. However, if in so doing the owner lowers the level of the water table so that a neighbor cannot get water, an injury is apparent, and the neighbor's cause may be actionable. Where a watershed supply is quite limited, the court will have a tough time assigning the rights to the water.

Irrigation has been the salvation of many areas of our country. If water must be pumped from a watershed, though, the lowered level of the water table may harm neighboring communities. Not all the water returns to the soil; much is lost by evaporation from the ground surface and through the leaves of the plants fed. The states have varying rules that govern the extent of a landowner's rights to such underground water.

A property owner often cannot raise the level of groundwater and do harm. The damming of a stream, for example, could result in flooding nearby. Similarly, rules often restrict a landowner's ability to cause water due to rain or the melting of snow to drain onto another's property.

♦ Prior Appropriation

Our western states have far more serious water problems than the eastern states. The common law rights mentioned above seem satisfactory where water is plentiful, but other rules have developed in the West.

The rights of *prior appropriation* (roughly—first come, first served) and prescription predominate in the West. Under this view, the first of two or more persons to appropriate water for his or her own use has the superior right to continued use. Continued use of water for an extended period of time, even by a nonriparian owner, gives one the right to further use. If a person wishes to appropriate a large amount of water for his or her own use, that person must first obtain a permit to do so. In this way, use of the water is controlled for the benefit of the public.

♦ ---

FAIRFAX V. VA. DEP'T. OF TRANSP.

247 Va 259, 29 ALR5th 759 (1994)

Lacy, Justice.

In this appeal of a condemnation award, we consider the appropriate basis for valuing land subject to a trust agreement providing that ownership of the land vests in a church if it is used for purposes other than a park.

In 1970, David Lawrence executed a trust agreement dedicating a 639-acre parcel to be used as a park. The trust agreement names Fairfax County Park Authority (FCPA) as the beneficiary in possession of the trust so long as the property in the trust is used as a public park dedicated to the memory of Ellanor C. Lawrence. If the property is used for any other purpose, title to the land passes to the trustees of the St. John's Episcopal church, who then own the land free from any restrictions. In the event of condemnation, the trust agreement requires the FCPA to contest the condemnation proceedings in "every fashion reasonably possible," but condemnation of a portion of the park land does not pass the remaining property to the Church.

In 1988, the Virginia Department of Transportation (VDOT) filed a certificate of condemnation for approximately 13 acres of the park property to expand and improve existing roadways. The certificate subsequently was amended to reduce the taking to 2.6497 acres in fee simple, 1.0584 acres for a temporary easement, and .3685 acre for a drainage easement.

VDOT filed a motion in limine seeking to establish the criteria for valuation of the condemned park property. The trial court ruled that the measure of compensation was the property's fair market value as restricted by the trust. The parties waived a hearing before the commissioners and presented evidence to the trial court of the value of the condemned property restricted to use as a park.

VDOT's appraisal witness, Edward S. Williams, III, testified that the value of the property used as a park or an open space was $2,125 an acre based on four sales he considered comparable. Clyde A. Pinkston, the appraisal witness presented by FCPA, testified that there was no market for park land and, therefore, the property, if restricted to use as a park, had no market value. If used for residential purposes, Pinkston testified that the condemned property would have a market value of $125,000 an acre.

The trial court, treating the land as park land, held that there was a market for the property as park land and accepted Williams's appraisal of $2,125 an acre. The court entered a final order setting the condemnation award of $6,450. We awarded FCPA an appeal.

We first consider FCPA's contention that the trial court erred in holding that the value of the condemned land must include consideration of its restricted use. In reaching this conclusion, the trial court acknowledged that cases in other jurisdictions have taken contrary positions. Some jurisdictions value the land based on the restricted use while others disregard such restrictions....The trial court elected to follow those cases which required consideration of the restricted use because of their factual similarities to this case. In each instance, the land taken comprised only a small portion of the condemnee's parcel. In our opinion, this factor—the amount of land taken relative to the amount left the owner—has little relevance to determining just compensation due as measured by fair market value, the standard mandated in condemnation proceedings in this commonwealth....Therefore, we must look to other grounds for determining the appropriate treatment of the use restriction in valuing the land taken in this condemnation proceeding.

In seeking support for their respective positions, the parties noted that we have not previously determined whether use restrictions such as those present in this case should be applied in determining the amount of the condemnation award. We have held, however, that similar restrictions should not be taken into account when valuing land for taxation purposes.

In a series of cases the Richmond, Fredericksburg and Potomac Railroad Company (RF&P) challenged the assessments of its railroad yard in Alexandria, the Potomac Yard....The assessments were attacked on various grounds, but as pertinent here, RF&P sought reduction in the appraised value because the land could only be used as a railroad yard. Furthermore, as in the instant case, price of the land was subject to an indenture agreement that provided that if the land was used for purposes other than maintenance or construction of railroad tracks, ownership of the land would revert to the United States government. We rejected on various grounds RF&P's argument that the restricted use as a railroad yard and the indenture agreement should be considered in determining the market value of the property....

As relevant here, we held that it is the fair market value of the land, not the value of the land *to the owner*, which is subjected to taxation, Therefore, the market value of the land is derived by considering the various uses to which the land is susceptible, not just those uses to which a particular owner may be restricted. If, however, the land is so committed to a particular use that it cannot be put to another use economically, we held that, under those circumstances, it is appropriate to take the committed use of the land into consideration when determining the market value of the land....

While these principles establish the circumstances under which use restrictions should be considered in calculating the fair market value of property in the context of real property taxation, we see no reason why they should not be applied in the context of the instant case. Fair market value of land is used not only for taxation purposes, but, as we have said, it is the prescribed method for determining the amount of "just compensation" due in condemnation proceedings....To adopt one set of principles for determining the fair market value of real property in a condemnation proceeding and another set to make the same determination for taxation purposes could result in a single parcel of land having more than one fair market value. Such a result would be inconsistent and inequitable and is unnecessary.

These principles are consistent with the condemnation jurisprudence of this Commonwealth. Condemnation is an *in rem* proceeding and, while the land is valued from the point of view of an owner rather than the condemnor, the value established is not the value to the owner personally....A determination of a particular owner's loss relative to that of others is only undertaken in the second step of the condemnation proceeding in which the condemnation award is allocated among those with interests in the property....

Finally, there is no evidence here that the condemned land was so committed to use as a park that it was not economically feasible to put the land to other uses. In fact, the trial court held that the highest and best use of the property was for residential purposes. There were no legal impediments to that use. Nor are future improvements required to adapt it for residential use....Therefore, applying the principles stated above, we find that fair market value of the property condemned in this case should be calculated without regard to the use restrictions placed on it by the trust agreement. In light of our disposition of this issue, we need not address the other errors assigned by FCPA.

Accordingly, we will reverse the order of the trial court and remand the case for further proceedings consistent with this opinion.

Reversed and remanded

ABOOD v. JOHNSON

200 N.W.2d 20 (Neb. 1972)

This is a boundary line dispute between adjoining landowners. The court found generally in favor of the plaintiffs, fixed the boundary line, and awarded damages to the plaintiffs for a fence destroyed by the defendant. The plaintiffs have appealed, contending that the boundary line fixed by the court did not include all the land enclosed by the fence. Defendant's cross-appeal contends that the line should be the lot line.

Plaintiffs were the owners of Lot 6, and the defendant was the owner of Lot 7, both located in Section 12, Township 8, Range 16, Buffalo County, Nebraska. The quarter-section line running north and south in Section 12 was the lot line between the two lots. Lot 6 was east of the line, and Lot 7 was west of it. The Platte River formed the north boundary of both properties, and a county road formed the south boundary. The distance from north to south was approximately 1,500 feet. The plaintiffs first acquired an interest in Lot 6 in 1926 and became the sole owners in 1953. Defendant acquired Lot 7 in 1947. The allegations and evidence of the plaintiffs were that a boundary line fence had been in existence for at least 50 years. It extended the entire distance of the lot line but was located some distance west of it. Plaintiffs' evidence fixed the location of the boundary line fence 61 feet west of the lot line at the south, 49.2 feet west of the line at a point near the center of the properties, and 28 feet west of the line at the north end of the properties.

The defendant contends that the correct boundary was the lot line; that there was no fence which was a boundary line fence; and that any use which plaintiffs had made of Lot 7 was with permission. Defendant has cross-appealed from that portion of the decree which fixes any part of the boundary line at any point other than the platted lot line.

Evidence convincingly established the existence of a fence between the two properties for more than 50 years. The fence was located at a point west of the quarter-section lot line, but the exact distance west of the line at particular locations in prior years was not surveyed or measured. The evidence was undisputed that the plaintiffs had dug and established an irrigation well approximately 40 feet west of the lot line on the south end of the properties and had used it continuously since 1957. The evidence was also undisputed that the plaintiffs had planted and cultivated crops on much of the south half of the disputed area. The evidence was convincing that plaintiffs had built and maintained a silo or ensilage bed at a point some 700 feet north of the south line of the properties. The west edge of the ensilage bed was 49.2 feet west of the lot line.

The evidence is undisputed that in 1967, the plaintiffs and defendant jointly removed approximately 400 feet of fence at the north end of the property to permit gravel operations to continue moving easterly along the south side of the Platte River and from defendant's to plaintiffs' property. In April of 1971, the defendant used a bulldozer and destroyed the fence on the remainder of the property. At the trial, the evidence established that there were some broken fence posts at the south edge of the property 61 feet west of the lot line. There was a fence post on the west side of the ensilage bed approximately 700 feet north of the south line and 49.2 feet west of the lot line. There was a broken concrete fence post some 400 feet further north. A line connecting these three specific locations and extended north for the remaining distance of approximately 400 feet ends 28 feet west of the lot line at the north border. Plaintiffs'evidence was that the fence had been located on that line. There is no evidence that a fence was ever located on the lot line at any point.

The court found generally in favor of the plaintiffs, fixed the boundary line, and directed the preparation of a survey accordingly. The boundary fixed by the court commenced 43.7 feet west of the platted lot line on the south, rather than 61 feet as shown by the plaintiffs' evidence; the next point was some 740 feet north, and at that point the line fixed by the court and the line established by plaintiffs' evidence were identical, both being 49.2 feet west of the lot line. From that point, the boundary fixed by the court ran generally north but toward the east where it joined the lot line at a point some 400 feet further north. The boundary then continued to the north end of the properties on the quarter-section lot line. The court also awarded damages to the plaintiffs for the fence destroyed by the defendant.

A thorough review of the record confirms the court's general findings in favor of the plaintiffs. The boundary line fixed by the court, however, does not fully conform to the evidence. The trial court did not indicate the basis upon which the boundary was fixed, nor make specific explanatory findings. The court determined that the dividing line on the north 400 feet of the property should be the quarter-section lot line rather than any fence line. Comments in the record indicate that determination was made because the plaintiffs and defendant had jointly removed the fence in that area for the sole purpose of removing gravel from both properties in 1967.

We believe this case is controlled by *McCain v. Cook*. . . . It is the established law of this state that, when a fence is constructed as a boundary line between two properties, and parties claim ownership of land up to the fence for the full statutory period and are not interrupted in their possession or control during that time, they will, by adverse possession, gain title to such land as may have been improperly enclosed with their own. See also *Ohme v. Thomas*. . . .

"After the running of the statute, the adverse possessor has an indefeasible title which can only be divested by his conveyance of the land to another, or by a subsequent disseisin for the statutory limitation period. It cannot be lost by a mere abandonment, or by a cessation of occupancy, or by an expression of willingness to vacate the land, or by the acknowledgement or recognition of title in another, or by subsequent legislation, or by survey." *McCain v. Cook*, supra.

Under the evidence here, the plaintiffs had acquired title to the property long before 1967, either by boundary line acquiescence or adverse possession, or both. Plaintiffs' title could be divested by nothing short of a validly executed deed, by adverse possession, or by other legal means not pertinent here.

The county surveyor who testified on behalf of both parties also prepared exhibit 57, describing the boundary fixed by the court. The surveyor also prepared a survey reflecting the fence line as established by plaintiffs' evidence. The survey was exhibit 50, but it was not admitted into evidence because a copy of it had not been furnished to the defendant within the time specified in the pretrial order. The surveyor did testify as to the location of the various points used as the basis of that survey and that he had observed the fence posts at the points shown. Exhibit 50 designates the line on which the dividing line fence was located in reference to the north-south quarter-section line. Exhibit 50 may not be complete insofar as a legal survey description of the property line is concerned, but subject to any such technical completion, it designates the boundary line established by the plaintiffs' evidence.

The decree of the district court was generally correct, but it should be modified by substituting the division line shown in exhibit 50 for the line shown in exhibit 57 and described in paragraph 2 of the journal entry. As so modified, the decree is affirmed.

Affirmed as modified.

REVIEW QUESTIONS

1. Distinguish between
   a. personal property and real property
   b. sale and bailment
   c. warranty deed and quit-claim deed
2. In what ways may personal property be lawfully acquired?
3. What are the requirements for a valid will?
4. Why must real estate transactions be recorded?
5. In what ways are a mortgagee's interests in real property similar to those of a landlord?
6. How are water rights controlled in your state?
7. For what purposes other than condemnation (as in *Fairfax v. Va. Dept. of Transp.*) is real estate valuation appraisal performed?
8. According to the logic in *Fairfax v. Va. Dept. of Transp.*, on what occasions, if any, should the market value of land not be determined by the various uses to which it is susceptible, but only to its restricted uses?
9. Return to the requirements for adverse possession and compare these requirements with the court finding for the plaintiff in *Abood v. Johnson*.

Chapter 20

Intellectual Property

INTRODUCTION

The phrase *intellectual property* is used to describe the types of intangible property that result from the creative exercise of the mind. There are four major legal systems in the United States which govern "intellectual property." These can be categorized as (1) the state laws of "unfair competition" which protect trademarks and trade secrets; (2) the federal patent laws; (3) the federal trademark laws; and (4) the federal copyright laws. In this chapter, we will consider these different legal systems. As we do, keep in mind that each legal system of protection is different and the rights and subject matter protected by each legal system may differ and, sometimes, overlap.

PATENTS

Everything in our civilization that distinguishes us from other forms of life results from man's creative efforts. Some of the most commonplace objects were marvels of invention only a few years back, and inventions keep coming at an ever increasing pace.

Engineers often find themselves in fortunate positions to invent things or to improve upon others' inventions. They are particularly favored by training in science, general temperament, and by the nature of engineering jobs for this endeavor. Inventions generally involve the application of scientific principles to practical problems with a practical solution as the objective. The relationship between invention and the training offered by engineering curricula seems quite obvious. Because of the engineer's training and talents, the engineer should be alert to the possibilities of making inventions, of obtaining protection for them, and of exploiting them.

Suppose that Mr. Black, an engineer, has just invented an entirely new type of internal combustion engine after many months of effort. The engine will run on a most readily available resource and promises to have a longer life than even the best competitors What should Black do to secure for himself the fruits of his creative efforts? A number of systems of legal protection are available, depending on the particular embodiment of his ideas that Black wishes to protect. For example, a patent might protect the engine, while trade secret protection might be useful for the method or process used to manufacture the engine. Probably the best course would be for him to contact an expert on intellectual property, such as a patent attorney or a patent agent. But even when the task of obtaining a patent is delegated to another, there is still considerable information Black should have. What, for instance, may be patented? What is the nature of a patent? Who is entitled to one? What protection does it afford? What are trade secrets and copyrights? What protection do they provide? How does one obtain them? What could prevent an inventor from getting a patent? How long will it take, and what delays can be anticipated? This chapter presents an effort to answer these and a host of other questions that are likely to arise.

♦ The Patent Right

One meaning of the word *patent* is "open or disclosed, obvious or manifest." These terms apply to the rights issued to an inventor by the Patent Office because the inventor must make a full and complete disclosure of his invention. No material feature or component may be withheld. The revelations must be such that a person skilled in the field could, by use of the patent, duplicate the thing patented.

The patent right is often regarded as a contract between the government and the inventor. In consideration for disclosing the invention, the government gives the inventor the right to *exclude others* from making, using, or selling his invention. The right extends for seventeen years from the date of issue, after which the patent enters the public domain. A patent only gives the right to exclude others from making, using, or selling the invention—*not* the right to make, use, or sell it. The reason for this is to avoid conflict with state or other federal laws that might prohibit making, using, or selling such things as those covered by the patent. For example, a statute might prohibit, making, using, or selling things in a particular *manner* proposed by the inventor. For instance, you might invent a truly fantastic insecticide, but the EPA and FDA may have something to say about its manufacture, sale, and use. Also, the inventor should not be able to impose upon the prior rights of another or of the public in general merely because he has obtained a patent for his invention. For instance, if use of the invention injures another, the presence of the patent should not allow the injury to continue.

Patent rights are property rights. Hence, patent rights may be bought or sold, mortgaged, licensed, or given or willed to another nearly as easily as any other personal property. Of course, patent rights are intangible rights.

♦ History of United States Patent Laws

The United States colonists brought the idea of patenting inventions to the new world. Prior to the adoption of the U.S. Constitution, patents were issued by the individual colonies, but a patent issued by one colony secured rights only in that colony. If someone in another colony could obtain the essence of the patent, he was free to make any use of it that he might desire. It is for this reason, among others, that the U.S. Constitution provides that Congress shall have the power "to promote the Progress of Science and useful Arts, by securing for limited Times to Authors and Inventors the exclusive Right to their respective Writings and Discoveries."

The first United States patent law was enacted in 1790. Since then there have been numerous changes of the law and of the governmental department which administers the law. Interestingly enough, the adoption of patent laws has been a consistent concern of Americans since the colonial days. For example, the Republic of Texas enacted patent laws shortly after its independence from Mexico in 1846. Similarly, the Confederacy adopted its own patent laws and issued its own patents while the Civil War raged.

The Patent Office (the correct name is the United States Patent and Trademark Office of the Department of Commerce, but is most commonly called the *Patent Office* or PTO) is now headed by a Commissioner of Patents and is part of the U.S. Department of Commerce. A 1952 Act of Congress (effective January 1, 1953) revised the law and brought it up to date. However, Congress continues to tinker with the laws, such as by passing the Process Patents Act of 1988, which was aimed at strengthening patents covering methods or processes.

♦ The Right to a Patent

The right to obtain a patent was neatly summarized in section 31 of the prior U.S. Patent Act. It stated that: Any person who has invented or discovered any new and useful art, machine, manufacture, or composition of matter, or any new and useful improvements thereof, or who has invented or discovered and asexually reproduced any distinct and new variety of plant, other than a tuberpropagated plant, not known or used by others in this country, before his invention or discovery thereof, and not patented or described in any printed publication in this or any foreign country, before his invention or discovery thereof, or more than one year prior to his application, and not in public use or on sale in this country for more than one year prior to his application, unless the same is proved to have been abandoned, may, upon payment of the fees required by law, and other due proceeding had, obtain a patent therefor.

The present act breaks this summary into separate elements intended to be clearer and more convenient. However, the meaning expressed is virtually unchanged.

> Any person (other than a Patent Office employee) may be granted a patent if the requirements are met. A corporation is a legal person, but it cannot effectively apply for a United States patent. The corporation could acquire patent rights from another and exploit those rights (and they often do), but the patent must first be applied for by a natural person or persons who are the actual inventors. Natural persons include those of foreign nationality as well as United States citizens.

A patent may be issued to two or more persons as joint inventors. Mere partnership in an enterprise or financial assistance does not make one a joint inventor. To be a joint inventor, one must collaborate with the other coinventor(s) and contribute to the conception of an invention.

Assignment

Because a United States patent has the attributes of personal property, its ownership may be assigned to another person. Assignment of the rights is accomplished using a written document, and may take place either before or after the inventor obtains his patent. If the Patent Office is not informed of an assignment (through specific procedures for recording assignment) within three months and the patent is subsequently assigned to another, the first assignee to record the transaction will be the new owner of the patent.

A patent may be issued directly to the assignee. When this is done, it is the patent application that is assigned by the inventor. The inventor must still execute the appropriate oath (discussed later), but any patent resulting from the application will be granted to the assignee.

It is, of course, possible to assign a portion of the rights acquired in a patent. But the interest assigned must be specified in the assignment. That is, assignment of a half interest in the royalties resulting from a patent could be determined and upheld in court.

Shop Rights and Contracts to Assign

Generally, patent rights belong to the inventor. If the inventor used his employer's time and equipment in his creative activities, it seems only fair that the employer should have some benefit from the result. The employer's right to benefit under these circumstances is well established in the law. Some difficult questions arise in this connection, though, in regard to contracts made in anticipation of invention, and inventions created wholly or partly on the employee's own time.

A shop right is the employer's nonexclusive, nonassignable license to use an employee's invention. It is limited to the particular employer involved and does not imply that compensation or royalties will be paid to the inventor. It arises from the employment relationship rather than from an express agreement to assign or license patent rights. Both the inventor and the employer have certain rights relating to the inventions; neither can totally exclude the other.

An employer's shop right to inventions created by his employees on company time and with the employer's facilities is an implied right; in some cases, the right may be based on the idea that it would be unfair to allow the employee to use the employer's resources and then try to refuse the employer any right to use the invention. However, the duty to assign patent rights may be made the subject matter of a contract. Often, an employee signs an agreement in which any patent rights are assigned to the employer. The duration of the agreement may be made to run for a period beyond termination of the employee's services; the scope of the agreement may be made to go well beyond things the employee might invent that would immediately benefit his employer. There is a limit, though, to the all-encompassing extent to which such agreements may be taken. An excessive length of time which may practically force a scientist or engineer to be tied to one employer is considered against public policy. It is common for such agreements to run for a year beyond termination of the employee's services. In essence, though, the question of an employer's rights to a patent obtained by a former employee is determined by when the creative work was done. If the work on which the patent was based was done for the former employer and if he had a right to the assignment of patents obtained by the employee, he may maintain an action for assignment of the patent.

It has become almost standard practice to require new engineers and scientists to agree to assign patent rights. Some companies even include an agreement to assign patent rights in their employment application form. In many instances this is merely an added precaution taken by the company. If an employee has been specially hired and retained to do research aimed toward obtaining patents, the patents so obtained must be assigned to the employer, even without an agreement to assign. On the other hand, the right of an employer to patents obtained by someone not hired to invent is a matter of shop rights, unless there exists a contract to assign. Because companies frequently move their technical personnel into and out of research as occasions demand, the precaution of an agreement as to patents seems well founded.

Consider Mr. Black and his engine. Assume him to be an engineer for the White Manufacturing Company. If Black developed an engine as a result of research and development endeavors for which the White Company paid him a monthly salary, Black's patent would have to be assigned to the company. If he was hired as a manufacturing engineer, but spent part of his time developing the engine with company facilities, the company would be entitled to at least shop rights in the engine. With shop rights the White Company could still make, use, and sell the engine even though the patent was granted to Black.

Mr. Black may have used his own facilities in developing the engine at home in the evenings. Would the White Company then have a right to his invention? Yes, if Black was hired to invent or if he had agreed to assign patents to the White Company as a condition of his employment. An oral contract to assign future patent rights may be enforced if the contract terms can be proved.

Invention or Discovery

To obtain a patent, the subject matter must constitute an invention. *Invention* and *discovery* are often used synonymously, despite differences in meanings. *Discovery* refers to the recognition of something in existence that has never before been recognized, whereas *invention* refers to the production of something that did not exist before. For ease, this chapter follows the more common usage of referring to an invention and a discovery as the same thing. To be patentable, an invention must not only be new or previously unrecognized, it must also be useful, and something which is not obvious to or normally predictable by a person of ordinary skill in the field involved.

There is no fixed yardstick of unobviousness in law. Each case is judged on its own merits in the light of similar cases and the present state of the art in the field of the subject matter of the proposed patent. That which might have been patentable at the turn of the century is now commonplace to journeymen in the field involved. In the final analysis, it is the judgment of the court that determines whether true invention or discovery is present. Because the judgment is a human opinion, it is open to the criticism that it could be fallible and, in some instances, arbitrary. True though this may be, no adequate substitute for human judgment on the question has yet been found.

New and Useful

A patentable invention cannot be something that is already known and used by others in the art involved. Description of the invention or discovery in a printed publication either in the U.S. or abroad can be a bar to obtaining a patent. Certainly, if the subject of the invention was published by another prior to the "inventor's" conceiving it, it cannot be said to be new, or that person's invention. Publication by the inventor more than a year prior to his application also operates to bar him from obtaining patent rights. Publication consists of making information available to the general public. It is not held to occur if the information is given to a restricted group of persons with the express or implied condition that the information is not to become public knowledge. Examples of prior art publications include previously issued patents, which are publicly available.

For a publication to bar patenting an invention, it must contain more than a mere reference to a new idea. Sufficient details must be present to enable the reader to make practical use of the idea. Black, for instance, might safely refer in a published article to a new internal combustion engine in very general terms. As long as he does not reveal the invention, the year's limitation "clock" does not run.

Public use of the invention more than a year prior to filing the patent application also defeats patentability. If Black started manufacturing his engines for public sale, time would start to run against the year at least from the date of his first sale (or the date he first offered to sell one of the engines). Of course, there are many shades of gray between black and white. For instance, manufacture of a limited number of machines for experimental use by certain persons is not likely to be held public use. This is especially true where the inventor has told the users of the need for secrecy regarding the invention. Court cases have held that use can be experimental even though the inventor may have made some profit in the transactions. Experimental use is determined in part by the intent of the inventor and the nature of the invention. Some things must be tried publicly. Experimental use of a new roadsurfacing material, for example, would almost necessarily require public use in a street or highway if truly typical conditions are to be encountered.

The invention must also be useful. (Note that a *design* patent, however, must be ornamental, not useful. Most patents are "utility" patents; hence, the requirement of usefulness.) The word useful is given a rather broad interpretation; essentially, it is required that no harmful effect on society would result from the invention. The primary purpose of the entire patent system is to benefit the public. It follows, then, that the inventor of something that offers the public no benefit or something that is harmful or immoral should not be entitled to a patent.

An invention must be operable or capable of use if is it to satisfy the test of usefulness. If there is a question of the ability of the invention to perform as claimed, the Patent Office may require a model to be built for demonstration purposes. Thus, if the examiner had difficulty understanding Mr. Black's use of internal combustion, he might demand a model from Black. However, the requirement of a model is exceedingly rare. It is much more likely that the examiner would require an affidavit from someone who had actually seen the engine in operation. Of course, "perpetual motion" machines are viewed with suspicion. Models of such inventions may be required.

What Is Patentable?

An idea is not patentable, but the machine, process, or thing into which it has been incorporated may be. That is, a physical law or principle, such as Newton's law of gravity, no matter how beneficial it is likely to be, is not patentable as long as it remains an idea or principle. Its physical embodiment might be patented, though, if the other tests of patentability are met.

There are seven categories of inventions which may be patented:

1. An art or a process or method of doing something—for instance, a new type of heat treatment of steel alloys to obtain certain physical properties.
2. A machine—an inanimate mechanism for achieving a certain result, such as transforming or applying energy (such as Black's newly conceived internal combustion engine), is a machine.
3. An article of manufacture—anything (other than a machine) made from raw material by hand, machine, and/or art. Many manufactured products serve as examples—a golf club, ash tray, paper clip, and paper pulp.
4. A composition of matter—examples include a new dental filling material to be used in place of dental amalgam or gold, a new drug and an insecticide.
5. A new and useful improvement of any of the above. Original patents in a field are often followed by numerous improvement patents. For instance, Morse's original telegraph patent was followed by over 5,000 improvements. Similarly, the basic patent on radio receivers, the automobile, and plastics each began a long parade of patented improvements.
6. A new variety of plant—roses, camellias, hybrid corn, and so on.
7. A design—a particular pattern, form, or contour of a product may be patented. Cloth, door chimes, soft drink bottles, and packages for goods have been subjects of design patents. Design patents run for fourteen years.

Debate continues over the proper legal scope of these categories, as well as whether the legal scope of the categories is appropriate as a matter of public policy. Should patents on man-made microorganisms be allowed? What about genetically engineered oysters or cloned mice? What limits are needed? What limits are realistic? Compared to such difficult ethical issues, the legal questions begin to look fairly straightforward.

THE PATENT APPLICATION PROCEDURE

The procedure involved in applying for and obtaining a patent ostensibly allows an individual inventor to obtain a patent on his invention without outside help. Why, then, should a prospective patentee spend money to obtain the services of a patent attorney or a patent agent? There are two main reasons for the expenditure: first, although the preparation and filing of a patent application (as outlined here) may appear fairly simple, technicalities and complications may arise which require knowledge that only a person with specialized knowledge and skill as to the applicable rules could possess. Second, unless the patent is properly prepared to begin with, the patentee is likely to find the wording he used (which was so clear to him) actually offers little or no protection.

Usually, the patentee hopes to make money on the potential patent by selling or licensing it to another, by using it himself, or by manufacturing and selling the product himself. An application for a patent, then, is usually based on economic motives. Inventors try to get the greatest possible coverage for their inventions. The extent of protection for an invention is largely determined by the wording of the claims. An experienced patent attorney or agent always tries to word claims in such a manner that the broadest possible coverage is obtained.

There is another prominent motive for obtaining patents—one in which the profit resulting from the patent is not so apparent. It stems from the fact that realistic limitations must be imposed upon the claims in any patent. The coverage in the claims of any patent can extend only so far as justified by the nature of the invention and its novelty. The novelty of the invention, in turn, is primarily a function of the amount of prior art on point. Hence, an improved widget will likely have narrower claims if widgets have been around for a long time. To prevent others from entering the field in which the patent is useful, some companies attempt to patent things similar to what is already covered. A simple example might be a patent covering a videocassette recorder, a programmable recorder, a recorder with several read/write heads, a recorder with stereo, and so on. One company might file a patent for each embodiment, thereby getting lots of patents and making it more difficult for others to enter the market. This practice, known as "blocking," is commonly used by many large enterprises. By obtaining numerous patents that cover lots of various features, the owner of the patents makes it harder for someone to enter the market without infringing. Because the practice causes frustration to outsiders, many criticisms have been leveled against it. It is lawful, however, unless the result would tend to give the patent holder an unlawful monopoly over an entire industry.

♦ Preliminary Search

One cannot assume that, just because something is not being manufactured and sold commercially, it is not disclosed by a patent or other publication. Somewhere in the more than 5 million United States patents, or in the multitude of foreign patents, there is quite likely to be something similar to the subject matter under consideration. The purpose of a prior art search is to ferret out any patents or publications relating to the subject matter of an invention to find out if a patent can be obtained on the invention, and, if a patent can be obtained, the potential limits of its claims. Such a search is not a legal requirement; rather, it is a practical expedient. It is quite possible to apply for a patent without a preliminary search, but it generally is not advisable.

♦ The Application

The key of any patent application is the specification (which includes the claims and drawings). The proper fee must accompany the application. Additional fees may be required for various amendments, responses, and the like, and still more fees are due for the allowance of a patent. Once issued, still more fees are due to "maintain" the patent in force.

Specification

The purpose of a patent specification is to clearly describe the invention. The clarity of description must be such that any skilled person in the field to which it pertains could, by using the specification, reproduce the invention and use it. One or more drawings (or possibly a model in unusual situations) may be required to make the invention clear.

Drawings are required in all cases in which they are meaningful. This includes almost every type of invention except compositions of matter and processes. In applications for processes, flow charts are often quite important to show the method clearly. There are special rules which are quite detailed pertaining to patent drawings and, for this reason, most applicants hire specialists to make them. Currently, applicants may submit "informal" drawings which, although not meeting all the detailed requirements, nonetheless provide an appropriate description of the invention during the examination of the patent application. A substantial portion of the body of a specification is usually devoted to describing the various views shown in the drawings and the functions of the components. Prior to the explanation of the drawing details, though, there should be a background of the invention (discussing the area to which it pertains) and a brief summary of the substance and nature of the invention and, possibly, its purpose. The claims follow the detailed description of the invention. The detailed description must set out the best mode of the invention contemplated by the inventor for accomplishing the goals of the invention.

The claims essentially state what is considered to have been invented and, therefore, what is reserved to the patent holder. From the applicant's point of view, the broader the coverage in the claims, the greater the rights which the patent will provide. It would be to Black's great advantage, for instance, to claim and patent "an internal combustion engine." Of course, such a claim would not be allowed; if it were, Black could force manufacturers of any kind of internal combustion engines to cease manufacturing the engines or pay him royalties. The principles of internal combustion engines are well known, however, so Black's claim would have to be restricted to a particular kind of engine. His claim might read "An internal combustion engine comprising..." with the remainder restricting the claim to the elements of the engine which are truly inventive. Copies of issued patents are available to the public. If a particular patent appears highly profitable, more people are likely to try to approach the invention as closely as possible without infringement. It is here that the scope of a patent's claims are really tested. Prospective producers of patented things may go to great lengths to avoid the payment of royalties and may try to "design around" a patent's claims. Of course, if a way around the claims exists, there is no reason why it should not be used.

Because it is difficult to tell in advance just what portion of an invention will turn out to be most important in the ensuing seventeen years, a considerable amount of imagination must be used in drafting the claims. Sometimes an apparently insignificant component of a device becomes more lucrative than the invention itself.

The original claims in the patent application should include some written as broadly as the preliminary investigation will allow. If they are too broad, such claims may be narrowed through amendment. Anything not claimed will be abandoned to the public when the patent issues.

Oath

Another essential component of every patent application is an oath taken by the applicant that he believes himself to be the originator of the thing for which he requests a patent. If appropriate, a number of persons may execute oaths stating themselves to be joint inventors. There are also a number of specific requirements for such oaths.

♦ Examination of the Application

After the patent application is prepared, the applicant submits it to the Patent Office. There, the application will be sent to one of the examining divisions (specializing in particular types of subject matter) and will be handled by a patent examiner. Generally, examinations take place according to the filing dates on the applications. An application may wait for six months or even a year before the examiner gets to it and sends a first "Office Action."

The application is examined for compliance with the law and the invention, as set forth in the specification and the claims, is closely scrutinized. If the form and content of the application comply with the law, the examiner turns next to the questions of novelty and obviousness. The patent examiner searches not only what already has been patented, but also trade publications, newspaper articles, and even mail order catalog descriptions and goods for sale in local stores. The result of the examiner's work is communicated to the applicant or his attorney in a letter known as an Office Action. If the examiner quarrels with the novelty or nonobviousness of the invention or claims, all of the claims may be rejected.

The Office Action may allow some or all of the claims. If all claims are allowed, the patent will be issued promptly. If some claims are allowed and others rejected, the applicant may obtain a patent including the approved claims or he may reword or further restrict the rejected claims to resolve the conflict between them and the prior art cited in the Office Action by the examiner. Of course, the applicant may choose to abandon the application.

Amendment

Amendment of an application that has been wholly or partly rejected as a result of its first examination is nearly always complex. Sometimes a visit to the examiner by the applicant's attorney can clear up misunderstandings or indicate rewording of certain claims to make them satisfactory. Much of the attorney's time will be spent in studying the claims in the application and comparing them with other patents and publications with which the examiner has noted a conflict. Amendments to the application are then made. A separate document must be prepared which answers all of the examiner's objections and makes the appropriate deletions or substitutions to cure the defects. The examiner then makes the corrections upon submission of the amendment. The original numbering of the claims is retained, although some of the claims may have been deleted. That is, if there were twelve claims in the original application and Number 11 were to be eliminated, the last claim would still be numbered "12."

Response

A response to an Office Action (such as the submission of the amendment) must take place no later than six months from the mailing date of the Office Action. If no response is made within the time limit, the application is considered abandoned. Another waiting period of from six months to a year may take place before the examiner again responds with an Office Action. These rounds of Office Action followed by response may continue indefinitely. If a year per round is assumed, considerable time may elapse between the original application and the final outcome. If the applicant persists in his pursuit of a patent, the rounds of Office Action and response will end either with the examiner finally allowing the claims and issuing a patent, or with the applicant's receipt of a final rejection.

♦ Appeal

A Patent Office "final rejection" is not always final. The examiner's decision may be appealed to a higher tribunal. Relatively few applications are finally rejected by the examiner and then appealed by the applicant. Generally, the applicant's attorney and the examiner can find common ground before an appeal is necessary.

There exists within the Patent Office a Board of Appeals and Interferences to which a finally rejected application may be taken. Although the Board of Appeals is a part of the Patent Office, there is no definite tendency for it to support the examiner's position. Frequently, the Board will overrule the examiner and allow claims that were previously rejected.

If the Board of Appeals upholds the examiner's position, a further appeal is still possible. The applicant may take his case to the United States Court of Appeals for the Federal Circuit, or he may file a civil action against the Commissioner of Patents in a United States District Court. If either action is successful, the court will order the patent to be issued. However, just as with court actions involving administrative boards, the Patent Office must be shown to be wrong for the court to overrule the examiner's decision.

♦ Interferences

The United States patent system is unique in that it is a "first to invent" system—the first inventor is entitled to the patent. Most other countries are based on a "first to file" system, where the inventor who files a patent application first is entitled to the patent. A few (very few, actually) of the patent applications filed in the United States become involved in interference proceedings. An interference arises when two patent applications claim the same thing. It may also arise as a conflict between an application and a patent that has been issued for less than a year. An interference is designed to determine who first invented the subject of the patent. As the result of the interference, one adversary wins and may proceed to prosecute the application; the other has nothing but experience as a reward for the trouble.

In an interference, just as in any other court case, the winner is determined by the evidence presented to the reviewing body (the Board of Patent Appeals and Interferences). Records of the conception of and the efforts to reduce to practice the invention must have been made and kept if a contestant is to have a chance of winning. This is one reason why nearly all companies require research staff members to keep notebooks of activities (with pages serially numbered and entries dated), and why the periodic witnessing of the contents is also required. Anyone capable of understanding the contents, by the way, may act as a witness to research notes. Interference actions are taken first to the Patent Office Board of Patent Interferences. Later appeals may be taken to the Court of Appeals for the Federal Circuit or to a United States District Court.

In determining who was the first inventor, two dates become important: the date of the conception of the invention, and the date of its reduction to practice. If one party proves that it first conceived the invention and also first reduced it to practice, that person is entitled to the patent.

Proof by one party that he or she first conceived the invention is not enough. The purported inventor may also be called upon to prove that he or she pursued the idea with reasonable diligence. If the other party was not the first to conceive of the invention, but was the first to reduce it to practice, the party may obtain the desired patent by proving that his adversary temporarily abandoned the idea. Inventions are often developed in secrecy, but too much secrecy can preclude obtaining a patent. Proof of reduction to practice requires testimony of someone who actually saw the invention in operation. (Remember that the filing of a patent application constitutes a "constructive" reduction to practice, as opposed to actually building an operable model, which constitutes an "actual" reduction to practice.)

Assume that Mr. Black's claims in his application for his engine have met with an interference. That is, one Ms. White filed an application for a similar engine and includes claims covering the same engine. The case is tried before the Board of Interferences. White proves by records that she has the earlier date of conception of the invention. If White can now prove an earlier date of reduction to practice, the patent will issue to her. Suppose Black built and ran his engine before witnesses while White was still at the drawing board. White may still get the patent if she pursued her invention with reasonable diligence, since she first conceived the engine. Black's hope for the patent rests on his ability to prove that White temporarily abandoned her endeavor for a substantial period—even four or five weeks might be enough. Here, White's records (if she made and kept them) will protect her rights; without records she could lose what is rightfully hers.

♦ Allowance and Issue

Few patent applications become involved in appeals; fewer yet in an interference. Most patent issues between applicant and examiner are settled satisfactorily in the early stages, and either a patent results or the application is abandoned. If the patent is allowed, the applicant is sent a notice of allowance and the patent will issue in due course.

INFRINGEMENT AND REMEDIES

The grantee of a patent has a lawful monopoly. His right would be virtually worthless, though, if he could not enforce it. A patentee may prevent others from encroaching upon the invention which is the subject of the patent. Often, the owner of the patent may recover damages and seek an injunction to prevent additional infringements of the patent.

In patent infringement actions, the defendant is charged with having made, used, or sold the invention in violation of the monopoly granted by the patent. In such actions, the claims are compared with the allegedly infringing device or process. It is here that unnecessary restrictions in claims are likely to prove costly; if the claim includes one or more elements which are absent from an accused device, there is no infringement. What the patentee has given up in the Patent Office he cannot get back in court; he has only the claims which were allowed. For instance, where a patent claim requires the use of a microprocessor, the use of a hard-wired control circuit in the defendant's device may mean that there is no infringement. If the original claims covered both the use of a microprocessor and a hard-wired circuit, and the patent owner deleted (or canceled) the hard-wired circuit claims in the face of their rejection by the Examiner, the hard-wired device clearly should not infringe.

♦ Defenses

The defendant in an infringement action essentially has two areas of defense. A defendant usually attempts to prove either that the patent is invalid or that the defendant's device or process is not an infringement. All patents are presumed valid. The defendant often attacks the validity of the patent, however, by arguing that the invention was not novel or was obvious or that, during the prosecution of the Patent Office, some "inequitable conduct" occurred, such as a failure to inform the Examiner of relevant prior art. The defendant may also claim there is no infringement. For example, the defendant may admit the acts complained of, but show that the claims are not sufficiently broad to cover these acts.

♦ Remedies

Patent infringement usually gives the patent holder a right to damages and an injunction against future infringements. The injunction is often an important remedy. If the defendant were not prohibited from infringing, an award of damages would essentially amount to forcing the patent owner to give up his or her exclusivity; it would be equivalent to forcing the patent owner to grant a license even if he or she did not wish to. Hence, the U.S. Courts often grant injunctions to prohibit patent infringement. In addition, the law allows the court to go beyond mere compensatory damages and assess up to triple damages against the defendant if there is "willful" infringement. The patent owner's attorney's fees can be covered if the case is "exceptional." Infringement may be considered "willful" if the defendant knew of the patent but did nothing to avoid its coverage or make sure that the defendant did not infringe.

The measure of compensatory damages in an infringement case is the plaintiff's losses. However, the statute requires that the patentee be compensated at least in the amount lost or which would have been lost in royalties because of the infringement. A question may arise as to the amount of royalties to be assessed if the plaintiff has not made previous royalty arrangements with others. The court then determines and awards a "reasonable royalty."

♦ Patent Markings

If the plaintiff in an infringement case is to get any damages for the injury, the plaintiff must have informed the defendant of the patent. Marking patented goods with the patent number serves this purpose. If the goods are not marked, the plaintiff must prove that the defendant was informed of the existence of the patent and that the infringement continued.

Goods sold are often marked "Patent Pending" or "Patent Applied For." These terms have no legal force so far as infringement is concerned. However, if the terms are used when, in fact, no application has been filed, it may be assumed that the purpose was to defraud the public. The patent statute prohibits such acts.

COMMON CRITICISMS

One common criticism is the time and money required to obtain a patent. At first, the criticism seems well founded. As to time, few patents are issued in much less than a year or two after the application is first filed; some have required twenty years or more. This sounds like a long time to wait for patent protection but, as a practical matter, the delay is seldom a hardship, and it may be beneficial to the applicant. The grater the delay in securing a patent, the greater is the period of practical patent protection. Theoretically, the applicant is not protected while his application is being processed. However, should the applicant be threatened with an infringement, the patent office, upon request, will usually expedite examination of the patent to permit legal action upon the infringement.

The monetary outlay required for a patent is extremely variable, depending upon the complications that may arise. It is possible for the patent to cost only a nominal amount. However, it is often likely to cost somewhere around $5,000 or more, just for the attorney's fees. A patent is somewhat like other business ventures; its probable cost in time and money should be weighed against the expected results. If economic analysis shows that the venture is likely to be profitable, it should be undertaken. If the likelihood of profit is remote, perhaps it is better kept confidential and used as a trade secret.

Occasionally, the objection is heard that a small inventor has no fair chance to compete with the research staffs of big business. It is true that research staffs are brought together to form a talented team and many times the effort pays off. But even in a research team the best ideas often flow from one individual or a small nucleus of people. Once the idea occurs it is often pursued more rapidly by teamwork, but if an independent inventor has the first date of conception and has pursued it diligently, the speed of a corporate research team is no help.

Members of industrial research staffs sometimes complain that they are not adequately compensated for patents they obtain and assign to their employers. Perhaps the criticism is valid, for many receive little or no compensation for the assignment. However, the opposite argument is that the staff member is paid not only for the time during which he had produced something worthwhile, but also for the other time when he was not so productive. It is also argued that the research job, considering only the activities involved, is more enjoyable to many people than the alternatives. The employer also has an investment in research facilities which must return a profit.

Corporations often are criticized because, with large funds available, they can indulge in long legal battles to win infringement or interference suits from less fortunate competitors. Sometimes the mere threat of a long legal battle is sufficient to intimidate a small inventor. While this criticism has merit, there is still an end to all legal battles. There is another aspect to this particular problem. Because of the large costs involved and the unpredictability of the outcome of such cases, large businesses frequently pay out of court settlements rather than undertake a long, drawn out court battle. That is, if an inventor claims that a manufactured product infringes his patent and demands a small royalty, the manufacturer may purchase the patent or pay the royalty rather than go to court, even though the particular patent might be only remotely related to his product.

One criticism which seems to have greater validity than most is concerned with the public welfare. Suppose someone invents something of great public benefit—perhaps a cure for cancer. Under the present law, it would be possible for the inventor to obtain a patent monopoly, make very limited amounts of the cure, and sell it at enormous prices. All but wealthy people would be deprived of the cure—a rather appalling prospect.

Other criticisms are directed at attempts to extend the scope of patent monopolies by pooling arrangements, tying clauses, marketing agreements, and the like. Some of these arrangements are lawful; others are not. The courts carefully scrutinize such arrangements under the antitrust laws. Despite shortcomings in the law, protection is afforded to the inventor, and the rate of technological advances testify to its success.

The U.S. Constitution succinctly explains why patent laws are so important—they promote the progress of science and engineering. In essence, the legal patent monopoly acts as an economic incentive to encourage innovation. Again, this view holds that our entire society benefits from innovation, such as by better medicines, cheaper goods, cheaper and more efficient means of production, and so forth.

TRADEMARKS

The good will attached to a trademark or a trade name is an example of intangible property. Trademarks are like other forms of property in that they can be bought, sold, and used by the owner. In the United States, the law regarding trademarks is based primarily on the common law. In addition, the federal Lanham Act of 1946 and state statutes provide trademark protection. Under the Lanham Act, a trademark is defined as including any word, name, symbol, or device or any combination thereof adopted and used by a manufacturer or merchant to identify his goods and distinguish them from the goods manufactured or sold by others. Similarly, a service mark may be any word, name, symbol, or device used to identify and distinguish services.

Trademarks are products of the Industrial Revolution. When goods were manufactured primarily for local sale, there was little need to identify the manufacturer. With specialization in manufacture and expanded marketing, successful advertising came to require that the manufacturer's products be identified in some way. A trademark differs from other property in that its value is in identifying the products of a particular manufacturer and distinguishing the goods of one manufacturer from similar goods of other manufacturers. Many marks and names are successfully used to identify various products, but the more common the mark or expression used, the more difficult it is to make a trademark out of it. It would, for instance, be difficult to obtain a trademark right for the word "pencil" or "automobile" because these words are generic of those types of goods.

A "descriptive" trademark is a mark that describes the goods, their characteristics, functions, uses, or their ingredients. Whereas *restaurant* may be generic, the mark "Steve's Great Chinese Restaurant" is descriptive. The distinction between generic marks and descriptive marks is important, because descriptive marks can, in some situations, be legally protected; generic marks are never protected.

Just as descriptive terms would be difficult to establish as trademarks, geographical names and family names present problems. Descriptive terms, as well as geographical and family names, may become protected as trademarks if they are used and advertised sufficiently so as to identify a source of the goods or services. A new company manufacturing automobiles could not use the Ford name to identify its products, even though the owner might be named Ford. Some states, however, still take the view that everyone has an "absolute" right to use their name in their business. New competitors in a field have a right to identify themselves and the location of their businesses, but not in such a way that the trademarks or trade names of others are infringed.

The strongest trademarks are newly coined terms that have nothing to do with the goods. Strong trademarks also often result from the use of words ordinarily not connected with a description of the specific product. The use of "Bluebird," for instance, would be an arbitrary or fanciful term for a television set. Many companies have made up their own trademarks or trade names. For example, "IVORY" is an arbitrary term for soap.

Rights in any trademark or trade name may be lost if the word, after use, becomes established as a term descriptive of many products. "Aspirin" was once a trademark of only one product; so was "Cellophane." For this reason, companies sometimes advertise in an attempt to persuade the public not to use the mark in ways which generically describe its goods or services.

Trademark rights are established by the first user of the mark; such rights arise out of the use of the mark in connection with the sale of the goods. A person who subsequently uses an established mark or one which is confusingly similar may be enjoined from continuing the use of the mark. If the use of the mark was an intentional violation of another's rights, the court may award damages in excess of actual damages, as well as the plaintiff's attorney's fees.

A mark is "confusingly similar" to another if there is a likelihood of confusion, mistake, or deception. The legal determination of whether a likelihood of confusion, mistake, or deception exists is a subjective determination by the trier-of-fact. The determination is made following the consideration of many facts. Elements considered in making such a determination include these:

1. The similarity of the names or marks in their appearance, sound, connotation, and commercial expression
2. The similarity in the nature of the goods or services of businesses in question
3. The similarity of the trade channels in question
4. The conditions under which and the buyers to whom sales are made
5. The fame of the prior mark including information about sales, advertising, and the length of time the mark has been used
6. The number and nature of similar marks
7. The nature and extent of any actual confusion
8. The length of time during which and the conditions under which there has been concurrent use without evidence of actual confusion
9. The variety of goods on which the mark is or is not used by the parties; and
10. The existence of any agreements between the parties regarding the use of the marks.

The relative "strength" or "weakness" of a mark also becomes important in determining whether a likelihood of confusion, mistake, or deception exists. If many different business entities exist which are using a particular word in connection with their goods and services so that the word has become very diluted or "weak," then a user of that word will have great difficulty convincing a court that some subsequent third party use of that same word should be prohibited on the basis that it is likely to cause confusion. The courts generally conclude that since so many people are already using the word, no harm can result from additional uses of the same word.

The common law established rights in the use of trademarks as well as penalties for abuses. Federal and state legislation (where statutes have been passed) have modified the established rules only slightly. The federal Lanham Act, which covers marks used in interstate commerce, provides for the registration of marks. Under the common law, trademark rights extended only to the geographic areas in which the mark was used. A state registration expands those common law rights to the boundaries of the state. Similarly, a federal registration expands those common law rights to the boundaries of the United States. Since 1989, it has been possible to apply for federal trademark registrations of marks that had not been used as of filing of the application. To obtain the registration, however, the applicant must begin using the mark and submit proof of such use to the Trademark Office. Registration of the mark continues for 10 years; if the mark is still in use after the 10-year period, it is renewable upon application by the party using it.

♦ Licensing and Franchising

The term "license" means a type of permission to do something or, in the case of a trademark (or other type of intellectual property), the right to *use* the mark (or other property). A "franchise" usually refers to a specific type of license. The complexities of license arrangements range from a basic license to use a trademark to a complex system in which the licensee receives the right to use the trademark, confidential know-how useful in the production of a good or service, and the benefit of widespread promotion. The latter arrangement probably would be considered a "franchise." Licensees (those who have a license from another) must submit to strict control of the quality of their product. Licenses and franchises are generally deemed beneficial to the public in that they tend to promote competition and assure a certain standard of quality. They provide an efficient means of combining centralized planning, direction, and standard-setting with a degree of local control and initiative.

The franchising coin has another side, however. If the use is such as to defraud the public, the trademark involved will not be protected. Consider the Brown Construction Company, a company so eminently successful in construction contracting in a particular city that it expands to new markets. Because its trade name and logo have become symbolic of high quality in construction, it expands by a licensing arrangement in which it reserves the right to dictate the company management and quality of construction, requiring a monetary return for the use of its well-promoted name. (Similar arrangements are common in the fast food, automotive repair, and numerous other industries.) Suppose the services provided by the Brown Construction franchisees come to be far different from what the original reputation and the promotion imply. Under such circumstances, Brown risks not only legal actions against it for the poor performances of its franchisees, but also the loss of goodwill of the trademark. Thus, Brown has legal as well as a practical incentives to be cautious in licensing its trademark and in monitoring the quality of its licensees.

COPYRIGHTS

The United States Constitution provides Congress with the right to enact copyright laws. Over the years Congress has exercised that right and created numerous different copyright acts. In 1976, Congress enacted the Copyright Revision Act of 1976, which took effect on January 1, 1978, and completely revised the law in many areas.

♦ The Subject Matter of Copyrights

Perhaps the easiest way to grasp the concept of a copyright is to visualize the distinction between the original physical copy of the author's work and imagine, hovering over that work, the intangible rights of copyright provided by the federal copyright system. While someone might purchase the original physical copy of the author's work, such as an original painting, and have the right to look at and enjoy the painting, unless they also own the intangible rights of copyright in that work, they do not have the right to make copies of the work, to use the work to prepare derivative works, or to distribute copies of the work to the public. They do not even have the right to publicly display the work except under certain conditions.

Under the new copyright system, the copyright laws automatically provide copyright protection to all original works of authorship which are fixed in any tangible medium of expression. The requirement of "originality" is met by most works, so long as they embody some creative effort. Occasionally, however, a court will hold that a work fails to meet the originality requirements, especially if the work consists of factual or historical information. As soon as the work is fixed in some tangible medium of expression, the copyright laws come into play. If the work is kept only in the mind of the author and is not fixed in some tangible medium of expression, the federal Copyright Act does not apply. But once the work is reduced to writing or recorded on a tape or in a photograph or the like, the federal Copyright Act automatically applies and the intangible rights of copyright exist.

The federal copyright laws do not govern the ownership and transfer of the original physical copy of the work. Normal state concepts of contract law govern those matters. However, the federal copyright laws do govern the ownership, transfer, and protection of the intangible rights of copyright that flow from the creation and fixation of the work in a tangible medium of expression.

Congress specifically provided in the new act that the federal laws apply to all "works of authorship" fixed in a tangible medium of expression. Congress defined the term "works of authorship" principally by example, including (1) literary works, (2) pictorial, graphic, and sculptural works, (3) motion pictures and other audio visual works, and (4) sound recordings. It appears likely that many activities will involve the fixation of original works of authorship, thus generating copyrights protected by the federal laws.

♦ The Idea/Expression Dichotomy

Congress specified in the new act that copyright protection for an original work of authorship does not extend to any idea, procedure, process, system, method of operation, concept, principle, or discovery, regardless of the form in which it is described, explained, illustrated or embodied in an original work of authorship. This statutory pronouncement by Congress maintained a long standing distinction evolved by the courts in their interpretation of the federal copyright laws, namely: the federal copyright laws do not protect ideas. They only protect the form of expression of an idea.

Perhaps the best way to understand this distinction is by example. Suppose that one came up with the idea of writing a novel about a debonair gentleman living in the South at the time of the Civil War. This would be an idea. But taking that idea and fleshing it out into a novel about Rhett Butler constitutes the form of expression of that idea. By way of another example, suppose that one came up with the idea for a new system of accounting. The idea for the new system would not be covered by the federal copyright laws. But descriptions in a textbook of the system constitute a form of expression of that idea and are protectable under the federal copyright laws.

Exclusive Rights

The rights afforded a copyright owner are exclusionary in nature; those rights allow the copyright owner certain exclusive rights with respect to the work and the right to prohibit others from engaging in those actions. Under the federal copyright laws, the owner of a copyright obtains five intangible exclusive rights:

- the exclusive right to copy or reproduce the work.
- the exclusive right to prepare derivative works based upon the work.
- the exclusive right to distribute copies of the work to the public by sale or other transfer of ownership, or by rent, lease, or lending.
- the exclusive right to perform the work publicly, if the work is a literary, musical, dramatic, choreographic, audiovisual work, a pantomime, or motion picture.
- the exclusive right to display the work publicly if the work is a literary, musical, dramatic, choreographic, pictorial, graphic, or sculptural work, or a pantomime. This bundle of rights can be broken up, and each right can be sold or licensed independently of the others.

Unlike the federal patent system, the rights afforded a copyright owner do not apply as against everyone in the world irrespective of how they may have happened to have created an identical work. For the copyright owner to recover against an alleged third party infringer, the copyright owner must prove that the infringer actually copied the copyrighted work, or actually distributed copies of the copyrighted work, or actually used the copyrighted work to prepare derivative works. If the alleged infringer can show that he independently created an identical work, there is no copyright infringement.

Moreover, once the copyright owner has sold a copy of the work, unless specifically provided otherwise in an enforceable contract or provided by the copyright laws themselves, there is no limitation on certain of the buyer's rights to "use" the work. For example, when a copyrighted novel such as *Gone With the Wind* is sold to a buyer, the buyer, although he cannot copy the work or use the work to prepare derivative works based thereon, can "use" the work by reading it as many times as he wishes, or he can sell, lease, or rent that copy of the work to another, or he can mutilate or destroy the work. The same is true of copyrighted drawings, manuals, or the product literature a company sells to third parties without any specific limitations on how the third parties may use those works.

Fair Use

One important limitation on the exclusive rights granted to the copyright owner is the doctrine of "fair use." The idea that certain uses of copyrighted works were "fair" and therefore should not be an infringement of copyright in a work first developed in the courts. However, as part of the comprehensive redevelopment of the federal copyright law, Congress codified the fair use doctrine in what is now Section 107 of the Copyright Act of 1976.

Section 107 provides that:

> ...the fair use of a copyrighted work, including such use by reproduction in copies or phonorecords or by any other means specified by that section, for purposes such as criticism, comment, news reporting, teaching (including multiple copies for classroom use), scholarship, or research, is not an infringement of copyright. In determining whether the use made of a work in any particular case is a fair use the factors to be considered shall include—
> (1) the purpose and character of the use, including whether such use is of a commercial nature or is for nonprofit educational purposes;
> (2) the nature of the copyrighted work;
> (3) the amount and substantiality of the portion used in relation to the copyright work as a whole; and
> (4) the effect of the use upon the potential market for or value of the copyrighted work.

Note that the four factors listed are only examples, not an exclusive list. For example, the United States Supreme Court has discussed at some length the notion that the limits of fair use are narrower when dealing with an unpublished work. The Court noted that the questions of when, where, and how a work was to be initially published were important questions reserved to the author of the work. Consequently, the Court held that the publication of excerpts of President Gerald Ford's memoirs in the *Nation*, prior to their scheduled publication in *Time*, constituted an infringement of copyright and was not a fair use.

Of course, the determination of what constitutes fair use varies from case to case, thereby lessening the precedential value of any particular case. In addition, none of the four factors set forth in Section 107 is determinative and the weight given to each of those factors varies from case to case. On the other hand, the effect of the use of the work on the commercial market for the work often is regarded as a most important factor. Because the purpose of copyright law is to provide an incentive for authors, uses which adversely impact the commercial market for an author's work are viewed as frustrating the intent of the federal copyright laws. Consequently, the parameters of fair use are narrowed when a particular use of a copyrighted work is commercial in nature and seems likely to eliminate or adversely affect the market for the copyright owner's work.

♦ Infringement

Because it is often difficult for the copyright owner to prove that the infringer actually copied the work or used the work to prepare derivative works, the courts have developed the doctrine that copyright infringement may be established by circumstantial evidence. The only evidence required to support a finding of infringement is that (a) the alleged infringer had access to the copyrighted work, and (b) the alleged infringing work is "substantially similar" to the copyrighted work. Determining whether or not two songs, books, movies, or other works are "substantially similar" may involve detailed expert testimony, as well as the impression that the two works have on the fact finder.

♦ Notice

For many years, U.S. laws required the use of a statutory notice of copyright to be placed on all publicly distributed copies or other media from which the work can be visually perceived. The copyright notice consists of the symbol © or the word copyright" (or the abbreviation "copr."), the year of first publication, and the name of the copyright owner or an abbreviation by which the name can be recognized. Thus, a copyright notice one might see is as follows:

© Copyright 1999 by Kendall/ Hunt
Publishing Company
All Rights Reserved

The inclusion of *All Rights Reserved* is a safety feature sometimes added to various works to satisfy the requirements of a treaty known as the Buenos Aires convention. Under this treaty, to which the United States and a number of Latin American countries belong, one way to protect a work is by including the phrase "All Rights Reserved." Most countries, such as the European and English-speaking countries, belong to other treaties which do not require this additional phrase. If foreign protection of copyright seems important, it is safest to use the symbol shown here, as well as the name of the owner and the year of first publication.

In 1989, the United States joined a large number of other countries by adopting the Berne Convention. The Berne Convention is the oldest copyright treaty. It was originally signed in 1886 and today includes eighty or so countries. To conform U.S. copyright laws to the requirements of the Berne Convention, Congress passed the Berne Convention Implementation Act of 1988. As of March 1, 1989, this act abolished the longstanding requirement of a copyright notice *for works first published after March 1, 1989*. This act also eliminated certain other formalities that were viewed as in conflict with the obligations of the United States as a member of the Berne Convention.

The act defines the term *publication* as the distribution of copies of a work to the public by sale or other transfer of ownership, or by rental, lease or lending. The offering to distribute copies to a group of persons for purposes of further distribution or public display also constitutes publication.

Under the prior U.S. law, the courts held that *any publication* of a work without the statutory copyright notice injected the work into the public domain, and thereafter anyone was free to copy the work. However, Congress provided that the omission of the copyright notice from publicly distributed copies did not invalidate the copyright in a work if (1) the notice was omitted from no more than a relatively small number of copies distributed to the public; (2) the registration for the work has been made before or is made within five years after the publication without notice, and a reasonable effort is made to add notice to all copies that are distributed to the public in the United States after the omission has been discovered; or (3) notice was omitted in violation of an express requirement imposed in writing upon the copyright owner's licensee or distributor. Because so many copyrighted works exist that were published before March 1, 1989, the prior laws regarding notice are still important.

♦ Registration and its Importance

The Copyright Act provides that at any time during the subsistence of copyright in any published or unpublished work, the owner of copyright in the work may obtain registration of the copyright claim by delivering to the copyright office the required deposit, together with the application and the required fees. The federal Copyright Act also provides that no action for infringement of the copyright in any work shall be instituted until registration of the copyright claim has been made. However, the Berne Act changed U.S. Laws to allow actions to be brought for the infringement of "Berne Convention works whose origin is" outside the United States even if no registration has been obtained. Thus, although rights under the Copyright Act are maintained for new works even if a statutory copyright notice is omitted, another legal action that *must* be taken for a U.S. author to file suit to enforce rights of copyright is registration of the work if it's of U.S. origin.

One advantage in obtaining early registration of copyright is the ability to obtain "statutory damages." The act provides that a copyright infringer is liable for either (1) the copyright owner's actual damages and any additional profits of the infringer, or (2) statutory damages. The act defines the statutory damages that the court may award to the copyright owner as a sum of not less than $500 or more than $20,000 as the court considers just for each infringing work. If the court finds that the infringement was "willful," the court may increase the award of statutory damages to the sum of not more than $100,000 for each infringing work.

The act also provides that the court may also award reasonable attorneys' fees to the prevailing party. No award of statutory damages or of attorneys' fees, however, may be made for any infringement of copyright commenced after first publication of the work and before the effective date of its registration, unless such registration is made within three months after first publication of the work. Consequently, if enforcing a copyright in the courts appears possible, registration is desirable to avoid losing the opportunity to recover statutory damages and attorney's fees.

♦ The Duration of the Intangible Rights of Copyright

The duration of the intangible rights of copyright in a work created on or after January 1, 1978, is as follows:

> (1) Except as provided in subparagraphs (2) or (3), copyright in a work subsists from its creation and endures for a term consisting of the life of the author and 50 years after the author's death.
>
> (2) In the case of a joint work prepared by two or more authors who did not "work for hire," the copyright endures for a term consisting of the life of the last surviving author and 50 years after such last surviving author's death.
>
> (3) In the case of an anonymous work (one in which no natural person is identified as author), a pseudonymous work (one in which the author is identified under a fictitious name) or a "work made for hire," the copyright endures for a term of 75 years from the date of the first "publication" of the work, or a term of 100 years from the year of its creation, whichever expires first.

Once the term of the copyright expires, anyone who lawfully obtains possession of the work or any copy thereof can slavishly copy it. For example, if one could lawfully obtain access to an original Remington western painting hanging in a museum, because the term of the rights of copyright therein has expired one could photograph the painting (thereby creating a derivative work that carries its own independent right of copyright) and sell the prints. For this reason, many museums are very careful to whom they loan their paintings and prohibit their patrons from photographing the works.

♦ Works Made for Hire

The federal Copyright Act contains specific provisions governing the ownership and transfer of the rights of copyright. The act generally provides that the individual who creates the work is deemed to be the "author" of the work and, as such, is the initial owner of all rights of copyright. There are only two situations when some other entity is deemed to be the "author" of the work. Both of those exceptions are referred to as *works made for hire.*

The first exception occurs when the individual who created the work did so in the course and scope of his employment. In that situation, the employer is deemed to be the "author" of the work and is the owner of all rights of copyright. The various federal courts that have addressed this issue recently have reached somewhat different results as to when certain persons are to be considered an "employee" for purposes of the copyright laws. The second exception occurs when the individual who creates the work was commissioned to create the work (on an independent contractor basis) for one of nine specified purposes and a written contract is signed by the independent contractor stating that the work is a "work made for hire." The nine specified purposes are (1) a contribution to a collective work, (2) a part of a motion picture or other audio visual work, (3) a translation, (4) a supplementary work, (5) a compilation, (6) an instructional text, (7) a test, (8) answer material for a test, or (9) an atlas.

Unlike the prior version of the law, the Copyright Act provides that the sale or other transfer of the physical object (such as the manuscript, tape, and so on) in which the copyrighted work is initially embodied (or the sale or transfer of any copy thereof) does not in and of itself transfer any of the rights of copyright. The Copyright Act provides that the only way to effect a valid transfer (other than by operation of law, such as when the owner dies) is by a written instrument of conveyance signed by the owner of the rights conveyed or his duly authorized agent.

Applying these principles, if an individual employed by a company creates a copyrightable work in the course and scope of his employment, then the company is deemed to be the "author" of that work, and is the owner of the work as well as the exclusive rights of copyright associated with the expression used in the work. This means that the company has the right to prepare derivative works from the original work, to make copies of the work, and to distribute those copies by sale, license, or otherwise. The employee who created the work retains none of these rights.

However, if an employee of the same company creates a copyrightable work outside of the scope and course of his employment, then the employee is deemed to be the "author" of the work and the initial owner of all of the exclusive rights of copyright. The company may, depending on whether the individual used the company's time, facilities and materials, have a right to use the *ideas* embodied in the work. However, the company will not have the right to reproduce, copy, or market the form of expression of the ideas covered by the rights of copyright.

The result is perhaps even more dramatic when a company commissions an independent contractor to create a work under his own control that is the subject matter of copyright. Assume that a corporation commissions a professional photographer to make photographs of the corporation's equipment for use in the corporation's company report. The photographer provides the corporation with the negatives. Assume that a written contract that included the magic words "work made for hire" was not utilized and that such photographs and the company report do not fall within any of the nine specific uses provided by the Copyright Act. In this case, because the photographer is not an employee of the corporation, the photographer is deemed to be the "author" of the works and the initial owner of all of the rights of copyright. While the corporation will have the right to use the photographs for the specific limited purpose authorized by the photographer, the corporation will not have the right in the future to reproduce the work, or to use the negatives to create additional works, or even to make copies of the work without the approval of the photographer. At an extreme, the photographer might be able to prevent the copying and distribution of the reports with his or her photographs.

♦ Architectural Works

More recently, Congress passed the Architectural Works Copyright Protection Act of 1990. For many years, our copyright laws protected architectural plans, drawings, and blueprints. However, protection for the design features of a building itself were not protected. Hence, you could freely copy the overall shape, style, or design of a building, but you could not copy, for instance, the blueprints for the same building.

When the United States decided to join the Berne Convention, the United States was required to protect "architectural works" to meet its treaty obligations. Thus, the Architectural Works Copyright Protection Act was adopted to meet the requirements of the Berne Convention. An "architectural work" is defined by our copyright law as

> The design of a building as embodied in any tangible medium of expression, including the building itself, architectural plans, or drawings. The work includes the overall form as well as the arrangement and composition of spaces and elements in the design, but does not include individual standard features.

Obviously, deciding what constitutes the "design" or "overall form" of a building sometimes is easy. In other cases, it will be difficult. Also, what are the "individual standard features" that should not be protected? Are these items things such as doors, windows, or stairs?

The owner of a copyright in an architectural work has the exclusive right to make and distribute copies and to prepare derivative works. However, these rights are subject to "fair use" by others. In addition, Congress adopted new laws to specifically provide that making, distributing, or displaying pictures, paintings, or photographs of a building that has been built is *not* an infringement. However, this provision applies only when the building is located in or is ordinarily visible from a "public place." For similar practical reasons, Congress adopted a provision to allow the owner of a building to make or authorize alterations or the destruction of the building.

The implications of this new act are far from clear. As is often the case with new laws, we can expect clarification of the scope of protection for "architectural works" to come from the courts.

TRADE SECRETS

Trade secrets laws serve two broad policies: They provide a vehicle for the maintenance of standards of commercial ethics and encourage innovation by providing protection that is supplemental to that offered by the federal patent and copyright laws. Trade secrets laws offer legal protection for an extremely wide variety of different types of information. Trade secret protection offers several advantages over other forms of legal protection. Unlike a patent or a copyright, a trade secret has no limited statutory life; trade secret protection may last forever. Moreover, trade secret protection extends to practically all types of information. Another advantage of trade secret protection over other forms of protection is that a trade secret is protected without any need to file any application for a registration or without any other formalities. Moreover, trade secret protection is secured without publicly disclosing the secret, in contrast to a patent, which discloses the information to the public.

In the United States, Section 757 of Restatement of Torts has been widely followed by the courts in trade secrets cases. More recently, a majority of the states have adopted the Uniform Trade Secrets Act (the "Uniform Act") in one version or another. Under both the Restatement and the Uniform Act, one who discloses or uses another's trade secret is generally liable if he or she discovered the secret by improper means or the disclosure or use constitutes a breach of confidence.

♦ Trade Secret

The Restatement defines a *trade secret* as follows:

> A trade secret may consist of any formula, pattern, device or compilation of information that may be used in one's business, and that gives him an opportunity to obtain an advantage over competitors who do not know or use it. It may be a formula for a chemical compound, a process of manufacturing, treating or preserving materials, a pattern for a machine or other device, or a list of customers....

As one would expect, some element of secrecy must exist to obtain relief. The information claimed to be a trade secret must not be a matter of public knowledge or generally known in an industry. If the information is obvious upon the viewing of an object embodying the "secret" or if the information is merely a base of experiential knowledge that is readily available to others in the same field, it appears unlikely that a court would protect the information as a trade secret. Generally, however, the availability of information through publicly available sources is irrelevant if the information is shown to have been acquired through a breach of a confidential relationship or through the use of improper means. Evidence of the unique and highly secret nature of the information at issue often increases the likelihood that the information will be held to constitute a trade secret.

♦ Confidence

The general rule of liability refers to a breach of "confidence." A confidential relationship may arise not only from technical fiduciary relationships such as attorney-client, partner and partner, and so forth—which as a matter of law are relationships of trust and confidence—but may arise informally from moral, social, domestic, or purely personal relationships. Courts have found that a confidential relationship may exist when the parties' relationship is that of partners, joint adventurers, licensor and licensee, and employer and employee.

Generally, not all employment relationships are confidential. When an employee acquires an intimate knowledge of the employer's business, however, the relationship can be deemed confidential. The law thus implies as a part of the employment relationship an agreement not to use or disclose information that an employee receives incident to the employment when the employee knows that the employer desires such information be kept secret or if, under the circumstances, the employee should have realized that secrecy was desired. As noted above, the use or disclosure of a trade secret learned through a confidential relationship constitutes a misappropriation of the secret. Thus, the common law prohibits an employee's use or disclosure of the employer's trade secrets, whether the use or disclosure occurs during or after the employment relationship ends. The courts are usually careful to ensure that the information at issue is indeed a secret. The courts thus have developed the rule that a former employee is free to use the general knowledge, skills, and experience gained during prior employment. This is true even when the employee has increased his or her skills or has received complex training during the former employment. Such knowledge, skills, and experience may even be used to compete against a former employer.

♦ Improper Means

In a famous case involving the question of *improper means*, a competitor hired a photographer to conduct aerial reconnaissance of a chemical plant under construction. In holding that the complaint stated a cognizable claim, the court emphasized the value of the secret and the efforts expended to develop it. The court noted that reverse engineering and independent development were proper ways of acquiring information, but

> one may not avoid these labors by taking the process from the discoverer without his permission at a time when he is taking reasonable precautions to maintain its secrecy. To obtain knowledge of a process without spending the time and money to discover it independently is improper unless the holder voluntarily discloses it or fails to take reasonable precautions to ensure its secrecy.

The court emphasized that the predicate of liability for such acts was the maintenance of standards of commercial morality.

♦ Remedies

Although injunctive relief is probably the most common remedy sought by trade secret owners, the courts also allow the recovery of actual damages and punitive damages in trade secret cases. There are essentially three measures of actual damages in trade secrets cases: (1) the plaintiff's actual losses due to the defendant's use or disclosure of the secret; (2) the defendant's profits resulting from its use of the secret; and (3) a reasonable royalty, which is to constitute the value of what has been appropriated.

DANN v. JOHNSTON
425 U. S. 219 (1976)

Respondent has applied for a patent on what is described in his patent application as a "machine system for automatic record keeping of bank checks and deposits." The system permits a bank to furnish a customer with subtotals of various categories of transactions completed in connection with the customer's single account, thus saving the customer the time and/or expense of conducting this bookkeeping himself. As respondent has noted, the "invention is being sold as a computer program to banks and to other data processing companies so that they can perform these data processing services for depositors." . . .

Petitioner and respondent, as well as various amici, have presented lengthy arguments addressed to the question of the general patentability of computer programs.... We find no need to treat that question in this case, however, because we conclude that in any event respondent's system is unpatentable on grounds of obviousness.... Since the United States Court of Customs and Patent Appeals (CCPA) found respondent's system to be patentable, *Application of Johnston,* supra, the decision of that court is accordingly reversed.

I

While respondent's patent application pertains to the highly esoteric field of computer technology, the basic functioning of his invention is not difficult to comprehend. Under respondent's system a bank customer labels each check that he writes with a numerical category code corresponding to the purpose for which the funds are being expended. For instance, "food expenditures" might be a category coded "123," "fuel expenditures" a category coded "124," and "rent" still another category coded "125." Similarly, on each deposit slip, the customer again, through a category code, indicates the source of the funds that he is depositing. When the checks and deposit slips are processed by the bank, the category codes are entered upon them in magnetic ink characters, just as, under existing procedures, the amount of the check or deposit is entered in such characters. Entries in magnetic ink allow the information associated with them to be "read" by special document-reading devices and then processed by data processors. On being read by such a device, the coded records of the customer's transactions are electronically stored in what respondent terms a "transaction file." Respondent's application describes the steps from this point as follows:

> To process the transaction file, the...system employs a data processor, such as a programmable electronic digital computer, having certain data storage files and a control system. In addition to the transaction file, a master recordkeeping file is used to store all of the records required for each customer in accordance with the customer's own chart of accounts. The latter is individually designed to the customer's needs and also constructed to cooperate with the control system in the processing of the customer's transactions. The control system directs the generation of periodic output reports for the customer which present the customer's transaction records in accordance with his own chart of accounts and desired accounting procedures....

Thus, when the time comes for the bank customer's regular periodic statement to be rendered, the programmed computer sorts out the entries in the various categories and produces a statement which groups the entries according to category and which gives subtotals for each category. The customer can then quickly see how much he spent or received in any given category during the period in question. Moreover, according to respondent, the system can "(adapt) to whatever variations in ledger format a user may specify." Brief for Respondent, at 66.

In further description of the control system that is used in the invention, respondent's application recites that it is made up of a general control and a master control. The general control directs the processing operations common to most customers and is in the form of a software computer program, i.e., a program that is meant to be used in a general-purpose digital computer. The master control, directing the operations that vary on an individual basis with each customer, is in the form of a separate sequence of records for each customer containing suitable machine instruction mechanisms along with the customer's financial data. Respondent's application sets out a flow chart of a program compatible with an IBM 1400 computer which would effectuate his system.

Under respondent's invention, then, a general-purpose computer is programmed to provide bank customers with an individualized and categorized breakdown of their transactions during the period in question.

II

After reviewing respondent's patent application, the patent examiner rejected all the claims therein. He found that respondent's claims were invalid as being anticipated by the prior art. 35

U.S.C. sec. 102, and as not "particularly pointing out and distinctly claiming" what respondent was urging to be his invention....

Respondent appealed to the Patent and Trademark Office Board of Appeals. The Board rejected respondent's application on several grounds. It found first that, under 35 U.S.C. sec. 112, the application was indefinite and did not distinctly enough claim what respondent was claiming to be his invention. It also concluded that respondent's claims were invalid under 35 U.S.C. sec. 101 because they claimed nonstatutory subject matter. According to the Board, computer-related inventions which extend "beyond the field of technology...are nonstatutory,"...and respondent's claims were viewed to be "nontechnological." Finally, respondent's claims were rejected on grounds of obviousness. 35 U.S.C. sec. 103. The Board found that respondent's claims were obvious variations of established uses of digital computers in banking and obvious variations of an invention, developed for use in business organizations that had already been patented....

The CCPA, in a 3-2 ruling, reversed the decision of the Board and held respondent's invention to be patentable. The Court began by distinguishing its view of respondent's invention as a "recordkeeping *machine* system for financial accounts" from the Board's rather negative view of the claims as going solely to the relationship of a bank and its customers." 502 F.2d, at 770 (emphasis in CCPA opinion). As such, the CCPA held, respondent's system was "clearly within the 'technological arts,'" id. at 771, and was therefore statutory subject matter under 35 U.S.C. sec. 101. Moreover, the Court held that respondent's claims were narrowly enough drawn and sufficiently detailed to pass muster under the definiteness requirements of 35 U.S.C. sec. 112. Dealing with the final area of the Board's rejection, the CCPA found that neither established banking practice nor the Kirks patent rendered respondent's system "obvious to one of ordinary skill in the art who did not have (respondent's) specification before him.".... In order to hold respondent's invention to be patentable, the CCPA also found it necessary to distinguish this Court's decision in *Gottschalk* v. *Benson,* supra, handed down some 13 months subsequent to the Board's ruling in the instant case. In *Benson*, the respondent sought to patent as a "new and useful process," 35 U.S.C. sec. 101, "a method of programming a general-purpose digital computer to convert signals from binary-coded decimal form into pure binary form".... As we observed, "(t)he claims were not limited to any particular art or technology, to any particular apparatus or machinery, or to any particular end use.".... Our limited holding, id. at 71, 93 S.Ct., at 257, 34 L.Ed.2d, at 279, was the respondent's method was not a patentable "process" as that term is defined in 35 U.S.C. sec. 100 (b).[1]

The Solicitor of the Patent Office argued before the CCPA that *Benson*'s holding of non-patentability as to the computer program in that case was controlling here. However, the CCPA concluded that while *Benson* involved a claim as to the patentability of an "apparatus" or "machine" which did not involve discoveries so abstract as to be unpatentable:

> "'The issue considered by the Supreme Court in *Benson* was a narrow one, namely, is a formula for converting binary-coded decimal numerals into pure binary numerals by a series of mathematical calculations a patentable *process*?' (emphasis added)...."(T)he instant claims in *apparatus* form, do not claim or encompass a law of nature, a mathematical formula, or an algorithm." 502 F.2d, at 771 (emphasis in CCPA opinion).

Having disposed of the Board's rejections and having distinguished *Benson* to its satisfaction, the Court held respondent's invention to be patentable. The Commissioner of Patents sought re-

1. "The term 'process' means process, art, or method, and includes a new use of a known process, machine, manufacture, composition of matter, or material." 35 U.S.C. sec. 100(b).

view in this Court, and we granted certiorari.... We hold that respondent's invention was obvious under 35 U.S.C. sec. 103 and therefore reverse.

III

As a judicial test, "invention"—i.e. " an exercise of the inventive faculty,"...has long been regarded as an absolute prerequisite to patentability.... However, it was only in 1952 that Congress, in the interest of "uniformity and definiteness," articulated the requirement in a statute, framing it as a requirement of "nonobviousness."[2] Section 103 of the Patent Act of 1952, 35 U.S.C. sec. 103 provides in full:

> A patent may not be obtained although the invention is not identically disclosed or described or set forth in section 102 of this title, if the differences between the subject matter as a whole would have been obvious at the time the invention was made to a person having ordinary skill in the art to which said subject matter pertains. Patentability shall not be negatived by the manner in which the invention was made.

This Court treated the scope of sec. 103 in detail in *Graham* v. *John Deere Co.,* 383 U.S. 1, 86 S.Ct. 684, 15 L.Ed.2d 545 (1966). There, we held that sec. 103 "was not intended by Congress to change the general level of patentable invention," but was meant "merely as a codification of judicial precedents...with congressional directions that inquiries into the obviousness of the subject matter sought to be patented are a prerequisite to patentability.". . . While recognizing the inevitability of difficulty in making the determination in some cases, we also set out in *Graham,* supra, the central factors relevant to any inquiry into obviousness: "the scope and content of the prior art," the "differences between the prior art and the claims at issue," and "the level of ordinary skill in the pertinent art." Ibid. Guided by these factors, we proceed to an inquiry into the obviousness of respondent's system.

As noted, supra, at 1395–1396, the Patent and Trademark Office Board of Appeals relied on two elements in the prior art in reaching its conclusion that respondent's system was obvious. We find both to be highly significant. The first was the nature of the current use of data processing equipment and computer programs in the banking industry. As respondent's application itself observes, that use is extensive:

> Automatic data processing equipments employing digital computers have been developed for the handling of much of the record-keeping operations involved in a banking system. The checks and deposit slips are automatically processed by forming those items as machine-readable records.... With such machine systems, most of the extensive data handling required in a bank can be performed automatically.

It is through the use of such data processing equipment that periodic statements are ordinarily given to a bank customer on each of the several accounts that he may have at a given bank. Under respondent's system, what might previously have been separate accounts are treated as a single account, and the customer can see on a single statement the status and progress of each of his "subaccounts." Respondent's "category code" scheme, see supra, at pp. 1394–1395, is, we think, closely analogous to a bank's offering its customers multiple accounts from which to choose for making a deposit or writing a check. Indeed, as noted by the Board, the addition of a category number, varying with the nature of the transaction, to the end of a

2. S. Rep. No. 1979, 82d Cong., 2d Sess., 6(1952); H.R. Rep. No. 1923, 82d Cong., 2d Sess., 7(1952).

bank customer's regular account number, creates "in effect, a series of different and distinct account numbers...." Petition for Certiorari, at 34a. Moreover, we note that banks have long segregated debits attributable to service charges within any given separate account and have rendered their customers subtotals for those charges.

The utilization of automatic data processing equipment in the traditional separate account system, is, of course, somewhat different from the system encompassed by respondent's invention. As the CCPA noted, respondent's invention does something other than "provide a customer with...summary sheet consisting of net totals of plural separate accounts which a customer may have at a bank."... However, it must be remembered that the "obviousness" test of sec. 103 is not one which turns on whether an invention is equivalent to some element in the prior art but rather whether the difference between the prior art and the subject matter in question "is a difference sufficient to render the claimed subject matter unobvious to one skilled in the applicable art...."

There is no need to make the obviousness determination in this case turn solely on the nature of the current use of data processing and computer programming in the banking industry. For, as noted, the Board pointed to a second factor—a patent issued to Gerhard Dirks—which also supports a conclusion of obviousness. The Dirks patent discloses a complex automatic data processing system using a programmed digital computer for use in a large business organization. Under the system transaction and balance files can be kept updated for each department of the organization. The Dirks system allows a breakdown within each department of various areas, e.g., of different types of expenses. Moreover, the system is sufficiently flexible to provide additional breakdowns of "sub-areas" within the areas and can record and store specially designated information regarding each of any department's transactions. Thus, for instance, under the Dirks system the disbursing office of a corporation can continually be kept apprised of the precise level and nature of the corporation's disbursements within various areas or, as the Dirks patent terms them, "Item Groups."

Again, as was the case with the prior art within the banking industry, the Dirks invention is not equivalent to respondent's system. However, the departments of organization and the areas or "Item Groups" under the Dirks system are closely analogous to the bank customers and category number designations respectively under respondent's system. And each shares a similar capacity to provide breakdowns within its "Item Groups" or category numbers. While the Dirks invention is not designed specifically for application to the banking industry many of its characteristics and capabilities are similar to those of respondent's system....

In making the determination of "obviousness," it is important to remember that the criterion is measured not in terms of what would be obvious to a layman, but rather what would be obvious to "one reasonably skilled in (the applicable) art."... In the context of the subject matter of the instant case, it can be assumed that such a hypothetical person would have been aware both of the nature of the extensive use of data processing systems in the banking industry and of the system encompassed in the Dirks patent. While computer technology is an exploding one, "(i)t is but an evenhanded application to require that those persons granted the benefit of a patent monopoly be charged with an awareness" of that technology.... Assuming such an awareness, respondent's system would, we think, have been obvious to one "reasonably skilled in (the applicable) art." There may be differences between respondent's invention and the state of the prior art. Respondent makes much of his system's ability to allow "a large number of small users to get the benefit of large-scale electronic computer equipment and still continue to use their individual ledger format and bookkeeping methods." Brief for Respondent, at 65. It may be that ability is not possessed to the same extent either by existing machine systems in the banking industry or by the Dirks system.[3] But the mere existence of differences between the prior

3. The Dirks patent does allow "the departments or other organizational users (i.e., the analogues to bank customers under respondent's invention) to retain their authority over operative file systems" and indicates that "programming is very easy and different programs are very easily coordinated."

art and an invention does not establish the invention's nonobviousness. The gap between the prior art and respondent's system is simply not so great as to render the system nonobvious to one reasonably skilled in the art.[4]

Accordingly, we reverse the Court of Customs and Patent Appeals and remand this case to that court for further proceedings consistent with this opinion. So ordered.
Reversed and remanded.

Mr. justice BLACKMUN and Mr. Justice STEVENS took no part in the consideration or decision of this case.

DUNLOP HOLDINGS LIMITED v. RAM GOLF CORP.
524 F.2d 33 (7th Cir. 1975)

Plaintiff sued Ram for infringement of its patent covering an unusually durable golf ball.[5] Ram convinced the district court that the patent was invalid because the invention had been made by a third party named "Butch" Wagner. Ram proved that Wagner had publicly used the new golf ball before February 10, 1965, the earliest date that plaintiff can claim invention;[6] however, Wagner had not disclosed his formula to the public. The questions on appeal are (1) whether the district court's findings on the prior invention issue are supported by the record, and (2) whether the nondisclosure of the method of making an article which is in public use is the kind of concealment or suppression that avoids the bar to patentability in Section 102(g).[7]

The patent covers the discovery that certain synthetic materials,[8] when fabricated by themselves (or with minor amounts of compatible materials), produces a golf ball cover with exceptional cutting resistance. An example of the material described in the patent is a DuPont product named "Surlyn." Golf balls made of Surlyn, with or without minor additives, infringe the claims in plaintiff's patent.

As noted, the date of invention claimed by plaintiff is February 10, 1965. In April of 1964, DuPont was trying to find a commercial use for its Surlyn, a recently developed product. Shortly thereafter, Butch Wagner, who was in the business of selling re-covered golf balls, began to ex-

4. Whereas "commercial success without invention will not make patentability,"*Great Atlantic & Pacific Tea Co. v. Supermarket Equipment Corp.,*...we did indicate...that "secondary considerations (such) as commercial success, long felt but unsolved needs, (and) failure of others" may be relevant in a determination of obviousness.... Respondent does not contend nor can we conclude that any of these secondary considerations offer any substantial support for his claims of nonobviousness.

5. U.S. Patent No. 3,454,280 covering "Golf Balls Having Covers of Ethylese-Unsaturated Monocarboxylic Acid Copolymer Compositions," issued to Dunlop Rubber Company Limited, a British company, as the assignee of the two individual inventors, pursuant to a U.S. Patent Application filed Feb. 2, 1966, and a British application filed on Feb. 10, 1965.

6. 35 U.S.C. Sec. 104 provides in part that "...an applicant for a patent, or a patentee, may not establish a date of invention by reference to knowledge or use thereof, or other activity with respect thereto, in a foreign country, except as provided in section 119 of this title." 35 U.S.C. Sec. 119 allows priority from the date of filing in certain foreign countries including Great Britain. Plaintiff is thus barred from establishing a date of invention by reference to its activities in Britain prior to Feb. 10, 1965, the date of its British application.

7. 35 U.S.C. Sec. 102 provides: "A person shall be entitled to a patent unless....

*****

(g) before the applicant's invention thereof the invention was made in this country by another who had not abandoned, suppressed, or concealed it...."

8. The parties use the term "copolymers" which we understand to refer to a kind of synthetic rubber or plastic material. The description in the patent is of "A golf ball comprising a core and a cover, said cover being formed of a composition comprising a copolymer of ethylese and at least one unsaturated monocarboxylic acid containing from three to eight carbon atoms, said copolymer containing up to thirty percent by weight of said acid." Patent No. 3,454,280 column 6, lines 45–50.

periment with Surlyn as a golf ball cover. He first made some sample balls by hand and then, using a one-iron, determined that the material was almost impossible to cut. He obtained more Surlyn and made several dozen experimental balls, trying different combinations of additives to achieve the proper weight, color, and a texture that could easily be released from an injection molding machine. By November 5, 1964, he had developed a formula which he considered suitable for commercial production and had decided to sell Surlyn-covered balls in large quantities. The date is established by a memorandum recording his formula, which Wagner wrote in his own hand and gave to his daughter for safekeeping on the occasion of her son's birthday.[9]

During the fall of 1964, Wagner provided friends and potential customers with Surlyn-covered golf balls. By the end of the year he had purchased enough Surlyn to produce more than 20,000 balls, and by February of 1965 he had received orders for ever 1,000 dozen Surlyn-covered balls. By the end of 1965, he had ordered enough Surlyn to produce more than 900,000 such balls. Without commenting further on the evidence, we note our conclusion that there is ample support in the record for the district court's findings that Wagner had discovered the use of Surlyn as a golf ball cover before November 5, 1964, had reduced the discovery to practice before February of 1965, and did not abandon the invention before his death in October of 1965.

We recognize that Wagner continued to experiment with different formulae after his decision to go into commercial production and that he encountered some problems with cracked covers. These facts do not undermine any of the district court's findings on the prior invention issue. The patent claims are broad enough to encompass any golf ball cover made principally of Surlyn and there is no doubt that Wagner had made a large number of such golf balls and successfully placed them in public use.[10] The only novel feature of this case arises from the fact that Wagner was careful not to disclose to the public the ingredient that made his golf ball so tough.[11] For that reason plaintiff argues that he " suppressed or concealed" the invention within the meaning of Sec. 102(g).

Since 1850 it has been settled that a patentee may be entitled to credit for making a new discovery or invention even though someone else actually made the discovery before he did.... That case established the proposition that an abandoned invention will not defeat the patentability of the rediscovery of "lost art."[12] The case has also been cited for the proposition that an inventor who had merely made a secret use of his discovery should not be regarded as the first inventor....

Gillman involved a patent on a machine which had previously been developed by a man named Haas; Haas had used the machine in his own factory under tight security. The output from the machine had been sold, but the public had not been given access to the machine itself.

---

9. In the district court's original findings this handwritten memorandum was erroneously described as a page in Wagner's formula book, whereas it was actually a separate sheet of paper. We attach no significance to this error, which the district judge subsequently corrected, since the testimony unequivocally established the genuineness of the document. We also find no substance in appellant's objection to the district court's making this minor correction in its findings after the notice of appeal was filed....

10. The evidence identifies at least three golfers who used Surlyn-covered balls during the fall of 1964 for rounds of golf played at Riveria Country Club in Los Angeles.... By February 1965, two of these golfers, both of whom were favorably impressed with the play of the new ball, had placed orders with Wagner for more than 1,000 dozen Surlyn-covered balls; they began to distribute them commercially although they both lacked knowledge of the Surlyn content in the cover construction....

11. In support of its contention that Wagner concealed his invention, plaintiff relies on (1) the deposition of an acknowledged expert on golf ball construction who failed to discover the Surlyn content of the cover in an analysis of Wagner's ball; and (2) the secretive manner in which Wagner gave the Surlyn formula to his daughter on her son's birthday "to keep in case something ever happened to him."...

12. "So, too, as to the lost arts. It is well known that centuries ago discoveries were made in certain arts the fruits of which have come down to us, but the means by which the work was accomplished are at this day unknown. The knowledge has been lost for ages. Yet it would hardly be doubted, if anyone now discovered an art thus lost, and it was a useful improvement, that upon a fair construction of the act of Congress, he would be entitled to a patent. Yet he would not literally be the first and original inventor, but would be the first to confer on the public the benefit of the invention. He would discover what is unknown, and communicate knowledge which the public had not the means of obtaining without his invention."...

In holding that Haas was not the first inventor, Judge Hand drew a distinction between a secret use and a noninforming public use.[13] There had been only a secret use of the Haas machine and therefore he was not regarded as the first inventor.

This case certainly involves neither abandonment nor a mere secret use, for the evidence clearly demonstrates that Wagner endeavored to market his golf balls as promptly and effectively as possible. The balls themselves were in wide public use. Therefore, at best, the evidence establishes a noninforming public use of the subject matter of the invention.

If Wagner had applied for a patent more than a year after commencing the public distribution of Surlyn-covered golf balls, his application would have been barred notwithstanding the noninforming character of the public use or sale.... For an inventor must exercise reasonable diligence if he is to be rewarded with patent protection.[14]

The question of diligence is especially significant in cases arising out of a dispute between two applicants for a patent on the same discovery. For in such a case, when the issue is which of the two applicants is entitled to the monopoly reward, it is often appropriate to weigh the later inventor's diligence in enabling the public to obtain the benefit of the concept more heavily than the earlier date of unexploited conception.... But in this case, although Wagner may have failed to act diligently to establish his own right to a patent, there was no lack of diligence in his attempt to make the benefits of his discovery available to the public. In view of his public use of the invention, albeit noninforming, we do not believe he concealed or suppressed the discovery within the meaning of Sec. 102(g).

We recognize, as appellant argues, that portions of Judge Rich's opinion in *Palmer* v. *Dudzik*...suggest that a public use which does not disclose the inventive concept may amount to concealment within Sec. 102(g). But that case, like *Gillman,* involved a patent on a machine; the benefits of using the machine were not made available to anyone except the inventor. Moreover, the case arose out of an interference proceeding in which the dispute was between two applicants for a patent, the earlier of the two having been less diligent than the later. In this case, Wagner was not only the first inventor, but also "the first to confer on the public the benefit of the invention"....

There are three reasons why it is appropriate to conclude that a public use of an invention forecloses a finding of suppression or concealment even though the use does not disclose the discovery. First, even such a use gives the public the benefit of the invention. If the new idea is permitted to have its impact in the marketplace, and thus to "promote the Progress of Science and useful Arts,"[15] it surely has not been suppressed in an economic sense. Second, even though there may be no explicit disclosure of the inventive concept, when the article itself is freely accessible to the public at large, it is fair to presume that its secret will be uncovered by potential competitors long before the time when a patent would have expired if the inventor had made a timely application and disclosure to the Patent Office.[16] Third, the inventor is under no duty to

---

13. "We are to distinguish between a public user which does not inform the art...and a secret user...."
14. The second sentence of Sec. 102(g) reads as follows:
In determining priority of invention there shall be considered, not only the respective dates of conception and reduction to practice of the invention, but also the reasonable diligence of one who was first to conceive and last to reduce to practice, from a time prior to conception by the other.
A conclusion that Wagner had concealed or suppressed his invention would have to be supported by a stronger showing than a mere lack of diligence. For it is less serious to hold that the first inventor has forfeited his right to a patent monopoly than it is to hold that he has forfeited any right to use his own invention without the permission of a subsequent inventor.
But we must bear in mind that it was not alone to reward the inventor that the patent monopoly was granted. The public was to get its reward and have the advantage of the inventor's discovery as early as was reasonably possible....
15. U.S. Const., art. 1, Sect. 8, clause 8.
16. In this case, for example, it is not unreasonable to assume that competing manufacturers of golf balls in search of a tough new material to be used as a cover, might make inquiries of Wagner's Surlyn supplier that would soon reveal his secret ingredient. After all, DuPont certainly had a motive to expand the market for Surlyn.

apply for a patent; he is free to contribute his idea to the public, either voluntarily by an express disclosure, or involuntarily by a noninforming public use. In either case, although he may forfeit his entitlement to monopoly protection, it would be unjust to hold that such an election should impair his right to continue diligent efforts to market the product of his own invention.

We hold that the public use of Wagner's golf balls forecloses a finding of suppression or concealment; that holding is consistent with both the decided cases and the underlying purposes of the statute.

Affirmed.

GUNTER v. STREAM

573 F.2d 77 (C.C.P.A. 1978)

This is an appeal from a decision of the Patent and Trademark Office Board of Patent Interferences (board) which awarded priority as to two counts of an interference to junior party, Stream. We affirm.

On August 11, 1975, an interference was declared between Gunter's application entitled "Heat Pipes for Fin Coolers," serial No. 507,314, filed on September 19, 1974, and Stream's application entitled "Method and Apparatus for Controlling the Viscosity of Glass Streams," serial No. 511,541, filed on October 3, 1974. The two counts of the interference correspond to claims 6 and 8 of Gunter's application and claims 28 and 29 of Stream's application.

The subject matter of the interference is a method and apparatus for employing heat pipe fins for cooling glass fibers as they are drawn through orifices of a glass fiber forming machine. Counts 1 and 2 define the subject matter:

1. A fiber glass bushing unit comprising in combination a container for the reception of molten glass, a plurality of orifices on the bottom of said container arranged in parallel rows, a plurality of plate-like fin members positioned between the rows of orifices by but below and out of contact with said container, said plate-like fins being mounted at one end in a header member, means to pass the fluid coolant through said header member, each of said plate members having a wick material affixed to the interior surfaces of said plate member and having a central cavity located therein, a vaporizable liquid on said wick capable of being vaporized from the surface of said wick and recondensed on said wick during operation.
2. A method of cooling glass fibers being drawn from a molten glass source from a plurality of glass orifices located on the bottom of said glass source, removing heat from said fibers by positioning a plurality of plate-like heat exchange members between said fibers to thereby absorb the radiant heat from said fibers on the surface of said platelike members continuously, maintaining the surface of the plate-like members receptive to heat absorption by vaporizing a volatile fluid on the interior surface of said plate-like members continuously from the surface of a wick contained therein and removing heat continuously from said plate-like members by indirect heat exchange with the mounting means for said plate-like members to thereby condense said volatile fluid in said plate-like members and thereby return it to the wick for further vaporization.

Gunter took no testimony and is restricted to his filing date of September 19, 1974, as the date of conception and constructive reduction to practice. Stream, an employee of Owens-Corning Fiberglass Corporation (OCF), assignee of his application, submitted testimony and documentary evidence to support a date of conception and reduction to practice prior to Gunter's filing date. Stream himself testified that, in August of 1970, he read an article on heat pipe technology published in the August 6, 1970, edition of *Machine Design,* which was admitted into evidence as Stream Exhibit 2. According to the testimony, Stream had a conversation, on or before August 27, 1970, with his supervisor, Mr. Hellmut I. Glaser,[17] in which he told Glaser of his idea to apply heat pipe technology to glass fiber making. Glaser corroborated Stream's testimony about their conversation and also testified that he reported on Stream's idea to Mr. Steven R. Gustafson, then patent attorney for OCF. Subsequent to this conversation with Glaser, Gustafson reduced Glaser's report to writing on September 2, 1970. Glaser testified that the contents of the Gustafson memo, which was admitted into evidence as Stream Exhibit 3, was an accurate summary of his report.

To support his reduction to practice of the invention, Stream produced employees of OCF who testified about contracting with Hughes Aircraft Company for construction of prototype fin shields which were eventually completed and shipped to OCF on November 11, 1973. Tests were performed on the prototypes by OCF at its Huntingdon Plant on April 17 and 18,1974.

In its opinion, the board defined conception as a disclosure of an invention which enables one skilled in the art to reduce the invention to a practical form without "exercise of the inventive faculty." The board was persuaded by the evidence presented by Stream on the question of conception. It found that appellee conceived the invention on August 27, 1970, when he understood it to the extent that he was able to disclose it to another who in turn understood the invention.

Opinion

On the question of reduction to practice, the board found that the Hughes prototype, successfully tested by OCF, embodied every essential element of the counts. The tests conducted on April 17 and 19, 1974, were proof of an actual reduction to practice, which is attributable to Stream. We agree with the board's finding on this issue, and we are not persuaded by appellant's argument that the reduction to practice does not inure to the benefit of Stream, since he took no part in this phase. Stream can prevail if he proves an earlier date of conception by a preponderance of the evidence.... The issue before the court, then, is whether Stream proved, as junior party, by a preponderance of evidence that he had conception on or about August 27, 1970. Gunter argues that Stream never had conception but only expressed an invitation to experiment with heat pipes in glass fiber forming machines. Stream is required, Gunter argues, to have conceived not only the invention, but the means to accomplish the invention. The conception was not completed by Stream because extensive research by Hughes Aircraft was necessary to achieve satisfactory performance of the invention. Gunter further argues that while Hughes Aircraft was working on the prototype, it suggested to OCF the use of cooling blocks in the construction of the prototypes. This, he argues, militates against Stream being in possession of a conception which included use of a "header member" which is recited in the counts. It is contended also that Stream had no connection with the invention from August of 1970 until September 27, 1974, when he reviewed and signed the application disclosing his invention and that there is no evidence on record to indicate that Stream was ever informed of the progress of Hughes Aircraft or of the tests of OCF.

17. Glaser is an engineer who had been with OCF since 1956 as a specialist in glass fiber formation. In 1970, he became a manager of the Process Technology Department.

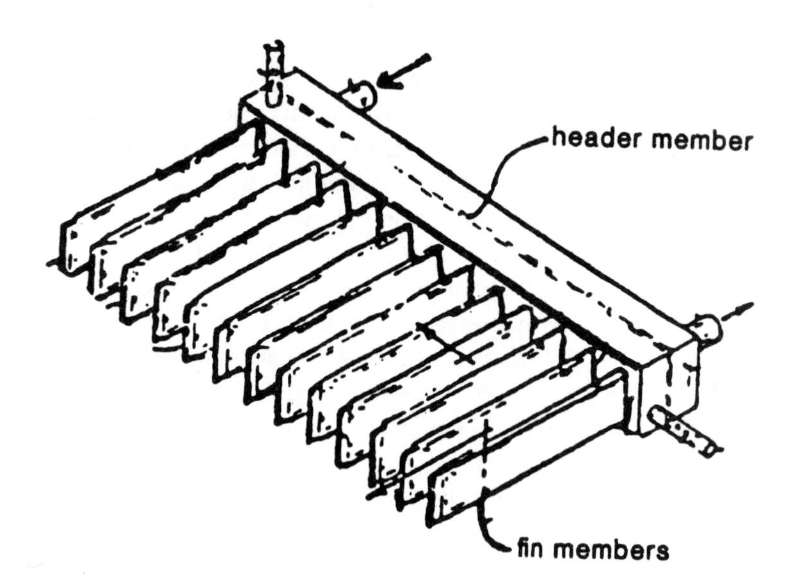

The board correctly cited the definition for conception initially stated in *Mergenthaler* v. *Scudder*:

> The conception of the invention consists in the complete performance of the mental part of the inventive act. All that remains to be accomplished, in order to perfect the act or instrument, belongs to the department of construction, not invention. It is therefore the formation, in the mind of the inventor, *of a definite and permanent idea of the complete and operative invention, as it is thereafter to be applied in practice, that constitutes an available conception, within the meaning of the patent law.*

We adopted this definition in *Townsend* v. *Smith*, supra. Reviewing the evidence supporting conception, we conclude that the conception of using heat pipe technology in glass fiber forming, as described in the counts, occurred on or about August 27, 1970. We hold, therefore, that Stream established by a preponderance of the evidence that he conceived the invention on or about that date. Having already shown that there was reduction to practice by the assignee, Stream's conception completes the showing necessary to award priority.

The facts show that on or about August 6, 1970, Stream read and understood the article on heat pipe technology. With his working knowledge of glass fiber forming and his understanding of basic heat pipe technology described in the article, Stream conceived the invention, applying heat pipe technology to glass fiber forming. The Gustafson memo is the only documentary evidence which corroborates the conception[18] It states, in pertinent part:

> On August 27,1970, Mr. Hellmut Glaser disclosed to Michael Mitcham and the writer that commercially available "heat pipes" might be advantageous substitutes for our present fin shields.
>
> The heat pipes cool the surrounding area through the vaporization of fluid contained inside the pipe. One such heat pipe is made by Electron Dynamics[19] and is

---

18. Both Gustafson and Glaser testified about the memo. Gustafson identified the document, whereas Glaser, on direct examination, testified that he understood the memo to be an accurate reflection of his conversation with Gustafson on August 27, 1970.

19. Electron Dynamics is a division of Hughes Aircraft Company.

> very similar in size and shape to a single fin shield. In operation, one end of the heat pipe is soldered or welded to a cooling manifold.
>
> According to Mr. Glaser the original suggestion for this incorporation of heat pipes was made by Ralph Stream.

The memo was the culmination of two conversations both involving Glaser, who corroborated Stream's testimony about the first conversation.

> Q. What do you recall about the conversation? A. Well, Ralph Stream came to my office with an article out of a trade journal, trade magazine on heat pipes, and disclosed to me that this might be application (sic) to be used on finshields; certainly might be something we could use there, that application.
> Q. If I hand you a copy of Stream Exhibit 2, do you recall if that's the article you just mentioned, that Ralph Stream came to you with? A. Yes.
> Q. Okay. Do you recall any of the details of the conversation between you and Ralph Stream which you just referred to? A. I think it's difficult to recall details about that. He brought the article into my office.
> Q. Sure. A. He brought the article into my office and I recall that we shortly discussed the concept of heat pipes, and that because of the advantages such a device offers, we should seek to utilize it in the cooling of forming cones, possibly as a substitute for finshields.
> Q. Based on your conversation, or at the time of the conversation with Mr. Stream, did you understand in general what a heat pipe is? A. I think the article pretty much describes it, yes.
> Q. Following your conversation with Mr. Stream, you felt that you understood the concept of utilizing heat pipe technology in the finshield environment? A. Yes. We were very intrigued by it, by that concept.

Glaser testified also to the second conversation with Gustafson:

> Q. Let's refer back to Stream Exhibit 3, and this is written by Mr. Gustafson, so you can't really testify as to the truth of what it says, but you could tell me if—does this refresh your memory as to whether or not it is an accurate reflection of your conversation with Mr. Stream? A. Yes.
> Q. Okay. And do you have any reason to believe that your conversation with Mr. Gustafson was after August 27, 1970? A. All I—it's difficult to remember dates, exact dates. We discussed, we, with Steve Gustafson, Ralph Stream's ideas, within a couple of days after Ralph Stream brought it in to me, because Steve came at that time almost weekly or bi-weekly to—
> Q. Okay. And Mr. Gustafson in Stream Exhibit 3 says in the third paragraph, "According to Mr. Glaser, the original suggestion for this incorporation of heat pipes was made by Ralph Stream." A. Right.
> Q. Do you agree with that statement?
> A. Yes.

With this evidence to support conception of the invention on or about August 27, 1970, the inquiry is directed to the substance of the conception. Did Stream conceive, at that point in time, all the essential elements of the counts? To answer this inquiry we compare the counts to the description in the Gustafson memo. The counts reproduced in their entirety, above, include an apparatus count and a method count. The apparatus count describes the fiber glass bushing unit as primarily a plate-like fin member, which has a central cavity for vaporization of a fluid coolant which travels through a wick, in combination with a header member, which possesses a means to pass the fluid coolant. The method count describes the cooling process as vaporization of fluid coolant and indirect heat exchange with a header member (cooling block).

The Gustafson memo describes the invention as a heat pipe, which replaces a standard fin-shield, connected to a "cooling manifold." This combination corresponds to the description of the apparatus in count 1. The plate-like fin member corresponds to the heat pipe, both cool through the "vaporization of fluid inside the pipe." The header member corresponds to the "cooling manifold," as both secure one or a plurality of heat pipes or fin members, and both provide a means for indirect heat exchange. Thus, the Gustafson memo describes the process of cooling as set forth in method count 2.

Gunter's argument that, in July of 1972, Hughes Aircraft suggested the cooling block and that, therefore, Stream was not in possession of a complete conception in August of 1970, is fully met by the suggestion of a "cooling manifold" in the Gustafson memo. The comparison of the counts with the description of the Stream invention in the Gustafson memo, we conclude, shows clearly that Stream conceived the invention with all essential elements of the counts at least by August 27, 1970.

Motion to Tax Costs

Appellant request the court to tax appellee for the cost of seventy-one pages of a ninety-three-page addition to the printed transcript. The ninety-three-page addition supplemented testimony of six of appellant's witnesses. Appellee, in opposition to the motion, argues that this testimony is relevant to the issues of conception and reduction to practice raised by appellant in his notice of appeal.

For purposes of allocating printing costs, appellant is required to bear the costs for that material necessary for the court to decide the issues he raised on appeal....

We conclude, after examining the testimony in question, that the material was necessary for consideration of the issues on appeal. Appellant's motion is accordingly denied. Having reached the above conclusion with regard to conception and reduction to practice, we *affirm* the decision below which awards priority to junior party, Stream.

Affirmed.

REVIEW QUESTIONS

1. Why is the patent right the right to exclude others from making, using, or selling the invention rather than the right to make, use, or sell it?
2. White, an electrical engineer, goes to work for the ABC Company to design automation circuitry. No patent agreement is required of him. Using company time and facilities (in part), he develops a new product and makes application for a patent on it. If he obtains a patent, what rights, if any, will the ABC Company have in the product?
3. In question 2, assume that White had signed an agreement to assign patent rights to the ABC Company but had developed the product at home, using only his own time and facilities. What rights, if any, will the ABC Company have in the patent?
4. Outline the steps of the patenting procedure.
5. Why are records of the development of an invention necessary?
6. Consider trademarks and copyrights. Can you conceive of something that might be the subject of either a trademark registration or a copyright?
7. In the case of *Dann* v. *Johnston,* the court refers to "the general patentability of computer programs." What arguments can you present for and against computer programs being patentable?
8. In *Dunlop Holdings Limited* v. *Ram Golf Corp.,* a) why was Ram the defendant? b) what is the practical result of this case?
9. In the case of *Gunter v. Stream,* what would Gunter be required to show to maintain his priority in the patent?
10. Gray, an engineer, learned a manufacturing process as a trade secret from White, his employer. Later, Gray quit and went to work for Black. Upon hearing that White had sold his plant to Brown, Gray quit Black's employ and went into business for himself, using the trade secret. Brown is suing for an injunction to prevent Gray from using the trade secret, claiming that he (Brown) bought the secret process with the rest of the business. Gray claims that he has respected the secret he learned from White because he did not use it until White sold the business. Would the court be likely to issue the injunction? Why or why not?
11. To what extent may an inventor publicly use or sell an invention prior to patent application without destroying its patentability?

Part Five

TORTS

Nearly all engineers are concerned with contracts and property; therefore, our time is well spent considering the rights and liabilities comprising the law in these areas. Just as important (perhaps even more important to some engineers), however, is the law concerning personal rights.

We all work with other people, so it follows that what we do may infringe upon their rights, and what they do may infringe upon ours. As is true for other professionals, an engineer's activities encounter *tort* risks. That is, what the engineer designs—a product, a process, or a system of some sort—may cause injury to someone's rights. For example, a design is based upon a mental image of the future. Without such plans, progress would cease. But designing a thing or a combination that has never before existed can be hazardous because one cannot possibly conceive of all ramifications of a new concept. Consequently, the price of progress in engineering is sometimes injury to someone—a tort.

A tort can be defined as a wrongful injury to another person or another's property rights. A plaintiff generally undertakes prosecution to obtain compensation for the injury suffered or to seek a decree that will prevent harm to personal or property rights. just as the law undertakes to prevent acts against society by prosecuting those who have committed crimes, it also provides procedures to redress wrongs to an individual. Such wrongs fall into two general categories—breaches of contract and torts.

Breach of contract cases arise when there has been an agreement of some sort between two or more parties. Such cases were treated in earlier chapters. In contrast, tort actions arise from the duties existing as a matter of law between parties. The driver of a car, for instance has a duty to avoid hitting others while driving. If the driver is negligent and, as a result, injures another, a tort action is available to the victim to make the driver pay for the damage. In chapter 21, we discuss a number of torts generally. The following chapter focuses on a particular type of tort action—products liability.

Chapter 21

Common Torts

As with our other laws of contract and property, a great deal of our tort law can be traced to the common law developed in England over the centuries. As a result, different types of torts sometimes overlap. For example, negligence is itself a kind of tort, yet other kinds of torts (against persons or against property) may also involve negligence. Torts can also be intentional (as opposed to negligent), and intent can be an element in torts against persons, property, reputation, and business. This chapter addresses the torts that arise most often.

Most tort cases are based on negligence—someone did something negligently, or neglected to do something he or she should have done. Although different torts often overlap, it is often helpful to distinguish between torts that involve personal injuries or death and those that involve property damage. Obviously, though, this distinction becomes blurry or nonexistent in situations (such as a car accident) involving both personal injury and damage to property. In the following sections, we will first explore the tort of negligence and its elements. The discussion then turns to other common torts, which will be treated under the following four general headings: (1) torts against a person, (2) torts against one's reputation, (3) torts against property rights, and (4) torts against economic rights.

NEGLIGENCE

The basic idea of *negligence* is that a failure to live up to one's duty is wrong and gives rise to liability for damages caused by that failure. *Black's Law Dictionary* defines negligence as "the omission to do something which a reasonable man, guided by those ordinary considerations which ordinarily regulate human affairs, would do." To establish the tort of negligence, the plaintiff needs to provide the following:

1. the existence of a duty
2. a breach of that duty
3. showing that the breach caused some injury
4. the damages

♦ Duty

Whether a legal duty exists under any particular set of facts and circumstances usually involves a question of law for the courts. Unless the actor's conduct creates some foreseeable danger to the victim, the actor generally does not owe a legal duty to the victim. Thus, the foreseeability of injury to a particular person or a group of persons is often considered an important basis for concluding that a legal duty does or does not exist.

Another factor sometimes considered by the courts is whether the conduct involved was "active" or "passive." Thus, a defendant who failed to prevent an injury by failing to act may be found to have no duty; on the other hand, a defendant who creates a dangerous condition by its conduct has a duty to do something to prevent any injury. However, the courts have eroded this particular distinction in more recent years. Some courts have adopted the view that "reasonable care" is required; the amount of care viewed as "reasonable" varies with the circumstances.

Another potential basis for the existence of a duty arises when one renders services recognized as necessary for the protection of persons or things. The service provider may have a legal duty if he or she fails to exercise care, and that failure increases the risk of harm to others, or if someone suffers harm because he or she relied on the service. A duty also can arise from the existence of contractual obligations. For example, contract duties may be considered tort duties in situations in which tort damages result from a breach of contractual duties, such as a breach of warranty.

Suppose Black Machinery Co. hires White to install a new lathe. White does so badly that he manages to tip the lathe over, destroying it and another piece of equipment. White's contract may provide the basis for imposing a duty to perform in a reasonably careful manner. Because White breached this duty by performing negligently, Black can sue for breach of contract and for negligence (in some states).

♦ Standard of Care

As noted, plaintiffs usually must show that the defendant owed them a duty of *care*. A general standard for this duty of care developed at common law over the centuries. In addition, many circumstances tend to modify the standard of care.

The general standard of care, as it has developed, is the degree of care that would be exercised by a *reasonably prudent person* in like circumstances. The average, or reasonably prudent, person is one possessing normal intelligence, memory, capacity, and skill. This person would have no handicaps, either physical or mental, that would set him or her aside as exceptional. It is, of course, easy to talk about an average person, but much harder to find one. Most people have something other than average intelligence or average physical structure. Reaction time (for a visual stimulus), for instance, averages somewhere around 0.19 seconds, but most people are either slower or faster. Nevertheless, a standard must be established, and then allowance must be made for the exceptions.

Modifications of the general standard of care exist in appropriate situations. The standard of care required of a surgeon in removing an appendix would be considerably more strict than the standard required of a person whose only claim to a knowledge of medicine came from a course in anatomy, but who, because of an emergency, had to attempt surgery. An engineer works under an exceptional standard of care when he or she designs or supervises the construction of a machine or structure. On the other end of the scale, people who have a physical or mental handicap cannot be held to the same standard as the average, or reasonably prudent, person.

♦ Proximate Cause

The main cause of most tort injuries is usually quite apparent when the facts are established. On occasion, though, more than one act or omission may cause the injury. For example, a motorist driving along a highway at night at a lawful speed may be so blinded by oncoming headlights that he will not see an object he is approaching. If he strikes another car in the rear, is *he* liable, or would the person who failed to dim the headlights be liable? The question is one of *proximate cause*. The failure of the oncoming driver to dim his lights would probably be posed as a defense, but with a probable lack of success. The driver who couldn't see properly should have slowed down.

To rise to the level of a proximate cause, something must have been a substantial factor in bringing about the injury. A factor may be characterized as either *cause in fact* or *legal causation*. *Cause in fact* is met if the *but for* standard is satisfied. Would the injury have occurred *but for* the defendant's conduct? The issue of *legal causation*, however, depends on the connection between the cause and effect; the injury must be part of a natural and continuous sequence.

For example, Black and White are engaged in the electrical repair of an overhead crane. Before starting the work, Black turned off the electricity at the switch box. White is working as the ground man of the pair. Gray, requiring electricity for a job he is doing, throws the wrong switch at the box. Black, receiving a shock from the conductor, drops a wrench on White, thereby injuring him. The immediate cause of White's injury was the force of the blow from the wrench dropped by Black. The proximate cause was Gray's action in throwing the wrong switch. If the case went to court, the probable result would be a finding for White against Gray.

Now suppose that when White is struck, he falls over, knocking over a can of gasoline. The gasoline spreads, evaporates, and when the crane is activated, a fire starts. The fire spreads, and sparks move to a neighboring house. The sparks move from that house to a nearby garage, in which a Rolls Royce is kept. The garage roof collapses and destroys the Rolls.

What was the proximate cause of the loss of the Rolls? At this point, the fire would probably be considered a proximate cause, but it seems difficult to go all the way back to Gray. Proximate cause is the doctrine by which the courts view certain factual causes as just too far removed to be the basic for legal liability.

♦ Res Ipsa Loquitur

In most tort cases, the plaintiff can point to specific negligent acts by the defendant. In some instances, however, it is difficult or impossible to show defendant's specific acts or omissions, but common sense tells us that someone was negligent. For example, a patient goes into an operating room and has surgery; a sponge is left in the patient's body. No one knows how it got there. In such situations, the plaintiff may still make a case based on the doctrine known as *res ipsa loquitur*. Essentially, this means "the thing speaks for itself." To use this doctrine, the plaintiff may show that the injury would not have occurred unless someone had been negligent, that the defendant had control of the instrumentality causing the injury, and that the plaintiff in no way contributed to the injury.

The result is not direct proof, but circumstantial evidence that suggests to the jury that someone was negligent. The defendant, of course, has an opportunity for rebuttal, which may take one of several forms, but usually consists of showing that he or she did exercise the requisite care.

♦ Gross Negligence

As stated earlier, doing something that should not be done or neglecting to do something that should be done, thereby causing injury to another, amounts to negligence. When the act is done or neglected intentionally or with a reckless disregard for the consequences, it ceases to be the common variety of negligence and becomes *gross negligence*. The victim is much more likely to recover punitive damages if he or she can show gross negligence. For example, many jurisdictions will not allow a plaintiff to recover punitive damages based upon a claim of negligence. Such damages, however, often can be recovered if gross negligence is shown.

♦ Negligence Defenses

As you may guess, there are many situations in which it is unclear just why or how something happened. In some situations, the injury was actually caused by the injured party. This discussion explains the defenses that developed in negligence cases.

Assumption of Risk

People do not always do what is best for them. Occasionally they assume risks for the experience or thrill of the very danger involved. If one is injured or dies as a result of the risk assumed (e.g., death of heart attack during a roller coaster ride), there can be no recovery. Since the assumption-of-risk defense acts to relieve the defendant of liability for negligence, it has usually been applied in a narrow fashion. For instance, the courts limited the doctrine to situations in which the plaintiff knew and understood the risks involved. It's one thing to risk a scratch; it's quite another to literally risk a limb. In addition, the courts required the assumption of the risk to be free and voluntary.

The picture is a little more complicated where a person accepts employment in a risky occupation. For many years, the courts prevented employees from recovering damages for injuries because of the holding of assumption of risk. Some courts rejected this view, however, instead holding that economic realities often compel employees to engage in risky conduct. These courts thus viewed the risks as involuntary. Under the Worker's Compensation laws,[1] though, the employer is deprived of this defense.

Assumed risks are only the risks normally and naturally involved with the undertaking. If, for example, the roller coaster suddenly became unsupported, with the resulting crash killing and injuring people, recovery would be quite possible.

A person often assumes a risk (of sorts) when he or she becomes a "good Samaritan" volunteer. If a person gives aid to another who is in distress, and this aid results in further injury to the distressed person, the volunteer is liable for such injury.

Contributory and Comparative Negligence

Assumption of risk ordinarily arises from a contractual situation of some sort, but *contributory negligence* comes from the negligent act of an injured party—that is, the victim failed to exercise ordinary care. At common law, if both the plaintiff and defendant were found

♦ ———

1. See chapter 25, Worker's Compensation.

to have acted negligently, the plaintiff could not recover. But this rule led to harsh results. Over time, the doctrine of contributory negligence gave way to today's rule of *comparative negligence*. Under comparative negligence, if both parties were negligent, and injury to the plaintiff would have occurred anyway, recovery may be allowed but diminished by an amount by which plaintiff's neglect contributed to the total damage.

TORTS AGAINST A PERSON

Assault and *battery* are two common personal torts. *False imprisonment* is also included as a tort against a person.

♦ Assault

The term *assault* is quite frequently used improperly. Its legal meaning refers to a *threat of violence*. It consists of one or more acts intended by the tortfeasor (the person who commits the tort) to create an apprehension of bodily harm in the victim. To rise to the level of assault, the tortfeasor must have an apparent, present means of inflicting the bodily harm. For instance, a knife or a pistol (it would make no difference that the pistol was not loaded if the victim had reason to believe that it was) would scare most people. The tort of assault requires that the threat be concerned with immediate injury—not next week or in the future, such as the case where someone threatens injury if they ever see you again. Harm from the tort of assault frequently occurs when the victim has a weak heart or when a pregnant woman receives threats.

♦ Battery

Assault ends and battery begins when the threat is carried out. *Battery* is the intentional and unlawful touching of another in an offensive manner. Battery is often incorrectly reported as assault; the two torts do often go together, but there is a distinction between them. A *battery* need not involve breaking a leg or hitting someone with a fist. An offensive contact amounting to battery could occur when someone gives another an unwelcome kiss or spits in another's face.

♦ False Imprisonment

False imprisonment occurs when one is intentionally confined within limits set by tortfeasor. The victim must be aware that he or she is being confined, and the victim must not have consented to the confinement. The means of imprisonment is incidental as long as the victim's personal liberty is restricted. For example, confinement of a person in a car that is traveling too rapidly for exit to be made safely would be imprisonment. A particular means of confinement might be imprisonment to one, but not to another; that is, an athletic young man might escape through a window, whereas a person confined to a wheelchair could not.

♦ Malicious Prosecution

The right to resort to the courts is sometimes abused. Occasionally, a plaintiff brings a suit against another merely as an annoyance or harassment. The tort involved in such an action is *malicious prosecution* (or *abuse of process* if based on a civil case). The tort hinges principally on the presence or absence of "probable cause." If there were reasonable grounds to believe that the facts warranted the action complained of, this is a defense against a suit for malicious prosecution. Usually, if a reputable attorney recommends an action at law after learning the facts, the plaintiff had "probable cause" to proceed.

TORTS AGAINST REPUTATION

A person has a right to whatever reputation he or she earns in day-to-day dealings with others. If false and malicious statements are expressed (orally or in writing), such statements may constitute *defamation*. Defamation occurs when false statements made about a person tend to expose him or her to public ridicule, contempt, or hatred. For example, defamation occurs when a statement falsely attributes a criminal act to a person. It also occurs in statements that tend to injure one in a job or profession. Defamation takes two forms. Oral defamation is known as *slander*. Printed or written defamation or defamation by pictures or signs is known as *libel*.

It is slander to falsely state that White is embezzling company funds; it is libel if the statement is written. Slander would not occur, however, if the statement were made only to White with no one else present. Someone other than White would have to hear the statement for slander to occur.

Libel results from printed matter. Even a radio or television broadcast of a speaker who reads a defamatory statement from a written article may constitute libel rather than slander. The damages recoverable for libel are usually greater than for slander because of the lasting impression created.

Truth and *privilege* constitute complete defenses to defamation suits. Regardless of the malicious manner in which the statements are made, if they are *true*, there is no defamation. *Privilege* refers to the right of one person to defame another. A judge has this right, and so does a sworn witness on the stand. Privilege is usually found when the otherwise defamatory statements are made in carrying out a judicial, political, or social duty. It arises from the necessity of making a full and unrestricted communication. At common law, slander and libel imposed *strict liability*. That is, the speaker was held liable if the statement was false, no matter how strongly the speaker believed in the truth of the statement. However, the Supreme Court has held that the First Amendment prevents this rule in situations where the speaker is part of the media, or the subject is a "public figure." Hence, when a newspaper (falsely) reports that a politician is known to have alcohol problems, the statements are not actionable unless the newspaper acted with malice in printing the story.

TORTS AGAINST PROPERTY

When one owns property, what he or she really owns is a set of rights. The owner has the right to possess the property, to use it (as long as the use does not infringe upon the rights of others), and to dispose of it. One can normally exclude others from using one's property or from taking possession of it. Tort actions result from the invasion of these rights.

♦ Trespass

The tort of *trespass* to land occurs whenever a person without license enters on the land of another. Even simply walking across a person's lawn is a tort. The law, however, does not concern itself with trifles, and a single instance of trespass such as the invasion of one's lawn probably would not be actionable. Even if it were, the result would likely be only nominal damages. An action for trespass is more likely when the trespass has been repeated numerous times or when material damage can be shown. Such damage to real property can be shown, for instance, where the foundation of a structure encroaches upon the property of another.

Consider this example. Black builds a pond near the edge of his property line (according to a survey). White, the owner of the adjacent property, later has another survey made. The later survey shows part of Black's pond to be on White's property. The court concludes that White's survey is correct. An equity court (where such a case would likely wind up) has the right to order the reconstruction or draining of the pond.

The person who is in possession of the land has the right to exclude others from trespassing. In other words, those who rent or lease property also have the right to exclude others from it (even the owner) as long as they lease the property. Still, personal rights take precedence over property rights. One does not ordinarily have the right to shoot trespassers. The force used usually must be no more than sufficient to remove the person from the property.

♦ Trespass Exceptions

At common law, a trespass involved no finding of intent or even negligence. If you entered or caused something to enter onto another's land, you were liable. Because this led to harsh results, the courts developed exceptions, whereby certain actions (which might otherwise amount to a trespass) did not have the harsh results that the old common-law rules sometimes caused.

Easement or License

If numerous members of the public use a person's property as they desire, an *easement may result*. If Black, for instance, owns lakeside property and the public crosses his property to reach the lake, he may eventually be prevented from excluding the public. The period of time for such a *public easement* to occur often runs from 15 to 20 years, depending on the state. To create the right of easement, the public use must be continuous. It is for this reason that one occasionally sees a road blocked off for one day per year.

The right to go upon another's land can be given by the person in possession. Such permission is known as a *license*. A caller at a home, for example, has the right to go as far as the door by a direct route. If a person must enter another's property to recover his or her own property, that person also has a right to do so.

Attractive Nuisance

Ordinarily, one who trespasses upon another's property assumes whatever risk may be inherent in the trespass. If the trespasser is injured by some hidden danger, that person has little chance of recovery against the owner. But, just as is true with many other general rules of law, this one has its exceptions. Probably the most prominent exception is known as *attractive nuisance*, pertaining to children of tender age. This principle is of recent origin as legal doctrines go, and it has been rejected by some courts, but the number and size of recoveries prompt its consideration. The doctrine of attractive nuisance began in the United States with a case involving an injury to a child playing around a railway turntable. In the century or so since then, a multitude of property conditions and instrumentalities (including swimming pools) have come to be considered attractive nuisances for children.

In jurisdictions where the attractive nuisance doctrine is applied, a property owner or occupant may be held liable for injuries sustained by children on his or her premises under the following conditions: (1) if the owner knew or should have known that the dangerous instrumentality or property condition would be attractive to children and failed to reasonably guard against injury to them or (2) if the owner had reason to expect children to play there (e.g., having seen them play in the vicinity) and did not warn them or take other suitable precautions.

The owner's (or occupant's) risk of attractive-nuisance liability is removed by taking reasonable precautions. For example, the owner is not expected to foresee very unlikely events—only those that might befall a normal, inquisitive child; and he or she would not be expected to guard something of danger obvious even to a child. The doctrine is aimed at conditions that would be inherently dangerous to a child but that the child could not be expected to foresee. Thus, an unguarded piece of machinery could easily be an attractive nuisance, whereas an open pit in a field would less likely be one. Generally, the attraction must be something unusual, uncommon, or artificial, as opposed to a natural hazard.

Attractive nuisance cases could involve children of any age, but children between 5 and 10 years old seem particularly susceptible. The courts also consider such things as the child's intelligence, state of mental health, and other conditions as significant in such cases. The largest factor, though, is the presence or absence of proper precautions by the owner or occupant of the premises.

♦ Conversion

The tort counterpart to the crime of theft is *conversion*. Conversion usually constitutes the wrongful retention of another's personal property. It also includes the wrongful alteration of property and the wrongful use of property by persons other than the owner. Conversion may arise in instances of bailment, where something left with another is used or sold by the bailee. For example, Black leaves a television set with White (as bailee) to be repaired. White sells the television set to Gray. White's tortious act is conversion, for which Black may maintain a conversion action in court. A successful suit in conversion normally nets the true owner of the property the market value of the converted property and vests title in the converter when the judgment amount has been paid. The owner, though, has two possible remedies available. He or she may sue on the tort of conversion or maintain an action *in replevin* to obtain the return of the property. In a replevin action, the owner sues to gain possession of the property (as opposed to suing for damages). If the owner wants the property back before the replevin action, he or she may usually obtain it by posting a bond to be forfeited in case the property is found to belong to the other party.

TORTS AGAINST ECONOMIC RIGHTS

In the United States, the government and the courts protect the right to compete with others in a business venture. Despite the likelihood that entrance into a particular field by an efficient newcomer may injure or even eliminate an established concern, such competition is favored. Usually, the result is healthy. The general tendency is to encourage efficiency, since the public benefits from it in lower consumer costs.

Competition, though, can lead to its own destruction. If a business drives all less efficient concerns from the field, a monopoly results. Since unregulated monopoly is usually associated with excessive prices, inefficient operation, and other undesirable effects, laws have been designed to preserve competition.

Since the field of law treating competitive practices is very large, no attempt will be made to cover the entire field here. Rather, a few of the most common torts against economic rights will be mentioned.

♦ Fraud and Misrepresentation

The tort of fraud occurs all too often. The elements of fraud are (1) a false representation of material fact (2) knowingly made (3) with the intent that the plaintiff rely on it, (4) reasonable reliance by the plaintiff on the representation by the defendant, and (5) damages due to the plaintiff's reliance.

The false representation must be about facts. Statements by a used car dealer, for example, that a car is "great" will probably not give rise to a claim for fraud. Courts tend to view such statements as opinion, and false statements of opinion generally are not actionable.

Other important concepts relating to fraud are its emphasis on the speaker's knowledge and intent and on the recipient's reliance. If the speaker truly believes what he or she says, there's no *intent* to mislead. In short, the speaker simply may not know that the statement is false. Second, the recipient must *rely* on the false statement before fraud can occur. If there's no reliance, then the recipient can't claim that he or she was harmed by the statement.

More recently, courts have recognized the tort of *negligent misrepresentation*. Like fraud, this tort protects against damages due to reliance on another's false statements. However, only the *negligence* of the speaker is required. If the speaker negligently made false statements, then there may be liability. The courts have been careful to limit this tort to situations where one party is clearly relying on the other in connection with some other transaction or relationship. In short, the courts are reluctant to impose blanket liability for any misstatement made negligently.

Suppose that Black, a civil engineer, is hired by White Mortgage to survey some land along an interstate highway. Black drinks too much at lunch one day and, as a result, his survey is flawed. Nonetheless, Black's survey is used by White Mortgage, which issues title insurance for the purchaser of the property. Because surveys are commonly relied on in connection with real estate transactions, Black probably could have foreseen that both White Mortgage and the buyer of the land would rely on the survey.

Consequently, surveyors have been held liable to purchasers of property. Note, however, that it would be practically impossible to show fraud by Black; Black's actions would instead be a negligent misrepresentation.

♦ Inducing Breach of Contract

Although breach of contract is treated under the law of contracts, inducing another to breach a contract is a tort. This is usually referred to as *tortious interference* with contract. According to the ancient common law, inducing a breach of contract was not actionable unless it was accompanied by violence or fraud. This concept was changed by the case of *Lumley* v. *Gye*,[2] in which an opera singer was induced to breach her contract and work for another. Though no fraud or violence had occurred, the court stated that a right of action against the person inducing the breach of contract existed.

Since that time, the courts have developed a significant body of law about when someone may encourage another to breach a contract. For example, merely advising a prospective buyer of the merits or properties of a product is not inducing breach of contract. The end result may be a breach of contract, but the seller must have actively persuaded the customer to breach if the seller is to be justly accused of having a hand in it. As a further example, Black has a contract to buy parts from the White Screw Machine Products Company. Gray offers to sell Black better parts at a lower cost. Black breaches his contract, but Gray cannot be said to have induced the breach of contract unless he actively advocated Black's breach.

Courts have broadened this tort to cover conduct where a person interferes with another's prospective contracts (i.e., an interference with the expectation of a contract). As you might imagine, however, the courts tend to view normal competition as privileged. In other words, a company is free to submit a lower bid to obtain a contract. However, doing more to advocate a breach, such as by disparaging a competitor's services or resources, runs a risk of moving from competition to a tortious interference. In situations where a contract already exists, though, the courts do not always view competition as a sufficient basis for interfering with the contract.

♦ False Advertising

The presence of false or misleading advertising is often apparent in our daily lives. Under common law, the only remedy afforded a person injured by such advertising was an action for fraud or deceit. However, federal and state statutes have modified the common law, and the courts have expanded the possible grounds for suits, as well as the remedies available. The Federal Trade Commission also seeks to *prevent* deceptive advertising, as do many state statutes. Enforcement, though, is often a major problem.

The federal Lanham Act governs federal trademark registration and trademark infringement actions. It also includes a provision that allows a company to sue a

2. Ellis & Blackburn 216, 118 Eng. Rep. 749 (1853).

competitor if the competitor engages in false advertising. (For historical reasons, the Lanham Act was viewed as preventing "unfair competition." One type of unfair competition is trademark infringement; another type is false advertising.) Not all false advertising is actionable, however. First, the statements must be false, misleading, or deceptive. Second, consumers must be confused or misled, or there must be a likelihood of confusion or deception. Third, the false statements must be material: They must be something that would influence the purchasing decisions of consumers. Under the Lanham Act, the court may enter an injunction to prevent further false advertising. In addition, a winning plaintiff may be able to recover damages and, in some situations, an accounting of the defendant's profits due to the false advertising.

Closely akin to false advertising is the *disparagement* of another's product. The tort resembles libel and is often called *trade libel*. Essentially, the law prevents a person from making false and misleading statements about a competitor's products. In addition to the preventive relief of injunction, damages for lost profits may be obtained if special damages (such as lost profits) can be shown. If any relief for disparagement is to be forthcoming, the plaintiff usually must prove the following:

1. The statements made by defendant were untrue.
2. The false statements were made as fact (rather than opinion).
3. The statements concerned the plaintiff's goods in particular—that is, defendant's statement that his or her goods were better than those of all competitors would not be disparagement.
4. The false statements were made with malice.
5. The plaintiff suffered some type of special damages, such as lost profits.

Disparagement is a little more difficult to establish than libel. In a disparagement case, it is the *plaintiff* who must prove all the elements. If the Plaintiff cannot prove each of the required elements, the plaintiff loses. In a libel case, however, the *defendant* must bear part of the proof burden. That is, in a case for disparagement, the plaintiff must prove that the statements made were *untrue*, whereas in a libel case the defendant would have the burden of proving statements true.

Just as in libel and slander, certain persons have the *privilege* of disparagement. If, for example, in the interest of preserving life and health, a doctor warns against the use of certain foods or drugs, the doctor does so with privilege. The same is true if a family member warns another member of the same family against using certain things. Furthermore, consumer organizations usually act without malice; they expect the public to benefit from their services. Each case involves a personal interest in benefiting others.

As you might imagine, some tension exists between the promotion of free speech under the First Amendment and the discouragement of speech that is misleading or deceptive. Hence, the courts are careful not to unduly prohibit advertising that is truthful.

NUISANCE

As you know, property owners have the right to use their property as they choose as long as they do not, in some way, injure the person or property of another. If one property owner does use his or her property in a way that injures the person or property of another, and the tort fits no other category, the tort of *nuisance* may cover it. A nuisance can be just about anything that interferes with the enjoyment of life and property. It may take the form of smoke or sulphur fumes, or pollution of a stream, or of excessive noise, to mention only a few types of nuisances.

Nuisances are either public or private. A public nuisance is one that interferes with the rights of a substantial number of the persons in a community. A private nuisance produces special injuries to the private rights in real property of one or a very few people. Any citizen may successfully lodge a complaint about a public nuisance, but only the person injured can maintain a successful action based on a private nuisance.

Consider this example. Black owns a factory in which semi-trailers are manufactured. Since the manufacture requires the use of rivets, the process is quite noisy. When Black first built the factory, several years ago, the building site was a cornfield, and the nearest neighbor was some two miles distant. With the passage of time, Black's trailer business expanded. Moreover, adjoining land was sold to a land development company, and houses have been built and sold. Recently, orders for trailers have forced Black to put on a third shift at the factory, from 11 p.m. to 7 a.m. Some of the new homeowners have complained to Black about the noise; one (White) has even instituted a nuisance suit. In answer, Black contends that he was there first, that an injunction would force him to close down his plant and deprive workers of jobs, and that the noise just isn't great enough to injure anyone anyway.

A variety of judgments would be possible in such a situation. Black's claim that he was there first and, thus, acquired a right to maintain the "nuisance" is doubtful to succeed as a defense. Proof by White that the noise had increased in time and intensity would be likely to defeat that defense. Such a defense might only succeed if White had full knowledge of the noise problem (as to degree and time of day) and bought his house in spite of the noise. This is often known as "coming to the nuisance."

As to Black's second claim—that an injunction would impose a hardship on him and his business—the response would differ from state to state. Certain state legislatures have adopted policies that encourage business migration into their states. In such places, the courts are reluctant to shut down an industry or company. The *balance of hardships* would probably also be considered here—for example, is it a greater hardship to the homeowners if operations continue, or is it a greater hardship to the company if it must eliminate the third shift (or perhaps move operations)? Hardship in terms of job loss and loss of income to the community would be weighed against the noise annoyance. The likely result of such a case in our example is a decree requiring Black to do all in his power to abate the noise problem. Many measures can be taken to attenuate such industrial noises.

JUSTIFICATION

Under certain circumstances, tortious conduct may be justified. For example, a person is justified in trespassing upon another's land if he or she must do so to regain possession of personal property. A person is justified in striking another in self-defense. License (permission) may be given to commit an act that is tortious in nature—such as a license to trespass upon the land of another in making a survey. Moreover, legal authority may be given to allow the commission of tortious acts—as the authority a police officer may exercise.

While a tort may be justified in some way, the person committing the tort still must exercise restraint and refrain from going beyond the limits justified. A person defending himself against the attack of another, for example, is justified in his defense up to the point where he becomes the aggressor. If he goes beyond the point of justification, he becomes answerable for the injury caused by his acts.

A defendant may justify a tort if he or she can prove that an inevitable accident caused the injury. Here, the defendant must show that he or she did everything reasonable under the circumstances to prevent injury. Accidents resulting from natural causes such as lightning, storms, and earthquakes, for example, are inevitable accidents.

TIME LIMITATIONS

An injured party should usually take prompt action against a tortfeasor. Most states have statutes of limitations for tort actions—the action must be instituted within so many years after the tortious act. If not timely filed, the action cannot be brought. Tort actions undertaken in a court having equity jurisdiction also run the risk of losing out to time. The equity term *laches* indicates a cause in which the plaintiff has "slept on his rights" too long. Stale causes are not popular, the feeling being that the court should not be more protective of the plaintiff's rights than the plaintiff was of his or her own.

DISCHARGE OF TORTS

A defendant may discharge the obligation to pay for damage from a tortious act in a couple of ways. First, not all causes of action find their way into court. Many are discharged by a simple agreement between the parties. Thus, the out-of-court settlement agreement (accord and satisfaction) is a common means of discharge. Rather than take the case to court, the tortfeasor agrees to pay the injured party for the damage done, thus avoiding court costs (in both time and money) and lawyer's fees.

If the case does goes to a jury, a judgment results. If the judgment is for the defendant, you might say the court has discharged the potential tort liability. The amount of the judgment in a tort case is generally made up of two elements: (1) the out-of-pocket cost to the plaintiff—such as medical costs, lost wages, and the like, and (2) compensation for pain and suffering (if any).[3] Here, bankruptcy of the tortfeasor may act as a discharge of sorts for tort obligations. If a tort action has been instituted, a judgment rendered, or the oblig-

3. Under common law, only the injured party was allowed to bring a tort action. If the injured party died, the cause of action ended. This has been changed by the almost complete adoption of survival statutes, which allow others to sue in the name of the deceased.

ation reduced to a contract before the bankruptcy proceedings are begun, the injured party shares in the bankrupt's estate as any other creditor. If no suit has been brought or contract made on the tort obligation, though, the tortfeasor's assets go to meet his or her obligations to creditors. The injured party's cause of action remains after bankruptcy, and he or she may elect to sue the tortfeasor for whatever remains.

STATE v. H. SAMUELS COMPANY, INC.
211 N.W.2d 417 (Wis. 1973)

Several issues are raised in the briefs, but the only one which is dispositive of the case is whether the repeated violation of a city ordinance constitutes a public nuisance which ought to be enjoined.

H. Samuels Company, Inc., has operated a salvage business in block 137 in the city of Portage since the early 1900s. In 1948 the junk business was expanded to include the salvaging of metals from automobiles and other machinery. Cranes were used after 1949, a guillotine shears after 1966, and a hammer mill after about 1971. At one time Samuels operated around the clock, but at the time of trial the operation at night had been reduced. In the processing of scrap metal, Samuels utilized railroad cars, trucks, heavy-duty cranes, guillotine shears, oscillators, conveyor belts, air tools, hammer mill, and metal-sorting equipment. Prior to 1966, block 137 was zoned commercial and light industry, but in 1966 the zoning was changed to heavy industrial. Block 137 is the only block so zoned in the developed portion of Portage. The areas immediately adjacent to block 137 are zoned either residential, single-family homes or commercial and light industry.

The defendant has a license to operate a junkyard. In its operation, the defendant unloads scrap metal from railroad cars with a magnetized crane and drops the metal into a steel guillotine shears which snaps the metal and drops it onto an oscillating conveyor belt, which in turn drops it on a pile or to a sorting house. Other operations involve a two-ton magnet lifting a car engine to the height of four feet and dropping it onto a large piece of steel wedged into the ground. Air tools are used to dismantle the engines, and the hammer mill is used to hammer metal into pieces in a large drum and to drop them on a conveyor belt where they are washed and sorted. The alleged nuisance consists of the air noise and ground vibrations created by the operation.

The city of Portage has an ordinance prescribing maximum permissible noise and vibration levels. The state contended the Samuels Company has repeatedly violated this ordinance and will continue to do so to the injury of the public. At the trial the state of Wisconsin attempted to prove the alleged nuisance by the testimony of two expert witnesses who monitored the noise and by the testimony of neighborhood homeowners of the disruption of their life patterns. The homeowners testified to their loss of sleep, domestic discord, added expense in remodeling their homes, suspension of home remodeling, moving from the neighborhood, rattling of windows, loss of hobbies such as working out of doors, loss of use of porches and yards for relaxation, shaking of pictures and furniture, shaking of beds, and rattling of dishes. The two experts testified their tests showed that at various times the sounds caused by the operation exceeded the maximum permissible decibel levels and sound frequencies established by the city ordinance of Portage. They also testified that the vibrations emanating from the salvage yard exceeded permissible displacement values prescribed by the ordinance for areas zoned heavy industrial.

The defendant's testimony consisted of the testimony of the chief of police who related that of the 32 complaints received in 1970, 24 were by one person and the rest by 4 persons; in 1971, of the 47 complaints, 38 were from 1 party and the balance from 7 other persons. No action on behalf of the city of Portage has been taken to enforce the city ordinance against

Samuels. The president of Samuels testified he had equipped his cranes with silencers, that he had reduced his operation, and intends to reduce the handling of automobiles in the future. In the area where defendant's plant is located, there are other industrial plants.

In its decision the trial court stressed that an injunction to enjoin a public nuisance was a drastic remedy, that the city of Portage had never brought an action for the violation of the ordinance against the defendant and "the defendant had taken considerable steps to improve the situation" and was operating a legitimate business where it had been carried on for many years. The court acknowledged the operation of the defendant's plant "is obviously an annoyance to the immediate neighbors." However, the court was impressed by the reasoning in the concurring opinion in *State ex. rel. Abbott* v. *House of Vision*...to the effect that before an injunction will issue when a statute has been violated, an effort must be made to prosecute for the violation of the statute, as this remedy is presumably adequate. The court also considered some economic factors, although it stated they were not directly involved, and commented that if an injunction to enjoin the nuisance required the defendant to stop operation, that would be taking of defendant's property without adequate compensation and therefore unconstitutional. The court concluded the case did not constitute such a case as called for the use of an injunction.

We think the trial court was in error. A public nuisance may be proved by a few witnesses. It is the extent and the nature of the acts and the resulting damage which are important, not the number of witnesses. The court questioned the accuracy of the tests performed by the state's experts by presuming noise from sources other than the Samuels plant contributed to the result of the tests. The presumptions are contrary to the testimony and the evidence. The fact the defendant has made some efforts to cut down the amount of noise does not go to the question of the existence of a nuisance. Neither the legitimacy of the business nor the length of time it has been in existence is controlling in determining whether a public nuisance exists. These factors are relevant to the question of whether the court should exercise its discretion to enjoin the nuisance. The reliance on the concurring opinion in *State ex. rel. Abbott* v. *House of Vision*, supra, is misplaced. The concurring opinion is really a dissent on the issue of whether a nuisance existed and whether a crime was enjoined because it is a crime. This dissent relied on *State ex. rel. Fairchild* v. *Wisconsin Auto Trades Association*.... But the language in *Fairchild* must be read in the context of its facts and the express statement, "There is no claim the acts of the respondent constitute a nuisance, public or private."

True, a court of equity will not enjoin a crime because it is a crime, i.e., to enforce the criminal law, but the fact the acts complained of cause damage and also constitute a crime does not bar injunctional relief. The criminality of the act neither gives nor ousts the jurisdiction of equity.... In such cases, equity grants relief, not because the acts are in violation of the statute, but because they constitute in fact a nuisance....

This view must be distinguished from the doctrine that the repeated violation of a criminal statute constitutes *per se* a public nuisance. This doctrine justifies the issuance of an injunction not to enforce the criminal statute but to enjoin illegal conduct which, of its repetition, constitutes a nuisance. Under this doctrine, a violation or a threatened violation of the statute does not constitute a nuisance. The violations of the statute must take place and be repeated to the extent their repetition affects such public rights as will constitute a nuisance.

In the majority opinion of *State ex. rel. Abbott* v. *House of Vision*, supra...this court said...:

"But the *Thekan* and *Cowie* cases received much study when they were before us and have been re-examined now and we conclude that the language of their opinions, broad as it is, expresses our view of the law and is applicable to acts repeatedly performed and with the avowed purpose of continuing, which do violate a statute whether or not they might be lawful under other and different circumstances. Consequently, we hold that if the statutes are violated as charged in the complaint a public nuisance is committed and the equitable remedy of injunction may be invoked."

The modern concept of injunctional relief is to use it when it is a superior or more effective remedy. This concept is illustrated by the many statutory provisions using an injunction to enforce sanctions and regulations in the commercial world. While the *Thekan* and *Cowie* cases might be distinguished, their broad language supported this court's view in *State* v. *J. C. Penney Company*...upon which this court enjoined repeated violations of a usury statute on the ground that the open, notorious, and flagrant violation of valid laws enacted for the benefit of the people of this state constituted a public nuisance. The court took judicial notice of the widespread use of the revolving charge accounts and the large number of Wisconsin citizens affected by these practices, and thus concluded the violations were a public nuisance which ought to be enjoined.

It would seem this court is now committed to the proposition that the repeated violation of criminal statutes constitutes *per se* a public nuisance. But whether such nuisance should be enjoined depends upon the amount of damages caused thereby and upon the application of the doctrine of the balancing of equities or comparative injury in which the relative harm which would be alleviated by the granting of the injunction is considered in balance with the harm to the defendant if the injunction is granted. If the public is injured in its civil or property rights or privileges or in respect to public health to any degree, that is sufficient to constitute a public nuisance; the degree of harm goes to whether or not the nuisance should be enjoined. The abatement of a nuisance by an *in rem* action is to be distinguished, as those cases generally involve a statutory declaration of what conduct constitutes a public nuisance and a statutory authorization for abatement against the property....

In *Jost* v. *Dairyland Power Cooperative*...this court reviewed the theory of public nuisance in the context of a suit for damages. The court relied on *Pennoyer v. Allen*...in which it was stated it was no defense to show that the business was conducted in a reasonable and proper manner and that the injury might have been done with more aggravation. The court emphasized that the destruction of the enjoyment of the plaintiffs of the comfort of their homes furnished the ground for action. It was pointed out that the lawfulness of the business did not justify the invasion of private rights. In *Dolan* v. *Berthelet Fuel & Supply Co.*...the court relied on *Pennoyer* v. *Allen*, supra, and concluded that even though a coal yard was operated properly and was a social and economic business, as was found by the trial court in this case, nevertheless the operation of the coal yard would be enjoined if it caused substantial damage to the adjoining plaintiff.

We need not decide whether defendant's operations in this case amount to a criminal-law nuisance because we think the repeated violations of the city ordinance constituted as a matter of law a public nuisance. We see no valid distinction between a city ordinance regulating business conduct and a criminal state statute. The cases involving city ordinances may be even stronger because ordinances in this state in their nature are not criminal. An ordinance is regulatory and prohibits undesirable conduct, but the consequence for its violation is a forfeiture rather than a fine or imprisonment.... A *fortiori*, if repeated violations of a public statute as in *Penney* constitute a public nuisance, then the repeated violations of an ordinance constitute a public nuisance....

It does not follow necessarily that the public nuisance resulting from the repeated violation of a statute or ordinance will be enjoined; this depends upon the degree of harm. In the instant case it may well be that the amount of harm caused by the repeated violations during the normal work hours of the day is insufficient to call forth an injunction, but the same degree of violation impairing the public's right to the enjoyment of their homes after normal working hours causes a greater injury and ought to be enjoined. If we were to consider the facts in this case in relation to establishing a criminal-law nuisance, it might well be that an operation less than that allowed by the ordinance would constitute a public nuisance which should be enjoined. However, the briefs ask for an injunction to limit the operation to what is permitted by the ordinance between 5:00 p.m. and 7:00 a.m. On the theory of our reversal, the injunction can only enjoin operations which constitute violations of the ordinance.

Judgment is reversed, with directions to enter a judgment enjoining the operation of the defendant from violating the city ordinance as to noise and ground vibration during the hours of 5:00 p.m. and 7:00 a.m. each day of the week.

MONARCH INDUSTRIES, ETC. v. MODEL COVERALL SERVICE

381 N.E.2d 1098
(Ind. Ct. App. 1978)

Plaintiff-appellant Monarch Industrial Towel and Uniform Rental, Inc. (Monarch), filed a complaint of tortious interference with a contractual relationship against Model Coverall Service, Inc. (Model). Monarch alleged that Model committed said tort by inducing Emmert Trailer Company (Emmert) to breach its uniform rental service contract with Monarch. (Emmert executed a similar contract with Model after terminating the contract with Monarch).

Following Monarch's presentation of evidence at trial, the trial court granted Model's motion for judgment on the evidence, pursuant to Ind. Rules of Procedure, Trial Rule 50. Monarch perfected this appeal arguing that the court erred in granting Model's motion for judgment on the evidence and in refusing to allow Monarch to reopen its case.

Indiana recognizes the tort of interference with contract relationships by inducing a breach of contract and has defined the essential elements which must be proved for recovery under such an action in *Daly* v. *Nau*..., as follows:

1. existence of a valid and enforceable contract;
2. defendant's knowledge of the existence of the contract;
3. defendant's intentional inducement of breach of the contract;
4. the absence of justification; and
5. damages resulting from defendant's wrongful inducement of the breach.

Model's T.R. 50 motion was grounded upon the lack of proper evidence as to damages. But it is well established that the judgment of the trial court will be affirmed on appeal if sustainable on any basis.... The evidence presented by Monarch failed to prove that Model intentionally induced Emmert to breach the contract with Monarch. Rather, the evidence shows that Emmert was dissatisfied with the service that Monarch was providing and had been looking at other rental services. The evidence is that Model approached Emmert to inquire as to whether Emmert was being supplied with a uniform rental service and that the Model sales representative was told that Emmert was not happy with the present supplier and intended to cancel the contract. And there was no evidence that Model offered any inducements to Emmert in the form of a better price or better services. Rather, the evidence is that there was no discussion of prices until Emmert had already given Monarch notice of termination of the contract and that the price Emmert was paying under its contract with Model was higher than it had paid under the contract with Monarch.

To sustain a judgment for defendant on the evidence, the evidence must be without conflict and susceptible of but one inference in favor of the moving party. If there is any evidence or legitimate inference therefrom tending to support at least one of plaintiff's allegations, a directed verdict should not be entered.... The evidence presented at trial clearly showed that Emmert elected to terminate the contract with Monarch and made the decision independent of Model's approach and subsequent contract with Emmert. There was no evidence from which an inference of intentional inducement to breach on the part of Model could be drawn....

Inasmuch as Monarch failed to prove inducement to breach, this Court need not discuss the sufficiency of the evidence as to the damages incurred by Monarch due to the breach of contract by Emmert or Monarch's argument that granting the T.R. 50 motion denied Monarch its rights to have the jury consider punitive damages.

Finally, whether the court erred in refusing to allow Monarch to reopen its case to present additional evidence as to the net profit it would have earned had the contract been completed is a question this Court does not decide. Error, if any, was harmless since that evidence could not have prevented the failure of the plaintiff's case on the essential element of intentional inducement to breach the contract....

The judgment is affirmed.

Affirmed:

STATON, J., and CHIPMAN, P. J., participating by designation, concur.

MENIFEE v. OHIO WELDING PRODUCTS, INC.

472 N.E.2d 707

Syllabus by the court

Under strict tort liability principles for the design of a product, a manufacturer need not anticipate all uses to which its product may be put, nor guarantee that the product is incapable of causing injury in all of its possible uses.

This appeal surrounds the death of appellant's decedent, William L. Menifee, from the inhalation of nitrogen gas while in the course of his employment at General Electric Company (hereinafter "General Electric").

The record establishes that the decedent was employed by General Electric to clean transformers and other equipment by using a pressurized stream of aluminum oxide. In addition to compressed air, nitrogen was used in the cleaning process. Liquid nitrogen was stored in a bulk tank and piped to the location where the cleaning took place. The compressed air and nitrogen were released through the same outlet in the grit blast building in which Menifee was working, and the flow of each substance was controlled by certain valves within a system of pipes intended to prevent the nitrogen from mixing with the compressed air.

On the day of the accident, the decedent was wearing a mask and receiving air to breathe from the compressed air supply line. However, the compressor overheated and automatically shut off and, thereafter, the valve system somehow allowed the nitrogen to enter his air supply. Menifee died four days later as a result of oxygen deprivation. Appellant, Ernestine Menifee, administratrix of the estate of William Menifee, subsequently filed this wrongful death action naming several defendants based on the following facts.

In November 1976, the management of General Electric had decided to expand its facilities due to additional sales and maintenance burdens associated therewith. General Electric entered into a contract with appellee KZF Incorporated, an architectural firm, for the design of the grit blast building and air supply lines which would extend from the existing facility to the new building.

General electric thereafter contracted with appellee J and F Harig Company to erect the building and air system as designed. In turn, the construction of the system was subcontracted to appellee Peck, Hannaford and Briggs Co. After the system's completion, General Electric contacted appellee Ohio Welding Products, Inc. for installation of a nitrogen gas holding tank and distribution system. The actual mounting, connecting and valving of the nitrogen gas system were performed by appellees A.L. Miller and A.L. Miller Plumbing, Inc. The air compressor was manufactured by appellee Cooper industries, Inc. and obtained from appellee Highway Rental Equipment Company, Inc.

The building and air system were completed on or about April 27, 1978. It is evident from a review of the record that all work was performed in the manner specified by General Electric. In addition, it is important to note that General electric did not inform any of the appellees that the compressed air would be used for breathing purposes.

The complaint filed on behalf of the appellant sounds in negligence and strict tort liability. Appellant alleges that Menifee died as a proximate result of the combined and concurrent acts by appellees who should have foreseen their consequences. The trial court granted summary judgment as to all appellees on the basis that no liability existed due to the lack of knowledge that the compressed air was used for breathing purposes. The court of appeals affirmed this ruling.

The cause is now before this court pursuant to the allowance of a motion to certify the record....

Holmes, Justice.

The issue presented is whether summary judgment in favor of appellees was proper on the theories of negligence and strict tort liability. For the reasons set forth below, we affirm the appellate court's ruling.

The crux of appellant's negligence argument is that it was foreseeable by appellees that the air supply system would be used by General Electric for breathing purposes. Appellant contends that due to the alleged foreseeable use of the system, a duty arose on the part of appellees to inquire of General Electric as to the system's contemplated use. Thus, argues appellant, appellees would have a duty to prevent any and all foreseeable injuries arising from such use.

It is rudimentary that in order to establish actionable negligence, one must show the existence of a duty, a breach of the duty, and an injury resulting proximately therefrom. ...The existence of a duty depends on the foreseeability of the injury....

The test for foreseeability is whether a reasonably prudent person would have anticipated that an injury was likely to result from the performance or nonperformance of an act....The foreseeability of harm usually depends on the defendant's knowledge....

In determining whether appellees should have recognized the risks involved, only those circumstances which they perceived, or should have perceived, at the time of their respective actions should be considered. Until specific conduct involving an unreasonable risk is made manifest by the evidence presented, there is no issue to submit to the jury.... Although each appellee raises various arguments in its defense, they all stand on one common ground: General Electric was the only entity with the knowledge required to prevent the decedent's injuries. It was clearly established by the parties before the trial court that only General Electric knew that the compressed air was going to be used for breathing purposes. In fact, General Electric represented to the appellees that the compressed air was going to be used to power air tools. Therefore, in the absence of the requisite knowledge, appellees could not have foreseen or reasonably anticipated the decedent's injuries and, as a matter of law, cannot be held liable for negligence....

Addressing her theory of strict tort liability, appellant argues that appellees may be held liable not only where the use of their product was one that was intended, but also where the use of the product for one specific purpose, among other conceivable uses, was, or should have been, perceived by the manufacturer. In support of this proposition, appellant argues that evidence was submitted to the trial court, by way of affidavit and deposition, that air compressors are commonly used in industry to supply breathing air. Appellant further contends that such use of these air compressors has been the subject of articles in industry and trade journals, and that this particular use is governed by safety standards established by the Occupational Safety and Health Administration and the American National Standards Institute. Accordingly, appellant argues that issues of liability have been raised pursuant to the principles of Section 402(A) of 2 Restatement of the Law 2d, Torts (1965) 346–347.

Even assuming that all of the appellees could be held responsible for the production, or the functioning, of the product in question, the facts before us do not permit a finding of strict tort liability. The design of a product cannot be held defective or unreasonably dangerous under Section 402(A) of the Restatement of Torts 2d unless the product is being used in an intended or reasonably foreseeable manner.... Furthermore, a manufacturer need not anticipate all uses to which its product may be put, nor guarantee that the product is incapable of causing injury in all of its possible uses....

In the case *sub judice*, appellees were told that the air system was to be used to generate power for air tools within the grit blast building. The system, including the air compressor, was specifically designed for this purpose. Appellant has failed to produce any evidence which would lead us to believe that appellees could have reasonably anticipated that the air system was intended for breathing purposes. Therefore, appellees are entitled to summary judgment as a matter of law on this issue.

Accordingly, the judgment of the court of appeals is affirmed.

Judgment affirmed.

REVIEW QUESTIONS

1. Name a tort that is not a crime.
2. What generally must be proved in a tort action?
3. Black, an engineer, is injured while visiting the White Manufacturing Company. Black was splattered in the face with hot metal from a die casting machine he was observing at the time. What complaint and reasoning might Black use in a tort claim? What reply would White be likely to use?
4. The Green Paper Company has responded to an invitation to set up a plant in a particular community. Several millions of dollars have been spent for buildings and equipment. However, in the first few months of operation, numerous complaints have been lodged and injunctions requested. It is claimed that the odors peculiar to the industry have lowered local property values and that the discharge of "black liquor" and dyes in the local stream has eliminated fishing. The plant employs approximately 1,000 people. How would the court be likely to treat the problem?
5. How does nuisance differ from attractive nuisance?
6. How does disparagement differ from defamation?
7. What practical effect does the court's decision have on the defendant in *State* v. *H. Samuels Company, Inc.*?
8. In *Monarch Industries* v. *Model Coverall Service,* what added evidence did Monarch need to win its case and recover from Model?
9. In the case of *Menifee* v. *Ohio Welding Products, Inc.*, Mrs. Menifee recovered nothing from the defendants for her husband's death. The reasoning by the court leads one to believe that perhaps General Electric, her husband's employer, may be at fault. Can she take successful legal action against G. E. in the case? Why or why not?

Chapter 22

Products Liability

As we have seen, tort law consists of a body of rules, statutes, legal theories, and principles designed to assure recovery for private injury. The plaintiff makes a case by showing that the defendant neglected a duty owed to the plaintiff and that this was the proximate cause of the plaintiff's injury. But now, suppose the defendant is the producer of a product that injured the plaintiff. To what extent may the plaintiff recover? What defense could the producer use? Can the plaintiff obtain evidence held exclusively by the defendant? Must the plaintiff show negligence? The answers to these and many related questions comprise the relatively new and currently expanding field of products liability.

HISTORY OF PRODUCTS LIABILITY

The idea of recovering from the producer for a product-related injury is of quite recent origin. A century or so ago, such a case would have been virtually inconceivable. Even several decades ago, products liability cases and recoveries were fairly rare. But in the 1960s, revolutionary changes in the field occurred; to understand how and why these changes came about, we must review the common law we inherited from England in this regard.

♦ Early History

Early English social and legal philosophy reflected the manufacturing nature of the economy. Producers of goods and services were held in high esteem. Their success meant success of the nation. The legal climate fostered their growth. Both logic and social philosophy supported the legal defense available if someone complained about a product—*caveat emptor* (the buyer beware). The logic was simple: people should examine what they are to receive before they buy it. If a purchaser is so negligent that he or she does not examine before buying, then the purchaser should live with this bad bargain.

Allowing a buyer to recover for a bad product was viewed as supporting a buyer's negligence, and the law usually will not aid those who are negligent. But then, of course, the products produced in those days were somewhat more easily examined than what we buy today. One can easily see defects in a shovel or wheelbarrow, but automobiles, television sets, automatic washers, and the like have created a rather different product climate. It is a rare consumer who would understand the internal components and functions of a new car even if that consumer were to dismantle it before buying it. So, is the consumer being negligent when he or she buys in reliance on those who produced the vehicle?

Another traditional defense available to producers was the requirement of *privity of contract*—the idea that one who is not a party to a contract should have no rights arising from it. In other words, if a consumer was injured by a product, but the consumer had not bought it directly from the manufacturer, he or she could not act against that manufacturer to recover for the injury. The producer or manufacturer only needed to interpose a middleman—a wholesaler or retailer—as an insulator.

Then, if the injured person could prove he or she bought the product, that person might sue the middleman but could not reach the "deep pocket." Although there is something to be said for the legal generality, the result does not seem quite right. After all, it was the producer and not the middleman who produced the faulty product. Nevertheless, such was the law in the past.

♦ The 20th Century

Around the turn of the century, the law of products liability began to change. Injured plaintiffs occasionally recovered from producers of faulty products. One decision stands out as being especially predictive of future events. This is the decision as written by Judge Cardozo in *McPherson v. Buick Motor Co.*[1] In that case, it seems that a wheel on a new Buick collapsed, and the plaintiff was injured in the resulting accident. In finding for the plaintiff in spite of the defenses mentioned above, Cardozo's reasoning is virtually a statement of products liability law as it evolved years later. However, during its development over the years, the law was not entirely predictable or "settled." Courts here and there followed the *McPherson* decision. During this time, also, the courts developed two legal philosophies that came to be used successfully in products liability cases: negligence and warranty.

Negligence

The idea in negligence here is the same as it is anywhere else in tort law. The plaintiff makes a case against the product's producer by showing (1) that the producer owed the plaintiff a duty to carefully design and produce the product, (2) that this duty was neglected, and (3) that this neglected duty was the proximate cause of plaintiff's injury. The plaintiff's problems often arose in attempting to prove the producer's negligence in design or manufacture, and the producer who could prove contributory negligence by the plaintiff had a formidable defense. Even so, showing negligence did win cases for plaintiffs.

Warranty

Plaintiffs won other cases based on a warranty theory. Specifically, the *warranty* is the promise that the thing bought will do the job for which it is intended. As we saw, the implied warranty of merchantability includes the promise that the product will not injure the product's purchaser while that job is being done. If injury occurs during the intended use of the warranted product, the injury gives the plaintiff a cause of action against the seller of the product; the implied warranty is breached, and this breach, in turn, amounts to a breach of contract.

But a problem occurs in taking a products liability action under warranty. Under the Uniform Sales Act, a predecessor to the Uniform Commercial Code, the only person protected by this warranty was the purchaser. Later, under the Uniform Commercial Code, this limitation was eased. As noted in Chapter 15, different states have adopted different versions of the UCC, which vary in just how broadly the warranty applies. But note also that the action was based on contract. The purchaser might act against the merchant who sold the product, but the purchaser still had no contract basis to act against the product's producer.

Thus, negligence was often difficult to prove, and warranty restricted the parties who might be plaintiff and defendant. Several influential judges and legal scholars believed that those injured by defective products deserved better treatment. A 1963 case[2] preceded a 1964 change of law[3] that, combined, improved the plaintiff's chances for recovery.

STRICT LIABILITY

Most states generally follow the rule of products liability as it appears in section 402A of the Restatement (Second) of Torts. As with statutes that set out legal rules, the wording of section 402A is important:

> (1) One who sells any product in a defective condition unreasonably dangerous to the user or consumer or to his property is subject to liability for physical harm thereby caused to the ultimate user or consumer, or to his property, if (a) the seller is engaged in the business of selling such a product, and (b) it is expected to and does reach the user or consumer in the condition in which it is sold. (2) The rule

1. 217 N.Y. 382 (1916).

2. *Greenman v. Yuba Power Prods*, 59 Cal.2d 57, 27 Cal. Rptr. 679, 377 P.2d 897.

3. Restatement (Second) of Torts, sec. 402A.

> stated in subsection (1) applies although (a) the seller has exercised all possible care in the preparation and sale of his product, and (b) the user or consumer has not bought the product from or entered into any contractual relation with the seller.

Although this statement of *strict liability*, as section 402A is often called, seems fairly clear, courts have interpreted and applied its provisions differently. But this is how things currently stand in the products liability arena—the setting where opposing attorneys (and expert witnesses) prove the defendant liable or not for a product and where "how much" is determined. Essentially, the battles could be pictured as shown below.

PLAINTIFF'S CASE

To win against the producer/defendant, the plaintiff must prove that the product was defective when it left the producer's control. Evidence must then prove a causal connection between the product defect and the injury to the plaintiff. Usually, the legal arena includes a jury to which the evidence is presented.

♦ Defective Product

Proving that a product was defective when it left the producer's control may be difficult. At times, it is very simple and so obviously true that the plaintiff may rely on the doctrine of *res ipsa loquitur* (the thing speaks for itself). At other times, it may or may not be obvious, with expert witnesses contending both ways and with plaintiffs trying to convince the jury to believe their expert rather than the defendant's expert.

Defects can be classified according to three basic types: (1) design defects, (2) manufacturing defects, and (3) marketing defects. The following paragraphs discuss these types of defects.

Design Defects

Of the three types of defects, often the most serious from the producer's point of view and the most difficult to define is the design defect. A design found to be defective usually involves not just one unit but an entire run or lot or, perhaps, many years' production. Furthermore, a defective-design decision made against one unit of product sets a precedent for recovery in subsequent cases. A great deal hinges, then, on determining when a design is defective.

Plaintiff's case		**Defendant's case**
1. Defective product and defendant's responsibility for it. 2. Defective product was proximate cause of plaintiff's injury.	*Defendant Liability?*	1. Product alteration. 2. Obviousness (contributory negligence). 3. Abusive use of product. 4. Functional necessity of design, product, or warnings, and state of the art. 5. Compliance with standards. 6. No proximate cause.
Plaintiff's proof (to maximize recovery)		**Defendant's proof (to minimize recovery)**
1. "Out of pocket" costs. 2. Lasting effects of injury. 3. Producer's neglect of safety considerations.	*Extent of damages (presuming defendant loses)*	1. Product abuse. 2. Comparative (or contributory) negligence. 3. Image of care.

The case of *Barker v. Lull Engineering Company, Inc.*,[4] is particularly thorough in its reasoning as to design defects and presents a logical approach to the definition. The *Barker* standard requires that for the design to be *defective*, (1) plaintiff must establish that the product failed to perform as safely as an ordinary consumer would expect when used in an intended or reasonably foreseeable manner, or (2) plaintiff must show that the product's design proximately caused the injury; and following this, the defendant must fail to establish that the benefits of the challenged design outweigh the risk of danger inherent in such design.

Although the risk-benefit, or cost-benefit, approach has often been a tacit component of products liability cases, *Barker* appears to have been the first case to advocate the risk analysis as a formal criterion. Accordingly, a producer should be able to avoid liability for omitting safety devices that would make products virtually inoperable and for failing to include production processes that would double or triple the product's cost for minuscule safety improvements. Adoption of the *Barker* case's reasoning in other states might provide a foundation for more uniform decisions in product design cases.

Manufacturing Defects

In contrast to a design defect, a manufacturing defect may exist in a properly designed product. *Manufacturing defects* are the defects that occur when a flaw is introduced in the manufacture of the product. For example, an airplane wing may be properly designed, but if flaws are introduced due to improper joining of components, the wing is defective. Such a flaw would be considered a manufacturing defect.

Marketing Defects

Some products are inherently dangerous to use. In this circumstance, the manufacturer has a duty to adequately warn prospective users of the nature of the product. A *marketing defect*, then, is a flaw that is due to a failure to adequately warn or instruct consumers. In one case of this nature,[5] a man was spreading mastic prior to laying a parquet floor. He had read the label and was aware of the message contained there, which cautioned that the mastic was flammable. The substance was actually more than merely flammable, however—it was explosive and should have been so labeled. The plaintiff was injured in the explosion and collected damages because of the inadequacy of the label.

Most products of more than a trivial nature are accompanied by user instructions of some sort. These may be installation instructions, maintenance procedures, repair details, or perhaps something more of a marketing nature. Obviously, following the instructions should not cause the product to injure someone. Suppose, however, that the instruction is meant to cure a known problem of the product. Suppose further that one who follows the instruction is, nevertheless, injured by the product.[6] The result can be devastating to the product's manufacturer, since the label or instructions probably refer to the dangerous nature of the product.

For example, suppose the manufacturer of a respirator provides instructions about its use for sandblasting. A court would no doubt conclude that the use of the respirator for sandblasting operations was entirely feasible. One can easily imagine that the instructions would emphasize the dangers if guidelines were not followed.

What should the producer do if a hazard is not discovered until a very large number of the products have been distributed? Very simply, the producer probably should do everything that can reasonably be done to remedy the situation. Automobile recalls are examples of the extent to which these remedies may be taken. Mailed warnings to all known users could be adequate if sufficient efforts were made to identify the likely victims. The producer's position is far from secure in such a circumstance, but the degree and sincerity of its efforts to warn people likely to be injured at least provides a reasonable defense.

♦ Hidden Defects

Numerous court cases indicate that our society has not completely abandoned the idea of *caveat emptor*. Use of an obviously defective product sets up a defense of contributory negligence. Thus, to bring a successful products liability case, the plaintiff must show that the product's defect was a hidden one.

In one case, the plaintiff was injured while using a fork-lift truck in a high stack area.[7] The court held it to be obviously dangerous to operate a fork lift in such a manner without a protective overhead guard and, there-

4. *Barker v. Lull Engineering Co., Inc.*, 573 P.2d 443 (Cal. 1978).
5. *Murray v. Wilson Oak Flooring Co., Inc.*, 475 F.2d 129 (1973).
6. See *Byrd v. Hunt Tool Shipyards, Inc.* 650 F.2d 44 (1981).
7. *Posey v. Clark Equipment Co.*, 400 F.2d 560 (1969).

fore, the manufacturer of the lift truck should not be responsible for the injury to the operator.

On the other hand, the presence of a hidden defect that causes an injury appears to give the plaintiff a right of recovery. In one such case,[8] the carbon content of steel with which the plaintiff was working was high enough to cause the steel to be brittle. In normal use, it shattered, causing injury to the plaintiff.

Sometimes, the plaintiff need not even point to the specific hidden defect in the product. In one case, for example, part of a hammer splintered off during normal use and struck the plaintiff in the eye. The hammer had been in use 11 months prior to the injury. Furthermore, the hammerhead hardness met military standards. The court found no specific fault in the manufacture or design of the hammer. Moreover, there was no proof that a defect existed when the hammer left the defendant's control. Yet the plaintiff recovered from the manufacturer of the hammer.[9]

Several cases have followed a similar line of reasoning. In one, for example,[10] the plaintiff could prove only that she was using her five-month-old car in a very reasonable manner in very reasonable circumstances, was an experienced driver, was not intoxicated, and did not use drugs. Still, the car malfunctioned, causing it to crash. She alleged, but apparently did not prove, that the drive shaft became disconnected and dropped to the roadway. Despite an inability to prove a specific defect, she recovered damages.

♦ Discovery

Very often, the evidence the plaintiff needs is in the defendant's hands. Products liability cases often involve evidence of quality control records, customer complaints, designs and design changes, research and product engineering records, and the like. In fact, documents like these are often vital to the plaintiff's case. As noted in an earlier chapter, document requests and other "discovery" procedures can and should be used to obtain such documents.

♦ Damages

Products liability cases are usually intended to obtain money from the defendant to compensate for plaintiff's losses. Occasionally, because of extremely hazardous or careless behavior by a defendant, a court may go further and award an additional amount to the plaintiff as a punishment or penalty against the defendant. Damages are discussed in greater detail in chapter 14.

♦ User or Consumer

The language of section 402A sounds as though there is a definite limitation as to who may bring a products liability case. What about injury to an innocent bystander? This question has been effectively answered in a California case[11] that courts in other states have followed where the question has arisen. Generally, the courts hold that the innocent bystander has the same right as the user or consumer to recover from the producer of the injurious product. Logically, the bystander should have even a higher right, since the user or consumer has at least a chance to look at the product before deciding whether to use it. In contrast, the innocent bystander has no such opportunity.

DEFENDANT'S CASE

Sometimes the plaintiff's case is so overpowering that the defendant has little alternative to settling the matter out of court. The defendant's expert witness may be extremely helpful in making the decision to settle. If a faulty design or manufacture is readily evident to the defendant's expert, a professional engineer and expert in the field, the fault would presumably be evident to the plaintiff's expert as well. In such a case, the best advice the expert can give a client is to settle out of court.

In contrast, when a defendant decides to go to trial, the defendant usually believes that it stands a high chance of winning or reducing the amount of damages to be paid. The decision as to the best course of action depends on what the evidence shows. The nature of the evidence may range from a cause for outright dismissal of the case to inconclusive opinions or speculation that a plaintiff could successfully counter.

♦ Product Alteration

The producer of a product should be held responsible only for those products he or she has produced. Close

8. *Moomey v. Massey Ferguson, Inc.*, 429F.2d 1184 (1970).
9. *Dunham v. Vaughan & Bushnell Mfg. Co.*, 247 N.E.2d 401 (1969).
10. *Hall v. General Motors Corp.*, 647 F.2d 175 (1980).

11. *Elmore v. American Motors*, 451 P.2d 88, 75 Cal. Rptr. 652 (1969).

examination of a component may indicate that it was repaired after manufacture—perhaps, for example, replacement brake linings or a rebuilt wheel cylinder failed. If a rebuilt part was the cause of the injury, it may no longer be the automobile producer's responsibility. The question is basically this: Who produced the product that failed?

Suppose it was the original brake linings (or other component) that failed and caused plaintiff's injury. Suppose further that these linings or other components had been supplied to the automobile producer by another manufacturer. Shouldn't the component supplier be held responsible for the injury rather than the producer who assembled these components into a finished product? The simple answer is no. It is the producer of the final product as sold to the consumer who is responsible for its safe performance. Whether the producer made the parts or bought them is irrelevant as far as responsibility to the injured plaintiff is concerned. The producer may be able to pass part or all of the loss along to the supplier, but the producer of the final product has primary responsibility.

♦ Proximate Cause

If no causal connection exists between the plaintiff's injury and the producer's product, the producer should have no liability. One might be able to show a faulty design in the location of the fuel tank in a car, for example, but if this flaw is in no way connected to plaintiff's injury, the faulty fuel tank location has no bearing on the case. Just as a showing of product alteration usually serves as a complete defense, a successful showing of the lack of any proximate cause is generally an effective defense.

♦ Obviousness

As noted above, a plaintiff will have difficulty winning if the defect causing the injury is obvious. The same is true where the plaintiff has ignored prominent labels attached to the product. Of course, what is obvious to a person experienced with the product in question might not be obvious to a novice, and a label written in English might not be much of a warning to one whose only language is Spanish. Such problems as these and many others are questions of fact for a jury to decide.

♦ Abuse

Producers must expect their products to be subjected to all sorts of abuses as well as the uses for which they were intended. Consider the uses to which you have put screwdrivers. Even television and newspaper advertisements show product abuse as persuasive proof of the rugged nature of products. Still, there is a logical limit to which anticipated abuse may be taken. For example, a plaintiff's injury resulting from the use of a typewriter as a hammer would probably not be tolerated as reasonable or foreseeable abuse. This sort of question is also often left for a jury to decide.

♦ Functional Necessity and State of the Art

In our present state of technological development, it is impossible to accomplish certain functions without some risk. A *completely* safe car would not move; a *completely* safe rotary lawn mower would not mow grass; a *completely* safe knife would not cut. But people need these pieces of equipment, so we cannot hold a manufacturer of such things to a standard of complete safety. Some danger is inherent in the operation of most products; perhaps the best we can expect is a condition we might consider *reasonably dangerous*. One way to state this standard is as follows: The manufacturer should produce a product as safely as the state of the art will permit. In other words, the manufacturer should review the state of the art at the time he or she produces the product. If at that time, a survey of the literature and a comparison with similar products show no safer way of accomplishing the function, it seems unreasonable to expect more from the defendant.

Proof of compliance with the state of the art, however, may not necessarily be a complete defense for a producer. The decision could favor the plaintiff for the simple reason that the court or a jury believes the state of the art should have advanced more rapidly. Consider, for example, Gray Harvester, a producer of agricultural machinery. White, a farmer, was injured while operating a Gray hay baler. Even though Gray successfully shows that he complied with the state of the art and industry standards, White may be able to win by showing the sluggish response by producers of farm machinery to injuries such as these. If the court feels the response should have been more rapid, White may win.

♦ Standards

Various federal agencies have become concerned with hazard reduction and safety issues, as have the military and various industrial organizations. Almost invariably, each of these organizations begins by formulating and publishing minimum standards to be met by whatever products are being considered. But suppose that Black Manufacturing Company, for example, has complied with all available standards in the manufacture of its product—say, a riding lawn mower. Green, in attempting to operate the lawn mower in very heavy grass, manages to injure himself. Green files a products liability action against Black Manufacturing. Is Black Manufacturing protected because it complied with the applicable standards?

The most common answer is no. The question of liability should probably still be answered according to the section 402A criteria. In fact, if the hazard were so improbable that those who wrote the standards completely missed it, this might suggest that what caused the injury was truly a hidden design defect. On the other hand, the defendant Black Manufacturing will argue to the jury that the standards show that problem was not foreseeable and the mower was not "unreasonably" dangerous.

If a producer fails to comply with published standards, and that failure results in injury to a plaintiff, the plaintiff certainly has strong evidence against the producer. Furthermore, such evidence might well be used as a basis for punitive damages. In addition to this, avoidance of standards published by federal agencies may lead to fines and/or other sanctions against the offending manufacturer.

♦ Plaintiff's Negligence

The courts treat a plaintiff's contributory negligence in connection with a strict liability case in various ways. Generally, though, proof of the plaintiff's contributory negligence will not relieve the producer of liability. That is, if Green proves a hidden defect in Black Manufacturing's lawn mower and a causal connection between the hidden defect and the injury, Black probably cannot win the case by proving that Green was careless. But that doesn't mean that Black shouldn't try the defense. Even if a claim of contributory negligence fails to win dismissal of the case, it is a valuable point to be made in minimizing damages. If the person injured was truly reckless, the jury may conclude that it was that person's conduct—not the defect—that caused the injury.

♦ Defense Image

Nothing in strict liability indicates that the character of the defendant has anything to do with plaintiff's case. In fact, section 402A would hold the manufacturer liable even though he or she has done a careful job in designing, making, and selling the product. The amount of damages must be assessed by a jury, by a judge, or by both. If the plaintiff can produce evidence of careless production, sloppy methods, fast and careless product-engineering changes, insignificant research, and misleading marketing, this may make an impressive case for punitive damages. On the other hand, if the defendant can show great care in design, manufacture, quality control, and attention to customer complaints, damages may well be held to a minimum. The producer may have to pay for the result of the faulty product, but the risk of punitive damages is minimized.

♦ State Statutes

No one should suffer because of faulty products. Truly defective products should not have been produced or should have been removed by the producer before they could injure someone. A products liability problem does exist today, of course, but some would contend that it suffers from "overkill." For example, cases arise where products were not really faulty, but were abused, or perhaps were produced many decades ago. The cost of defending products liability suits and paying judgments and out-of-court settlements for such cases is added to production costs, and the public pays by way of higher prices. In recognition of such problems, some states have enacted products liability laws.

State products liability acts vary considerably in their content and their restrictions. However, such laws generally set out the following standards: (a) that after a certain period of time the producer's responsibility for the product ends, (b) that the "state of the art" can be used as a defense, (c) that compliance with recognized standards (e.g., military standards) is a reasonable defense, (d) that alteration and/or abuse by the plaintiff is a defense, and (e) that a failure to warn is not a component of strict products liability.

Such restrictions as these may become more common in the future. In addition, advocates have suggested that there should be a federal products liability law. The primary advantage of a federal law would be its uniformity. Thus, a company making products in Illinois, for example, and selling them throughout the United States would not have to be concerned with the laws (and related risks) of every state. Opponents tend to argue that each state should be free to choose for itself what is appropriate for protecting people within its territory from injuries.

DUNHAM v. VAUGHAN & BUSHNELL MANUFACTURING COMPANY
247 N.E.2d 401 (Ill. 1969)

A jury in the circuit court of Macoupin County returned a verdict in the sum of $50,000 in favor of the plaintiff, Benjamin E. Dunham, and against the defendants Vaughan & Bushnell Mfg. Co. and Belknap Hardware and Mfg. Co. Judgment was entered on the verdict, and the Appellate Court for the Fourth judicial District affirmed. . . . We allowed the defendants' petition for leave to appeal.

The injury that gave rise to this action occurred while the plaintiff was fitting a pin into a clevis to connect his tractor to a manure spreader. He had made the connection on one side, using a hammer to insert the pin. To insert the second pin he lay on his right side underneath the tractor and used the hammer extended about two and one-half feet above his head. The hammer moved through an arc which he described as about eight inches. He testified that as he undertook to "tap" the pin into the clevis a chip from the beveled edge of the hammer, known as the chamfer, broke off and struck him in the right eye. He lost the sight of that eye.

The hammer in question is a claw hammer of the best grade manufactured by the defendant Vaughan & Bushnell Mfg. Co. It bore the "Blue-Grass" trademark of its distributor, the other defendant, Belknap Hardware and Manufacturing Co. The plaintiff had received the hammer from a retailer, Hayen Implement Company, located near his home. He received it as a replacement for another "Blue-Grass" hammer, the handle of which had been broken. Before the accident occurred the plaintiff had used the hammer for approximately 11 months in connection with his farming and custom machine work. He had used it in repairing a corn crib and had also used it in working upon his farming implements and machinery.

Each party offered the testimony of an expert metallurgist. Neither expert found any flaws due to the forging of the hammer, or any metallurgical defects due to the process of manufacture. The experts agreed that the hammer was made of steel with a carbon content of "1080." The plaintiff's expert testified that such a hammer was more likely to chip or shear than one made of steel with a lower carbon content of "1040," which would not be so hard. The defendant's expert disagreed; it was his opinion that a hammer made of harder steel, with the higher carbon content, would be less likely to chip or shear than one made of steel with a lower carbon content. Both experts testified that use of a hammer produced a condition described as "work hardening" or "metal failure," which made a hammer more likely to chip or shear. The defendants apparently suggest that the plaintiff should not have used a claw hammer to tap the pin into the clevis because the mushroom head of the pin was made of steel of "Rockwell" test hardness of C57, which was harder than the head of the hammer, which tested Rockwell C52. But as the appellate court pointed out, the specifications of the General Service Administration used by all Federal agencies, call for a Rockwell "C" hardness of 50–60 in carpenter's claw hammers and a Rockwell "C" hardness of 50–57 for machinists' ball peen hammers. Those specifications also require that sample carpenter's claw hammers and sample ball peen hammers be subjected to identical tests by striking them against another hammer and against a steel bar, to determine their tendency to "chip, crack, or spall." The specifications thus negate the defendant's suggestion that the plaintiff should have used a ball peen hammer, rather than the hammer in question, in tapping the pin into the clevis.

The basic theory of the defendants in this court is that the requirements of strict liability, as announced in *Suvada v. White Motor Co.*, . . . were not established, because the testimony of the experts showed that the hammer contained no defect. Suvada required a plaintiff to prove that his injury resulted from a condition of the product which was unreasonably dangerous, and which existed at the time the product left the manufacturer's control. But the requirement that the defect must have existed when the product left the manufacturer's control does not mean that the defect must manifest itself at once. The defective "aluminum brake linkage bracket," with which the court was concerned in ruling upon the legal sufficiency of the complaint in *Suvada*, was alleged to have been installed in the tractor not later than March of 1957; it did not break until June of 1960.

Although the definitions of the term "defect" in the context of products liability law use varying language, all of them rest upon the common premise that those products are defective which are dangerous because they fail to perform in the manner reasonably to be expected in light of their nature and intended function. So, Chief Justice Traynor has suggested that a product is defective if it fails to match the average quality of like products. . . . The Restatement emphasizes the viewpoint of the consumer and concludes that a defect is a condition not contemplated by the ultimate consumer which would be unreasonably dangerous to him. Statement, Torts (Second) section 402A, comment g. Dean Prosser has said that "the product is to be regarded as defective if it is not safe for such a use that can be expected to be made of it, and no warning is given." (Prosser, The Fall of the Citadel, 50 Minn. L. Rev. 791, 826). Dean Wade has suggested that apart from the existence of a defect "the test for imposing strict liability is whether the product is unreasonably dangerous, to use the words of the Restatement. Somewhat preferable is the expression 'not reasonably safe.'".

The evidence in this case, including both the General Services Administration specifications and tests and the testimony of the experts as to "work hardening" or "metal failure," shows that hammers have a propensity to chip which increases with continued use. From that evidence it would appear that a new hammer would not be expected to chip, while at some point in its life the possibility of chipping might become a reasonable expectation, and a part of the hammer's likely performance. The problems arise in the middle range, as Chief Justice Traynor has illustrated: "If an automobile part normally lasts five years, but the one in question proves defective after six months of normal use, there would be enough deviation to serve as a basis for holding the manufacturer liable for any resulting harm. What if the part lasts four of the normal five years, however, and then proves defective: For how long should a manufacturer be responsible for his product . . ."

The answers to these questions are properly supplied by a jury, and on the record that is before us this case presents only the narrow question whether there is sufficient evidence to justify the jury's conclusion that the hammer was defective. The record shows that it was represented as one of "best quality" and was not put to a use which was regarded as extraordinary in the experience of the community. The jury could properly have concluded that, considering the length and type of its use, the hammer failed to perform in the manner that would reasonably have been expected, and that this failure caused the plaintiff's injury.

Strict liability, applied to the manufacturer of the hammer, Vaughan & Bushnell, extends as well to the wholesaler, Belknap Hardware and Mfg. Co., despite the fact that the box in which this hammer was packaged passed unopened through Belknap's warehouse. The strict liability of a retailer arises from his integral role in the overall producing and marketing enterprise and affords an additional incentive to safety. . . . That these considerations apply with equal compulsion to all elements in the distribution system is affirmed by our decision in *Suvada v. White Motor Co.* . . .

The defendant's objections to the instructions to the jury were adequately disposed of in the opinion of the appellate court. The judgment of the appellate court is affirmed.

Judgment affirmed.

GIERACH v. SNAP-ON TOOLS CORPORATION
255 N.W.2d 465 (Wis. 1977)

This action for negligence is brought by a user of a ratchet wrench against the manufacturer, Snap-on Tools Corporation. Peter Gierach, the Plaintiff, was an employee of the Riebe Oldsmobile Garage in Grafton, Wisconsin. On May 1, 1973, Gierach was working on an Oldsmobile automobile when the ratchet wrench slipped. Gierach sustained injuries to his face, mouth, and teeth when the ratchet wrench struck him. The evidence showed that Gierach was using the wrench properly at the time the accident happened. After the accident, the wrench was placed in the tool box of Karrels, the service manager at Riebe, to whom the wrench belonged. About a week later, Dick Augers, a salesman for Snap-on Tools, made one of his regular sales calls to the garage and was told about the accident. He took the wrench into the back room and disassembled the gear housing. It was apparent to those present that a gear tooth was sheared. The plaintiff testified that Augers said, "Well, this is what the problem is, that the gear is sheared and this would cause it to slip." Karrels, the service manager, also testified that the sheared gear was discovered when the Snap-on salesman disassembled the housing. Karrels also testified to the salesman's statement.

The action was brought by the plaintiff on the theory that the accident was caused by the manufacturer's improper design or manufacture of the ratchet wrench.

At the trial, Karrels, who was a master mechanic with extensive experience, gave as his opinion that the gear teeth were not properly hardened, and, as a consequence, the gear tooth sheared off. On the other hand, there was testimony by Godlewski, Snap-on's quality assurance inspector, that the gears were properly hardened, that they were built and designed to withstand 5,000 pounds of pressure, and that there was nothing wrong with the manufacture or design of the tool. Snap-on's witness acknowledged, however, that the maximum pressure that could be exerted on the wrench by a worker would be less than 1,000 pounds.

It was Godlewski's testimony that the wrench had not been cleaned and maintained properly, that grease and debris had built up within the gear housing, and the gear pawl could not seat properly to fully engage the gear teeth. He said that it was as a consequence of this improper maintenance that the wrench slipped.

The parties, prior to trial, stipulated that the damages for personal injuries were $4,500.[12] The jury, accordingly, was required only to return a verdict in respect to negligence. Ninety percent of the negligence was attributed to Snap-on Tools Corporation and ten percent to the injured plaintiff, Gierach. Judgment for the plaintiff was entered for the total sum of $4,249.69, including costs and disbursements. The appeal was taken from that judgment.

The trial judge instructed on the elements of negligence, and, in addition, instructed in respect to the duty of a manufacturer to exercise ordinary care in the design and manufacture of its product so as to render that product safe for its intended use. He also instructed that the manufacturer had the duty to exercise ordinary care to give adequate warning of the dangers attendant upon the proper use of the product. A *res ipsa loquitur* instruction was given in the following form:

12. The record is confusing, for the judgment recites that the parties had, prior to trial, stipulated to a $4,500 damage ceiling. Our review of the record indicated, however, that, both In plaintiff's motion after verdict for entry of judgment and in verbal representations by both parties immediately prior to trial, the damage ceiling was stipulated to be $5,000. However, for purpose of this appeal, we accept the statement in judgment that the stipulated ceiling was $4,500.

> You are further instructed that if you find that the plaintiff was properly using the ratchet wrench in question at the time of the accident and that nothing occurred to cause the ratchet wrench to become defective after it left the defendant's control, and if you further find that the accident in this case ordinarily would not have occurred if the ratchet wrench had not been defective in its design, you may then infer from the accident itself and the surrounding circumstances that there was negligence on the part of the defendant in the design or manufacture of the ratchet wrench.

The court instructed the jury that Gierach had a duty to use ordinary care for his own safety.

The defendants, insofar as the record shows, submitted no proposed instructions to the court and failed to object to any of the instructions that were given. Yet, on this appeal, defendant argues for the first time that, because this is an action for negligence and not for strict liability under sec. 402(A) of the Restatement of Torts, a *res ipsa loquitur* instruction was inappropriate. This objection comes too late, because the defendant fully acquiesced in this instruction. In addition, a *res ipsa* inference is permissible in respect to any tort which occasions unintentional personal injuries if the elements necessary for instructing on that inference are placed in evidence. The evidentiary preconditions for the doctrine were summarized in *Utica Mut. Ins. Co. v. Ripon Cooperative. . . .*

> (1) The event in question must be of the kind which does not ordinarily occur in the absence of negligence; (2) the agency or instrumentality causing the harm must have been within the exclusive control of the defendant.

The instruction given in the instant case was in substantial conformance with these standards and was not objected to. The evidentiary preconditions for the *res ipsa* inference were satisfied in this case. There is no dispute that the slipping of the gears would not have occurred in the absence of negligence. Exclusivity of control does not mean that the instrumentality be in the physical possession of the defendant at the time of the occurrence. . . . As we noted in *Ryan v. Zweck-Wollenberg Co.,* . . . quoting from Prosser, Torts, sec. 43, p. 298:

> All that is necessary is that the defendant have exclusive control of the factors which apparently have caused the accident; and one who supplies a chattel to another may have had sufficient control of its condition although it has passed out of his possession. . . .

Defendant also asserts that *res ipsa* is inapplicable in a products liability case where negligence is alleged. That assertion is completely without foundation in the law; and we conclude that, under the circumstances, it was within the sound discretion of the trial judge to give that instruction.

The defendant, on this appeal for the first time, impliedly objects to the instruction in respect to the manufacturer's duty to warn in respect to the use and care of the ratchet wrench. There was no objection to that instruction when it was offered by the court. However, the point which the defendant now seeks to assert was arguably made when the defendant objected to the special verdict question which inquired:

> Was the defendant negligent with respect to:
> (b) Failing to warn of dangers with respect to the ratchet wrench?

The objection to the question was based, however, not on the lack of an evidentiary basis at the trial, but rather on an answer given by the plaintiff to a question asked by defendant's counsel.

Defendant's counsel asked, "(C)an you think of any further instruction or warning that could have been given to you. . . ." The plaintiff responded that he could not think of any. From this, defense counsel concludes that it was conceded that no warnings were necessary. This position, however, is contrary to the entire tenor of the defendant's case.

It was the position of the defendant throughout the trial that the plaintiff and his fellow employees had failed to clean and maintain the wrench properly and that the consequent buildup of grease and dirt within the gear housing of the wrench caused the slippage. The response appropriately made by plaintiff's counsel to this assertion is simply that, if such hazard existed, the purchasers and ultimate users of the wrench should have been warned by the manufacturer of the necessity of periodically opening the gear box and cleaning it. No warning was ever given by the manufacturer. The record shows that it was the defendant's theory of the case and its evidence that made it reasonable for the trial judge to instruct on the manufacturer's duty to warn of the necessity for the periodic cleaning of the wrench.

It was Godlewski, the defendant's expert witness, who testified that grease and dirt could build up within the gear housing and cause it to slip. It is apparent from the defendant's own testimony that it was aware of the slippage danger and yet it failed to give any warnings or instructions with respect to the maintenance or cleaning of the wrench.

Snap-on Tools, knowing that someone could be injured if the tool were not properly maintained, had a duty to warn, and it failed to discharge that duty.

In view of the fact that the defendant failed to object to any of the instructions, and because the instructions given were not legally erroneous, no objection to the instructions can be raised on this appeal.

The defendant on this appeal argues that a verdict question should have been submitted to the jury in respect to whether the service manager, Karrels, was negligent in the supervision of Gierach in respect to the use and care of the ratchet wrench. The trial judge refused that requested question because there was no evidence that Karrels was negligent. The trial judge, therefore, properly excluded a question about Karrel's negligence. Although, as we held in *Connar v. West Shore Equipment of Milwaukee, Inc.*, . . . a special verdict question in respect to the negligence of an individual who is not a party may be included in the verdict, it is necessary that there be "evidence of conduct which, if believed by the jury, would constitute negligence on the part of the person or other legal entity inquired about.". . The trial judge properly concluded that there was no evidence upon which a jury could have found Karrels negligent for the failure to warn or supervise Gierach.

The evidence was sufficient to support the verdict, whether the verdict be viewed in the light of the plaintiff's initial theory that the wrench was negligently designed and manufactured or whether it be viewed in light of the defendant's theory that the accident occurred because of the buildup of grease and dirt within the wrench housing.

There were sufficient facts to support the jury's verdict in respect to negligent manufacture and design, because the evidence showed that the manufacturer represented that the gears were designed to withstand pressures up to 5,000 pounds. Yet, the testimony presented by the defendant's expert indicated that at most a maximum force of 1,000 pounds could have been exerted by Gierach.

Applying the inferences that are appropriate under the *res ipsa* instructions, the jury could reasonably conclude that the instrumentality was within the control of the manufacturer and that the breaking of the gear would not have occurred were it not for some negligence in the manufacture or design. Looking at the jury verdict in connection with the defendant's theory that the buildup of grime and dirt occasioned the slippage, it is apparent that the evidence showed that the manufacturer was aware of this hazard and yet failed to give any warning or instruction whatsoever in respect to the cleaning or maintenance of the sealed gear housing.

The most difficult question in respect to negligence is one not posed on this appeal at all. Our examination reveals no evidence of negligence on the part of the plaintiff Gierach. Apparently, however, plaintiff's counsel, as a matter of strategy, concluded not to cross-appeal in light of the relatively small amount of reduction of damages occasioned by the jury's attribution of ten percent of the negligence to Gierach.

The question of the sufficiency of the evidence was posed on motions after verdict; and the trial judge, upon re-examination of the record, concluded that there was ample evidence to sustain the findings of the jury.

While there was evidence submitted by the defendant which, if believed by the jury, would tend to eliminate or minimize any negligence on the part of the defendant, the trial judge correctly pointed out that the jury was not obliged to believe such testimony. It did not.

Our review of the record reveals no error of law, and the evidence was sufficient to support the verdict.

Judgment affirmed.

SAUPITTY v. YAZOO MFG. CO., INC.

726 F.2d 657 (1984)

Logan, Circuit Judge

Defendant Yazoo Manufacturing Company, Inc. appeals from a jury verdict awarding plaintiff, James Saupitty, $560,000 compensatory and $440,000 punitive damages. Plaintiff was injured while operating a lawnmower manufactured by defendant. Plaintiff brought suit on the theory of manufacturer's product liability, alleging that design defects rendered the mower unreasonably dangerous to the user. Oklahoma law applies in this diversity case.

Plaintiff was a civilian employee of the United States at Fort Sill, Oklahoma, in the Grounds Maintenance Department. One of his responsibilities was to cut the grass on the Fort Sill grounds. To do so he used a six-year-old Yazoo YR-60 riding lawnmower that defendant had manufactured. Plaintiff was riding the mower down a hill when the mower began bouncing and shaking. He attempted to slow or stop the mower by placing the machine into reverse gear. The gear shifting momentarily locked the mower's drive wheels, bucking plaintiff forward over the top of the machine. In his attempt to stop his fall his thumb and two fingers of his left hand were severed and his arm was injured.

Plaintiff alleged that the mower was defectively designed and unreasonably dangerous in a number of respects: It did not have a proper rear weight, a stabilizer bar, a dead man switch, or adequate brakes; the operator's seat was in an unsafe position; the controls allowed the operator to shift directly from forward to reverse; and the mower lacked sufficient warnings.

On appeal defendant asserts that the trial court should have directed a verdict in its favor because the mower's brakes and belt guard had been removed before the accident. Oklahoma cases have adopted the rule of *Restatement (Second) of Torts* sect. 402A (1)(b) (1965), which imposes liability only when the product "is expected to and does reach the user or consumer without substantial changes in the condition in which it is sold." Thus, a manufacturer is not liable when an unforeseeable subsequent modification alone causes the plaintiff's injury. *Texas Metal Fabricating Co. v. Northern Gas Products Corp.* The manufacturer is liable, however, if the subsequent modification was foreseeable...or if it was not a cause in fact of the injury. See *Blim v. Newbury Industries, Inc.*... In *Blim*, a plastic injector press was originally equipped with mechanical drop bars designed to prevent injuries to operators' hands. The plaintiff's employer testified that he removed the bars because the press did not operate properly with the drop bars in place. In a suit against the manufacturer the jury found for the plaintiff, who was injured when the press closed on her hand. The manufacturer, relying on *Texas Metal*, contended that the removal of the drop bars constituted a material alteration of the product and thus was an intervening and independent cause of the injury as a matter of law. This court disagreed, stating:

"Appellant's reliance upon *Texas Metal* is misplaced. In that case there was no demonstrated relationship between the rattle (the alleged defect) and the ultimate explosion. ...Here, the mechanical drop bars were safety features designed to prevent just such an injury as that sustained by appellee. Since evidence demonstrated that they were already ineffective, their removal could not even exacerbate the hazard; *a fortiori*, it could not, as a matter of law, constitute a superseding, intervening cause of the injury...."

Similarly, in the instant case, plaintiff's coworkers and his expert witness testified that the scrubber brakes would not stop the mower. Even the manufacturer's brochure declared that the foot brake was ineffective while the mower was in gear. Thus, the jury could find that to slow or stop the mower plaintiff would have had to shift into reverse even if the brakes had not been removed. Furthermore, plaintiff testified that his hand was injured by the mower's cutting blades rather than by the belts exposed by the removal of the mower's belt guard. Plaintiff did testify that he disengaged the mower's blades shortly before the accident, but his expert testified that the model of mower involved in the accident had a history of failing to disengaged the blades when the operator placed the blade control in neutral. Thus, plaintiff presented evidence from which the jury could reasonably have concluded that the cutting blades continued to turn and cut plaintiff's hand, as he testified, and that the removal of the belt guard was not a cause of plaintiff's injury. Therefore, construing all of the evidence and the inferences therefrom in the light most favorable to the plaintiff, we cannot conclude that the removal of either the mower's brakes or its belt guard constituted a superseding, intervening cause of plaintiff's injury as a matter of law.

We see no merit in defendant's other arguments for reversal. The plaintiff produced evidence from which the jury could reasonably have concluded that the mower possessed the alleged design defects when defendant sold it to the government. Defendant's argument that the district court improperly instructed the jury is unmeritorious, particularly since its counsel did not object to the instructions at the time they were given. The Oklahoma Supreme Court has held that punitive damages are recoverable in products liability cases.... Plaintiff produced sufficient evidence of the manufacturer's reckless disregard for public safety to justify submitting the issue of punitive damages to the jury.... The trial court did not err in refusing to instruct the jury on the theory of comparative negligence, since the Oklahoma Supreme Court has said that comparative negligence statutes have "no application to manufacturers' products liability.... Finally, although the award of $560,000 compensatory and $440,000 punitive damages is high, the amounts do not shock our conscience.

AFFIRMED.

Seth, Chief Judge, dissenting:

I must respectfully dissent. The majority concludes that the modifications to the six-year-old mower made by the government should be disregarded, and would apply tort standards and tort doctrines. Reading the record in the light most favorable to the plaintiff, I disagree. The plaintiff was awarded one million dollars for the loss of three fingers from his left hand and for punitive damages.

The government made two material alterations to the mower, either of which should constitute grounds for a directed verdict for the defendant on the strict liability claim under applicable doctrines. First, the mower's belt guards had been removed. At trial an issue was whether the mower's blades had cut Mr. Saupitty's fingers or whether his hand had become entangled in the swiftly moving belt and his fingers severed when they were pulled into contact with the belt's pulley. The majority accepts the plaintiff's assertion that his fingers were cut by the mower blades. The plaintiff testified that he had disengaged the mower's blades before starting down the hill. the majority assumes that the disengagement was unsuccessful because "his expert testified that the model of mower involved in the accident had a history of failing to disengage the blades when the operator placed the blade control in neutral." However, that theoretical possibility was directly refuted by Mr. Saupitty's own testimony that the blades were off and had stopped rotating. In these circumstances it is hard to understand why theory and not the actual facts were used. If the blades had indeed stopped turning, then the only other possibility was that Mr. Saupitty's fingers were severed when they were caught in the belt. Since the belt guard had been removed by the government after it bought the mower it must be concluded that a material alteration had occurred preempting strict liability.

The government also removed the mower's brakes sometime during its six years' use of the machine. That removal necessitated stopping the mower by pulling it into reverse. The machine could not even be otherwise slowed down because both the brakes and the throttle control had been removed. The degree of efficiency of the brakes before they were removed is not necessarily pertinent because although inefficient they nevertheless would have provided an alternative to the necessity of throwing the mower into reverse. In any event, the brakes originally on the mower met the standards for the industry. The plaintiff did not show that he had to stop the mower to avoid an accident. The mere slowing of the machine by use of the brakes could easily have prevented the accident. Their removal must constitute a material alteration. The proof of the plaintiff on both the brake and belt guard removals did not meet the requirements of *Stuckey v. Young Exploration Co.*,...to permit the trial judge to submit the case to the jury.

Even were strict liability to be imposed in this case, there is no evidence whatever in the record to support an award of punitive damages.

Since the defendant did not raise a contract specification defense, we cannot consider it on appeal.

I thus dissent.

REVIEW QUESTIONS

1. Why was *caveat emptor* an appropriate defense 200 years ago but is not appropriate today?
2. What is the meaning of *privity of contract*? Under what circumstances does it make sense? Why isn't it an appropriate defense in a products liability case?
3. Contrast strict liability as stated in the Restatement (Second) of Torts, section 402A, with negligence and warranty as they would be applied to products liability.
4. The White Company produces a combination tool for home use that may be converted to a lathe, drill press, grinder, and many other power tools. Gray bought one of these tools and was injured while using it as a lathe. The injury occurred because the lathe chuck loosened, and the piece on which Gray was working flew out of the chuck, striking him in the head. Gray was out of work for several weeks and has lost partial sight of his right eye. What would you as plaintiff's expert witness look for to make a case for Gray to recover? What would you do as defendant's expert witness in the case?
5. Can you make the cost-benefit ideas of *Barker v. Lull Engineering* agree with the section 402A definition of strict liability? Does it agree better with negligence or warranty?
6. In the case of *Dunham v. Vaughan & Bushnell Manufacturing Co.*, the manufacturer was found responsible even though no negligence in design or manufacture was proved. If such is the trend under strict liability, what could a manufacturer do to protect against such losses?
7. In the case of Dunham v. Vaughan & Bushnell Manufacturing Co., is there any indication that the tractor manufacturer or the manure spreader manufacturer could have been made party defendant rather than the producer of the hammer?
8. The *Gierach* case involves the absence of written instructions. How should the producer of a product with known hazards warn a distant and unknown user of those hazards? Suppose the hazard is discovered after many products have been sold and distributed. What measures should be taken then?
9. After reading both the majority opinion and the dissent in *Saupitty v. Yazoo Mfg. Co.*, decide this million dollar case for yourself and justify your decision.

Part Six

GOVERNMENTAL REGULATION

Political considerations often dictate engineering activity to a considerable degree. This is obviously true in projects run by the federal government (for example, engineering for NASA and the military) and by state and local governments (such as engineering for roads, bridges, and water supplies). Political intervention is an expected component of such projects. A major problem in the political control of engineering activities, however, is the lack of effective communication between politicians and engineers. Unfortunately, many politicians lack training in the sciences or else are not concerned with the technical aspects of a project. On the other hand, many if not most engineers find political involvement and motivations difficult to understand and reconcile with engineering requirements. The results are frequently less than desirable.

Administrative agencies exercise a great deal of control over engineering activities pursuant to administrative rules and laws. The basic idea is that an administrator with some knowledge of the engineering and business aspects involved in the project will act as the control mechanism according to established rules. Of course, most regulatory statutes are politically inspired (and the administrative rules are often made according to that inspiration). Unfortunately, too, like politicians, many regulatory personnel lack the background and the training that would allow them to do an effective job. Under these circumstances, it is not surprising that engineers often complain about the controls imposed on them.

Here, we will look at the law and activities of administrative agencies in general. Then we will discuss how administrative agencies affect three industrial functions: labor, worker's compensation, and safety.

Chapter 23

Administrative Law

We have always had administrative agencies, bureaus, or commissions. For the first century and a half (or so) after our government was formed, agencies were few in number. Since then, however, their number has grown and grown. The federal government has established the ICC, the FCC, the SEC, the NLRB, the IRS, and the FTC to name only a few. The state government often has some sort of industrial commission, a governing board for the universities, a public utilities commission, a tax commission, numerous licensing boards, and many other agencies. Local agencies include zoning boards, school boards, welfare agencies, and others. All these entities have powers to investigate, make rules and rulings, and supervise activities in some limited geographical area and some sphere of activity. All have powers delegated to them by a legislative body or a chief executive (such as the president, governor, or mayor). These bodies have an ever-increasing influence on what we may do and how we may go about doing what we do. It seems only reasonable, then, that we should consider what or who controls these agencies, why they exist, and how one learns to live with them.

WHY AGENCIES?

Agencies often do not reflect the separation of powers so carefully assigned in our Constitution. Rather, they blend the powers to govern a segment of the population or business world in the name of the public. Administrative agencies thus tend to have judicial, legislative, and executive powers.

Administrative agencies act like courts when they rule for or against an individual or a business. For example, Brown may be granted or denied a license to operate a business, or his license may be taken from him because he failed to follow an agency rule. Gray's tax return may be investigated by the Internal Revenue Service, and deductions allowed or denied. Black may not be allowed credits for the courses she took at another university when she tries to enter the University of Ames. White Company may be required to cease and desist its pollution of a stream. Green Trucking Company may need agency approval of its rates before it can offer its services to customers. Whenever an agency renders a decision such as these, its actions resemble those of a court. Unlike courts, however, agencies do not have to wait until an injury has occurred before they act.

Agencies are also legislative in nature. They have the power to investigate situations that come to their attention and then make rules based on what they find. For example, the Internal Revenue Service may find it desirable to require people like Green to keep another kind of record. After following the appropriate rule-making procedure, the IRS's new rule requiring taxpayers (like Green) to keep such a record would appear in an issue of the *Federal Register*. Publication in the *Federal Register* is deemed by law to provide adequate notice to anyone affected by the proposed rule.

In many ways, agencies also act like an executive or supervisor. They generally have the right not only to investigate, but to act to prevent occurrences that they anticipate will harm someone. Specifically, many agencies have the power to prosecute violations of the rules adopted by the agency. As a practical matter, the threat of active supervision may be sufficient to obtain the objective sought by the agency. For example, companies will sometimes voluntarily withdraw a product from the market rather than try to meet the requirements of a new rule from the FDA.

The suggestion is sometimes made that courts could do better what agencies are supposed to do. But one need only consider the traditional nature of courts to see the magnitude of changes that would be required for courts to replace agencies.

The main function of a court is to decide specific disputes. Our traditional court procedures are designed to obtain all pertinent information regarding some past event. Then, armed with this information, the court makes a judgment. Only rarely, and then in the nature of an equity action, will courts concern themselves with things that have not yet occurred. In numerous cases, motions to dismiss are granted because the action is *moot*, meaning no real and present issue remains to be decided. Often a court's judgment refrains from going beyond the minimum reasoning required to answer a present question. Anything further is considered *dicta*, language that is not binding on other courts in future disputes.

For example, court supervision of the nuclear reactors throughout the United States seems less comforting than that provided by the Nuclear Regulatory Commission (NRC). The NRC includes experts, can investigate issues, and can make and enforce rules for future application. Thus, an agency's value tends to be its ability to expertly handle situations as they arise and, indeed, sometimes before they arise.

Another point in favor of administrative agencies is that they often include experts in a particular area of activity. The point is simply this: Judges are experts in law, and usually not expert in the technicalities of a given kind of activity.

Judge Brown, for example, probably was an attorney for a period of years before being elected or appointed to her present position. She may have specialized within the law, but even so, she might be classed as expert or semi-expert in almost any area of civil or criminal law. By brushing up on recent cases in a given area, she could make intelligent judgments on questions arising in those areas. But now, suppose we place her in a position where a knowledge of atomic energy or electrical power distribution questions arise. Suppose further that we ask her to supervise activities by companies involved with these questions. By so doing, we are probably asking too much of the judge. A better choice might be Black (a nonlawyer), who has been involved with the field for a substantial period. By choosing Black, we presumably reduce the chances of horrible and costly errors and increase the probability of sound decisions.

♦ Administrative Abuses

Despite the advantages of having our government operate through agencies, newspaper reports, investigators' comments, legal cases, and comments from friends who have been victims often reveal problems with administrative agencies.

Consider Black, above, who was just appointed to a six-year term as a member of the seven-person committee heading the XYZ Commission. He was appointed because he is an expert in the XYZ field. Of course, he is also likely to be a member of the appointment-maker's political party or at least sympathetic toward it. Now, Mr. Black may be a very ethical sort of a person. But even if he is, opportunities and pressures in administrative agencies can bring about changes in this facet of his personality.

As a government employee, Mr. Black's travel costs are covered by submitting an expense voucher at the end of a trip. If he needs to visit a supervised company, however, he might ask the company to pick up the tab. In fact, if his ethical nature is beginning to fray, Mr. Black might try to get expense money from both sources. Moreover, if he can work in visits to other supervised companies at the same time, he may, perhaps, obtain multiple travel payments. If one supervised company is engaged in the business of transporting people, Mr. Black may obtain free trips to "examine" first hand the nature of transportation it offers, And if a company deals in vacations, Mr. Black must, of course, investigate the nature of the vacation in person. Mr. Black may very soon learn that because of the supervisory power he wields, he may successfully call upon almost any company he supervises for favors.

Mr. Black's appointment to the commission lasts only six years. If Mr. Black intends to live longer than six years, he will want to be eligible for another job at the end of that period. His most likely future employer is one of the companies he supervises. Therefore, a significant practical pressure may induce him to treat at least one large company favorably in trying to secure future employment.

Of course, not all commissioners and administrative agency officials have erodible ethics. But then, neither are they all paragons of virtue. Instead, they tend to be human with human strengths and weaknesses.

ADMINISTRATIVE AGENCY ACTIVITIES

Administrative agencies are primarily responsible for (a) holding hearings, (b) making rules, and (c) supervision. In carrying out these functions, the agencies often have broad discretionary powers. If an agency steps beyond the boundaries of these powers as outlined by the statute that created the agency, it may be called upon to engage in a fourth activity—defending itself in court.

♦ Hearings

In acting in their judicial role, agencies generally hold "hearings," not trials. In doing so, they often follow their own rules. Proceedings before an agency usually involve less formalities than court proceedings. The result of a hearing is a *finding*, which the hearing officer submits to the agency for appropriate action. Most hearings are *adjudicative*—that is, they involve a plaintiff (often the agency itself) and a defendant. Some, though, are *legislative*—that is, they are simply a device for collecting evidence prior to making a general rule of some sort. The legislative hearing is appropriately a part of the rule-making procedure.

Adjudicative hearings are conducted (often before an *administrative law judge*) to determine what happened in some instance or set of instances in question. For example, the Gray Company has been accused by the National Labor Association (NLA) (a union) of an unfair labor practice. Specifically, the NLA says that just before the recent representation election, one or more members of management made statements amounting to threats if the NLA became the elected representative of the employees. (Obviously, the NLA lost the election or there would be no hearing.)

The unfair labor practice complaint was made to the National Labor Relations Board, which reacted to the complaint by appointing a hearing officer to investigate. A time and place for the hearing was set, the date came, and a hearing was held to elicit the facts of the case. Witnesses were called, sworn in, examined, and cross-examined. Hearsay may have been allowed, and sometimes, something less than the best evidence may have been used.

The presiding officer (again, this is usually an administrative law judge) sent the results of the hearing on to the NLRB, along with his or her recommendations for action (or inaction). The NLRB is not bound by the presiding officer's recommendation—that is, it may or may not order a new election, regardless of the recommendation it receives. If it does order an election, the Gray Company or the union that won the election may wish to appeal the order. If the NLRB rules against the NLA, *it* may wish to appeal. Depending on the nature of the question and other circumstances, appeal may be taken to a U.S. district court or a U.S. court of appeals.

The hearing procedure described above is similar to that required by most federal administrative agencies and some state agencies. In a local or state government agency, however, the hearing is likely to take place before the entire agency (that is, before all of the appointed members of the agency) rather than a hearing officer, and the appeal procedure will go through the state or a local court system rather than the federal system.

♦ Rules and Rule Making

Rule making is a legislative procedure. Administrative agencies make rules (presumably in the public interest) to regulate the activities of the entities controlled. One problem of the procedure is that sometimes the public interest seems to get lost in the process. In some ways, this is not surprising: The daily operations of administrative agencies bring the agencies in close contact with the businesses they supervise. Thus, it is often the businesses themselves who propose the rules. In such circumstances, one should not expect the resulting regulations to bring excessive hardships to the businesses. A rule so made may very well benefit the business community supervised more than it does the public.

Rules are usually made pursuant to directives in the statute creating the agency and delegating certain powers to the agency. If a rule is one called for by statute, it is known as a *legislative rule*. Such legislative rules, if adopted by the agency pursuant to the required procedures, have the same force as laws. The other general category of rules is referred to as *interpretive rules*—rules that interpret something. An interpretative rule may interpret a statute, a policy, another rule, a rate schedule, or something else. Interpretive rules generally do not have the full force of law, but they usually are treated with deference by the courts.

Procedural Fairness

Rules could be made in such a way as to be arbitrary and completely unfair to one or more of those supervised or to the public in general. A federal law (some states have similar laws) known as the *Administrative Procedure Act* was designed to prevent such unfairness. This law, together with the act creating the agency, sets forth general procedures to be followed by an agency in making rules. The primary means of assuring fairness exists in the requirement that agency actions be open to public scrutiny.

In keeping with procedural fairness, agencies usually do not make rules without first giving notice that they are considering such rules. They then give an opportunity for those who would be adversely affected to attempt to counter the rule. Generally, an agency publishes a proposed rule in the *Federal Register*, along with an invitation to interested parties to make written comments.

The agency's staff reviews the comments it receives and then may revise the proposed rules. The finalized rules are then published. In many situations, the finalized rules are published with a summary of the comments and, in some cases, a response by the agency's staff as to its views on the issues raised by the comments. Generally, such "finalized" rules become effective 30 days after their publication. Legislative hearings are sometimes scheduled, and those who would be affected are invited to attend. In addition, an agency may publish the proposed rule and make its effective date an extended period of time from publication. Such a publication acts as a notice. If no one objects, the rule may then become effective on the published date.

Rule making may have both beneficial and adverse effects. Consider Brown, who is licensed by the FCC to operate a television station reaching a given audience. Brown is very conscientious in his occupation. He has taken numerous surveys of his audience, and they all show a strong preference for old movies and game shows. Suppose that the FCC, in an effort to "clean up" television, passes a rule banning the types of programs Brown televises. Or suppose the FCC rule simply requires all network stations to televise only network programs during the prime viewing times. Such rules might help the television networks to sell greater amounts of advertising, since they would reach larger audiences. But Brown and his audience might well suffer because of the rule. Brown's audience would be depleted, and his ability to sell local advertising would be injured because of his smaller audience. Obviously, it is the smaller businesses that often cannot afford to hire lawyers to monitor proposed rules, provide comments, and lobby the agency.

Ex Post Facto Laws

The United States Constitution prohibits Congress from passing *ex post facto* laws. Such laws make an action or event unlawful and set the law's effective date *before* the law was passed. Accordingly, to be "unlawful," an act must have violated a law that existed when the act occurred. The courts have interpreted this ex post facto provision to apply only to criminal laws, and our courts generally view administrative rules as civil in nature. So Gray, with his income tax problem, may feel that his troubles are over when he pays what the IRS requires. But suppose the Internal Revenue Service passes a new regulation two years from now that would require Gray to pay a penalty in addition to the tax he has already paid. Gray may have to pay that additional amount or suffer whatever sanctions are specified in the law. "Final" statements by administrative agencies are not always final.

♦ Supervision and Coercion

Administrative agencies generally have the right to supervise those entities with whom they are concerned. In other words, the agency may find fault with one of those entities and then discipline it for whatever fault was found. Or they may simply require those supervised to "cease and desist" a practice found faulty. The means that agencies use to accomplish their supervisory function vary from one agency to another. The administrative "investigation" is one commonly used device.

Investigation

Investigating a person or a company may sound like a perfectly innocent and necessary activity, and it is in many situations. In some hands, however, the investigation turns into something far from innocent. Such an investigation can be very much akin to arresting and trying someone for a crime. The person arrested may be innocent of any wrongdoing, but to many who hear of his or her plight, the very fact that the person was arrested implies guilt. Similarly, the fact that a company or a person is under investigation can cause customers to drop away, lenders to balk, and stockholders to flee. Publicized investigations can become a means of coercion; the threat to "straighten up or we'll investigate you" often has teeth.

Of course, one who is subjected to such coercion may resort to the courts. A person or business may even be successful in getting the agency to drop the investigation. The chances of such success are usually low, however. The court is likely to uphold the agency's investigation if the agency can show a lawful purpose in conducting it. The court may even support the investigation if no reasonable purpose can be shown. After all, agencies exist to supervise activities; how can one effectively supervise without being aware of the facts in a given situation? To learn the relevant facts, the agency will argue, it must investigate. The agency may even have cause to investigate just to make sure everything is operating as it should.

Publicity about an investigation can be harmful, but so can some of the tactics used in the investigation. Consider the discovery procedures we mentioned earlier. An agency can serve a set of written interrogatories, or questions, on the subject of an investigation. When used, such *interrogatories* may ask a number of broad questions. Truthful answers often require the recipient to reveal detailed, private, and sometimes confidential information. Answering a properly phrased interrogatory can be both time-consuming and expensive. Failure to answer may subject the recipient to extreme sanctions. If one must answer many interrogatories phrased in rather broad fashion, he or she may find it necessary to hire a staff simply for this purpose, and of course, accept responsibility for paying the staff salaries. Similar problems arise when agencies subpoena numerous documents.

Discovery procedures by an agency may turn up unexpected (and sometimes trivial) violations that lead to further investigation. Gradually, the company's attention is diverted to the defense of the investigation rather than to the production of whatever goods or services it deals in. The imposition of such a burden by an agency may possibly be necessary and desirable—but it can also be abusive.

Supervisory Variations

The threat of investigation is one way agencies control those supervised. The use of *cease and desist* orders is another. Such an order usually amounts to an order entered by an administrative law judge that requires the defendant to stop (that is, to cease and desist) actions specified in the order.

But other, more practical techniques exist as well. Certain agencies, the Environmental Protection Agency (EPA), for example, are empowered to levy fines against offenders. Suppose the law gives the EPA the right to assess a fine of $10,000 a day for a given violation. Suppose further that fines of this nature are frequently levied, but that the agency nearly always reduces them to a much lower figure—perhaps in the neighborhood of $1,000—when payment is made. We now have a handy device not only for supervision, but also to keep those supervised from complaining. If a company refuses to pay the assessed fine, a court action is the natural result, with the amount to be paid $10,000 a day, not $1,000.

Another practical means of supervising and avoiding court conflicts is known as the *consent decree*. On discovering a violation, an agency may threaten to take court action against the alleged offender and, at the same time, offer an alternative—a means to avoid the action. All the alleged offender needs to do is to, essentially, admit guilt and agree not to violate the rule again. Since court actions in which an agency is a party are often extended, costly, and easily lost (even if one is innocent), there is a pronounced tendency to negotiate and enter into a consent decree. In this way, the agency can avoid the time and expense of protracted court battles and still create an image of success in doing its supervisory job.

Of course, as noted earlier, with so much power available for supervisory use, there is a possibility of abuse. If the agency itself is not effectively controlled, it may choose arbitrarily which of its supervised entities to investigate and which to leave alone. An agency may well decide to overlook infractions by large companies with numerous governmental and political connections and concentrate its supervision on others.

Advisory Opinions

Since an agency's sanctions may be extreme, sometimes it is better for a business to ask the agency's advice, in the form of a document called an *advisory opinion*, when a question arises. Suppose Green Trucking Company, for example, has been engaged in a certain practice for an extended period. Then, after discussing the practice with others, Green is led to believe that this practice may violate a particular rule. Would it be better to clear the doubt by asking the advice of the appropriate agency?

Some agencies are willing to give advisory opinions, but others are not. However, even if the agency is willing, at least two significant risks arise in asking for an advisory opinion. First, if Green's past practice is truly a transgression, the inquiry would attract attention to it and may lead to an investigation of the practice. Second, even if Green's inquiry results in a stamp of approval by someone in the agency, advisory opinions generally are not binding on the agency. Even if the agency were bound, it could still issue an interpretative rule holding the questionable practice to be a violation. So perhaps the less said about Green's customary behavior the better. Of course, this leaves Green in the unenviable position of perhaps continuing to violate an existing rule.

CONTROL OF AGENCIES

Courts exercise a limited degree of control over agencies, and legislative committees add a bit more. What an agency may or may not do is spelled out by legislation, and the legislation presumably responds to a public need. Most agencies are created by a legislative enactment; then, as a need becomes apparent, the behavior of agencies is modified by further legislation.

♦ Standing

One who is injured usually has a right to take action to recover for the wrong that has been done. For example, an applicant who applies for a license and whose application is denied has an "interest" to be protected and can show injury to that interest. But how about members of the public who might have benefitted if the license had been issued? Can they bring an action against the agency? Since actions by agencies affect large segments of the population (at least to a remote degree), the question of *standing*, or the degree of interest required to bring an action, often arises.

A few years ago, according to the courts, the people in our license-denial example would not have had sufficient interest to give them standing to act. Recent cases, however, have revealed a tendency of the courts to lower this barrier. Still, courts are usually careful to avoid "opening the floodgates" of litigation by allowing too large a group to have standing.

♦ Ripeness

Ordinarily the harvesting of an agricultural product must wait until that product is ripe. Roughly the same idea applies to court proceedings in which an administrative agency's actions are challenged. The action must be ripe for review—that is, there must be a concrete problem for the court to decide before it will evaluate the agency's action (or inaction).

The doctrine of ripeness was designed to "prevent the courts, through avoidance of premature adjudication, from entangling themselves in abstract disagreements over administrative policies, and also to protect the agencies from judicial interference until an administrative decision has been formalized and its effects felt in a concrete way by the challenging parties. The problem is best seen in a twofold aspect, requiring us to evaluate both the fitness of the issues for judicial decision and the hardship to the parties of withholding court consideration."[1] Generally, an agency's action is not *ripe* for a challenge unless it substantially threatens or injures someone.

♦ Scope of Review

An issue that often arises is whether an agency's action is subject to review by a court. The degree of "reviewability" of an agency action depends upon the laws under which the agency operates. As a general rule, if the legislature has entrusted a regulatory function to the agency, a court has no right to replace the action taken by the agency. Usually, judges are reluctant to decide such cases, anyway, in recognition of the agency's expertise in an area in which the judge is not likely to be an expert.

Boundaries and limits are either established by statutes or practice, and when an agency exceeds these limits, a call for review of the agency's action may be successful. Some statutes expressly state that certain agency decisions are not subject to review by any court. Challenges based on alleged procedural unfairness often have a better chance for success, assuming that the agency did not follow the required procedures.

One standard basis for review of agency action alleges that the agency action was not supported by *sub-*

1. Abbott Laboratories v. Gardner, 387 U.S. 136,148–49, 87 S. Ct. 1507, 1515 (1967).

stantial evidence. When a court examines this type of challenge to agency action, it is reviewing in its own area of expertise—the use of evidence and drawing inferences from it. The court is not substituting its judgment for the agency's in an area delegated by statute to the agency. Rather, the court is in a position to criticize the *manner* in which the agency did its job and require the agency to give a better performance.

The court's control of agency actions can only be limited, since a court cannot review those features of agency action delegated exclusively to the agency. Courts are thus left reviewing agency actions to see if they are *arbitrary, capricious, or an abuse of discretion*. The only remaining legal control of unwise agency action is by the legislative body that created the agency—either by legislative "oversight" committees or by legislative enactment.

In creating agencies we have created controlling devices necessary in our society, but devices that themselves must be controlled to limit or eliminate their abusive practices. The courts offer only a partial answer to agency control. Perhaps the most important control is requiring that agencies act via hearings open to the public and by maintaining a free press to bring injustice to the attention of the voters.

BOWLING GREEN-WARREN COUNTY AIRPORT BOARD v. C.A.B.

479 F.2d 553 (6th Cir. 1973)

This is a petition for review of the decision and final order No. 72-7-25, rendered by the Civil Aeronautics Board (hereinafter the Board) on July 7, 1972. By this order the Board granted the application of Eastern Air Lines, Inc., for amendment of its certificate of public convenience and necessity for Route 10 so as to delete the point Bowling Green, Kentucky, from that route.

The following facts, as adopted by the Board, appear in the initial decision of the examiner. Bowling Green is the county seat of Warren County, Kentucky, and the trade center for several counties in south-central Kentucky. In recent years Bowling Green and Warren County have experienced a trend of industrialization and relatively rapid growth. Thus Bowling Green's population has increased from about 18,000 in 1950 to about 38,000 in 1970, and the County's population increased from about 43,000 to about 55,000 during the same period.

Bowling Green is located approximately 60 miles north of Nashville, Tennessee, and approximately 100 miles south of Louisville, Kentucky, both of which cities are classified by the Federal Aviation Administration as medium air traffic hubs. From the Nashville airport, five trunk lines and four local service air carriers provide single-plane service to over 65 cities throughout the United States. From the Louisville airport, four trunk lines and three local service air carriers provide single-plane service to over 70 cities across the country. Louisville, Bowling Green, and Nashville are linked by Interstate Highway 65, a major north-south highway, as well as a major bus line providing frequent service between these points and beyond.

Since its certificate was amended in 1947 so as to add Bowling Green as a point on its Nashville-Louisville segments of Route 10, Eastern provided continued daily service to Bowling Green up to September 1969.[2] In fiscal year 1969, the number of origin and destination passengers for Eastern at Bowling Green had reached 11,852 per year, or over 32 per day on its daily schedules (southbound in the mid-morning and northbound in the late afternoon), and traffic had increased at a rate of about 2,000 passengers per year.

2. Claiming an insufficient number of passengers at Bowling Green to support the two daily round-trip flights it was then providing to Nashville and Louisville, in 1961 Eastern applied for suspension and deletion of Bowling Green and replacement of its Nashville-Bowling Green-Louisville service by Ozark Air Lines, Inc. The Board denied this application in a decision rendered in 1964.

On May 1, 1969, Eastern filed an application requesting suspension of its service at Bowling Green and approval of an agreement between Eastern and Air South, Inc., whereunder Air South would be employed by Eastern as an independent contractor to perform replacement service to Bowling Green. Air South was to provide three daily round trips between Bowling Green and Louisville and two daily round trips between Bowling Green and Nashville, and bear its own expenses subject to Eastern's agreement to underwrite Air South in amounts of no more than $50,000 the first year, $10,000 the second year, and $5,000 the third year. In support of its application, Eastern cited its competition from numerous strong trunk lines which are able to devote their full attention to dense and/ or long-haul routes upon which Eastern is dependent for economic survival. Eastern noted that it would save $300,000 annually from suspension at Bowling Green and that Bowling Green would benefit from the more frequent service to Nashville and Louisville under the replacement service.

In August 1969, the Board approved Eastern's application for temporary suspension at Bowling Green and the agreement for replacement service by Air South. The Board's approval was made subject to the condition that the suspension granted to Eastern would immediately terminate if Air South should cease to satisfactorily provide the specified replacement service.

Air South provided reliable service from September 2, 1969, through March 31, 1970, when it terminated its replacement service due to a lack of financial resources and support. Notwithstanding Air South's morning, afternoon, and evening service from Bowling Green to Louisville and Nashville, the number of origin and destination passengers at Bowling Green dropped to an average of 21.9 per day during this seven-month period—a decline of about one-third from the traffic Eastern had carried just prior to its replacement by Air South.

Faced with the legal obligation of providing air service to Bowling Green when Air South terminated its replacement service, Eastern on March 30, 1970, applied to the Board for continuation of its temporary suspension at Bowling Green and for approval of a one-year agreement between Eastern and Northern Airlines, Inc., which was then providing scheduled air service between Louisville, Bowling Green, and Nashville. Under the agreement Eastern would underwrite Northern in the amounts of $8,000 per month for the first three months of operation and no more than $4,000 per month thereafter for the latter's replacement service at Bowling Green. Pending the Board's consideration of the above application, Northern's service was admittedly unreliable and unsatisfactory, and Northern in fact terminated service at Bowling Green on or about May 8, 1970. Bowling Green was thus without reliable and satisfactory air service from April 1, 1970, until May 25,1970, and with no service at all during the last 17 days of this period.

On May 25, 1970, Eastern entered into a third contractual agreement for replacement service to Bowling Green, this time with Wright Air Lines, Inc. Whereas Air South and Northern in essence had been employed as independent contractors, retaining all revenues and bearing all expenses subject to underwriting payments by Eastern, Wright was essentially employed to provide substitute air service on a cost-plus-10 percent basis, subject to an initial ceiling of $28,000 per month. Under the agreement, Wright was to provide Bowling Green with at least three daily round trips to Louisville and at least two daily round trips to Nashville.

On September 24, 1970, the Board approved Eastern's application for continued temporary suspension at Bowling Green, subject to Wright's performance of the specified replacement service, and the agreement between Eastern and Wright for such replacement service. In so ap proving Eastern's application, the Board acknowledged that replacement service at Bowling Green had not been satisfactory. The Board was nonetheless of the belief that Wright's replacement service, sponsored and sustained by Eastern, would be superior to prior replacement service and superior to the service which would be provided directly by Eastern were the suspension denied.

On December 31, 1970, Eastern filed the present application for deletion of Bowling Green from its certificate for Route 10.

In his initial decision rendered after a public hearing, Examiner Sornson apparently recognized that there is no reasonable prospect that reinstitution of direct service at Bowling Green by Eastern would be economically feasible. The Examiner nonetheless concluded that Bowling Green should not be deleted from Eastern's certificate, but rather that Eastern should continue to be responsible for the community's air service needs through support of air taxi replacement service. The Examiner found Eastern's request for total relief by way of deletion of Bowling Green to be unwarranted in that "Eastern (had) not shown that a substitute service arrangement at Bowling Green cannot be successful...." In support of this conclusion, Examiner Sornson cited what he found to be inadequate subsidies paid by Eastern to Air South and Wright and Eastern's failure to effectively promote and advertise the replacement service.

Reviewing the decision of the Examiner, the Board rejected his finding that Eastern had failed to show that a substitute service arrangement cannot be successful at Bowling Green and his conclusion that Eastern should remain responsible for the community's air service needs through support of air taxi replacement service. Instead, the Board concluded from the evidence that:

> Continued air taxi operations have only a limited chance of eventual success...there are no unusual or extenuating circumstances which would warrant support of the magnitude which would be required to continue such operations.

In support of its conclusion, the Board reviewed the decline in the number of passengers experienced by both Air South and Wright and the resulting operating losses suffered by these replacement carriers, notwithstanding the substantial subsidies paid by Eastern. Thus Air South, with its fully satisfactory schedules and performance, was able to generate only 22 daily origin and destination passengers, resulting in a loss of over $127,000 during its seven months of operation. And during its first year of operation, Wright experienced traffic of under 20 daily origin and destination passengers, resulting in a loss of $242,000 or $35 per passenger. The Board also relied on the evidence that after approval—without civic objection—of reduced service by Wright in May 1971 to stem continuing losses, traffic dropped to about 12 passengers per day.

The Board further rejected the Examiner's finding that the decline in passenger traffic is attributable to Eastern's failure to promote and assist the substitute service and Eastern's inadequate subsidies. In addition to providing support services for both Air South and Wright, Eastern spent about $3,500 in the year ended May 1971 advertising and promoting Wright's substitute service. Eastern paid Air South $50,000 in cash subsidy for the latter's seven months of operation and paid Wright almost $300,000 for the first year of its replacement service.

The Board therefore concluded that the public convenience and necessity require the amendment of Eastern's certificate for Route 10 so as to delete Eastern's authority to serve Bowling Green.

Under our narrow scope of review as set forth in 49 U.S.C. sec. 1486(e), the Board's findings of fact are conclusive if supported by substantial evidence. Beyond this scope of judicial review, we must defer to the Board's expertise and discretion in weighing the many factors and policy considerations which enter into Board decisions respecting public convenience and necessity....

From the record before us we find that the Board's findings of fact are supported by substantial evidence. Relying upon the undisputed rates of traffic generation at Bowling Green between 1962 and 1969, the Board found that there is no reasonable prospect that a reinstitution of direct service by Eastern could be profitable in the future. This finding—which also appears to be a conclusion of the Examiner—is supported by evidence that during its peak year of 1969 with 11,853 origin and destination passengers at Bowling Green, Eastern achieved load factors of only about 30 percent and suffered an operating loss of almost $300,000 (about $25 per Bowling Green passenger) on its Bowling Green-Nashville and Bowling Green-Louisville flights.

We also find substantial evidence on the record to support the Board's findings that continued replacement air taxi operations supported by Eastern have an extremely limited chance of success and that traffic response to these operations does not warrant their continuation. As reviewed above, the declining traffic under Air South's and Wright's operations, and the substantial operating losses suffered by these replacement carriers despite Eastern's subsidies fully support these findings by the Board.

We find no merit in Petitioner's argument that the Board's order must be set aside because the Board failed to give sufficient weight to the several factors prescribed in 49 U.S.C. sec. 1302.[3] Petitioner asserts that the Board failed to adequately consider the future commercial needs of Bowling Green as well as the effects of its decision upon the Postal Service and the national defense.

With respect to Bowling Green's future commercial needs, its opinion discloses that the Board was fully aware of the factors favoring the maintenance of air service not only in Bowling Green but also in all smaller communities across the nation. On balance, however, the Board found that Bowling Green's future commercial needs and the community benefits from certified air service do not warrant continued requirements that Eastern—as a trunk line carrier-bear the responsibility and expense of the unsuccessful operations at the City. Finding no abuse of the Board's discretion, we are powerless to reweigh these factors which were before the Board....

Moreover, we find in the record no evidence respecting the effects of the Board's decision on the present and future needs of the Postal Service and the national defense. Even if such evidence had been introduced, however, we do not believe that the Board is required in every case to make express findings on every factor set forth in section 1302....

The order of the Board is affirmed.

MARSHALL v. NORTHWEST ORIENT AIRLINES, INC.

574 F.2d 119 (2d. Cir. 1978)

Northwest Airlines, Inc., ("Northwest") seeks to prevent representatives of the Occupational Safety and Health Administration ("OSHA") from inspecting one of its airport facilities, claiming that the safety aspects of the hangar in question are already regulated by the Federal Aviation

3. 49 U.S.C. sec. 1302 provides as follows:

in the exercise and performance of its powers and duties under this chapter, the Board shall consider the following, among other things, as being in the public interest, and in accordance with the public convenience and necessity:

(a) The encouragement and development of an air-transportation system properly adapted to the present and future needs of the foreign and domestic commerce of the United States, of the Postal Service, and of the national defense;

(b) The regulation of air transportation in such manner as to recognize and preserve the inherent advantages of, assure the highest degree of safety in, and foster sound economic conditions in, such transportation, and to improve the relations between, and coordinate transportation by, air carriers;

(c) The promotion of adequate, economical, and efficient service by air carriers at reasonable charges, without unjust discriminations, undue preferences or advantages, or unfair or destructive competitive practices; (d) Competition to the extent necessary to assure the sound development of an air-transportation system properly adapted to the needs of the foreign and domestic commerce of the United States, of the Postal Service, and of the national defense;

(e) The promotion of safety in air commerce; and

(f) The promotion, encouragement, and development of civil aeronautics....

Administration ("FAA"). It adopts this position notwithstanding the fact that an administrative record concerning the alleged overlap in jurisdiction has not been assembled, and discovery is yet to commence. We agree with the determination of the district court that the challenge to OSHA's jurisdiction cannot be brought until Northwest has more adequately developed the record by exhausting its existing administrative remedies.

I.

On October 27, 1976, Efraim Zoldan, an OSHA compliance officer, attempted to inspect Northwest's Hangar No. 1 at John F. Kennedy International Aport. After being denied entrance by the agent in charge, he sought an inspection warrant from Magistrate Vincent Catoggio on November 4, 1976. Relying on affidavits establishing that Northwest's hangar had in the past been cited for 13 violations of OSHA standards, the magistrate issued a warrant.

The following day, Northwest again turned Zoldan away when he attempted to enter its facility, ostensibly because it would soon be moving to quash the warrant. It did so that very afternoon, arguing that section 4(b)(1) of the Occupational, Safety and Health Act ("Act") explicitly withdrew from OSHA any jurisdiction over safety matters already regulated by other federal agencies. Northwest contended that the FAA oversaw the occupational safety of hangar workers, and cited an opinion by Administrative Law Judge Jerome C. Ditore as support for this principle.

The opinion relied upon concerned the airline's challenge to a previous OSHA citation for failing to provide adequate protection to employees changing landing lights on aircraft. After a 24-day hearing, Judge Ditore determined that the FAA had preempted this area by promulgating regulations "affecting the occupational safety and health of ground maintenance personnel changing landing lights through the leading edge flap cavities of Boeing 747 aircraft." A member of the Occupational Safety and Health Review Commission subsequently granted the Secretary of Labor's petition for discretionary review of Judge Ditore's decision by the full Commission, but such plenary consideration has been delayed pending the appointment of a third commissioner.

Upon being apprised of Judge Ditore's decision, Magistrate Catoggio vacated the search warrant. He interpreted the earlier ruling to indicate that the FAA had preempted "at least part of the area (of airline safety)," and thus had, for section 4(b)(1) purposes, prevented OSHA from exercising jurisdiction over any safety hazards within the hangar. The Secretary of Labor then appealed to the district court, which reinstated the warrant, finding that the challenge to OSHA's jurisdiction was premature, and requiring the airline to first exhaust all administrative remedies.

II.

Section 4(b)(1) of the Occupational
Safety and Health Act, 29 U.S.C. sec. 653
(b)(1), provides that:

> Nothing in this chapter shall apply to working conditions of employees with respect to which other Federal agencies...exercise statutory authority to prescribe or enforce standards or regulations affecting occupational safety or health.

In essence, this provision is designed to eliminate any duplication of the efforts of federal agencies to secure the well-being of employees. At the same time, however, section 4(b)(1) was not intended by Congress to eliminate OSHA's jurisdiction merely on the basis of hypothetical conflicts. Employees should not lightly be denied the protection of OSHA for, as the Senate Report accompanying the Act noted, "(t)he problem of assuring safe and healthful workplaces for our working men and women ranks in importance with any that engages the national attention today."...Thus, as the section's wording makes clear, a sister agency must actually be *exercising* a power to regulate safety conditions in order to preempt OSHA. Were this not the case, many occupational health hazards might remain uncorrected merely because some agency whose prime concern is not safety has failed to take notice of that area.

In a similar vein, section 4(b)(1) is not designed to provide wholesale exceptions for entire industries....The exact standard to be applied in determining the scope of preemption, however, is less than crystal-clear. The *Southern Pacific* court concluded that agency rule-making directed " at a working condition—defined either in terms of a 'surrounding' or 'hazard'—displaces OSHA coverage of that working condition."...The Fourth Circuit, on the other hand, has established a somewhat different test, concluding that "working conditions" encompass:

> an area broader in its contours than the "particular discrete hazards" advanced by the Secretary, but something less than the employment relationship in its entirety....(They include) the environmental area in which an employee customarily goes about his daily tasks....

As these vague standards demonstrate, a determination of preemption requires an inquiry into complex issues of law and fact. Accordingly, it is proper for a court to defer examination of such difficult questions of agency jurisdiction until a party has fully exhausted its administrative remedies....The exhaustion requirement is particularly apt in cases such as this, where administrative review is forthcoming and the agency will be given a chance to apply its expertise....

The airline asks this Court to conclude that an entire hangar, and the multitude of possible hazards within it, are excluded from OSHA protection. Given the breadth of the problem, our inquiry would of necessity be amorphous, unwieldy, and lacking in expertise. In contrast, following an OSHA inspection and its administrative review, the issues will have been greatly refined, and we will be made aware of the specific types of hazards that OSHA officials believe are regulated by the FAA. Certainly, given Congress's determination to create a "comprehensive, nation-wide" approach to improving occupational safety, the agency charged with that mandate should be accorded an initial opportunity to inspect potential safety hazards.

III.

In addition to disputing OSHA's jurisdiction, Northwest contests the constitutionality of the warrantless searches conducted by the agency's inspectors. The airline notes that in most instances, a warrantless search is inconsistent with the protections of the fourth amendment....In fact, the greater number of courts to consider this issue have found unconstitutional the warrantless search procedure authorized by section 8(a) of the Act....That question is now before the Supreme Court in *Marshall v. Barlow's, Inc.*

We need not decide, however, whether Congress could constitutionally authorize warrantless searches of all business premises. Nor need we determine whether the "pervasive regulation" of the airline industry creates an exception to the warrant requirement. Assuming *arguendo* that warrantless searches are unconstitutional, it would then be proper for a court to imply a warrant requirement to preserve the statute's constitutionality....The Supreme Court followed exactly this course of action in *Camara*, supra.

Adhering to these precedents, the district court issued a warrant, relying on Northwest's previous OSHA violations for a showing of probable cause. On this appeal, the airline does not dispute the existence of factual support for this finding. Accordingly, having determined that a warrant requirement can properly be implied if the statute, as written, were unconstitutional, we affirm the order of the district court reinstating the inspection warrant.

♦

REVIEW QUESTIONS

1. Why must we have agencies? Why, for example, must we have an agency to control air traffic? What could be a better way to accomplish this control?
2. What is the *Federal Register*?
3. What could an agency (i.e., a local zoning board) do to a person or company whose actions it found undesirable?
4. How are agency hearings similar to and different from court trials?
5. How does rule making differ from a legislative procedure?
6. What coercive tactics are available to agencies? Consider a local liquor control commission.
7. Who controls agency activities? Is the control adequate?
8. The case of *Bowling Green-Warren County Airport Board* v. *C.A.B.* posed the kind of problem the C.A.B. commonly handled. Think of some reasonable solutions to the problem posed by the case. What could be done besides discontinuing service to and from Bowling Green?
9. In *Marshall* v. *Northwest Orient*, the problem seems to center on whether OSHA or FAA should inspect the Northwest Orient facility at JFK International Airport. In your opinion, which agency would do the better job of safety inspection? What problems would this agency likely encounter in carrying out this function?

Chapter 24

Labor

A necessary component of everything we buy or sell is labor. According to one view, what we really pay for when we buy something is labor in one form or another. Capital, it is said, merely represents stored compensation for past labor; management, too, is just another form of labor. Thus, labor is a vital and, sometimes, volatile portion of all projects. For this reason, we will consider here some of the problems posed when people's services are bought and sold.

In this discussion, problems such as the cost, availability, and transportation of labor are not addressed—instead, this chapter addresses some of the legal problems involved in labor conflicts. The discussion will center around unions because without unions, or the threat of them, there is usually little conflict. Two powers must be fairly evenly matched for a conflict to occur. A single worker usually cannot successfully dispute a management edict, however wrong it may seem. By leaving, the employee can deprive a company of his or her services but, in most cases, others can be hired to do the job as well.

The laborer's answer to this rather unbalanced state of affairs is organization. A concerted refusal to work, in which all employees participate, can be a powerful weapon. To accomplish such organized activity, workers formed *unions*. Occasionally, the balance of power shifts very heavily to the union side. When this occurs, management cannot afford the conflict that might be caused by refusing a union request, especially a refusal likely to promote action that could destroy the company.

HISTORICAL BACKGROUND

Working people now have the right to form unions. Through unions, they may petition their employer for improvements in wages or working conditions. Under certain circumstances, employees may resort to economic "warfare," in the form of strikes, to enforce their demands. Although strikes are generally lawful today, such was not always the case.

♦ Criminal Conspiracy

Under the English common law existing at the time of the American Revolution, the formation of an association of employees engaged in similar employment was unlawful. To a considerable extent, the judge—made common law reflected the economic beliefs of the time—the *law of supply and demand* and the *wage-fund theory*, for example. Reasoning from basic notions of supply and demand, the courts held that it was unlawful to "artificially" regulate the price of labor. Such decisions viewed the only fair and lawful price as that established by competition for a particular type of labor. Moreover, the wage fund theory essentially held that a worker could only improve his or her wages or working conditions by stealing from another worker. Since organizations of working people clearly intended to improve the lives of their members, such organizations were viewed as unlawful "criminal conspiracies."

Over the years, the courts gradually came to recognize the right to organize. In the meantime, organizations of employees made some improvements in the working conditions and wages of their members. The demise of the criminal-conspiracy doctrine in its application to labor unions came about during the second half of the nineteenth century. This demise also brought with it a rise of union activities. New unions were formed, and attempts were made to organize on a national scale. Gradually, the use of economic force to bring about improvements in working conditions began to win recognition in the courts.

♦ Antitrust Laws

In 1890, the Sherman Antitrust Act was passed with the apparent purpose of preventing monopolistic practices of "big business."

However, the act's application to unions was unclear. In an early case involving a union boycott, the United States Supreme Court held that the boycott was a violation of the antitrust laws, and the union was fined triple damages. Subsequent use of the antitrust laws, and the threat of their further use, stifled unions.

♦ Other Union Constraints

By this time, employers came to rely on two legal devices for minimizing union activity. The *yellow-dog contract* prevented employees from joining a union or from forming a union among themselves. And the *labor injunction* could be used to prevent anticipated damage from threatened union activities.

Yellow-Dog Contract

The *yellow-dog contract* was a clause in an employment contract stating, in effect, that the employee agreed, as a condition of continued employment, not to engage in union activities. Under contract law, the courts generally enforced such agreements. If a union tried to persuade employees to become members, it was inducing the employees to breach their contracts and was subject to either an injunction to prevent the activity or a claim for damages if it was successful. Moreover, the employees were terminated if they breached their contracts.

Labor Injunction

Generally, anyone who is threatened with harm to his or her property has a right to request an injunction from a court having equity powers. This right, of course, applies to employers, too. In the case of employers, not only are their physical assets, such as the plants and machines, considered their property, the concept extends to an employer's good will in relations with customers and the public as well. Thus, for example, delayed filling of customers' orders could be such an injury. Therefore, an employer who was threatened with a strike or other economic action could ask a court for an injunction to prohibit the strikes or to stop the union from picketing. Courts responded to these requests quite freely with temporary injunctions or *labor injunctions*, as they came to be known. Some courts even went so far as to direct the injunctions against "whomsoever" might harm the employer.

♦ Union-Supportive Legislation

Unions saw the courts as a friend of management. Hence, the unions began to lobby for legislation to curtail abusive legal tactics by management. The Clayton Act, passed in 1914, appeared to be the answer. Many advocates viewed the Clayton Act as the key to eliminating injunctions against unions under the antitrust laws. Its wording seemed clear. However, the Supreme Court interpreted the Clayton Act as applying only to strikes and picketing by employees against their employer. In other words, a court could still issue an injunction against non-employees who picketed. Thus, the courts were still able to issue injunctions against secondary economic pressure or boycotts (discussed later).

In 1932, Congress passed the Norris-LaGuardia "anti-injunction" Act. The Norris-LaGuardia Act eliminated nearly all labor injunctions by federal courts. Employee organizations could now strike and picket much more freely. Essentially, the balance of legal power had shifted toward the unions.

♦ National Labor Relations Act

Congress went even further in 1935, with the passage of the National Labor Relations Act (NLRA), sometimes called the *Wagner Act*. The only unfair labor practices (i.e., unlawful practices) spelled out in the act were acts of employers. The Wagner Act set up the National Labor Relations Board (NLRB) to determine whether the federal law should be applied to a particular case, to hold elections to certify employee representatives, and to investigate and prevent unfair labor practices.

With the new protections afforded by the Norris-LaGuardia and Wagner acts and the sympathetic attitude of government and the courts, the labor movement thrived. Old unions expanded, and new ones were born. Collective bargaining and enduring labor-management contracts became almost a way of industrial life.

♦ Taft-Hartley Act

Unions became unpopular in the period right after World War II—in part because they had called strikes in important industries. In an effort to more evenly balance the parties' respective bargaining powers and balance the power back toward management, Congress passed the Taft-Hartley Act in 1947.

The Taft-Hartley Act prohibited certain union actions as unfair labor practices. Among other things, Taft-Hartley (and later NLRB rulings) made unlawful the creation of a *closed shop* arrangement with an employer—an arrangement in which a new employee had to be a union member before he or she could be employed. Moreover, the secondary boycott, jurisdictional strikes, and featherbedding were all made illegal. Furthermore, a 60-day notice of intent to strike was required in most cases before a union could lawfully strike. National emergency procedures were also set up in an attempt to avoid national disasters and still preserve free collective bargaining. A federal mediation and conciliation service was created to aid in reaching collective bargaining settlements. Finally, the act made unions responsible for their torts and allowed them to sue and be sued in court actions.

It is an understatement to note that unions objected to the new legislation; one of the more complimentary descriptions for it was the "slave labor act." Despite union objections to the act, unions continued to expand during the period following its passage. They grew and combined, becoming nationwide and even international in their affiliations. The American Federation of Labor (an association of trade unions) and the Congress of Industrial Organizations (built along industry lines) merged to present a united front.

Unions themselves became big business. With the increased size came certain abusive practices. For example, people who handle money belonging to others are often tempted to divert some of it to personal gain, disguising the diversion or obscuring it by simply not keeping records. Here, accumulating union dues, initiation fees, and assessments provided such a temptation. A senate investigating committee, headed by Senator McClellan, uncovered proof of such practices in some unions and strong suspicions of them in others. In at least one union, many officers were found to have long prison records. The work of the McClellan Committee was a main cause of passage of a second National Labor Relations Act amendment, the Labor-Management Reporting and Disclosure Act of 1959, known as the *Landrum-Griffen Act*.

♦ Landrum-Griffin Act

While the main purpose of the Landrum-Griffin Act was to cause employers, labor organizations, and their officials to adhere to high standards of responsibility and ethical conduct, it also amended many sections of the Taft-Hartley Act. It imposed, for example, a $10,000 fine or imprisonment up to one year, or both, on those who did not comply. Enforcement of the Taft-Hartley Act, on the other hand, had been achieved mainly through cease and desist orders and contempt of court for noncompliance. In addition, the Landrum-Griffin Act restricted the ability to picket for purposes of recognition or organization.

CONTEMPORARY LABOR RELATIONS LAW

Today, the main source of law governing the relationships between labor and management is the *National Labor Relations Act*, or NLRA. Of course, other federal acts such as the Fair Labor Standards Act, the Walsh-Healey Public Contracts Act, the Work Hours Act, and the Equal Pay Act address labor relations as well. Moreover, in the last few decades, employment discrimination based on race, sex, age, and disabilities has been made unlawful. Numerous state and local laws also have a bearing on the employment relationship. Still, the central law of labor relations remains the National Labor Relations Act.

Over the years, decisions by the courts and the NLRB have refined and interpreted the act. Many of today's controversies are decided upon the basis of the " settled law" of that particular area of labor relations. But, then, with technological change and the seesaw bargaining of union-management contracts, new problems continue to arise. Thus, although the general rules are fairly well settled, the law continues to evolve.

♦ Purpose, Policy, and Jurisdiction

The avowed purpose and policy of the National Labor Relations Act is to minimize industrial strife and thereby promote the full flow of commerce. The means chosen to accomplish this was to define certain rights of employees, rights of employers, and orderly procedures for settling disputes between them.

Employees are guaranteed the right to organize and to bargain collectively with their employers. They are also guaranteed the right to refrain from such activities unless they are subject to a union-management contract requiring membership in a union as a condition of their employment.

The agency charged with resolving conflicts involving such organizing and collective bargaining is the NLRB. It functions to prevent and remedy unfair labor practices and to conduct secret-ballot elections to select employees' bargaining representatives. In doing so, it engages in a wide variety of administrative procedures and practices. It must receive petitions and allegations, determine its own jurisdiction, investigate, hold hearings, issue orders, and make rules, and it often becomes a party to controversies in the federal courts. In short, it functions in much the same manner as do most federal administrative agencies.

The NLRB decides whether it has authority over a matter when it responds to petitions either claiming an unfair labor practice or requesting an election. In fact, its first order of business in response to such a petition is to decide whether it has jurisdiction in the case. According to its rules, the employer in the petition must be involved in interstate commerce for the NLRB to take the case; but involvement with interstate commerce does not necessarily mean that the NLRB must assert jurisdiction. Usually, the NLRB will not assert its jurisdiction unless the employer's volume of business exceeds set dollar limits for a particular industry. If the NLRB refuses to take jurisdiction, the petitioner may be able to turn to a state act for relief.

♦ Employee Representation

In industries affecting interstate commerce, certification by the NLRB as the bargaining agent is largely a matter of winning an election. Section 9 of the Taft-Hartley Act (as amended by the Landrum-Griffin Act) describes the procedure. The following material gives an overview of that procedure.

Petition

The process begins with a petition to the NLRB for an election to be held. The petition may come from an employee, a labor organization, or the employer.

Investigation

On receipt of the petition, the NLRB must determine whether interstate commerce is affected and determine the appropriate *bargaining unit*. Various types of bargaining units are used to represent the employees involved. The unit might be an *employer unit*—that is, it would consist of all the employees of a particular employer or a substantial portion of them. The unit also could be composed of the employees of *one plant* or of one *department* within it. Alternatively, the choice might be a *craft unit* if the employees involved in a particular trade or craft desire such representation. A group often chooses the unit that it believes will generate a majority vote. If the employer challenges the union's choice of a unit, review by a hearing officer and, possibly, the courts may occur. The bargaining-unit determination should be made in such a manner as to assure employees the fullest freedom in exercising their rights.

In determining bargaining units, special consideration is given to two groups of employees: professional employees and guards. Professional employees (such as engineers, accountants, buyers, and the like) are not to be included in a bargaining unit with other employees unless the majority of the professional group votes for inclusion. Guards employed for protection of company property cannot be included in the same bargaining unit with other employees of the company under any circumstances. Generally, the guards' labor organization may not even be affiliated with unions representing other employees. Combining representation could partially defeat the guards' functions.

Elections

The NLRB regulates elections to determine who will be the employees' bargaining representative. If the only choice is between a particular union and no union at all, a simple majority of the votes determines the outcome. If the union fails to win more than half, the results of the vote will be certified as showing that the union was not the choice of a majority of the voting employees. If more than two choices exist and none receives more than half the votes cast, a runoff is held between the two choices receiving the most votes. Once a union has been selected and certified, that choice of a bargaining representative settles the issue for at least a year; after a year, certification can be removed by petition and another election.

Who May Vote

Each employee in the appropriate bargaining unit is entitled to one vote to determine the bargaining representative. Shortly after passage of the Taft-Hartley Act, the question arose as to whether employees on strike could vote in an election. Rulings on this question proved difficult to follow in some situations, and the law was clarified in the Landrum-Griffin Act. Employees on a valid economic strike against their employer retain the right to vote in an election for a year after the beginning of the strike. Seasonal or temporary employees generally are not eligible to vote unless they have a legitimate expectation of being reemployed and an interest in the working conditions.

Campaigning

In ensuring the free election of a bargaining representative, the NLRB is often concerned with preelection conduct by both the employer and the union. Neither the employer nor a union may use coercion or threats of reprisal to attempt to influence an employee's vote. The employer cannot promise benefits to be given if the union is unsuccessful. Furthermore, the employer must be particularly careful under NLRB rulings during the 24 hours directly preceding the election. In short, the employer cannot do anything to create an atmosphere that would discourage an election.

The limitations imposed on the employer, however, are not intended to inhibit the normal conduct of business. The employer can still hire new employees and discipline employees, even to the point of firing an employee for cause. The employer can speak to the employees on company time, or on their own time if attendance is voluntary. He or she can voice an opinion of unions or of the result to be expected from joining a union as long as threats or promises of benefit could not be implied from such statements. The employer can state the company's legal position, the dangers and costs of union membership, and what the union can and cannot do for employees.

In the context of a campaign, it is useful to consider just who is an employer. The term *employer* applies to the management of a company. This usually includes not only the president and general manager, but generally all salaried line or staff employees outside the bargaining units. Thus, what a foreman, an engineer, or a production control clerk says or does prior to an election may be interpreted as the words or deeds of the employer.

♦ The Contract

The contract between the company and the bargaining representative of the company's employees is not an employment contract; no one holds a job by virtue of it. Rather, it is an agreement that sets forth the conditions under which those who are employed will work.

It has become almost standard practice to include certain provisions of importance in collective bargaining agreements. Besides addressing wages and items such as sick days, holidays, and vacations, for example, it usually contains a statement to the effect that the union agrees not to strike and the company agrees not to lock the union out during the life of the agreement. Most agreements set up a step-wise grievance procedure—an orderly process for settling disputes without resort to coercive tactics or to the courts.

If, for example, the paint gang at Black Manufacturing believes a newly issued incentive rate is unfair, its recourse is to petition for a change pursuant to the agreement and follow the route of the grievance procedure. A strike by the paint gang as a result of the rate issuance would constitute an unlawful *wildcat strike*. Such a strike would probably allow Black Manufacturing to permanently replace any or all of the paint gang members. However, if the paint gang members are allowed to return to work, they come back with the same rights they would have had if they had not engaged in the wildcat strike.

In a similar vein, Black Manufacturing must bargain with its employees; it cannot lock them out. If, for example, Black's bargaining sessions with its employees' representative reveals what Black considers to be unreasonable demands, Black must still bargain. Black may not even be able to go out of business to avoid bargaining. For example, if Black is a multi-plant organization, it could not go out of business at, say, Dayton, where it has labor problems and move its Dayton operations to one of its other plants. To lawfully go out of business at Dayton to avoid the union, it would have to go completely out of business—at all its plants. Otherwise, its behavior constitutes an *unlawful lockout*—an unfair labor practice. Still, the duty to bargain is *not* a duty to agree to terms perceived as unreasonable.

♦ Unfair Labor Practices

Section 8 of the NLRA lists a number of *unfair labor practices*. It is, for example, an unfair labor practice for either a union or a company to restrain or coerce employees as the employees exercise the right to form or join labor organizations or to choose not to join. It is also an unfair labor practice for the union and the employer to refuse to bargain with the other. In this regard, it is noteworthy that neither party is required to accept any proposal by the other or to agree to any concession. Bargaining simply means meeting at reasonable times and conferring in good faith on employment-related subjects. Other unfair labor practices are elucidated below.

CLOSED SHOPS

Since 1947, it has been unlawful to form a new closed shop—a shop in which a new employee has to be a union member *before* he or she can be employed. the reasoning is that in a closed shop, the union rather than management selects prospective employees. If an employee proves unsatisfactory, the company has little or no right to replace that employee; generally, the company must keep whomever the union sends. Since 1947, the only lawful, newly bargained provisions of this kind allow for a *union shop*. In a union shop the employer hires new employees and has a period of at least 30 days (except for building construction industries) as a "trial period." The employer may replace the new employee if he or she does not perform satisfactorily. After the probationary period, the new employee must join the union to retain his or her employment. If Black Manufacturing operated under a closed shop provision prior to 1947, it might be forced to retain that arrangement; but if the original contract provision was made after the Taft-Hartley amendment, Black might have a union shop or an *open shop* (where union membership is not required), but not a closed shop.

One version of the open shop is an *agency shop*. Union membership is not required in this arrangement, but employees eligible to be union members are required to pay periodic dues (usually equivalent to union dues) whether they are members or not.

SECONDARY BOYCOTTS

In this country and elsewhere in the world, people may choose freely between dealing with one competitor or another. If, for example, the Black Company chooses to buy a die cast machine from Gray Enterprises rather than from Green, It is Black's choice. In fact, if Black wishes to deal exclusively with Gray rather than Green, that is Black's prerogative. The problem arises when someone else forces Black to make a particular decision. The situation is shown graphically below. If Union A wishes to receive some concession from Management C, A may bring added coercive force against Management C by forcing Customer B to refuse to deal with Management C. This is known as a *secondary boycott*, and it takes many forms. Such conduct is secondary in the sense that it is directed not at management C (the primary target of the union), but at some other company.

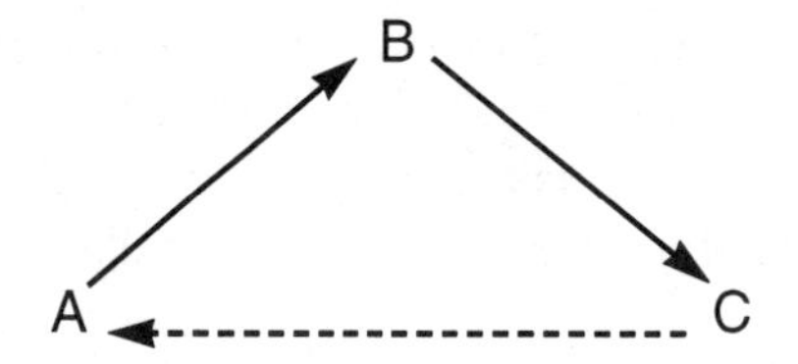

Under the NLRA, it is an unfair labor practice for a union to engage in or induce its members to strike where the objective is to force one company to cease dealing with another. However, not all conduct that may influence a second company to curtail business with the employer amounts to a *secondary boycott*. For example, picketing directed at the target employer is probably acceptable.

It is an unfair labor practice for a union to force or threaten to force a company to assign work to members of one trade union rather than another (a "jurisdictional dispute"). For example, a contractor might be required to assign the job of hanging metal doors to metal workers rather than carpenters, or the reverse.

Hot Cargo

It is an unfair labor practice to force or threaten to force an employer to refuse to handle the products of some other company. A union-employer agreement of this nature is known as a *hot cargo agreement* and is unlawful for both parties. Such agreements were sometimes made as an attempt to get around the restrictions on secondary boycotts. Specifically, unions tried to negotiate agreements with the employers to the effect that the employees could not be required to handle or transport products from an "unfair" company.

Suppose, for example, that Black Company usually buys products from White Manufacturing. A strike against White occurs. If a hot cargo agreement exists, the union at Black may insist on enforcing Black's agreement to stop dealing with White to help force a favorable settlement with White. Here, the union could argue that its dispute with Black was "primary" (and therefore not a secondary boycott) because it was an attempt to enforce its contract with Black. Such agreements are unfair labor practices under the NLRA.

Allies

Sometimes, what appears to be a secondary boycott may not really be one. For example, a union may have a legitimate strike against Black, but it chooses to picket White, who is Black's main customer. This sounds like a secondary boycott against White. But suppose Black and White are owned and controlled by a common parent company. Suppose they have formed a pact to the effect that they will assist each other in opposing union pressures. What if their products form an integrated link in a production network? In short, such situations or acts would make it apparent that White is no longer a "neutral." In such circumstances, the union usually has a right to act against White as Black's *ally*.

Common Site

Another problem involving secondary pressure may arise when a company hires a contractor to do work on its premises. Suppose this contractor has labor problems while the contracted work is being done. The union's primary and, presumably, legitimate target is the contractor, but the company for which the work is being done may well get splattered in the fray too. The *common site* situation occurs frequently enough so that special rules exist to deal with it. The rules are intended to minimize injury to the second company without depriving the contractor's employees of their rights. If the contractor's employees have no way to exercise their rights without injuring the second company, that company may well expect to feel some pressure.

Subcontracting

Suppose the Black Company, facing a strike, subcontracts some of its work to Green. Certainly, subcontracting is ordinarily lawful. But if Green takes the work involved, Green may very well find Black's employees' union on its doorstep, lawfully picketing the plant and otherwise applying pressure. By taking the subcontracted work under the circumstances, Green has abandoned its neutral position.

Lawful Union Coercion

Of course, in certain situations, a union has a right to take coercive action against an employer; and in certain situations, what seems like a secondary boycott is really a lawful primary action. However, for a union's coercive action to be lawful, it must meet three tests:

1. the objective sought by the union must be lawful (the objective of forming a new closed shop, for example, would not be lawful)
2. the means used must be lawful (use of the secondary boycott, for example, would not stand this test)
3. the union must be the legitimate bargaining agent of the employees involved (for example, picketing by a union that had lost a representation election at the picketed plant would be unlawful)

Suppose the Black Company has a labor-management contract with the representative of its employees, the UWW (the United Wood Workers). Sixty days before the contract is due to end, Black receives a strike notice from the UWW. This does not necessarily mean that Black will have a strike on its hands in two months. Instead, it indicates that the union will enter negotiations for a new contract with a weapon it has a right to use when the current contract expires—the strike. The union also may picket Black's place of business and attempt to persuade employees of other employers to refuse to cross the picket line. In addition to the strike and such picketing, the UWW could use other means of informing the public of the strike against Black and attempt to persuade members of the public not to do business with Black.

Strikes

The NLRA specifically preserves a union's right to strike, but it limits this right. Generally, the limitations depend on the objective sought, the means used, and the existence of a labor-management contract with a no-strike provision.

If a no-strike provision in a contract is currently in force, the employees may still strike, but doing so may be a basis for discharge. Sometimes, the right to strike is contingent upon the presence of an immediate hazard to the employees. If a real danger threatens—for example, the assigning of a welding job near a paint booth—union members may not have time to resort to the grievance procedure. Under such a circumstance, it would be unreasonable to require the employees to continue to work.

Economic Strikes and Unfair Labor Practice Strikes

If the union has no contract provision barring it from a strike or if the contract time has run out, a strike may be quite lawful. In such a case, the strikers' rights depend to a considerable extent upon the objectives sought. An *economic strike*, for example, is one in which the objective sought is an increase in wages, improvement of working conditions, a change in overtime policy, or some other such concessions from the employer. In contrast, an *unfair labor practice strike* occurs in response to an unlawful labor practice by the employer. Economic strikers are still employees, but they may be replaced by their employer. If the employer has hired permanent replacements by the time the economic strikers indicate a desire to return to work, the strikers are not entitled to reinstatement.

Strikers participating in an unfair labor practice strike are in a somewhat stronger position. If they are replaced by other employees, they retain their rights to reinstatement even though the replacement workers may have to be discharged.

Of course, strikers can act in ways that will bring their dismissal in any case. If strike activities go well beyond those allowed by the NLRA and other laws, the employees involved may lose their rights, whether they are economic strikers or unfair labor practice strikers, whether they are suffering from a mass "illness," refusing to work customary overtime, or simply engaging in a mass slowdown. Misconduct by the strikers—violence or threats of violence, for example—could be the cause of such a loss of rights.

The employer, just as anyone else, has the rights of dominion over and use of his or her property. The strike known as a *sit-down strike*—in which employees remain at the employer's premises but refuse to work—deprives the employer of these rights and is, therefore, unlawful.

Picketing

We all have the right to picket. Whether we are employees, union members, or just ordinary consumers, we may inform others of our grievances; whether they are real or fancied makes little difference. Removal of this right would require a drastic change in the First Amendment of the U.S. Constitution. Thus, picketing at an employer's place of business is generally seen as a constitutional right. However, the courts traditionally have not viewed picketing as speech that receives the full protection of the First Amendment. Of course, when people exercise their rights, they often run headlong into others' rights—this is what legal conflicts are made of.

Ordinarily the right of a union to picket an employer's place of business is a right reserved to the representative of the employer's employees. Particularly, organizational picketing by a noncertified union is barred as unlawful *blackmail picketing*. Sometimes, unions picket to inform the public that the employer does not employ union members or have a contract with a union. Such picketing is lawful, at least up to the point where it interferes with the employer's business—preventing pick up and delivery of goods, for example.

♦ Methods, Changes, and Protected Work

The management of a company selects the methods by which the company's product is produced. When a company is formed, it sets up whatever processes are needed to produce the goods and/or services it sells. It then arranges for labor to operate the equipment that performs the processes. From the management point of view, it seems right and reasonable that the company should also be able to change the processes when economic analysis indicates the desirability of such change. But unions sometimes attempt to block or delay such changes.

What happens if two processes are merged into one by a methods change? Or suppose it becomes more economical to subcontract a process or a set of processes. What if economics dictate closing a plant? Perhaps management wishes to buy or build a new plant in another state to house some of the present processes. Maybe a merger with another company seems desirable. Many people assume that all these decisions are a management prerogative. However, while management generally has the right to make such changes (unless it has bargained the right away), it usually must be willing to bargain over their effects on the employees.

For example, the merger of two companies into one is a management decision. Black Company, for instance, may merge with White Company to form Gray Enterprises without consulting any of the unions involved. Some mergers would have little or no effect on employees. Even if employee status is seriously affected by the merger, though, management is not obligated to bargain with the union *about the merger*—only about its *effects on the employees*. In such situations, conflicts may arise about just which union has the right to bargain on behalf of the successor company.

Subcontracting work is usually a management decision as well, so long as it can be shown that the sole inspiration for farming the work out was economic. There are some exceptions to this rather broad generality, however. In one case,[1] farming the work out to another precipitated a layoff, and the court supported the NLRB's contention that this was going too far. In another case,[2] the subcontracted work (maintenance) previously done by union members on the company premises was to be done by nonunion people. The NLRB decision, which was affirmed by the U.S. Supreme Court, indicated that this action was beyond reasonable limits of the management prerogative.

In the construction industry, union-management contract provisions restricting subcontracting are permitted and are usually upheld. One version of such provisions has come to be known as a *work preservation clause* and has the full sanction of both the NLRB and the U.S. Supreme Court. Such clauses are designed to limit or prevent the subcontracting of work. Generally, a clause that limits subcontracting on grounds unrelated to the presence or absence of a union at the subcontractor is upheld. A clause that restricts subcontracting to other companies who have signed or contract with the union violates the NLRB. Such clauses are viewed as attempts to induce the subcontractors to unionize.

1. Weltronic Co. v. NLRB, 419 F.2d 1120 (1969).
2. Fibreboard Paper Products Co. v. NLRB, 379 U.S. 203 (1964).

Featherbedding

One form of union resistance to change is known as *featherbedding*, or the requiring of payment for work not done or not to be done. For an example, suppose the Black Company completely automates a process involving 10 people, and because of the change, the 10 jobs no longer exist. A reasonable move by management would be to retrain the people for other jobs and perhaps wait for normal attrition to take care of any excess personnel. The company might, of course, try to lay the people off or try to get them employment with another company. As long as Black Company treated the employees as fairly as it could under the circumstances, the union probably would have little to oppose. (Depending on the magnitude of the changes, the employer may need to bargain with the union about the changes.) The union could not require Black to retain and pay the employees as though they were still performing the eliminated jobs. Such a demand would be considered featherbedding, and it is unlawful.

♦ Conclusions

While labor relations remains an extremely important area of the law, many other laws govern the employment relationship as well. A state's contract laws, for example, generally shape much of the relationship. As already noted, federal (and often state) statutes prohibit discrimination based on factors such as race, sex, age, handicaps or disabilities, etc. Other statutes govern pension funds, employee stock ownership plans, health insurance plans, and so on. Thus, the sum total of the laws relating to employment relationships is quite complex. Nonetheless, an understanding of these issues is important, because engineers often find themselves affected by collective bargaining and the laws of labor relations.

SOUTH PRAIRIE CONSTRUCTION COMPANY v. LOCAL 627 I. U. OP. ENGINEERS
425 U.S. 800 (1976)

Per Curiam

Respondent Union filed a complaint in 1972 with the National Labor Relations Board alleging that South Prairie Construction Co. (South Prairie) and Peter Kiewit Sons' Co. (Kiewit) had violated secs. 8(a)(5) and (1) of the National Labor Relations Act, as amended, 61 Stat. 140, 29

U.S.C. secs. 158(a)(5) and (1), by their continuing refusal to apply to South Prairie's employees the collective-bargaining agreement in effect between the Union and Kiewit. The Union first asserted that since South Prairie and Kiewit are wholly owned subsidiaries of Peter Kiewit Sons', Inc. (PKS), and engage in highway construction in Oklahoma, they constituted a single "employer" within the Act for purposes of applying the Union-Kiewit agreement. That being the case, the Union contended, South Prairie was obligated to recognize the Union as the representative of a bargaining unit drawn to include Prairie's employees.[3] Disagreeing with the Administrative Law Judge on the first part of the Union's claim, the Board concluded that Prairie and Kiewit were in fact separate employers and dismissed the complaint.

On the Union's petition for review, the Court of Appeals for the District of Columbia Circuit canvassed the facts of record. It discussed, *inter alia,* the manner in which Kiewit, South Prairie, and PKS functioned as entities; PKU's decision to activate South Prairie, its nonunion subsidiary, in a state where historically Kiewit had been the only union highway contractor among the latter's Oklahoma competitors; and the two firms' competitive bidding patterns on Oklahoma highway jobs after South Prairie was activated in 1972 to do business there.

Stating that it was applying the criteria recognized by this Court in *Radio and Television Broadcast Technicians Local Union 1264* v. *Broadcast Service of Mobile, Inc.*, and decided that on the facts presented, Kiewit and South Prairie were a "single employer." It reasoned that in addition to the "presence of a very substantial qualitative degree of centralized control of labor relations," the facts "evidence a substantial qualitative degree of interrelation of operations and common management—one that we are satisfied would not be found in the arm's length relationship existing among unintegrated companies". . . . The Board's finding to the contrary was, therefore, in the view of the Court of Appeals "not warranted by the record...."

Having set aside this portion of the Board's determination, however, the Court of Appeals went on to reach and decide the second question presented by the Union's complaint which had not been passed upon by the Board. The court decided that the employees of Kiewit and South Prairie constituted the appropriate unit under sec. 9 of the Act for purposes of collective bargaining. On the basis of this conclusion, it decided that these firms had committed an unfair labor practice by refusing "to recognize Local 627 as the bargaining representative of South Prairie's employees or to extend the terms of the Union's agreement with Kiewit to South Prairie's employees.". . . The case was remanded to the Board for "issuance and enforcement of an appropriate order against . . . Kiewit and South Prairie. . . ."

Petitioners South Prairie and the Board in their petitions here contest the action of the Court of Appeals in setting aside the Board's determination on the "employer" question. But their principal contention is that the Court of Appeals invaded the statutory province of the Board when it proceeded to decide the sec. 9 "unit" question in the first instance, instead of remanding the case to the Board so that it could make the initial determination. While we refrain from

---

3. The relevant portions of the act, secs. 8 and 9, 29 U.S.C. secs. 158 and 159, provide in part: "Section 8(a) It shall be an unfair labor practice for an employer—

"(1) to interfere with, restrain, or coerce employees in the exercise of the rights guaranteed in section 7:

"(5) to refuse to bargain collectively with the representatives of his employees, subject to the provisions of section 9(a).

"Section 9(a) Representatives designated or selected for the purposes of collective bargaining by the majority of the employees in a unit appropriate for such purposes, shall be the exclusive representatives of all employees in such unit....

"(b) The Board shall decide in each case whether, in order to assure to employees the fullest freedom in exercising the rights guaranteed by this Act, the unit appropriate for the purposes of collective bargaining shall be the employer unit, craft unit, plant unit, or subdivision thereof...."

On the facts of this case, the Union first had to establish that Kiewit and South Prairie were a single "employer." If it succeeded, the existence of a violation under sec. 8(a)(5) would then turn on whether under sec. 9 the "employer unit" was the "appropriate" one for collective-bargaining purposes.

disturbing the holding of the Court of Appeals that Kiewit and South Prairie are an "employer, ..."[4] we agree with petitioners' principal contention.

The Court of Appeals was evidently of the view that since the Board dismissed the complaint, it had necessarily decided that the employees of Kiewit and South Prairie would not constitute an appropriate bargaining unit under sec. 9. But while the Board's opinion referred to its cases in this area and included a finding that "the employees of each constitute a separate bargaining unit," ..."its brief discussion was set in the context of what it obviously considered was the dispositive issue, namely, whether the two firms were separate employers. We think a fair reading of its decision discloses that it did not address the "unit" question on the basis of any assumption, *arguendo*, that it might have been wrong on the threshold "employer" issue.[5]

> Section 9(b) of the Act, 29 U.S.C. sec. 159(b), directs the Board to: decide in each case whether, in order to assure to employees the fullest freedom in exercising the rights guaranteed by this Act, the unit appropriate for the purposes of collective bargaining shall be the employer unit, craft unit, plant unit, or subdivision thereof. . . .

The Board's cases hold that especially in the construction industry a determination that two affiliated firms constitute a single employer "does not necessarily establish that an employer-wide unit is appropriate, as the factors which are relevant in identifying the breadth of an employer's operation are not conclusively determinative of the scope of an appropriate unit...."

The Court of Appeals reasoned that the Board's principal case on the "unit" question, ...was distinguishable because there the two affiliated construction firms were engaged in different types of contracting. It thought that this fact was critical to the Board's conclusion in that case, that the employees did not have the same "community of interest" for purposes of identifying an appropriate bargaining unit. Whether or not the Court of Appeals was correct in this reasoning, we think that for it to take upon itself the initial determination of this issue was "incompatible with the orderly function of the process of judicial review...." Since the selection of an appropriate bargaining unit lies largely within the discretion of the Board, whose decision, "if not final, is rarely to be disturbed," ...we think the function of the Court of Appeals ended when the Board's error on the "employer" issue was "laid bare."...

> As this Court stated... ,
>
> (i)t is a guiding principle of administrative law, long recognized by this Court, that "an administrative determination in which is imbedded a legal question open to judicial review does not impliedly foreclose the administrative agency, after its error has been corrected from enforcing the legislative policy committed to its charge...."

In foreclosing the Board from the opportunity to determine the appropriate bargaining unit under sec. 9, the Court of Appeals did not give "due observance (to) the distribution of authority made by Congress as between its power to regulate commerce and the reviewing power which it has conferred upon the courts under Article III of the Constitution...."

The petitions for certiorari are accordingly granted, and that part of the judgment of the Court of Appeals which set aside the determination of the Board on the question of whether Kiewit and South Prairie were a single employer is affirmed. That part of the judgment which

---

4. "Were we called upon to pass on the Board's conclusions in the first instance or to make an independent review of the review by the Court of Appeals, we might well support the Board's conclusion and reject that of the court below. But congress has charged the Courts of Appeals and not this Court with the normal and primary responsibility for granting or denying enforcement of Labor Board orders."

5. The ALJ's decision in favor of the Union included a conclusion that the pertinent employees of Kiewit and South Prairie constituted an appropriate unit under sec. 9(b). But that conclusion was of course preceded by the determination that the two firms were a single employer. In disagreeing on the "employer" issue, the Board was not compelled to reach the sec. 9(b) question in order to dismiss the complaint.

held that the two firms' employees constituted the appropriate bargaining unit for purposes of the Act, and which directed the Board to issue an enforcement order, is vacated, and the case is remanded to the Court of Appeals for proceedings consistent with this opinion.

It is so ordered.

Affirmed in part, vacated in part, and remanded.

PEAVEY CO. v. NLRB

648 F.2d 460 (7th Cir. 1981)

Petitioner, Peavey Company, seeks review of an order of the National Labor Relations Board ("NLRB"). The NLRB cross-petitions for enforcement. We grant enforcement in part and deny it in part.

I

The Board found that Peavey violated section 8(a)(3) when it discharged Mellinda Snider. The Administrative Law Judge ("ALJ") found that Snider's discharge "was motivated at least in substantial part by her protected activities." While affirming the ALJ's conclusions of law, the Board labeled Peavey's reasons for the discharge a "pretext." The Board then found that "Snider was discharged solely because of her concerted and union activities." Peavey claims that legitimate business reasons justified Snider's discharge.

Since its decision in this case, the National Labor Relations Board issued its *Wright Line* decision.... *Wright Line* set forth definitive rules for resolving cases in which a "dual motive" discharge is alleged. The Board expressly rejected the "in part" test relied upon by the ALJ here. Instead, it applied the Supreme Court's test in *Mt. Healthy City School District Board of Education v. Doyle*, ... to dual-motive discharge cases. Under the new test, the General Counsel must first make a prima facie showing that the employee's protected conduct was a motivating factor in the employer's decision to discharge the employee. Once this is established, the burden shifts to the employer to demonstrate that he would have discharged the employee even in the absence of the protected conduct....The Board ruled that the *Mt. Healthy* test aimed to determine the causal relationship between the employee's protected activities and the employer's action. Once found, a causal relationship justifies liability under section 8(a)(3), without any quantitative label such as "in part" or "dominant motive." *Id.*

This is the first "dual-motive" case to reach us for decision since *Wright Line*. At least one other circuit has adopted the *Mt. Healthy* approach.... We have reviewed the decisions and have decided to follow the *Mt. Healthy/Wright Line* test in "dual motive" cases in this circuit.

Our review of the record as a whole convinces us that Peavey Company met its burden under the *Wright Line* decision. The evidence showed that, although a good typist, Snider was a sloppy worker and had a history of disputes with her supervisors. Peavey disciplined her in writing and advised her of possible discharge unless her work record and attitude improved. A few days before her discharge, she refused to insert notices of new pay scales into paycheck envelopes she was told to prepare. Snider's discharge was ultimately prompted by her refusal to retype some poorly typed letters.

The Board backed away from the ALJ determination that Snider's discharge was motivated "in part" by her union activity. In labeling Peavey's reasons as "pretextual," however, the Board relied on the same rationale as did the ALJ. It agreed that Peavey had "tolerated Snider's poor job performance for over eighteen months" until she began concerted activities. The Board also relied on Peavey's demonstrated "animus" towards the union and the timing of Snider's discharge.

Peavey's reasons here cannot be labeled pretextual. As *Wright Line* held, a pretext can be found to exist when "the purported rule or circumstances advanced by the employer did not exist, or was not, in fact, relied upon (sic)." ...Here, however, it is undisputed that Snider had been disciplined, for cause, prior to her contact with the union. Snider testified, in fact, that she went to the union as a result of her discipline. Her prior discipline also undermines the Board's claim that Peavey had tolerated Snider's poor performance without complaint for eighteen months. Unlike in *St. Luke's Memorial Hospital, Inc.* v. *N.L.R.B.*, . . . there were independent acts of misconduct at the time of Snider's discharge which justified Peavey's action. Moreover, an employer's silence does not extinguish its right to discipline an employee whose conduct continues to worsen....

Once Peavey's reasons for discharge are stripped of the label "pretext," it is apparent that Peavey met its burden under the *Wright Line* decision. Even though Snider engaged in some protected activity, Peavey showed that she would have been discharged even in the absence of the protected conduct.... The substantial evidence on the record as a whole does not support the Board's finding of an 8(a)(3) violation. We therefore deny enforcement of that portion of the Board's order calling for reinstatement of and back pay for Snider.

II

The Board also found five violations of section 8(a)(1) of the National Labor Relations Act. 29 U.S.C. Sec. 158(a)(1). The Board found that Peavey violated the Act when (1) its General Manager told Snider that she was a confidential employee and thus not entitled to participate in union activities; (2) Supervisor Sandy Noe followed employees and eavesdropped on their conversations; (3) it announced, after the union began organizing, that employees could see their personnel files; (4) its General Manager told an employee that persons who joined the union would be considered disloyal; (5) its General Manager promised the employees at a meeting that "he was going to see that things" got better.

After a review of the record as a whole, we conclude that substantial evidence supports the Board's finding of section 8(a)(1) violations. Peavey's contentions to the contrary are without merit. Accordingly, we enforce the rest of the Board's order.

Enforced in Part, Enforcement Denied in Part.

NLRB v. BROWN AND CONNOLLY, INC.

593 F.2d 1373 (lst Cir. 1979)

This is an application for enforcement of an NLRB order arising out of an employer's alleged failure to abide by its oral recognition of a union in violation of sections 8(a)(1) and (5) of the Act. 29 U.S.C. Secs. 158(a)(1) and (5)(1976). Before addressing the facts we recite briefly the course of the Board's position on recognition short of an election. At one time the Board ruled that an employer faced with a demand *prima facie* supported, as with authorization cards apparently signed by a majority of its employees, was obligated to bargain, absent some good faith reason to doubt the union's showing. We accepted this principle....Thereafter the question arose in a modified form, where, after rejecting the union's demand supported by a showing of cards, the employer engaged in unfair labor practices. Although even here some circuits did not, we again agreed with the Board, and held that the employer's conduct making a fair election impossible destroyed any right it might otherwise have had to demand an election.... On certiorari to resolve the conflict, we were affirmed....

After *Gissel*, perhaps recognizing the criticisms there expressed of the consequences of organization without the protection of secret elections and without permitting the employer to be heard, see..., the Board took the position that, absent other violations, the employer had a right to require an election regardless of the union's showing. In this it was supported by the Court....On the other hand, if the employer once recognizes a majority union, no matter how informally, the Board holds that the right is lost.... This is the present case.

> On September 14, 1976, 11 of the respondent's 17 employees,[6] together with two union organizers, called on respondent's president, Brown, at his office, unannounced, and followed what would appear to be a predetermined script to obtain an admission from Brown that he "recognized" the union. ("Do you recognize that we are a majority of your employees?" "Do you recognize that we all wear union buttons?" etc., etc.) Caught unprepared, with no prior knowledge of organizational activity, and quite possibly unaware of an employer's right to require an election, Brown ultimately, according to the credited testimony of the employees, said the magic word. Although it took a little longer, even on General Counsel's case the whole meeting might remind chess players of the fool's mate.

We could have some sympathy with respondent's attempt to require an election,[7] asserted promptly, and before any further steps had been taken, and when, at best, Brown had been the victim of an orchestrated encounter, 13 to 1, with considerable variation among those testifying against him. However, even if we were the policy-maker, we would not quarrel with the Board's rule that once an employer had affirmatively agreed to recognize a union, it cannot change its mind. The facts in this case might suggest the desirability of a rule that recognition must be clearly and convincingly evidenced, and not dependent upon a staged performance and resolution of disputed, possibly perjured, testimony, which may have fostered here, at least in part, the very strife the Act is intended to prevent.[8] While voluntary recognition may be favored, *NLRB* v. *Broadmoor Lumber Co.*, ...recognition versus election does not seem to us that important. The argument made to us, that if an employer could change its mind it would lead to unfair labor practices in an attempt to cause the union to lose the election, would also undercut the Board's policy recognized in *Linden Lumber Division*, ante, an inconsistency counsel for the Board avoided by not citing the case. We consider such policy matters, however, to be outside our province....Equally, we reject respondent's request that we reverse the Board's evidentiary findings.

The order of the Board is to be enforced.

---

6. The company concedes that these employees constitute an appropriate unit.
7. No sympathy can be expected, however, in regard to respondent's conduct subsequent to its refusal to bargain, which resulted in the Board's finding several unfair labor practices which respondent does not here contest. In light of our resolution, it is unnecessary to remand to the Board for a determination whether, given these violations, a fair election became impossible under NLRB v. Gissel Packing Co., ...
8. *Cf.* NLRB v. Gogin, ante, where the court accepted the Board's resolution of a telephone conversation, that the employer's attorney had perjured himself, without "intend(ing) adverse reflection upon counsel whatsoever." ... A disputed telephone call seems an unhappy way of resolving the important question of whether a union represents the employees.

REVIEW QUESTIONS

1. Show graphically how Adam Smith's supply and demand theory might be used by a court to prove "unlawful" union action. Show by another graph how the wage-fund theory might be used for the same purpose.
2. What do the following terms mean? What is their present legal status?
   a. Yellow-dog contract
   b. Labor injunction
   c. Closed shop
   d. Picketing
   e. Strike
   f. Wildcat strike
   g. Featherbedding
   h. Jurisdictional dispute
   i. Agency shop
   j. Collective bargaining
3. What are the nature, purpose, function, and jurisdiction of the National Labor Relations Board?
4. Compare the effects on employers of the union shop and the closed shop.
5. What are the limits on the efforts that may be taken to persuade employees either to join or not to join a union prior to a union election?
6. Describe how a secondary boycott works. Give an example of a secondary boycott involving a union. Give an example of one not involving a union.
7. Distinguish between an economic strike and an unfair labor practice strike. What are the rights of employees involved in each?
8. Reasoning from the Supreme Court opinion in *South Prairie Construction Co.* v. *Local 627 I.U. OP. Engineers,* what are the respective functions of the U.S. Court of Appeals and the U.S. Supreme Court?
9. According to the decision in *Peavey Co.* v. *NLRB,* what employee activities justified discharge? What specific acts by the company were censured?
10. In the case of *NLRB* v. *Brown and Connolly, Inc.,* does it seem that the U.S. Court of Appeals has supported an activity that is unfair, if not illegal? What would be a better policy?

11. For quite some time the Amalgamated Embalmer's Union tried to win a representation election to represent the employees of the Explosion, Iowa division of the Black Mfg. Co. Last July, just before the election, "Whitey" Black, the company president, came in from Stitch-in-time, Nebraska to speak to the employees. In his speech he mentioned that since the Explosion plant was operating so close to the break-even point, the added overhead caused by the union would make it necessary to reduce all wages by $.30 per hour if the union got in. Despite this, the union won by 122 votes to 121. Contracts were signed and the wage rate stayed the same. One day last March the eight employees of the paint shop walked off their jobs as a complaint about their piecework rates; a few minutes later they were joined by the other hourly employees. The group milled around the parking lot committing various injurious acts against management members' vehicles. An accountant had taken several pictures of the occurrence when he was accosted by group members and beaten and his camera was smashed. The following morning all the workers returned, but two of the paint shop employees were told they were fired. More work stoppages then occurred and interferences with production have continued to the present. The two employees who were fired have lodged a grievance which is pending final settlement. "Whitey" Black has stated flatly that if these disturbances do not cease he will close the Explosion plant. The union has replied that if he does this it will strike the materials supplier which he uses at the home plant and force him out of business there, too. Identify the acts and proposals of the union and the company as either lawful or unlawful according to the interpretations of the federal labor law.

Chapter 25

Worker's Compensation

People who work for a living risk being injured on the job. Some of the machines and equipment they must use are inherently dangerous. Sometimes the building or the work situation itself is hazardous. Occasionally, it is dangerous to get to the job site or to return from it. Some degree of danger is associated with almost any human activity. Such being the case, who should bear the risk of injury on the job—the employee or the employer?

THE HISTORICAL PERSPECTIVE

The answer a century or so ago was that the employee bore all these risks. In other words, the law heavily favored employers. The injured worker might sue an employer for compensation for a work-related injury, but the chances of winning were virtually nonexistent. The employer had three strong defenses: assumption of risk, contributory negligence, and the fellow servant rule. Because of these three defenses, employees hardly ever won a suit against their employers. On the other hand, there was no monetary limit on what an employee could win. Thus, the occasional win by an employee usually involved a large amount.

♦ Assumption of Risk

As you know, in tort law, one who voluntarily assumes a risk and is injured as a result cannot sue another and recover for that injury from the other person. Logically, then, one who accepts employment accepts the risks that are part of that employment. If Green, for example, applies for work at the Brown Company, and the Brown Company hires him as a machine operator, doesn't Green assume the risk of being injured? A century ago, the answer to such a question was clearly, yes. Therefore, if Green were injured during his employment in the operation of a Brown Company machine, Brown might express sympathy or even offer Green a payment as a gratuity, but Green probably could not win an award of compensation from Brown.

♦ Contributory Negligence

If the assumption-of-risk defense should fail, an employer could use the same or a similar set of facts to establish the employee's contributory negligence. That is, if Green lost a finger or a hand in one of Brown Company's machines, Brown could simply argue that working around machines was obviously dangerous and that Green was negligent in not keeping his hand out of the machine. If Green was found at all negligent, the doctrine of contributory negligence would bar his claim against Brown.

♦ Fellow Servant Rule

Assumption of risk and contributory negligence usually covered the employer in all worker-injury cases. But if they did not, the employer had a third line of defense. Returning to our example above, let's assume Green can establish that the machine that injured him was faulty, perhaps because of poor maintenance. Brown Company's answer could include as an added defense that a maintenance man or some other employee had been negligent, and this caused the mechanical fault. Brown's next step, then, would be to allege that the real defendant should be the negligent fellow employee—not the employer for whom both employees worked. Under the common law, any fellow employee's negligence, was implied to be the cause of the injury. This was known as the "fellow servant rule."

♦ Employers' Liability Acts

Still, the specter of maimed and injured workers caused public concern. In various jurisdictions, legislatures changed the law in the late 1800s to what were known as *employers' liability acts*. Although these acts benefited the injured worker, he or she still had to sue the employer for compensation. This often required the worker to find an attorney willing to take a case on a contingency-fee basis. The contingency fee, in turn, would reduce the plaintiff's recovery by 25 to 50 percent of the amount (if any) recovered from the employer.

Presuming the injured employee could find an attorney willing to take the case, the employee usually had a better chance to win under employers' liability acts. Such acts often eliminated two of the employer's defenses and substantially changed the third one. Assumption of risk and the fellow servant rule were eliminated; contributory negligence was changed to *comparative* negligence. The resulting legal battles frequently left it up to a jury to decide the question of who was more negligent. Jury awards were often tempered to reflect a sharing of the negligence.

For example, if a jury found Brown and Green to be equally negligent, and the damages were $20,000, Green might be awarded $10,000. After paying an attorney, 40 percent of the award, Green would be left with $6,000 to compensate him for a $20,000 injury. Other expenses, such as expert witness fees, could reduce Green's award even further.

Under the acts, employers began to lose more cases and pay higher awards to injured employees. Employers developed ways to try to avoid liability. For example, suppose Green was severely injured while working. As he recovered in a hospital, he would receive a visit from a Brown Company representative, urging him to accept an amount (say, $500) from his employer to "help him in his time of need." Green would be required to sign a "receipt" (which turned out to be a release from liability) to obtain the cash.

The employers' liability acts were a significant improvement over the common law of torts, but they still left something to be desired. The injured worker still had to find an attorney and initiate action. The employee's award might be trimmed by comparative negligence and other expenses. Attempts by employers to obtain a release after paying only a paltry sum often were viewed by other workers and the public as reprehensible. In addition, the injured worker often had difficulty finding fellow workers who would risk their jobs by testifying for him. The system known as worker's compensation had been adopted in most European countries by 1900 or so. After a couple of false starts here, the federal government and a few states began to enact such laws. Between 1908 and 1948, all of our states adopted some version of the law.

WORKER'S COMPENSATION LAWS

It would certainly be surprising if the various federal and state versions of worker's compensation laws agreed on the details. just the fact that the laws were enacted over a 40-year period suggests a fair amount of variance. Even so, the coverage among the various laws is similar enough that we may discuss a general concept of worker's compensation, pointing out prominent exceptions along the way.

The laws attempt to reach the same general objective—to spread the risk of loss for injuries. Hence, these laws provide compensation to a worker for an injury arising "out of and in the course of employment." At the same time, these laws usually limit an employee's ability to recover from the employer. Specifically, they set up an insurance scheme through which employers pay "premiums." If an employee is injured in the course of employment, he or she will usually qualify for coverage and a payment.

♦ Coverage

In almost all the laws, provision of coverage is either mandatory, with penalties for noncoverage, or is permissive, with a provision that those companies choosing not to cover workers are rendered virtually defenseless when sued. Such laws may provide that the employers who choose not to be covered may be sued in tort for an employee injury and cannot rely on the defenses of assumption of risk, contributory negligence, or the fellow servant rule. Some laws, however, require coverage and state that each day of operation without coverage is a separate violation of the act, with a fine or imprisonment to result from each violation. At the same time, the statutes usually provide the employer with immunity to suits from employees, except for injuries caused by gross negligence or intentional misconduct. In either case, a company is under pressure to adopt worker's compensation for its employees.

The worker never has to pay for worker's compensation coverage. The employer bears the cost of the coverage in some way. Generally, an employer may contract with an insurance company for worker's compensation coverage. In many states, though, the employer may elect to self-insure. In a few states, the employer is required to buy this insurance from a state agency. In short, the employer must take the required steps to assure that he or she is capable of paying an injured worker's medical expenses and a weekly compensation for lost wages.

Each of the worker's compensation acts includes certain exceptions to general coverage. Many of the laws exempt agricultural employees, domestic servants, independent contractors, casual workers, and/or workers in religious or charitable organizations. Some require that the company have a minimum number of employees before the act applies. Sometimes owners are counted as employees. Sometimes an employer will voluntarily elect to obtain worker's compensation coverage for employees not automatically covered by the act. As time passes, the laws have tended to expand the coverage to more and more occupations.

♦ Premiums

The cost to an employer for worker's compensation reflects the employer's insurance company's premium requirements. Generally, an employer pays the premium for coverage on a given job set forth in the insurance company's rate manual. Such premiums are usually based on $100 of payroll cost. If Green operates a rather hazardous machine for the Brown Company, the premium Brown has to pay for worker's compensation coverage may be as high as $10 for every $100 Green earns. On the other hand, if Green works in the office, where the only prominent hazard might be the possibility of tripping over a wastebasket, Brown's premium cost might be $.25 or $.50 per $100 of Green's earnings.

After the initial period of coverage, the premium cost is raised or lowered according to the employer's experience rating. Few severe injuries over the period may result in reduced premiums. Numerous and very severe injuries probably will increase the premiums.

♦ Second Injury Funds

Suppose that Green has lost the sight of an eye. Would the Brown Company hire him? Suppose he lost the sight of the other eye while working for Brown. He would then be permanently, totally disabled. The cost of such a permanent, total disability is far greater than the cost of losing the sight of one eye, and Brown might well hesitate to risk such a heavy loss. For this reason, nearly all worker's compensation laws limit the second employer's responsibility to injuries suffered while the employee was with that second employer. If Green lost the sight of his other eye, Brown would be responsible only for the loss of sight in one eye. But Green, of course, would be totally blind, and this is where the *second*, or *subsequent, injury fund* comes in. That fund is used to make up the difference between what Brown would be required to pay and what Green would receive for being totally disabled. Second injury funds are created in a variety of ways, the most common being legislative appropriation.

♦ Amount of Compensation

The amount of compensation an injured worker is to receive for a work-connected injury depends upon the nature of the injury and the specifications of the worker's compensation law. Generally, worker's compensation laws require the payment of all medical expenses. Compensation for lost wages, though, usually depends upon the nature of the injury. Four categories of disability exist: (1) temporary total disability, (2) permanent partial disability, (3) permanent total disability, and (4) death.

A temporary total disability usually requires the employer to pay compensation that is some percentage of the employee's wage—For example, two-thirds of Green's customary weekly wage. Upper and lower limits nearly always accompany this provision. That is, while Green is waiting in the hospital for his broken legs to heal, he might receive two-thirds of his weekly wage providing this amount was no more than $300 per week and no less than $50 per week. Also, the law often has a holdback feature so that if Green is out only a few days, he receives little or no compensation for lost wages. If he is out two or three weeks, the law may require Brown to compensate Green for the first few days in addition to the continuing payments until Green is ready to return to work.

A *permanent partial disability* usually involves the loss of function of some part of the worker's body. When Green lost the sight of one eye, for example, that was a permanent partial disability. Loss of a hand, a finger, a foot, a leg, or the hearing in one ear would all constitute permanent partial disabilities. Worker's compensation laws generally have a schedule of the time period for which the worker will be paid a portion of the weekly wage for the disability. Green's loss of an eye, for example, might require his employer to pay Green two-thirds of his weekly wage (between the listed maximum and minimum) for perhaps 150 weeks. In addition to this, many of the worker's compensation laws provide for some form of rehabilitation for both permanent partial and permanent total disabilities.

When Green lost the sight of both eyes he had a *permanent total disability*. Under many of the worker's compensation laws, all his related medical costs would be paid, and he would receive some compensation as lost wages for as long as he was disabled. If rehabilitation cannot be effective, this means wage compensation for life. Under other states' laws, some limitation (usually as to a maximum compensation total) is imposed. However, laws of this type tend to become more liberal over time.

If Green lost his life while working for the Brown Company, Brown would have to pay all Green's medical expenses and then make payments to Green's wife and children, or to Green's other beneficiaries. This wage compensation often continues until a statutory limit had been reached or until Green's widow has remarried. The amount of compensation might be modified when Green's children reach a certain age or are married. Brown usually must also pay for Green's burial up to a statutory maximum.

♦ Determination of Compensability

As you have just seen, worker's compensation laws clear up the question of whether a worker injured while working will be compensated for the injury. They even go on to specify *how much* this compensation will be. Beyond that, the laws usually set up procedures for the filing and handling of claims. These schemes also set up procedures for appeals. The reason for appeals, of course, is that the laws still leave unanswered questions. A common question from which many, many cases arise is whether an injury actually *was* work connected. In other words, did the injury arise *out of and in the course of the worker's employment?* If the worker's injury occurred while he or she was at home mowing the lawn, painting the house, or playing tennis on a weekend, it is difficult to see why worker's compensation coverage should apply. If, instead, the worker is injured at work, there is usually little question as to the application of the law. But what if the worker is injured while going to work or while on lunch hour, for example? These are only two of numerous questions that the statutes left unclear.

Going and Coming

Generally, an employee is not covered by worker's compensation on routine journeys to and from work. Still, there are enough exceptions to this generality that a rather imposing body of case law has developed on this topic alone. The general statement is still true, but numerous special circumstances have caused decisions for coverage.

For example, Green, traveling in his car at his customary time of travel over his customary route to or from work probably is not covered. However, if Green is a sales representative and is required to use his car in his work, he probably would be covered by worker's compensation in many jurisdictions. If Green is responding to a call to come in early or is returning home later than usual because of some unusual work requirement, he has a sound case for coverage. If he has taken an unusual route in response to a work requirement (even a mere suggestion by his boss), he also is very likely to be covered. Furthermore, if his work has required him to be awake so long that he falls asleep at the wheel and is injured as a result, worker's compensation should cover him. In short, anything work-connected that causes him to alter his customary routine of travel to and from the job gives him a reasonable cause to claim worker's compensation. Such injuries can be viewed as resulting from a *special mission* given to the employee. In such situations, the employee's travel (and injury) may have arisen out of the employment and in the course of the employment.

Lunch Hour

Suppose Green leaves the Brown Company premises for lunch. Generally, Green is not covered. Again, however, there are plenty of exceptions. If Green can show a connection between his job and his lunch hour activity, he has a case for worker's compensation coverage. Suppose that Green and Gray, Green's supervisor, go to a restaurant for a meeting with a vendor. Even if Gray suggested that Green ride with him, and Green is injured in an accident in Gray's car on the way, Green has a case for coverage. (For that matter, so does Gray.) The coverage is also likely where the lunch involves customers or co-workers on a business lunch, or where Green is asked to pick up materials or make a delivery in conjunction with his lunch period. In all these situations, Green has a reason to allege that injury arose out of and in the course of employment.

Intoxication

If Green comes to work so inebriated that he cannot properly control his actions and, as a result, becomes injured at his job, should this injury be compensated? The question has been raised many times, and compensation frequently has been denied. The logic for denial is that the injury was really caused by the employee's condition, into which he voluntarily placed himself. Thus, many worker's compensation acts provide that coverage does not extend to an employee who is intoxicated.

But let us now change Green's situation a little. Green is now a traveling salesman. His job requires him to entertain customers, and this often involves having a number of drinks. Brown Company actively encourages Green's drinking habit, or knows of it and tolerates it. Green is involved in an auto accident after having entertained customers and is injured. Should he be covered by worker's compensation or not? The answer to this is usually in the affirmative.[1] The entertainment of Brown's customers was part of Green's job, so Green's injury arose out of and in the course of his employment.

Injury by Co-worker or Third Person

In many situations, an error of another employee or an outsider causes the injury. If the injury arose out of and in the course of the employee's job, the employee has a right to worker's compensation. But this may not be the end of it. If strong evidence of a tort exists, the injured employee or the insurer (or both) may be justified in acting against the one who caused the injury. For example, Green is injured because a machine on which he was working malfunctioned. Green has a right to recover worker's compensation from Brown Company, his employer. He may also have a tort action, perhaps for products liability, against White Company, the manufacturer of the machine. Brown Company's insurer, Gray Insurance, also has an interest in the tort action. Under subrogation (i.e., if the contract provides that after making the payments, the insurance company stands in Brown's shoes), the insurer may seek to recover whatever has been paid to Green in worker's compensation payments.[2]

For worker's compensation coverage, injury by a co-worker or outsider must have arisen from the injured employee's work. That is, if Green were injured in a strictly personal altercation with Black, coverage would not be appropriate. If the Black-Green altercation had nothing to do with Green's work, Green's only appropriate recourse is a tort action against Black.[3] The injury did not arise out of and in the course of Green's employment.

Other Situations

It would be nearly impossible to describe all the cases that have occurred in the context of worker's compensation. A few more general categories of cases might be mentioned, however, simply as examples of problems that arise. Suppose Green is injured while at work by "an act of God," such as lightning or a tornado or other violent storm. Customarily, a person such as Green would be covered by worker's compensation, but only if his job exposed him to the hazard in a greater degree than members of the general public were exposed.

Now suppose Green disconnects a safety device on equipment with which he must work. Cases on this sort of situation go both ways. If the company suggested or else knew of Green's added hazard and did nothing about it, the chances for coverage are good. Otherwise, under most worker's compensation laws, coverage is likely to be denied.

1. See, for example, *Boyd v. Francis Ford, Inc.*, 504 P.2d 1387 (Or. 1973).

2. A worker's compensation case in which a set of payments of this nature occurred is *Moomey* v. *Massey Ferguson, Inc.*, 429 F.2d 1184 (10th Cir. 1970).

3. One case of this nature is *Highlands Underwriters Ins. Co.* v. *McGrath*, 485 S.W.2d 593 (Tex. 1972).

In another work setting, now, Green may *intentionally* injure himself. If this can be shown, worker's compensation coverage is likely to be denied.

What if Green's employment requires him to live for short periods away from home, in an apartment near the location of his work? Suppose further that Green suffers an injury in that apartment—the stove blows up, or a wall collapses. As in other types of circumstances, these cases tend to be decided on their own unique facts: The courts try to examine all relevant facts and decide whether the injury arose "out of and in the course of" the employment.

TIETZ v. HASTINGS LUMBER MART, INC.

210 N.W.2d 236 (Minn. 1973)

Relators seek review of a decision of the Workmen's Compensation Commission awarding benefits to respondent, widow of the deceased employee. The only issue is whether a drowning, resulting from a boating accident during a company picnic, was work-related and, therefore, compensable. We affirm.

The picnic was an annual outing sponsored and financed by the Hastings Lumber Mart, Inc., for the benefit of all of its fulltime male employees. Full attendance was actively encouraged, and actual attendance was usually close to 100 percent. The outing was held on a workday afternoon chosen in advance to provide as little conflict as possible with other obligations of the employees. Those who attended received a full day's pay. Those who did not attend were not required to work, as the business premises were closed at noon on the day of the outing. However, some wage adjustment was made for those not attending, either by a reduction in the employee's sick leave or, in one case, by a direct docking of wages.

Although this occasion was not used to make speeches, present awards, or to otherwise enhance the vocational abilities of the employees, the commission found that the personal injury did arise out of and in the course of decedent's employment within the guideline of whether or not the employer derived a direct and substantial benefit from the employees' attendance at the outing "beyond the intangible value of improvement in the employee's health or morale that is common to all kinds of recreation and social life."...

We have weighed the factors which we regard governing,[4] and are convinced that the commission's findings are not manifestly contrary to the evidence and that they have followed the guidelines of *Ethen* v. *Franklin Manufacturing Co.,*....in reaching their decision, although this case is factually distinguishable and a different result is mandated.

Attorneys' fees in the amount of $400 are allowed respondent on this appeal.

Affirmed.

ATKINSON v. LITTLE AUDREYS TRANSFER CO., INC.

212 N.W.2d 350 (Neb. 1973)

Richard B. Atkinson, plaintiff-appellant, prosecutes this appeal from an order of the District Court for Douglas County, sustaining the order of dismissal of the Nebraska Workmen's Com-

4. See 1 Larson, Workmen's Compensation Law, sec. 22.23.

pensation Court. The cause was dismissed by a one-judge court, and on rehearing this finding was reaffirmed by the Nebraska Workmen's Compensation Court *en banc*. We affirm.

Plaintiff by vocation is an over-the-road truck driver. This action is predicated on his claim of a preinfarction angina on May 31, 1968, leading to a more severe myocardial infarction on June 3, 1968. He claims his heart condition arose out of and in the course of his employment with the defendant. Plaintiff's acute myocardial infarct occurred on June 3,1968, while he was on a cross-country journey from Fremont, Nebraska, to the west coast. The basis of the present action, however, is plaintiff's contention that he sustained a preinfarction angina on the previous trip, which originated at Fremont, May 25, 1968, and terminated on May 31, 1968, at Fargo, North Dakota.

On the trip in question plaintiff was working on a refrigerated unit with a cargo consisting of lettuce, carrots, oranges, and celery which had been picked up in California. The separate items of produce were segregated in the truck and were packed in cartons and crates which varied in weight between 35 and 70 pounds. At Bismarck, North Dakota, one-half of the load, or 20,000 pounds, was unloaded onto skids furnished by the consignee. The unloading of the trailer was done by the plaintiff and the driver, individually picking up cartons of produce and placing them on the skids. The skids were then taken into the consignee's place of business by his own employees. Work commenced at the end of the trailer. As the trailer was unloaded the plaintiff and the driver worked their way into the interior. To unload this 20,000 pounds required 2 1/2 hours of work inside the refrigerated trailer. This same process was repeated with the balance of 20,000 pounds of produce at Fargo, North Dakota. Plaintiff worked another 2 1/2 hours in the trailer at that point.

Plaintiff testified that while unloading at Bismarck he had a shortness of breath and a feeling like he had drawn in too much cold air. He then experienced pain in his chest. These symptoms first appeared when he was inside the refrigerated trailer unloading produce. When plaintiff was unloading at Fargo, North Dakota, the pain increased. He didn't know what caused the pain and thought it was just cold air in his lungs. Two symptoms persisted, shortness of breath and the pain across his chest on both sides. This pain continued after he had unloaded the cargo at Fargo and when he arrived at Pipestone, Minnesota, where he left the truck. It continued during his return home to Valley, Nebraska, and was still present the next day, Saturday morning. It is plaintiff's contention that he was under unusual strain in temperature and work and that these conditions combined to produce the myocardial infarction.

On cross-examination the plaintiff admitted that he did not have any pain in his chest while he was in Bismarck. He did, however, have shortness of breath and thought he had sucked in too much air.

The testimony of the driver of the unit on which plaintiff was working does not support his position. He testified that the plaintiff made no complaints to him of chest pains. If he had made such complaints the driver would have remembered them because he liked the plaintiff.

The testimony of plaintiff's doctors tended to support his contention that there was a causal connection between his employment activities and his subsequent heart attack. Defendant's medical witness, however, testified that during the period in question the plaintiff had been engaged in ordinary work to which he was accustomed and there was no direct relationship between the work he had performed on May 31 and his subsequent heart attack on June 3.

It would serve no useful purpose to further detail the evidence herein. It is apparent that the evidence is irreconcilable and in direct conflict. A review of the record convinces us that we cannot say that the findings of fact are not supported by the evidence. This case therefore is controlled by *Gifford* v. *Ag Lime, Sand & Gravel Co.*...in which we said: "Upon appellate review of a workmen's compensation case in the Supreme Court, the cause will be considered *de novo* only where the findings of fact are not supported by the evidence as disclosed by the record."

Here, the triers of fact in each instance found against the plaintiff. The Workmen's Compensation Court *en banc* said in part: "...plaintiff failed to prove a causal connection between the work he did on May 31, 1968, and his subsequent heart attack which he suffered on June 3, 1968....The burden was on the plaintiff to prove this causal connection.

We said in *McPhillips* v. *Knox Construction Co., Inc.*..."Where the claimant in a workmen's compensation case fails to show with reasonable certainty that the disability of which he complains arose out of and in the course of his employment, the proceeding will be dismissed."

We agree that the plaintiff has not maintained his burden herein. The judgment of the District Court sustaining the order of dismissal is affirmed.

Affirmed.

ANDERSON'S CASE

370 N.E.2d 692 (Mass. 1977)

In this case we are called upon to construe the meaning of G.L. c. 152, sec. 7A, and to determine whether that statute was correctly applied by the reviewing board (board) of the Industrial Accident Board. We conclude that, since the findings and decision of the board leave it conjectural whether it correctly applied the statute, the case must be remanded to the board for further proceedings consistent with this opinion.

The plaintiff, widow of Albert S. Anderson (employee), filed a claim for compensation under G.L. c. 152, sec. 31. After hearing, a single member of the Industrial Accident Board ordered the insurer to pay the claimant-widow weekly dependency compensation in accordance with sec. 31 from March 28, 1972, to date and continuing at the rate of $45 a week. There were further orders for payments under other sections of c. 152.

The insurer claimed a review to the board. The matter was tried without witnesses, and the parties filed a statement of agreed facts. It was agreed that on March 28, 1972, the employee, then forty-nine, reported for work between 7 and 7:15 A. M. to perform his regular duties. As he was office manager, those duties were, in general, supervisory and clerical, and involved only occasional physical exertion, although his workweek almost always considerably exceeded forty hours. It is unknown what the employee did from the time of his arrival, but at 7:30 A.M. he was found by fellow employees at his desk, choking and semi-conscious. He kept breathing but never spoke. A police ambulance arrived and removed him to Worcester City Hospital, where he was pronounced dead on arrival. A copy of the death certificate is attached to the record and was incorporated therein by reference.

By agreement, a report of John P. Rattigan, M.D., was submitted. He gave as his opinion, based upon the available medical data, that the employee's sudden collapse and subsequent death on March 28, 1972, occurred in the natural progression of coronary artery disease, and was not causally related to his employment. Dr. Rattigan regarded it as particularly significant that the employee died shortly after he arrived at work before he had engaged in any of his routine activities; and further, Dr. Rattigan gave as his opinion that the employee's general activities on the day of death were so completely void of suggestion of emotional stress that it was not possible to causally relate his collapse and death to work.

The board reversed the decision of the single member, and denied and dismissed the widow's claim. In its "Findings and Decision," after a statement of the agreed facts, the board stated: "Based upon the foregoing facts and the medical opinion of John P. Rattigan, M.D., whose opinion the Reviewing Board adopt, the Reviewing Board find that the deceased employee was found choking and semi-conscious at his place of employment; that his duties were supervisory and clerical; that there was no direct or inferential evidence of emotional or physical stress on the day of his death or involving his duties. Accordingly, the Reviewing Board find and rule that there is prima facie evidence that the employee was performing his regular duties on the day of his death but that the claimant-widow has failed to establish by a fair preponderance of the affirmative evidence that the deceased employee sustained a personal injury arising out of and in the course of his employment. The Reviewing Board further find the claim does not come within the provisions of Chapter 152. The widow's claim for dependency compensation is hereby denied and dismissed." A judge of the Superior Court affirmed the decision.

General Laws c. 152, sec. 7A, as appearing in St. 1971, c. 702, reads as follows, in its entirety: "In any claim for compensation where the employee has been killed, or found dead at his place of employment or is physically or mentally unable to testify, it shall be prima facie evidence that the employee was performing his regular duties on the day of injury or fatality or death or disability and that the claim comes within the provisions of this chapter, that sufficient notice of the injury has been given, and that the injury or death or disability was not occasioned by the willful intention of the employee to injure or kill himself or another."

The plaintiff argues that sec. 7A should be construed as establishing, inter alia, prima facie evidence that the employee's death was causally related to his duties as an employee. She also argues that, given that construction of the statute, the board's findings and decision do not demonstrate that the board correctly applied the statute to this case.

The plaintiff does not argue that the evidence did not support the board's denial of her claim. In this she is correct; the opinions expressed by Dr. Rattigan clearly warranted the board in concluding that the death was not causally related to the employment.

We look first to our standard of review in a case such as this. It has been well established that this court must sustain the findings of the reviewing board and they are final unless they are wholly lacking in evidential support or tainted by error of law....

For a claim to be compensable it must arise out of and in the course of the employment. Clearly a causal relationship is required between the employment duties and the injury or death. In a case such as this one, where the employee was found dead at his place of employment, we construe the statute, sec. 7A, as establishing, inter alia, prima facie evidence of causal relationship between the employment and the injury or fatality. We believe that was the meaning intended by the legislature, particularly in its use of the words, "and that the claim comes within the provisions of this chapter."[5]

We turn next to the definition of "prima facie evidence," as we construe that term as used in the statute, sec. 7A. Prima facie evidence, in the absence of contradictory evidence, requires a finding that the evidence is true; the prima facie evidence may be met and overcome by evidence sufficient to warrant a contrary conclusion; even in the presence of contradictory evidence, however, the prima facie evidence is sufficient to sustain the proposition to which it is applicable.[6]....

---

5. Cases construing the statute, sec. 7A, as it existed prior to its amendment in 1971, are not apposite, since the statute in its earlier wording was more limited in scope, and in its effective assistance to the claimant.

6. In *Commonwealth* v. *Pauley*... see the following language: "According to the Massachusetts view, when by statute or common law, one fact probative of another is denominated prima facie evidence of that second fact, proof of the first or basic fact requires a finding that the second, the inferred or presumed fact, is also true. The finding is mandatory. To avert this result, the opponent must assume the burden of production (the burden of persuasion remains with the proponent). It is only when the opponent has introduced sufficient evidence, which, cast against the natural inferential value of the basic fact, creates an issue of the fact for the trier, that the opponent has satisfied his burden and the mandatory effect disappears. In a case tried by jury where the opponent does not assume his burden, the judge should charge that if the jury find the basic fact, they are required to find the inferred fact; if the basic fact is admitted or otherwise undisputed, the judge should charge that the jury must find the inferred fact, and if the inferred fact encompasses the substance of the case, the judge should direct a verdict.

In light of what we have said as to the meaning of the statute, we examine now whether the board has correctly applied the statute in this case. It is the duty of the board so to deal with cases before it that when a certified copy of the record is presented to the Superior Court, that court can determine with reasonable certainty whether or not correct rules of law have been applied to facts which could properly be found....

It is possible that the board interpreted the statute, sec. 7A, as we have construed it, *supra*, and applied it correctly in this case.

The result reached, denial of the claim, is consistent with a correct application of the law, since the board might have determined that the prima facie case which arose from the statute was controlled and overcome by other evidence, particularly the opinion of Dr. Rattigan. Nevertheless, the board's legal reasoning is far from clear in the record. Ambiguity is particularly shown in the board's conclusion that "the claimant-widow has failed to establish by a fair preponderance of the affirmative evidence that the deceased employee sustained a personal injury arising out of and in the course of his employment, and that his death was causally related to a personal injury arising out of and in the course of his employment."

There is no question in this case that the board was on clear notice that the legal effect of the statute, sec. 7A, was at issue before the board, because the statement of agreed facts by the parties posed, as an "issue presented," the question whether the statute warranted a finding for the plaintiff widow without actual medical evidence of causal relationship. As we have shown, that was a correct statement of the law as we have construed it, *supra*. It is at best conjectural whether the board applied the prima facie effect of the statute to the issue of causal relationship. The plaintiff is entitled to a clear and unambiguous statement of the board's reasoning.

The decree is reversed and a new decree is to be entered in the Superior Court remanding the case to the Industrial Accident Board for further proceedings consistent with this opinion.

So Ordered.

REVIEW QUESTIONS

1. What three defenses did early employers have when sued by injured employees? How was this changed under the employers' liability acts? What further changes were made under worker's compensation laws?
2. Why should an employer be rendered virtually defenseless when an injured employee sues because of a work-connected injury, whereas the employer would have all the defenses normally available in a tort action by a nonemployee? Why should the employee be placed in such a preferred position?
3. Why were employers' liability acts unsatisfactory?
4. What basic proof is necessary for an injured employee to recover from an employer under worker's compensation laws?
5. What generally happens if an employer refuses to be covered by the state's worker's compensation system?
6. What effect does experience rating have on an employer's worker's compensation coverage purchased from an insurer?
7. What is a second (or subsequent) injury fund? Why is it necessary, and how does it work?
8. Give two examples for each: (a) temporary total disability, (b) permanent partia disability, and (c) permanent total disability.
9. Black, an engineer for the White Company, is injured while returning home from work. What circumstances might cause this injury to be covered by workers' compensation?
10. In the case of *Tietz* v. *Hastings Lumber Mart, Inc.*, list the reasons why worker's compensation coverage was allowed.
11. Heart attack cases form an important and very difficult set of worker's compensation cases. Considering the *Atkinson* case, suppose Atkinson had returned home and suffered a heart attack (myocardial infarction) on the evening of May 31. Do you think he would have been covered by worker's compensation? Why or why not? If you think he would have been covered, suppose further that between finishing his workday and arriving home, he had stopped at a local tavern for "a few beers." Would this alter his chances of coverage? What further proof do you think Atkinson would have needed in the actual case to turn the balance in his favor?
12. Based on your reading of *Anderson's Case*, do you believe that an engineer found dead at his or her desk died out of and in the course of employment? Suggest a scenario under which coverage might reasonably be denied.

Chapter 26

Safety

It seems only reasonable to expect that the environment in which people work should not be hazardous to them. Furthermore, since each employee often represents a significant investment in training (either formal or informal), an employer suffers a sizable economic loss when an employee is absent due to illness or injury. The actual monetary costs of work-related injuries go beyond those of the employer, however. The wealth of a society is measured by the goods and services it produces. Losses due to an employee's illness diminish that wealth. In addition, such injuries increase the public burden. Many people who rely on welfare and similar assistance can trace their plight to industrial accidents. Industrial injuries, then, are a source not only of personal misery, but also of social and economic problems.

In Chapter 25, we discussed worker's compensation and its historical perspective. Worker's compensation is a remedy—a device to partially compensate an injured employee for partial or total temporary or permanent loss. It is not and was not intended to be a total replacement for the loss suffered. For example, it includes no compensation for pain and suffering, despite the fact that the loss of a limb causes lots of pain. The fact that there *is* a loss should itself deter industrial injuries for both employee and employer. Still, the strength of the deterrent (or the incentive for safety) does not seem to have been sufficient.

In 1994 there were 6,210 work-related deaths, and 2,236,600 illnesses and injuries. To increase the incentive for industrial safety, the United States Congress in December 1970, passed the Williams-Steiger Occupational Safety and Health Act (OSHA). It took effect in April 1971. Congress's purpose in enacting OSHA was "to assure so far as possible every working man and woman in the Nation safe and healthful working conditions."

Of course, *complete* safety in a working environment is a virtual impossibility and probably would be undesirable even if it were achievable. That is, few of us would be likely to accomplish much in a "padded cell" atmosphere. The work environment should be safe and healthful, yet not so safe as to represent a stultifying atmosphere. The real, practical goal, then, is to provide working conditions that are as safe and healthful as is technologically and economically feasible. In addition, OSHA required a federal determination of an appropriate, minimum level of safety. The policy statement of the act acknowledges the practical limits of providing a safe working environment.

OSHA

The legal essence of OSHA is encapsulated in the following statement of employer duties:

> Section 5:
>
> a) Each employer—
>
> 1) shall furnish to each of his em ployees employment and a place of employment which are free from recognized hazards that are causing or are likely to cause death or serious physical harm to his em ployees;

2) shall comply with occupational safety and health standards promulgated under this Act.

b) Each employee shall comply with occupational safety and health standards and all rules, regulations, and orders issued pursuant to this Act which are applicable to his own actions and conduct.

♦ Applicability

Congress passed OSHA in part because of the drain unsafe working conditions created on the national economy. As a federal law, OSHA is limited in its jurisdiction to companies that affect interstate commerce. Of course, involvement with interstate commerce is often a matter of judgment. Suppose White Company operates only in Oregon. If it uses or handles goods produced in Kentucky, for example, it may be said to operate in interstate commerce. Or Green Company may be a road builder building roads only in the vicinity of Helena, Montana; still, the argument could be made that the roads are intended for interstate travelers as well as those who live nearby. The courts tend to take a very expansive view of what constitutes interstate commerce.

OSHA was not intended to replace coverage by other federal laws or the worker's compensation acts. OSHA therefore states that it does not apply to working conditions where other federal agencies exercise statutory authority affecting occupational safety and health. The meaning of this provision is not altogether clear, however, as may be noted from the cases.[1]

Consider this example. Black Trucking is involved in interstate commerce. Its activities are, therefore, supervised by several federal agencies, including the Interstate Commerce Commission (ICC). Black might well think that it already has sufficient governmental supervision and resist what it sees as an attempt by OSHA to generate more paperwork and problems. A reviewing court, on the other hand, might conclude that the ICC is insufficiently involved with safety and require Black to also comply with OSHA. If a serious safety problem is found or suspected, some device for exercising OSHA jurisdiction may also be found.

1. See, for example, *Marshall v. Northwest Orient Airlines, Inc.*, at the end of Chapter 23, "Administrative Law."

♦ The OSHA Agency

When Congress passed OSHA, the Occupational Safety and Health Administration was created to enforce the safety standards. Congress delegated the power to set appropriate standards to the Department of Labor. The secretary of labor is the executive head of the OSHA. (Since the acronyms for the act and the agency are the same, we use *the OSHA* to mean the Occupational Safety and Health Administration and simply *OSHA* to refer to the statute.) A company's main contact with the OSHA usually comes from the local inspector (or compliance officer). If a company has a problem with the local officer, however, various appeals are possible. OSHA provides for a three-member review commission (the Occupational Safety and Health Review Commission, or OSHRC) to reassess contested citations for violations issued by the officer.

If White Company, for example, files an appropriate notice contesting an OSHA citation (discussed later), White will voice its complaint against the citation at an administrative hearing before an administrative law judge of the commission. If White Company is unhappy with the outcome there (or if the secretary of labor is unwilling to accept the result), a request may be made for a hearing before the full commission. The commission may, at its discretion, review any hearing. When all commission review possibilities have been exhausted, the next step is review of the case by a United States court of appeals. The final step available is to request review by the United States Supreme Court.

♦ NIOSH

In 1978, OSHA created a new entity, the National Institute for Occupational Safety and Health (NIOSH). The director of NIOSH is immediately responsible to the secretary of the Department of Health and Human Services (HHS), but is required to work closely with the secretary of labor. NIOSH is charged with research and education functions, particularly as these functions are concerned with toxic substances. Of course, toxic substances also concern the Environmental Protection Agency (EPA) (which suggests that cooperation between the two agencies is desirable).

By itself, NIOSH cannot adopt or enforce standards. However, the NIOSH findings as to safety requirements (for example, maximum levels of toxic substances) are proposed to the secretary of labor as prospective industry standards. Approval and publication of the requirements in the *Federal Register* by the department of labor often lead to their adoption as rules. One example of a toxic substance with which both the EPA and OSHA are concerned is lead.

Section 20 of OSHA requires the HHS secretary to respond to requests by an employer or employee for information about the possible toxic nature of substances in the workplace. White Company or its employees may suspect a harmful concentration of lead in the air in the work environment, for example. In response to a request from either White Company or its employees, NIOSH would investigate the environment and report its results, assessing whether the lead would have potentially toxic effects in the concentrations found.

♦ State Programs

Section 18 of OSHA provides for state occupational safety and health programs to replace those of the federal government. Essentially, if a state adopts a plan that is as effective as OSHA, the state will have jurisdiction over safety matters, not the OSHA. Of course, a proposed state program must meet certain requirements. Typically, the state prepares a plan and submits it to the OSHA. For three years after the approval of a state plan, the state program is considered developmental. During this time, the state must show it is capable of doing an effective job. If the state shows adequate legislative backing, standard-setting ability, standard-enforcing ability, and competent personnel, the OSHA may enter into an "operational status agreement" with the state. Final approval of the state plan may occur after an additional year or more of effective operation. One inducement for states to initiate their own programs is the provision that the OSHA will pay up to half the operating costs of state programs.

Provisions exist for decertification of state programs, and occasionally states voluntarily give up their programs.

♦ On-Site Consultation

Much of the OSHA's efforts focus on work-environment standards and penalties for noncompliance, but this is not the complete story. Suppose, for example, that White Company is sincerely concerned about the safety of its employees and turns to the OSHA for advice. White Company's request will result in a visit by a consultant to identify hazardous conditions and recommend corrections to be made. The consultant sent by a state agency is likely to be a state employee, but if the OSHA is called upon, the consultant will be a private firm or individual. The consultation costs White Company nothing.

The procedures vary somewhat from state to state and in the federal agency, but the following provides an overview: The consultant arrives, has an opening conference, and then tours the workplace, identifying any hazards and possible violations. The consultant also may interview employees. After the tour, the consultant usually has a closing conference to review what he or she has seen and has been told. A written report follows. The report should identify observed hazards and instruct the employer about how to correct them. The consultant, however, *cannot* issue citations. For employers, the fact that the consultant's visit will not result in an investigation or enforcement proceeding by the OSHA is quite important. Thus, White Company could benefit from a free safety consultation without fear of being penalized by the consultant's findings.

♦ Standards

The question of safety in a workplace is a relative one. As noted earlier, absolute safety is virtually impossible to achieve in any kind of practical circumstance. Thus, a decision must be made as to what is "safe enough." Congress directed that the standards should adequately assure "to the extent feasible,...functional capacity even if such employee has regular exposure to the hazard dealt with by such standard for the period of his working life."

When Congress created the OSHA, it gave the agency the first two years of its existence to examine and adopt (promulgate) so-called consensus standards. The result was wholesale adoption of existing industry standards, state standards, ANSI (American National Standards Institute) standards, and standards from various other sources. Most of these standards have been highly useful, some have had to be modified, and others have been fairly useless.

As an example of the standard-setting process, consider the OSHA's *ground-fault protection standard*. The OSHA adopted this standard from the 1971 National Electric Code (NEC), which existed at the OSHA's inception. In addition to requiring proper grounding of all electrical tools, the 1971 NEC provided the following: "All 15- and 20-ampere receptacle outlets on single-phase circuits for construction sites shall have approved ground-fault circuit protection for personnel. This requirement shall become effective on January 1, 1974." This meant that the exclusive protection device was to be the ground-fault circuit interrupter. This device detects low levels of current leakage and trips the circuit breaker when this leakage exceeds some preset level, commonly 5 milliamperes. Since the time required for the device to react is quite short (as short as 1/40 of a second), electrocution of the worker is avoided, but the worker may still suffer a severe shock.

Shortly before the effective date of the ground-fault protection standard, the OSHA advisory committee recommended a delay. The reason for this was a competing concept, known as an *assured equipment grounding program*. This latter program requires unusually stringent inspection and testing of each cord set, circuit, receptacle, and equipment for grounding before each day's use, plus other periodic tests. If this program is carried out, the risk of shocks of any sort is diminished to an insignificant level. The OSHA standard for protection against shocks, effective February 22, 1977, included the option of either method of protection.

Standard Setting and Rule Making

When the OSHA determines that a particular toxic material or harmful physical agent must be handled in a certain way for the sake of safety, it creates a standard. As with rules promulgated by other federal agencies, it does this by publishing the proposed standard or rule in the Federal Register along with a request for objections, data, or comments by interested parties. Those interested usually have a 30-day period in which to respond. If no one responds, the rule may be finalized by a second publication in the *Federal Register*. If objections or adverse comments do appear, there is often a hearing before the OSHA.

Variances

Sometimes an employer may legally avoid compliance with the OSHA standard. Exceptions can be made under both temporary and permanent variances. Suppose, for example, that White Company (a) has a better (or equivalent) means of protecting its employees than the OSHA standard provides or (b) cannot comply with the standard within the time allotted. As an example of the first situation, assume OSHA had passed the ground-fault electrical standard as first proposed and White Company already had an assured equipment grounding program. This would appear to be sound reason for White Company to request a permanent variance, which, if granted, would permit it to avoid the standard.

As an example of the second kind of variance (a temporary variance), White Company may find itself in the position of being required to comply with a standard in an impossibly short time. Reconstruction may be necessary, or personnel or materials to make the change may be hard to find. White Company's request for additional time to comply (i.e., temporary variance) may be honored by OSHA, but this is usually contingent upon White doing all it can to protect its employees from the hazard involved in the meantime.

Of course, other reasons exist for granting variances. For example, suppose an OSHA standard conflicts with an employee's religious views. For example, the OSHA requires carpenters to wear hard hats, but the Old Order Amish are required by their religion to wear wide-brimmed black felt hats. In this case, the OSHA issued a permanent variance to allow Old Order Amish carpenters to comply with their religion. Of course, if issuance of the variance might have endangered others, the problem would have been more difficult to resolve.

Standards Reforms

One of the major complaints about the OSHA involves the issuing of citations about trivia. It is difficult for a company to understand why the design of a toilet seat, a 2" variation in the location of a fire extinguisher, or a 1"variation in the distance between the rungs of a wooden ladder (12" for some ladders, 13" for others) might be important to someone's safety. The explanation is that such regulations often came to be part of the OSHA during its wholesale adoption of consensus standards at its inception. Thus, many trivial regulations were adopted along with the truly important. Furthermore, Congress required inspectors from the OSHA to issue a citation for *any* violation they observed.

To avoid wasting resources on enforcing outmoded and trivial regulations, OSHA began a purge of such rules in 1978. The editing efforts have now resulted in the deletion of many unnecessary regulations, and OSHA's efforts to "clean up its act" are continuing.

♦ Inspections

Probably the greatest single source of complaints against the OSHA is the unannounced inspection. White Company, for example, may have the best safety practices in its industry and, therefore, believe that any reasonable inspection would find no serious faults. But inspections have many disturbing aspects (many of which are discussed in the investigation section of Chapter 23 on administrative law). To make matters worse, the OSHA suffers from an apparently well-deserved adverse image acquired in the early years of its operation.

For example, many early investigators or compliance officers apparently had little or no industrial experience. They committed blunders of ineptness and engaged in investigations and arguments over trivia. Occasionally, a primary concern of such compliance officers was neatness—whether or not the floor was swept, for instance. Thus, the last thing White Company may desire is an unannounced interruption of its operations by a compliance officer who has no knowledge of the practical effects of inept inspection upon the production facilities.

Congress apparently wanted the OSHA to make surprise inspections. OSHA allows a compliance officer to "enter without delay and at reasonable times any factory, plant, establishment, construction site, or other area, workplace, or environment where work is performed by an employee of an employer" and to "inspect and investigate during regular working hours, and at other reasonable times, and within reasonable limits and in a reasonable manner, any such place of employment and all pertinent conditions, structures, machines, apparatus, devices, equipment, and materials therein, and to question privately any such employer, owner, operator, agent, or employee."

The reasoning behind these rules was that surprise investigations are the only way to catch industries violating a rule. If the investigator must first ask, be refused, and return with further authority, the employer has an opportunity to change the conditions of the employment environment.

The Fourth Amendment to the United States Constitution indicates that warrantless searches are not allowed. The question thus arose as to whether a warrantless and unannounced inspection violated the Fourth Amendment. Such controversies as this generate work for the U.S. Supreme Court. In the case of *Marshall v. Barlow's, Inc.*[2] (an electrical and plumbing installation business in Pocatello, Idaho), the issue was decided. Barlow's, it seems, had twice refused the OSHA inspector access to its premises. Some of the court's statements in finding the act unconstitutional in regard to warrantless searches are interesting: "...the businessman, like the occupant of a residence, has a constitutional right to go about his business free from unreasonable official entries upon his private, commercial property. The businessman, too, has that right placed in jeopardy if the decision to enter and inspect for violation of regulatory laws can be made and enforced by an inspector in the field without official authority evidenced by a warrant."

Thus, in *nonconsensual* instances (where the business refuses entry to the OSHA), the *authority to inspect* is not removed, but some of the intimidation from the threat of a warrantless search is gone. The OSHA officer can still get in to inspect, but he or she must first obtain a search warrant if the attempt to inspect is refused. The Court also held that OSHA provides statutory authority for the issuance of such warrants.

Citations

Just as an avid police officer could probably find something wrong with the way each of us lives, a zealous inspector from the OSHA can probably find rule violations present in almost any establishment. Just as traffic citations *could* result if you are observed speeding or failing to stop at a stop sign, citations *could* result from the OSHA inspections. OSHA and the OSHA's regulations require the compliance officer to issue citations for all observed violations.

Of course, the employer is not required to meekly submit to the citation, posting it for the employees to see and paying the penalties involved. After receiving a citation, the employer has 15 working days to contest the citation, providing the employer has a disagreement with it and is willing to spend the necessary time and money to fight the citation. The employer's first step is to file a *notice of contest*. Generally, the employer does this by sending a written notice to the area director.

Employers appeal OSHA citations on many bases. One of the more common reasons is objection to the lack of specificity in the citation: Occasionally it is difficult to tell from the citation just what the inspector's complaint was. The penalties may also be challenged as excessive, or the time period allowed for abatement of the hazard may be inadequate. Even when the citation is specific, it may be that the employer's safety practice is superior to the agency's rule.

2. 436 U.S. 307 (1978).

Penalties

Citations may or may not include proposed penalties for the cited violation. If the inspection reveals a *serious violation*, one involving a substantial probability of physical harm or death, there is a mandatory penalty assessment of up to $1,000. A *willful violation* runs the risk of a considerably greater penalty. In such cases where the employer knew of the hazard and did nothing to correct it or to protect against it—the employer risks substantial penalties. The maximum penalty for these willful violations runs up to $10,000 per violation and/or six months in jail. A second violation can double these maximum penalties.

Safety hazard citations usually state an *abatement period*, a time limit within which the employer is expected to remove the hazard. A failure to meet the time limit (or appeal the citation in this regard) may prompt the assessment of up to $1,000 per day of abatement-period violation.

Other penalties exist for falsifying records, failing to post notices of citations, and assaulting OSHA inspectors.

♦ Record Keeping

Under OSHA, all but fairly trivial work-related injuries or illnesses must be recorded, and the records must be available for inspection by compliance officers. The purpose, of course, is to assemble a body of knowledge of work-related human problems, an apparently worthy and desirable motive. However, one wonders whether such record keeping is truly a cost-effective way of gathering such information.

With some exceptions, this requirement pertains to all employers having more than 10 employees. The Bureau of Labor Statistics, however, may ask even small businesses for a sampling of such information, so each such employer risks being required to keep those records. Besides small businesses, certain service industries are exempted, as are farmers with respect to records for members of their families.

When a company must keep records, those records are to include all occupational illnesses (assuming a determination can be made as to environmental cause). Also to be recorded are all occupational injuries resulting in (a) death, (b) lost work-days, (c) restriction of motion, (d) loss of consciousness, (e) transfer to another job, or (f) medical treatment other than first aid.

♦ OSHA and the Political Environment

OSHA began in a healthy business climate. In such a climate people, including those who make laws, tend to be concerned with the health and well-being of workers. In an economic down-turn, people tend to be more concerned about the health and well-being of businesses and the unemployment rate. As the economic climate varies, the OSHA emphasis changes to reflect the change in public attitude. The OSHA's actions, thus, can be viewed as resulting not only from the law under which it was created, but also from the prevailing economic and political environment in which it operates.

WHIRLPOOL CORP. v. OCCUPATIONAL SAFETY AND HEALTH REVIEW COMMISSION

645 F.2d 1096 (D. C. Cir. 1981)

The Occupational Safety and Health Act of 1970 (OSHA) employs two devices to protect workers from the unconscionably high risk of tragic death or injury: regulations, which define safety standards for specific industrial environments; and, a catch-all "general duty" clause which requires employers to abate "recognized hazards" in the workplace. As this court recognized in *National Realty & Constr. Co. v. OSHRC* the laudable and sweeping mandate of the general duty clause must be focused through clear notice of any specific hazard, in order to ensure fairness to employers and open, reasoned decision making by OSHRC (the "Commission").

As the commission acknowledges, the Secretary of Labor's (the "Secretary") citation here provided Whirlpool Corp. (the "petitioner") with inadequate written notice of the alleged general duty violation, and the evidence presented at the administrative hearing was not a "model of precision." The predictable result of these haphazard procedures was a poorly developed record which fails to support OSHRC's findings. Because we cannot discern whether this lack of evidence reflects the confusion caused by the Secretary's procedures rather than an unmeritorious charge, we must remand this record for further development.

I. The General Duty Clause

A. *Substantive Elements*

(1, 2) Section 654(a)(1) of OSHA, the general duty clause, provides that

(a) Each employer—

(1) Shall furnish to each of his em ployees employment and a place of employment which are free from recognized hazards that are causing or are likely to cause death or serious physical harm to his employees.

As OSHA's legislative history makes clear,[3] this subsection does not impose strict liability on employers, but instead limits their liability to "preventable hazards." The three elements of a general duty violation are: (1) a hazard likely to cause death or serious bodily harm; (2) recognition of the hazard either by the specific employer or generally within the industry; and (3) existence of a feasible method of abatement.

The Secretary has the burden of coming forward with evidence on the feasibility issue. This procedural burden is closely related to the broad sweep of the clause, for proof of the specific method of abatement, perhaps more than the other substantive elements, helps provide the employer with notice of the precise hazard at issue.

B. *Notice requirement*

Section 658 of OSHA provides that an employer charged under the general duty clause must be given notice in the form of a citation which "shall be in writing and shall describe with particularity the nature of the violation...(and) shall fix a reasonable time for the abatement of the violation."

Ideally, the citation should provide the employer with notice of the Secretary's contentions pertinent to each of the three elements underlying a general duty violation. Where detailed prehearing notice has been lacking, the reviewing court must carefully scrutinize the Secretary's presentation of evidence to ensure that the cited employer has been afforded a fair opportunity to address the specific violation charged. As this court found in *National Realty*, the virtues of adequate notice extend even beyond due process: "To assure the citations issue only upon careful deliberation, the Secretary must be constrained to specify the particular steps a cited employer should have taken to avoid citation, and to demonstrate the feasibility and likely utility of those measures."

II. Procedural History of the Instant Violation

Petitioner manufactures appliances at its Marion, Ohio, plant. Overhead conveyors move parts through the manufacturing process. To protect employees working beneath the conveyors from falling parts, petitioner maintains a huge protective guard screen, which consists of multiple steel mesh panels secured by metal clips to angle iron frames, which are joined together by bolts. Maintenance personnel routinely traverse the guard screen to retrieve fallen parts.

In years prior to the instant violation, maintenance personnel fell partially through the guard screen. In 1974, an employee fell to his death when the bolts joining two frames failed.[4] After an OSHA compliance officer's inspection, petitioner was cited for: "failure to

---

3. See 116 Cong. Rec. 38377 (Nov. 23, 1970) (recognition requirement intended to mitigate employer liability);...(employer's duty "not absolute," but instead limited to protecting workers from "preventable dangers"). Although the employer's duty under OSHA extends beyond that imposed under common law...it must be "achievable and not a mere vehicle for strict liability."

4. J.A. 36. Because the instant citation was precipitated by this tragedy, petitioner had reason to believe, absent further clarification from the Secretary, that it was the bolts, not the tensile strength of the mesh, that was the hazardous condition. This ambiguity was compounded by the citation itself. See J.A. 43.

provide a safe walking and working surface on the screens under the conveyor." The citation further required "immediate" abatement. Petitioner contested the citation.

At the hearing, the Secretary offered the testimony of employees who described occasions on which workers had fallen partially through the screen. The Secretary's compliance officer refused to specify whether the source of the hazard was the bolts linking the panel frames or the tensile strength of the screen mesh. A civil engineer testified for the Secretary, and suggested a third theory of the hazard: the absence of a heavy steel catwalk. Petitioner's engineer testified on cross-examination that one-third of the screen's panels had been replaced by heavier-gauge mesh, but he added that complete replacement was architecturally infeasible.

The Administrative Law Judge ("ALJ") vacated the citation without deciding whether the Secretary had proven a general duty clause violation. OSHRC remanded the matter to the ALJ for findings under the general duty clause. One Commissioner dissented from the remand order, however, on the grounds that the Secretary had failed to specify or prove a feasible abatement method.

On remand, the ALJ again dismissed the citation, this time on the grounds that the record did not reveal a feasible abatement method. The ALJ assumed that the hazard contemplated by the Secretary was the tensile strength of the screen panels. However, the ALJ accepted the unrefuted testimony of petitioner's expert, who stated that complete screen panel replacement was infeasible.

OSHRC again overruled the ALJ. The Commission was admirably candid in acknowledging the substantial deficiencies in the notice provided petitioner.

> [T]he citation failed to specify [what condition rendered the screen unsafe] and accordingly facially lacked particularity. [In addition] a foundation had been laid in the citation for [the Secretary's] own confusion of focus and for [petitioner's] possible misconception of the case, not only by the absence of reference to the [tensile strength of the mesh] as the hazard, but by the requirement of immediate abatement, which most likely suggested abatement by replacing the bolts rather than by replacing a substantial part of the guard screen panels.

The Commission concluded that, notwithstanding the confusion precipitated by its citation, the "purposes of the particularity requirement [were] fulfilled in subsequent stages of the proceeding." Although the "Secretary's evidence at the hearing was not a model of precision in identifying the hazard," the Commission found that repeated reference to the strength of the steel mesh gave petitioner constructive notice of the hazard. Moreover, according to the Commission, petitioner apparently perceived that the tensile strength of the screens was the hazard, for it introduced rebuttal evidence on this point.

The Commission also rejected the ALJ's findings regarding the lack of evidence on the feasibility issue.[5] The Commission relied on the "record as a whole," and specifically: (1) the testimony of petitioner's engineer that heavier-gauge screens were safer; and, the testimony of petitioner's em ployees, who acknowledged that heavier screen panels were typically deployed to replace screens that had torn. The Commission concluded

> Because Whirlpool systematically was replacing the mesh panels with heavy-duty wire throughout the plant, the feasibility of the heavy-duty wire to eliminate the hazard and its likely utility in Whirlpool's plant is *apparent.*

5. In addition to the absence of evidence on the feasibility issue offered by the Secretary, the ALJ noted disparities in the credibility of the parties' witnesses. The compliance officer did not test the tensile strength of the various panels used in the guard screen, and the Secretary's engineer admitted that he had no experience with guard screen systems. On the other hand, petitioner's engineer was not only familiar with the special characteristics of petitioner's screen, but was also generally acquainted with guard screen structures. J.A. 28–29.

The commission reversed the findings of the ALJ, and amended the citation to provide an abatement period of six months.[6] This appeal followed.

III. Notice and Evidence of a Feasible Abatement Method

A. Notice

OSHRC points to employees' references to the guard screen panels and to rebuttal testimony on the tensile strength of the screen as evidence that petitioner had adequate notice of the hazard. In so doing, the Commission ignores the importance of detailed notice to the fair administration of the general duty clause.

Assuming *arguendo* that petitioner's rebuttal suggests a general awareness of the hazardous condition, it hardly follows that the Secretary ever gave petitioner a fair opportunity to address a necessary and independent issue: how the hazard could have been feasibly abated. The Commission acknowledges that there was no written notice. Nor was there constructive notice, for no witness outlined "the Secretary's own theory of what steps the (petitioner) should have taken to abate the hazard." The Secretary's compliance officer addressed the subject of abatement only on cross-examination, and he pointedly refused to reveal the Secretary's abatement plan .[7] The Secretary's engineer did not even address abatement. Of course, the Commission's post-hearing alteration of the citation clarified the hazard charged, but came far too late to provide petitioner with a fair opportunity to meet the Secretary's contentions.

B. Substantial Evidence

The substantial evidence requires a court reviewing an administrative adjudication to accept findings of fact supported by "such relevant evidence as a reasonable mind might accept as adequate to support a conclusion." As tolerant as this standard is, the vague notice and unfocused evidence offered by the Secretary at the hearing has resulted in a record devoid of support of OSHRC's conclusion.

Although some witnesses called by the Secretary testified to the *efficacy* of the heavier-gauge screen, none addressed the feasibility of complete screen panel replacement. On cross-examination, petitioner's engineer, after noting that one-third of the panels had been replaced before the fatality, stated that *complete* panel replacement was architecturally infeasible. In view of his unrefuted testimony, it is by no means "apparent" that complete replacement was a feasible means of abating the hazard .[8]

Accordingly, the order of the OSHRC is vacated and the record remanded for further development on the issue of a feasible abatement method.

So ordered.

---

6. J.A. 49. Post-hearing alterations of the citation can signal deficiencies in the notice afforded respondents.... In this instance, alteration of the citation resolved an ambiguity in the precise hazard charged.... It would appear, however, that this ambiguity should have been resolved before or at least during the hearing to provide petitioner with a full and fair opportunity to meet the Secretary's charge.

7. Q. Are you saying that the Secretary would agree that the wire, the heavy-duty wire type does comply...?

A. No, Sir, I am not saying that.

Q. You're saying it does not comply?

A. I'm not saying it does, and I'm not saying it does not comply.

J.A. 73.

8. J.A. 48. At oral argument, counsel for petitioner noted that subsequent to the hearing *all* of the guard screen panels have been replaced.

OSHRC does not contend, however, that this disposes of the feasibility issue. As the Commission indicated in response to an initial remand from this court, eventual compliance with an abatement order does not moot the original citation, and our review of that citation is based on the record which purportedly supported it. The general duty clause's focus on preventable hazards, [citation omitted] limits liability to dangers actually or constructively recognized at the time of the violation.... It would appear, then, that the feasibility of abatement must be judged as of the time of the violation as well. We cannot discern from this record whether petitioner's apparently recent success in abating the hazard contradicts the testimony on feasibility of abatement recorded at the 1975 hearing.

IRWIN STEEL ERECTORS v. *OCCUPATIONAL SAFETY & HEALTH REVIEW COMMISSION*
574 F.2d 222 (5th Cir. 1978)

Per Curiam
The Occupational Safety and Health Act of 1970, 29 U.S.C. sec. 651, et seq., created the Occupational Safety and Health Review Commission (OSHRC), and authorized it to adopt standards,...make investigations,...and issue citations for violations. OSHRC may obtain judicial review.... In such review, The findings of the Commission with respect to questions of fact, if supported by substantial evidence on the record considered as a whole, shall be conclusive....

The employer who here seeks review was engaged in steel erection work. On the job in question, its welders were working on steel beams or joists in a single-story building about 23 feet above the ground. They were required to move along the beams in straddle position to do the necessary welding. The OSHRC ordered safety belts, lifelines, and lanyards to be used.[9] The employer contended that compliance with the standard would expose its welders to a greater safety hazard than non-compliance and, therefore, urged that the citation issued to it should be set aside and the proposed penalty of $600 be vacated.

While the evidence adduced before the administrative law judge was conflicting, there was substantial evidence to support the conclusion that the welders were subjected to the danger of falling, a serious violation of 29 CFR 1926.28(a), and that the use of the safety equipment ordered would not create a safety hazard.[10] The employer asserts that the preponderance of the evidence showed the requirement to be unwarranted and to create other hazards. We are prohibited by the statute and the authorities cited from reweighing the evidence. Accordingly, these findings must be affirmed.

The employer contends that the OSHRC increased the burden required for establishing the affirmative defense of "greater hazard" by requiring proof of (1) unavailability of alternative means of protecting employees, and (2) inappropriateness of a variance application, and cites *Secretary v. Industrial Steel Erectors, Inc.* 1974...for the proposition that the only requirement of proof of the greater hazards defense is that the safety or health of employees would be endangered rather than protected by compliance. *Industrial Steel Erectors,* supra, and *American Bridge*, supra, involved fact situations, however, where employers used safety devices in compliance with statutory requirements except at specific points in their work. In both of these cases, a preponderance of the evidence showed that the employees were safer in not complying with the standard than if they had. In fact, the basis of the opinions was only that an employer would be permitted to assert an affirmative defense when the safety of employees would be endangered rather than protected by compliance with a standard.

Later, however, in *Secretary v. Russ Kaller, Inc.,*...the OSHRC spelled out specific requirements for the greater hazards defense:

9. The citation alleged a violation of 29 CFR 1926.28(a), which states: Personal Protective Equipment. (a) The employer is responsible for requiring the wearing of appropriate personal protective equipment in all operations where there is an exposure to hazardous conditions or where this part indicates the need for using such equipment to reduce hazards to the employees.

10. Under the Act, a serious violation shall be deemed to exist in a place of employment if there is a substantial probability that death or serious physical harm could result from a condition which exists, or from one or more practices, means, methods, operations, or processes which have been adopted or are in use, in such place of employment unless the employer did not, and could not with the exercise of reasonable diligence, know of the presence of the violation. 29 USC sec. 666(j). Under Section 666(c) a "nonserious" citation involves a determination that the violation in question was "specifically determined not to be of a serious nature."

> The scope of the defense recognized in those cases is, however, narrow. It is not enough that compliance with literal terms of the standards would create new hazards.... The record must show that the hazards of compliance are greater than the hazards of noncompliance...that alternative means of protecting employees are unavailable...and that a variance application under section 6(d) of the Act would be inappropriate....

These are the standards that the Commission applied. It did not change the rules to the petitioner's prejudice in mid-course. Moreover, we need not decide here, in any event, the propriety of applying the *Russ Kaller*, supra, standards, for the employer did not establish by a preponderance of the evidence the defense under the criteria it considers applicable.... The question for us is not whether the preponderance of the evidence test has been satisfied. That test is for the OSHRC, just as the preponderance of the evidence test in a civil jury case is for the jury. Once the OSHRC concludes that the evidence demonstrates a proposition by a preponderance, the only review available is whether there is substantial evidence on the record as a whole to support its conclusion.

The first criterion for a successful defense, admittedly valid, is for the employer to show that the hazards of compliance are greater than the hazards of noncompliance. There was substantial evidence to support the Commission's conclusion that the employer failed to show that the welders were in need of mobility and that the danger of dangling from a safety belt was greater than the danger of falling.

For these reasons, the petition for review is DENIED.

CHAMPLIN PETROLEUM CO. v. OSHRC

593 F.2d 637 (5th Cir. 1979)

An employer appeals an order of the Occupational Safety and Health Review Commission (Commission) finding a violation of the general duty clause of the Occupational Safety and Health Act (OSHA) which obligates the employer to

> furnish to each of his employees employment and a place of employment which are free from recognized hazards that are causing or are likely to cause death or serious physical harm to his employees....

The recognized hazard to which employees were exposed was the escape of hot oil. The Commission's order was based on the employer's failure to prevent such exposure by effectively communicating a rule against opening valves, through which the oil might escape, unless they are equipped with handles. Because we fail to find substantial evidence on the record considered as a whole,...that either the use of handles on valves or the effective communication of a rule against opening valves without them would have materially reduced the hazard of oil fire injuries, we reverse.

A pipeline control valve malfunctioned at the Corpus Christi, Texas, oil refinery of employer, Champlin Petroleum Company (Champlin). Isolation of the pipeline section and draining of its contents were necessary before the defective valve could be removed. A block valve was closed on each side of the control valve to permit the isolated crude oil, which flowed at a temperature hot enough to ignite on contact with air (auto-ignition temperature), to cool before draining through the bleeder valve. Several hours later, the unit operator, Cobb, and two maintenance employees, Benson and Bennett, prepared to bleed the control valve. Bennett stood up on the pipes a few feet to the side of the bleeder valve. Cobb stood to the right of and below the valve with the valve at waist or shoulder height. Benson stood up on the pipes on the same level as the valve. The valve had been newly installed with a handle four days before. On this occasion, however, the handle, a circle 2-1/2 inches in diameter, was missing. Benson reached down and opened the valve by turning the valve stem with a crescent wrench. A thin stream of oil ran into the bucket which had been suspended from the valve. Shortly thereafter, smoke emerged from the valve indicating that auto-ignitable oil would follow. Benson was unable to close the valve, and in the ensuing flash fire the three employees were injured, Cobb fatally. Foreign matter had apparently been trapped in one of the block valves, preventing it from completely closing. The pressure created by opening the bleeder valve dislodged the sediment, permitting hot oil to flow into the isolated section.

Following inspection of the refinery by an OSHA compliance officer, a citation was issued charging a general duty clause violation in that:

a. There was no fixed handle on the bleeder valve to control the flow of liquid hydrocarbons from the line. Employees were using a pair of pliers on the valve stem since the handle was missing.
b. Liquid hydrocarbons were not piped away from the bleeder valve located approximately 3 feet from an inservice heater unit.

An $800 penalty was proposed.

The administrative law judge vacated the citation and proposed penalty, finding the hazard of exposure to auto-ignitable oil unpreventable because the employer could not have foreseen that its employee would open the valve without a handle in violation of the company's well-established safety policy. Reversing the administrative law judge, the Commission affirmed the citation and penalty, concluding that the company's safety policy was ineffectively communicated to employees.

We observe that in reversing the decision of the Commission we need not and do not assess the sufficiency of support for the administrative law judge's conclusions. The relationship between the ALJ and Commission differs from that of trial and appellate courts in that OSHA contemplates that the Commission be charged with fact-finding responsibility, 29 U.S.C.A. sec. 659(c), and the ALJ is merely an arm of the Commission for that purpose...and may weaken the contrary conclusion of the Commission,...we disturb the Commission's decision only because it lacks the support of substantial evidence....

To establish a general duty clause violation, the Secretary must prove "(1) that the employer failed to render its work-place 'free' of a hazard which was (2) 'recognized' and (3) 'causing or likely to cause death or serious physical harm.'"...The general duty obligation, however, is not designed to impose absolute liability or respondeat superior liability for employees' negligence. Rather it requires the employer to eliminate only "feasibly preventable" hazards.... It is the Secretary's burden to show that demonstrably feasible measures would materially reduce the likelihood that such injury as that which resulted from the cited hazard would have occurred.... The Secretary must specify the particular steps the employer should have taken to avoid citation, and he must demonstrate the feasibility and likely utility of those measures....

At the hearing it was suggested by the OSHA compliance officer that such flash fires are preventable by attaching a drain pipe to the valve so hot oil would be piped away from the area of employee exposure and by maintaining handles on bleeder valves so they may be turned off quickly when smoke appears. Where handles are missing, the compliance officer suggested, an effective training and safety program discouraging opening of handleless valves could reduce injury. The administrative law judge found that use of a drain pipe would have been infeasible under these circumstances, and the Commission, assuming that an adequate safety rule regarding handles would have abated the hazard, restricted its discussion to the effectiveness of communication of the policy against opening handleless valves.

The conflicting evidence on effectiveness of company policy communication need not be confronted in this case. The decision properly turns on the lack of evidentiary support for the underlying assumptions of both the administrative law judge and Commission that either the use of handles or the strengthened communication of the company's policy against opening valves without handles would materially reduce the hazard of injuries from auto-ignitable oil fires.

We consider first the proposed improvement of the company's communication of its rule against opening handleless valves. The effectiveness of this communication is declared by the Commission to be the critical issue in the preventability of this violation. The emphasis of the Commission is misplaced, however, in its focus on what the company has taught rather than what the employees have learned. Improved education will effectively prevent repetition of a mishap only where ignorance was at least the partial cause of that mishap.

Close scrutiny of the record reveals no evidence that any Champlin employee has ever opened or is likely ever to open a handleless valve out of ignorance that it is contrary to company-approved procedures. Benson's opening of the valve on this occasion, the only known instance in which a handleless valve has ever been opened by a Champlin employee, was clearly done with knowledge that he was following an improper and unsafe procedure. Bennett, the other maintenance employee present, testified that he had never run into a handleless bleeder valve and suggested that he would know to get the handle replaced if he noticed its absence. He recalled being taught proper valve-opening procedure while coming up through training. Pascal, a unit operator, testified that standard company procedure was to open and close only valves with handles on them. It is incumbent upon the Secretary to demonstrate exactly how the company should and could improve communication of its policy so as better to reduce auto-ignitable oil fires. Because the record reveals no proof that any employee has ever opened a handleless valve before, in ignorance or otherwise, and because no witness testified that he was unaware of the correct procedure for opening valves, we conclude that the Secretary has failed to show that wider or louder broadcasting of the company's policy would have had any material effect on the likelihood of injuries of the type sustained here.

Even if the record revealed sufficient evidence from which to conclude that improved policy communication would reduce the opening of valves without handles, the record fails to demonstrate, as it must, that the use of handles themselves would effectively reduce the hazard that caused these injuries.

The administrative law judge stated

> It does not take a conscientious expert, familiar with the refinery industry, to recognize the factor of safety provided by the presence of a handle on a bleeder valve. Its utility enables the operator to crack this valve ever so gently, and if smoke is encountered from the flow the source of ignition can be halted abruptly.

The Commission too concluded that once smoke is observed, the valve "could be closed sufficiently quickly by means of the handle to prevent the hot oil from emerging." While common sense suggests that valves can be closed more quickly by use of handles, close examination of the record suggests that these conclusions are beyond the record evidence....

The first witness, Champlin's chief supervisor, Cooper, testified that use of handles to open valves is standard company policy. He did not assert the efficacy of closing by handle in a case like this one. Rather Cooper commented that hanging the bucket from the valve instead of placing it on the deck below reduced the time available to close before flashing and speculated that the valve had been opened dangerously wide in this instance, perhaps because the valve itself was plugged. The OSHA compliance officer testified that increased maneuverability of a device with a handle is obvious, but he expressed doubt that use of a handle alone would have eliminated the hazard:

> Possibly if people were quick, without the hose there, if they were quick enough they might be able to control it. Who is to say, but at what point you are going to have a bleeder valve, assuming it is not plugged, in normal operations, who is to say at what point there might be a possible washout? And so therefore, a handle may or may not in itself solve a particular problem.

Bennett, when asked if the valve could have been shut in time with a handle, replied

> I would have to say that is an individual preference. I can't say that I would. I think from what I witnessed in this particular instance, the first thing I thought about was self-preservation.... I wouldn't have taken time to shut it, personally.

Bennett testified it would be standard procedure to try to close the valve "if you thought you could get it closed safely." Benson testified

> My first reaction was to try to close the valve and I reached for it and then there was so much smoke I realized I couldn't do it....

Benson admitted that common sense directs the use of handles as the safest method to open valves, but explained

> In the case of this valve, there is no place to be that would be—how do I say this? The valve is so located that you can't get in a position that you can stay in, in order to close the valve and stay at the valve handle at all times.

Another maintenance employee testified that the standard procedure on seeing smoke would be to shut the valve off and speculated, "You wouldn't think to have enough oil to go out at one time to flash without you could shut it off immediately." In this particular accident there may have been no direct causal connection between the lack of handle and the flash fire. Evidence suggests, for example, the valve may have been opened so far that even a handle would not have closed it in time and that the employees may have been in unusually awkward positions. Because OSHA is designed to encourage abatement of hazardous conditions themselves, however, rather than to fix blame after the fact for a particular injury, a citation is supported by evidence which shows the preventability of the generic hazard, if not this particular instance.... We conclude, however, that the record fails to show substantial evidence that an employee, under any operating conditions, would be able to close this valve in response to the appearance of smoke signaling auto-ignitable oil if the valve were equipped with a handle. For example, there is no evidence of the time involved between appearance of smoke and ignition of the oil which followed. Similarly, the Secretary adduced no evidence that other Champlin employees had encountered smoke in bleeding valves and had been able to close valve handles in time to block the escape of hot oil.

Because the Secretary has failed to shoulder its burden of proof of the violation, the Commission's decision must be

REVERSED.

REVIEW QUESTIONS

1. Why was the OSHA created?
2. In the business of the OSHA, what are the functions of the inspector, administrative law judge, the Occupational Safety and Health Review Commission, and a U.S. Court of Appeals?
3. How is NIOSH related to the OSHA?
4. In what ways could a state program for occupational safety and health be superior to the federal program? In what ways could the federal program be expected to be superior?
5. Consider the ground-fault protection standard. Which of the two standards discussed would be better under a particular set of factual conditions? Which should be better in general?
6. Consider a power press (punch press or brake press). List at least five ways to improve the safety of its human operator.
7. The U.S. Supreme Court decision in *Marshall v. Barlow's, Inc.* indicates that the OSHA must obtain a search warrant if an employer refuses to submit to inspection upon the inspector's first visit. What arguments can you propose for and against the warrant requirement?
8. In the *Whirlpool* case, why did the court require that the OSHRC order be vacated and remanded? What would you expect the OSHA to do next in regard to its citation against Whirlpool?
9. According to your reading of the *Irwin Steel Erectors* case, should structural steel welders be required to wear safety belts, lifelines, and lanyards? Compare the dangers of such a requirement with the danger if such equipment is not required.
10. In the *Champlin* case, it seems obvious that the dangers of oil at an auto-ignitable temperature are life threatening. Why, then, did the court reverse the OSHRC finding and order regarding valves through which such oil flows?

Glossary

Administrative law: Body of laws pertaining to the regulation of administrative agencies.
Advisory opinion: Nonbinding response by an agency to questions having to do with the functions supervised.
Agency shop: Shop in which union membership is not required, but union dues are collected from nonunion members as well as those who are members.
Agent: Representative of the principal in dealing with other persons.
Apparent authority: Implication of authority to deal with others in an agency relationship based upon principal's act or negligence, despite presumed agent's real lack of such authority.
Arbitration: Substitute for a trial in which the dispute is submitted to an impartial entity whose decision is binding.
Assault: Threat of harm to an individual.
Assumption of risk: Voluntary assuming of the risks normally associated with an activity.
Attractive nuisance: Instrumentally attractive and dangerous to children.
Bargaining unit: Determination by the National Labor Relations Board of what portion of a work force is to constitute a group of workers for collective bargaining purposes.
Battery: Unlawful touching of another person in an offensive manner.
Caveat emptor: Let the buyer beware. People are assumed to be cognizant of obvious hazards to their own safety.
Citation: Notification of a violation. Here, specifically a safety violation.
Closed shop: Shop in which a prospective employee must be a member of the union *before* being hired.
Coercion: Tactics or devices used to cause another's compliance with the wishes of the entity bringing the pressure to bear.
Community property: Property ownership shared by husband and wife.
Competent party: Mature person with no identifiable incapacity to make a contract.
Condition precedent: Condition, the occurrence of which causes an obligation under a contract to become effective.
Condition subsequent: Condition, the occurrence of which terminates an existing contract.
Consent decree: Device used by administrative agencies in supervising entities.
Contempt of court: Enforcement device based upon failure to comply with a court order.
Contingencies: Unexpected occurrences in a project.
Contract: Agreement enforceable at law.
Contract documents: Array of documents setting forth the work to be done and the relationships of the parties involved.
Contributory negligence: Reasoning that plaintiff's injury was caused by his own negligence as well as that of defendant.
Conversion: The tort counterpart to the crime of theft.
Copyright: Protection of the copyright holder's exclusive right to copy, use, sell, perform, or display the subject matter.
Corporation: Artificial person created according to state statute to act according to the provisions and limitations of its charter.
Cost-plus contract: Arrangement in which the contractor's costs are paid by the owner, with the contractor being paid for his services by a special arrangement, usually a fixed fee or a percentage of the costs.
Criminal conspiracy: A bit of reasoning in early courts by which the combining of workers into unions was outlawed.
Deposition: The taking of evidence at a time other than when the legal action is to be held and/or at a location other than the courtroom. All the formal rules of evidence taking are customarily observed.
Design defect: Product defect caused by faulty design; may pertain to many products.
Discovery: Right and procedure to obtain evidence in the hands of other people; especially pertains to defendants.
Easement: Right in land which belongs to another.
Eminent domain: Right of the government to take private property for public use upon payment of the market value for it.
Employer's liability acts: Acts allowing employers to be sued in tort by employees injured by work-connected events.
Engineering ethics: Ethical code or set of moral standards governing the activities of engineers in the functions they are called upon to perform.
Engineering management: Management of people who perform engineering tasks.
Equity: Legal system developed to handle "unusual" cases with extraordinary remedies.

Estoppel: Legal forbidding of one from denying his former act or statement by pleading or alleging the truth.

Exemplary damages: Often called "punitive" damages, meant to punish the defendant for reprehensible behavior, thus making him or her an example to deter others from the conduct.

Ex post facto: Laws or rules made after occurrence of the act or event to which they pertain.

Express warranty: Promissory statement in which a seller agrees to assume a risk normally borne by the buyer.

Feasibility study: Preliminary examination of a proposed project to determine whether it is reasonable to proceed to its conclusion.

Featherbedding: Requirement of payment for work not done or not to be done.

Federal Register: Daily publication of the Federal Government Printing Office giving the most recent activities of the United States Government agencies.

Fellow servant rule: Employer's defense of injured worker's lawsuit, claiming that another worker really caused the injury.

Fiduciary relationship: Relationship involving trust and loyalty, as between principal and agent.

Grand jury: Jury of inquiry convened to hear accusations and evidence in proposed criminal cases to determine if an indictment is indicated.

Hearing: Device somewhat similar to a trial for collecting evidence.

Hearsay evidence: Indirect evidence; frequently, testimony as to evidence experienced by another.

Hot cargo: Refusal to handle the products of another company.

Implied authority Agent's authority to represent the principal based on circumstances surrounding the arrangement; e.g., implications based on previous dealings, trade customs, etc.

Implied warranty: Warranty from the seller to the buyer in any sale unless expressly disclaimed by the seller.

Indemnity: In agency, payment by the principal of costs encountered by the agent in carrying out his or her duties.

Infringement: Making, using, or selling an invention protected by a patent.

Injunction: Equity remedy requiring a person to do or refrain from doing some specific thing. Permanent: to last as long as the cause of the injunction exists. Temporary: generally used to hold the *status quo* until further action by the court.

Insurable interest: Interest in which the beneficiary risks an economic loss if the event insured against occurs.

Intellectual property: Intangible property resulting from the creative processes of the mind.

Interference: Occurrence of two patent applications claiming the same thing.

Investigation: Device for obtaining evidence, often used in a coercive manner by agencies.

Joint tenancy: Shared ownership of property in which each joint owner has the right of survivorship.

Joint venture: Joint undertaking by two or more persons limited in scope and duration. Often, a partnership formed to accomplish an objective, terminating when that objective is accomplished.

Labor injunction: Temporary injunction to preserve employer's property when threatened with coercive labor activity.

Labor law: Body of rules and laws pertaining to union-management relations.

Land contract: Security arrangement in which the seller retains ownership of the property until contract provisions are met (usually, a set of payments).

Libel: Defamation of another by writing, printing, pictures, or signs.

License: Temporary right to use the property of another.

Liquidated damages: Damages of an amount predetermined by the contract.

Mandatory injunction: Injunction requiring someone to do a specific thing.

Manufacturing engineering: Design of processes to produce products (goods and/or services).

Master-servant relationship: Employment relationship in which the owner or principal may exercise control over the means of accomplishment of tasks assigned to employees.

Mechanic's lien: Interest in real estate established by unpaid workers who worked upon the real estate involved.

Mortgage: Security interest in property which is in effect until the property is paid for according to the terms of the mortgage contract.

Negligence: One's failure to do one's duty to avoid injury to another.

Novation: Replacement of one of the parties to a contract; requires the consent of all parties to be binding.

Nuisance: Almost anything that interferes with life or property, e.g., excessive noise, smoke, or fumes.

Open shop: Shop in which union membership is not required.

OSHA: Federal Occupational Safety and Health Act (1971).

Owner-independent contractor relationship: Relationship in which the independent contractor agrees to produce a result with little or no control to be exercised by the owner or principal.

Partnership: Arrangement, similar to a single proprietorship, in which two or more persons agree to carry on a business for profit.

Patent: Right to exclude others from making, using, or selling an invention.

Patent pending: Essentially a warning that an invention may soon be protected by a patent.

Personal property: All property other than real property, such as money, goods, accounts receivable, patents, contract rights, etc.

Permanent partial disability: Permanent loss of function of some part of a person's anatomy.

Permanent total disability: Complete, permanent disability.

Petit jury: Jury to decide issues of fact as a result of presentations made during a trial.

Picketing: Informing of others (the public) of grievances against someone or something.

Principal: Person an agent represents in dealing with others.

Privity of contract: Rights of parties to a contract; a legal action based upon a contact presumes that the parties involved are in a contractual relationship.

Products liability: Liability of a producer or seller of a product which causes injury to someone.

Professional engineer: Qualified engineer certified by his or her state's registration board as capable of performing engineering services in his or her specialization to a professional standard.

Prohibitive injunction: Injunction requiring someone not to do a specific thing.

Proximate cause: Legal causation; a substantial factor in the cause of an injury.

Public policy: Beneficial public interest.

Quit-claim deed: Deed transferring whatever rights the grantor has in the property to the grantee.

Ratification: Agreement by a principal to be bound by arrangements made on his or her behalf by a nonagent. Also, agreement by employees to be bound by a labor contract arranged between a labor union and an employer.

Real property: Land and things attached to it which were intended to become part of the real property.

Res ipsa loquitur: The things speaks for itself.

Respondeat superior: Employer is responsible for acts of his employees involved with the employees' employment.

Ripeness: Necessity that there must be a real problem to be solved by the court before a legal action may be taken.

Rule: Rules are agency counterparts to laws.

Safety: Absence of danger from an activity.

Sale on approval: Seller retains ownership of the goods even though the buyer may have possession of them until approval is registered, thus completing the sale.

Sale or return: Sale in which the merchant has the right to return the goods to the seller.

Secondary boycott: Indirect coercion against a disinterested party to gain some concession from another party.

Security interest: Rights or interest in the thing sold that the seller has until contracted payment is made.

Shop rights: An employer's nonexclusive, nonassignable right to make, use, or sell an employee's invention.

Slander: Oral defamation of another in the presence of other people.

Specification: Contract: description of work to be done or thing to be purchased. Patent: description of the invention.

Specific performance: Order of a court having equity powers requiring a person to perform some specific act required by a contract.

Standing: Degree of interest required to take legal action.

Strict liability: Liability for product injury based on product being defective in some way and a causal connection being shown.

Strike: Employee work stoppage.

Subcontractor: Entity retained by a contractor to accomplish specific parts of a contract.

Summary judgment: Court judgment, based on all the preliminary materials, that there is no reason to have a trial and that the moving party should be entitled to the judgment.

Taft-Hartley law: Federal labor-management relations act passed in 1947.

Temporary total disability: Temporary condition, such as a curable disease, causing a person to be unable to work.

Tenancy in common: Shared ownership in property in which the share of ownership may be sold or willed or given to another.

Tort: Injury by one person to another's person or property.

Trademark: Identification of goods distinguishing them from goods of another.

Trade secret: May be any secret formula, pattern, device, or compilation of information used in a business.

Ultra vires act: Act by a corporation which is beyond the limits set by its charter.

Undisclosed principal: Agency event in which the agent does not reveal the existence of the principal to the third party.

Unfair labor practice: Unlawful practice by either a management or a labor organization.

Uniform Commercial Code (UCC): General statute pattern modified and enacted by the various states setting forth the rules governing commercial transactions.

Union shop: A shop in which a new employee must join the union after a probationary period.

Value: Worth of something; it may vary substantially in the various contexts in which it is established.

Warranty deed: Deed transferring property rights in which the grantor warrants that the title is good.

Worker's compensation: Payment to workers injured out of and in the course of their work.

Work preservation clause: Limitation or prevention of subcontracting of work.

Yellow-dog contract: Contract clause in which a new employee agreed not to become a union member.

Index

A

D

E

F

M

T

X

Y